Elementary Linear Algebra with Applications

Second Edition

Elementary Linear Algebra with Applications

Second Edition

Richard O. Hill, Jr.
Michigan State University

Harcourt Brace Jovanovich College Publishers
Fort Worth Philadelphia San Diego
New York Orlando Austin San Antonio
Toronto Montreal London Sydney Tokyo

ISBN: 0-15-521092-0

Library of Congress Catalog Card Number: 90-82256

Printed in the United States of America

PREFACE

Elementary Linear Algebra with Applications, Second Edition, covers the basic material of linear algebra, giving students both an understanding of the theory and an appreciation of the applications which motivate it, while giving the instructor considerable flexibility in the way many topics can be covered in the course. The primary purpose of this book is to present the core material of linear algebra in a student-oriented, highly readable way. The many examples and the large number and variety of exercises in each section will help the student learn and understand the material.

Most elementary linear algebra students are technically oriented; surveys indicate that most are majoring in engineering, in the physical, biological, or mathematical sciences, or in economics, geography, psychology, and the like. Most of these students will be using linear algebra in later courses and in their professions. Such students can well use a course that provides them not only with the theoretical foundations of linear algebra but also with a feeling for a variety of applications and an appreciation of some numerical considerations required for implementing the theory on a computer. This textbook is designed for such a course and allows the instructor to choose the applications and numerical topics to be covered according to his or her tastes and the students' needs.

Surveys indicate that most elementary linear algebra students have had at least one term of calculus, and this book is written at that level. However a number of schools have successfully used this textbook in a course *without* a calculus prerequisite. The few examples and exercises in this book that require calculus are clearly marked and can either be briefly discussed (for breadth) or skipped. (And, of course, these problems can always be assigned for extra credit.)

As has been said, applications are a key feature of this book, and following this preface is a list. Please note that the applications are scattered throughout the text—many of them are short, and most of them are there to illustrate or to motivate the mathematics. For sound pedagogical reasons the applications are integrated into the text and not just tacked on to the ends of chapters or (worse) collected at the end of the book.

Determinants will always play a role in linear algebra, and traditionally this role has been considerable. The role of determinants is, however, mainly theoretical, and they should usually be avoided in order to solve problems efficiently on a computer. Moreover, many instructors, especially when teaching a ten-week course, prefer to spend a minimal amount of time on determinants. *Elementary Linear Algebra with Applications*, Second Edition, gives the instructor the flexibility of presenting a traditional coverage of determinants (Chapter 2, which can be presented any time after Chapter 1 and before eigenvalues) or (by skipping Chapter 2 completely) of presenting only the amount of material on determinants necessary for eigenvalues (Section 5.1).

Chapter 1 presents the traditional material on Gaussian elimination and offers an introduction to matrix algebra. Applications are interspersed throughout, beginning with an elementary, but comprehensive, background example in Section 1.1. One of the special features in this chapter is the discussion of solving $AX = B$, where A is square and invertible. A mathematical answer is $X = A^{1}B$, but in the real world it is usually not done this way. Section 1.5 explains why not and what is really done, using LU decompositions. Only a simplified discussion of LU decompositions is given in Section 1.5, although it is enough to give students an idea of the principles involved. More details are given in Appendices A and B. For more advanced technical students, a more sophisticated application of the material in Chapter 1 is given in Appendix C. Chapter 1 ends with an optional section on various numerical considerations, including ill-conditioned systems, a topic *all* technical students should be aware of.

Chapter 2 presents the traditional development of determinants. This chapter, Chapter 5 in the First Edition, may be assigned at any time after Chapter 1 and before eigenvalues. A new section containing applications of determinants has been added. Both the change and the addition were made at the encouragement of users and reviewers.

Chapter 3 starts with a review of $\mathbb{R}^2$ and $\mathbb{R}^3$ and then develops the traditional material on vector spaces, subspaces, linear independence, dimension, rank, and so on. The emphasis is on subspaces of $\mathbb{R}^n$; but "abstract" vector spaces of matrices, polynomials, and continuous functions are also discussed. A new section on coordinates and change of basis has been added, the second part of which is optional. The chapter ends with an optional section on coding theory. This somewhat nontraditional topic is fascinating and relevant, and the section can be easily read by students on their own; the material illustrates well the subject matter of the chapter.

Chapter 4 introduces the concept of inner product spaces and other associated material. It also discusses the least-squares problem, an important application, and uses least squares to motivate orthogonality and the Gram-Schmidt process. This in turn is reorganized as the QR decomposition, which nicely simplifies solving the least-squares problem numerically. The chapter ends with two optional sections. The first deals with a nontraditional topic for the course, using Householder transformations to obtain a *QR* decomposition. This is a beautiful interaction between geometry, algebra, and numerical considerations. The second optional section deals with similarity and general matrices of linear transformations. This traditional, but slightly higher-level material, was asked for by a number of users.

Chapter 5 starts with brief coverage of determinants (which would be skipped if Chapter 2 is covered). The core material on eigenvalues and eigenvectors is presented in Sections 5.2–5.4, and applications are presented in Sections 5.55.7. For flexibility, Sections 5.5–5.7 are completely independent of one another. The chapter ends with an optional section that introduces numerical approaches to the eigenvalue problem.

Chapter 6 concludes the body of the text with a potpourri of topics that are described in that chapter's introduction. The topics were chosen to provide a bridge between the introductory material of Chapters 1–5 and material in future courses in applied and pure mathematics. Here too, the sections are completely independent of one another. There are two new sections here, whose inclusion was suggested by several users. Indeed, Professor Alberto Delgado of Kansas State University was even kind enough to send me his class notes on iterative solutions; they proved most helpful in writing Section 6.3.

Altogether the core material for a course might include the following: Chapter 1, Sections 1–6; Chapter 3, Sections 1–7 and the first part of Section 8 (Section 1 may be skipped or only the last third reviewed, according to students' background); Chapter 4, Sections 1–4 and the first part of Section 5; either Chapter 2 or Section 5.1; Sections 4.2–4.4, and some of Sections 4.5–4.7; and other topics and sections according to time and the instructor's preferences.

Changes for the second edition came from several sources. But by far the largest number came from suggestions of users. Some of these suggestions are acknowledged elsewhere in the preface and occasionally in the text. But since most of the suggestions are simply gratefully used, I would like to thank all of you who phoned (about a dozen) or wrote (about two dozen—often including a list of misprints) with suggestions of interesting problems or of topics you felt should be included. I hope that anyone who comes up with improvements to this edition will feel free to contact me.

Here are some of the changes for the second edition. About 200 exercises have been added, some elementary to help clarify concepts, some a little more challenging to help students put together two or three concepts. Five new sections, described earlier, have been added. Much of the exposition has been retained, but scattered throughout are sentences, paragraphs, examples, and subsections added to improved clarity or expand the exposition. For example, Section 4.1 contains a new Example 1 using computer-aided design to help motivate linear transformations, and Section 5.2 now has a small optional subsection on trace.

In Section 1.3, the discussion of the definition of product has been expanded to include a brief introduction to linear combinations. Students usually see this very important concept for the first time along with linear independence, bases, dimension, etc. My hope is that this very early introduction will help reduce the confusion when other new topics are introduced later.

One final addition is the inclusion of cumulative review exercises at the ends of Chapters 3, 4, and 5. Answers have been placed in the Instructor's Manual to provide flexibility. Instructors might, for example, assign these exercises as minitests and then post the answers.

I have grown to like the idea of using supplemental projects ln linear algebra courses, so that students can see an in-depth examination into various topics. I particularly like to use, where appropriate, a computer and PC Matlab for these projects. PC Matlab is a wonderful, easy to-use, professional quality piece of software for linear algebraic problems. Several projects I have used, together with a few others, are given in the Instructor's Manual.

Many people have contributed to the current form of this book. First, of course, are all the people who contributed to the first edition, to whom I repeat my gratitude. I would also like to thank the reviewers of this edition and those who provided extensive responses to questionnaires. They are Gerald Armstrong, Brigham Young University; Gladwin E. Bartel, Otero Junior College; Frank Battles, Massachusetts Maritime Academy; L. Becerra, University of Houston–Downtown; Bruce H. Bemis, Westminster College; Martha L. Bouknight, Meredith College; Dr. Eddie Boyd, Jr., University of Maryland–Eastern Shore; George W. Brewer, Odessa College; David C. Brooks, Seattle Pacific University; John Dawson, Pennsylvania State University–York; Mark Christie, Louisiana State University; Alberto L. Delgado, Kansas State University; John Fink, Kalamazoo College; Susan Friedman, City University of New York–Bernard Baruch College; Dewey Furness, Ricks College; Genaro Gonzalez, Texas A&I University; Eric Gossett, Bethel College; Ray Haertel, Central Oregon Community College; Dr. Jack E. Hofer, California State Polytechnic University–Pomona; Daniel A. Hogan, Hinds Community College; Dr. David Horowitz, Golden West College; Dr. K. L. Huehn, California State Polytechnic University–Pomona; David L. Hull, Ohio Wesleyan University; James A. Hummel, University of Maryland; W. Kahan, University of California–Berkeley; Steven Kahn, Anne Arundel Community College; Anne Landry, Dutchess Community College; Ira Lansing, College of Marin; Dean S. Larson, Gonzaga University; Mauricio Marroquin, Los Angeles Valley College; Mary McCarty, Sullivan County Community College; David Meredith, San Francisco State University; R. S. Montgomery, University of Connecticut; Maurice Ngo, Chabot College, G. F. Orr, University of Southern Colorado; Robert W. Prielipp, University of WisconsinOshkosh; Norman

Richert, University of Houston–Clear Lake; Jo E. Smith, GMI Engineering and Management Institute; Dr. Henrene E. Smoot, Alabama A&N University; Ward A. Soper, Walla Walla College; Jay Treiman, Western Michigan University; Michael H. Vernon, Lewis Clark State College; Paul Vicknair, California State University–San Bernardino; Ken Weiner, Montgomery College; and W. Thurmon Whitley, University of New Haven.

Finally, I would like to thank the editorial and production staff at Harcourt Brace Jovanovich and the production coordinator, Lynn M. Edwards, for their contributions to the project.

Richard Hill
Department of Mathematics
Michigan State University
E. Lansing, MI 48824

LIST OF APPLICATIONS

This is a list of the applications in the text and the purpose each application serves. These applications are integrated into the text as a motivational tool. They illustrate the material as it is developed (and are not separated into special sections or chapters at the end of the text).

Section 1.1, Example 4. A demographics problem: provides motivating background for several principal types of problems treated in the text.

Section 1.3, Example 6. An inventory problem: motivates the definition of matrix multiplication.

Section 1.6, Example 3. A bridge structure problem: justifies studying matrices with special forms and further justifies the use of *LU* decompositions.

Appendix C. A temperature distribution problem using finite differences: justifies studying matrices with special forms and further justifies the use of *LU* decompositions; also illustrates an important tool of applied mathematics based on linear algebra.

Section 2.5. Wronkians; cross products; area and volume; change of coordinates in multiple integrals: shows applications of determinants.

Section 3.1, Example 1 (includes prior discussion on page 132). Navigation problems: motivate addition of vectors.

Section 3.1, Figure 3.15. Force problems: illustrate projections.

Section 3.9. Coding theory: a marvelous nonstandard application; illustrates many basic concepts and relationships developed in Chapter 3.

Section 4.1, Figure 4.2. Computer-aided design: illustrates the geometric action of linear transformations.

Chapter 4 Introduction and Section 4.3, Example 6. Fitting a line to data: illustrates a typical least-squares problem.

Section 4.3, Example 7. Electronic imaging: illustrates a typical least-squares problem.

Section 4.4, Example 6. Fitting a line to data:; motivates orthogonalization and the need for the Gram–Schmidt process.

Section 5.5. Fibonacci sequences: a fascinating application; illustrates the usefulness of diagonalizing symmetric matrices.

Section 5.5. Markov processes: solution to the deomographics example (Example 4 on page 4), Section 1.1: an important application; illustrates diagonalization and limits..

Section 5.6. Differential equations and a population-interaction model: illustrates diagonalization.

Section 5.7. Quadratic forms: illustrates the usefulness of diagonalizing symmetric matrices by orthogonal matrices.

Section 5.8. Direct and inverse iteration: illustrates the theory of diagonalization.

Section 6.1. Orthogonalizing polynomials: illustrates the Gram–Schmidt process

Section 6.2. General case of the least-squares problem: illustrates the singular value decomposition.

CONTENTS

---

*Chapter 2 and Section 5.1 provide alternative presentations of determinants. Either one can be covered anytime after Chapter 1 and before Section 5.2.

†The applications in section 5.5, 5.6, and 5.7 are completely independent of one another.

Chapter 1

INTRODUCTION TO LINEAR EQUATIONS AND MATRICES

Systems of linear equations have many applications. They are found in economics, social sciences, and medicine as well as the biological and physical sciences and mathematics. The method we will use to solve such systems is called *Gaussian elimination*, after Carl Friedrich Gauss (1777–1855), one of the most prolific mathematicians in history. However, the method was known well before Gauss; indeed it appeared in a Chinese text, *Chui-chang Suan-shu*, around 250 B.C. We study this method not only because it has historical significance, but also because it is the basis for the best direct methods for programming a computer to solve linear systems. Equally important, the mathematical study of this method opens the door to a fascinating and widely applicable branch of mathematics, linear algebra.

1.1 Introduction to Linear Systems and Matrices

The word "linear" comes from the word "line." An equation for a line in the xy-plane is any equation that can be put in the form

$$a_1x + a_2y = b$$

where a_1, a_2, and b are constants, a_1, a_2 not both zero. Such an equation is called a linear equation in the variables x and y. In general a **linear equation** in the variables $x_1, x_2, \ldots, x_n$ is one that can be put in the form

$$a_1x_1 + a_2x_2 + \cdots + a_nx_n = b \tag{1.1}$$

where $a_1, \ldots, a_n$, and b are constants, $a_1, \ldots, a_n$ not all zero. In this form all the variables appear only to the first power and are not arguments for logarithmic, trigonometric, or other kinds of functions. There are also no products or square roots of variables.

Example 1 The equations

$$2x - 3y = 4 \qquad x_1 - x_2 + x_3 - x_4 = 6$$
$$z = 5 - 3x + \tfrac{1}{2}y \qquad x_1 + 2x_2 + 3x_3 + \cdots + nx_n = 1$$

are all linear equations, whereas the equations

$$x^2 + y^2 = 1 \qquad \sin x + e^y = 1$$
$$xy = 2 \qquad 7x_1 - 3x_2 + \frac{9}{x_3} + 2x_4 = 1$$

are not linear. ■

A **solution** of a linear equation $a_1x_1 + a_2x_2 + \cdots + a_nx_n = b$ is a sequence of numbers $t_1, t_2, \ldots, t_n$ such that if we substitute $x_1 = t_1$, $x_2 = t_2, \ldots, x_n = t_n$ into the equation, we obtain a true statement. To **solve** an equation means to find all solutions; the set of all solutions is called the **solution set**.

Example 2 Solve the following

(a) $3x = 4$ (b) $4x - 5y = 3$ (c) $2x_1 - 7x_2 + 10x_3 = 8$

Solution (a) Divide both sides by 3 to obtain $x = \frac{4}{3}$ as the unique solution.

(b) One way is to solve for x:

$$4x = 3 + 5y \qquad \text{so that} \qquad x = \tfrac{3}{4} + \tfrac{5}{4}y$$

Notice that for every value of y we obtain a corresponding value for x. Thus there is an infinite number of solutions. We need some way to indicate them all. One way to do this is to let y equal an arbitrary variable, t, so that $x = \frac{3}{4} + \frac{5}{4}t$ and all solutions are of the form

$$x = \tfrac{3}{4} + \tfrac{5}{4}t, \quad y = t, \qquad t \text{ any real number} \tag{1.2}$$

This is representative of the way we shall describe infinitely many solutions when there are more unknowns than equations.

(b) Another way is to solve for y:

$$4x - 3 = 5y$$
$$y = \tfrac{4}{5}x - \tfrac{3}{5}$$

Then let $x = s$, which yields

$$y = \tfrac{4}{5}s - \tfrac{3}{5}, \quad x = s, \qquad s \text{ any real number} \tag{1.3}$$

Although the forms of the answer in Equations (1.2) and (1.3) are different, they in fact represent the same answers. For example, letting $t = 1$ in Equation (1.2) gives $(x, y) = (2, 1)$, which is obtained in Equation (1.3) by letting $s = 2$.

(c) Solve the equation for x_1:

$$2x_1 = 8 + 7x_2 - 10x_3$$
$$x_1 = 4 + \tfrac{7}{2}x_2 - 5x_3$$

Here we need two arbitrary variables, say s and t. We let $x_2 = s$, $x_3 = t$, and obtain $x_1 = 4 + \frac{7}{2}s - 5t$. Thus all solutions are of the form

$$x_1 = 4 + \tfrac{7}{2}s - 5t, \quad x_2 = s, \quad x_3 = t, \qquad s \text{ and } t \text{ any real numbers}$$

Note that we could have solved for x_2 or x_3, obtaining alternate forms of the solution. ■

Very often we wish to solve several linear equations at the same time. A finite collection of linear equations in the variables $x_1, x_2, \ldots, x_n$ is called a **system of linear equations**, or more simply a **linear system**. A **solution** of a linear system is a sequence of numbers $t_1, t_2, \ldots, t_n$ that is a solution of each of the linear equations simultaneously.

A system of two linear equations in two unknowns can have no solutions, one solution, or infinitely many solutions. A system with no solutions is called **inconsistent**; a system with at least one solution is called **consistent**.

Example 3 Graph each of the following systems and determine the number of solutions.

(a) $\begin{aligned} 2x + y &= 8 \\ 2x + y &= 4 \end{aligned}$ (b) $\begin{aligned} 2x + y &= 8 \\ x - y &= -2 \end{aligned}$ (c) $\begin{aligned} 2x + y &= 8 \\ 4x + 2y &= 16 \end{aligned}$

Solution The graph of the equation $ax + by = c$, a and b not both zero, is a straight line in the xy-plane. Thus the graph of each linear system that is composed of two linear equations consists of two straight lines. The solutions of each such system correspond to the intersection of the lines (see Figure 1.1).

Figure 1.1

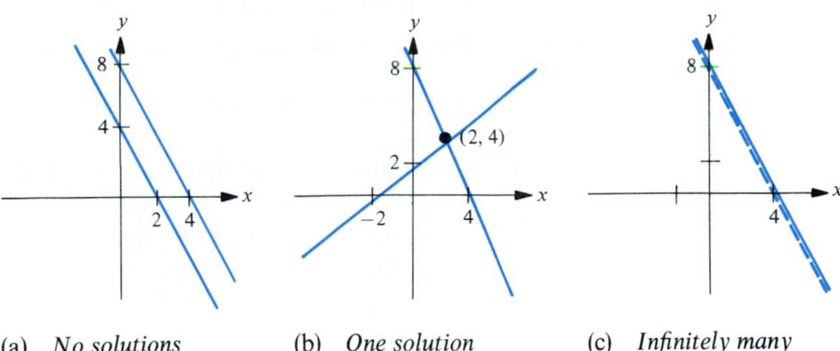

(a) *No solutions* (b) *One solution* (c) *Infinitely many solutions*

(a) The lines are parallel and distinct and do not intersect. The system has no solutions and is inconsistent.
(b) The lines intersect in one point, (2, 4). The system has one solution.
(c) The two lines coincide and hence intersect in infinitely many points. Thus the system has infinitely many solutions. ■

In general, every linear system has either no solutions, one solution, or infinitely many solutions.

Background Example

Linear systems arise in many different contexts and give rise to several different types of problems. The next example will give you an idea of a few of the types of problems you will be able to solve *by the end* of the course. So do not panic over not being able to solve the whole problem now; just absorb the discussion of the solution, the beginning part of which you already know how to do, and look forward to knowing how to do the rest later. The example is centered around a population flow model, but there are similar models in biology, chemistry, and economics.

Example 4 Suppose between 1990 and 1995 the population of the United States is roughly constant at 270 million people. Suppose, in addition, that during each of these years 20% of the people living in Texas move out of Texas and 10% of those outside move into Texas. Assume that the population of Texas at the beginning of 1991 will be 50 million people (see Figure 1.2).

Figure 1.2

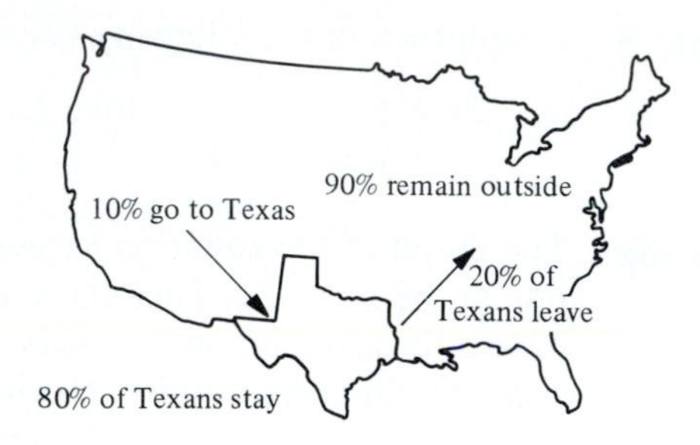

(a) Set up a linear system describing this situation.
(b) What will be the population inside and outside of Texas at the end of 1991?
(c) What will be the population inside and outside of Texas at the beginning of 1990?
(d) Is there a population distribution inside and outside of Texas for which there would be no total change from one year to the next (even though 20% of those inside moved out and 10% of those outside moved in)?
(e) What happens to the population distribution over those six years? Does it tend to stabilize or vacillate?

Discussion of Solution (a) Let

$x =$ the population of Texas at the beginning of a year
$y =$ the population outside of Texas at the beginning of a year
$u =$ the population of Texas at the end of a year
$v =$ the population outside of Texas at the end of a year

Then the hypothesis of the problem gives the system

$$\begin{aligned} u &= 0.8x + 0.1y \\ v &= 0.2x + 0.9y \end{aligned} \tag{1.4}$$

and, of course, $x + y = 270$ million people and $u + v = 270$ million people.

(b) We let $x = 50$ million and $y = 220$ million in the system (1.4), solve, and obtain $u = 62$ million and $v = 208$ million. This is a straightforward **linear substitution** problem.

(c) We let $u = 50$ and $v = 220$ in the system (1.4), solve, and obtain $x \approx 33$ million and $y \approx 237$ million. This is an **inverse linear** problem. We shall study how to solve such problems, especially when they are large, later in this chapter.

(d) We let $u = x$ and $v = y$ in the system (1.4):

$$\begin{aligned} x &= 0.8x + 0.1y \\ y &= 0.2x + 0.9y \end{aligned}$$

This system has infinitely many solutions, as we have already discussed, but there is only one solution such that $x + y = 270$ million, namely, $x = 90$ million and $y = 180$ million. This is, in fact, an **eigenvalue problem**; eigenvalues are discussed in Chapter 5.

(e) This is a **Markov process** problem, and the surprising answer is that the populations tend to the answer of part (d)! We explain why this is true in Section 5.5. ■

One of the main purposes of this chapter is to find all solutions of linear systems. The method we use to do this is called **Gaussian elimination**. It is efficient, simple, and easy to program on a computer. Consider the following system.

$$\begin{aligned} 2x + 3y - 2z &= -7 \\ 3y + 2z &= 3 \\ 5z &= 15 \end{aligned}$$

Such a system is said to be in **triangular form**, or in **(row) echelon form**, because of the shape of the system on the left-hand side of the equal signs. A system in this form is easy to solve by a process called **backsubstitution**: solving the last equation for z and then working backward, each time substituting the values that have been found into the preceding equation.

Example 5 Solve the system

$$\begin{aligned} 2x + 3y - 2z &= -7 \\ 3y + 2z &= 3 \\ 5z &= 15 \end{aligned}$$

Solution First solve the last equation for z:

$$5z = 15, \quad z = 3$$

Next substitute this value for z into the preceding equation and then solve for y:

$$3y + 2(3) = 3, \quad y = -1$$

Finally, substitute the known values for y and z into the first equation and solve for x:

$$2x + 3(-1) - 2(3) = -7, \quad x = 1$$

Thus the solution is $(x, y, z) = (1, -1, 3)$. ■

We shall see that large triangular systems can be solved easily by backsubstitution.

Gaussian elimination is used to reduce a general system to triangular form, which we can then solve by backsubstitution. We now introduce a new concept that will make the reduction process quicker and easier. A **matrix** is a rectangular array of constants. (The plural of matrix is **matrices**.) We fully develop matrices in Section 1.3; here we briefly illustrate their role in Gaussian elimination. If we consider the system

$$\begin{aligned} 2x - 3y - z &= 4 \\ 7x + y \phantom{{}+9z} &= 3 \\ -3x - 2y + 9z &= 0 \end{aligned}$$

there are three different matrices associated with the system

$$\begin{bmatrix} 2 & -3 & -1 \\ 7 & 1 & 0 \\ -3 & -2 & 9 \end{bmatrix}, \quad \begin{bmatrix} 4 \\ 3 \\ 0 \end{bmatrix}, \quad \begin{bmatrix} 2 & -3 & -1 & 4 \\ 7 & 1 & 0 & 3 \\ -3 & -2 & 9 & 0 \end{bmatrix}$$

The first and last have special names. The first is called the **matrix of coefficients** of the system and the last, the **augmented matrix** of the system. The use of matrices not only eliminates the need to write the variables and + and = signs, but also leads to a better understanding of the mathematical relationships involved in a problem and facilitates the use of computers.

WARNING When finding these matrices, the variables must be written in the same order in each equation and zeros must be inserted for "missing" variables (as in the second equation).

A linear system can be reduced to triangular form by applying the following three types of operations in a systematic way. These operations are reversible. Thus no solutions are introduced or lost when these operations are applied to a system.

(1.5) **ELEMENTARY OPERATIONS ON A LINEAR SYSTEM**

(a) Add a multiple of one equation to another.
(b) Interchange two equations.
(c) Multiply an equation by a nonzero constant.

Since the equations of a system correspond to rows of the augmented matrix, elementary operations on a linear system correspond to the following.

(1.6)

ELEMENTARY ROW OPERATIONS ON A MATRIX

(a) Add a multiple of one row to another.
(b) Interchange two rows.
(c) Multiply a row by a nonzero constant.

In step (c), multiplying a row by c means multiplying *every entry* in that row by c.

In the next section, we discuss Gaussian elimination more thoroughly. In particular we cover how to choose operations and decide upon their order. At this point you should *not* worry about *why* operations are performed in the order given; you should just try to understand *how* the operations work.

Example 6 illustrates how these operations can be used to reduce a system to triangular form.

Example 6

Linear System	*Associated Augmented Matrix*
$\begin{aligned} 3y + 2z &= 7 \\ x + 4y - 4z &= 3 \\ 3x + 3y + 8z &= 1 \end{aligned}$	$\begin{bmatrix} 0 & 3 & 2 & 7 \\ 1 & 4 & -4 & 3 \\ 3 & 3 & 8 & 1 \end{bmatrix}$
Interchange the first and second equations.	Interchange the first and second rows.
$\begin{aligned} x + 4y - 4z &= 3 \\ 3y + 2z &= 7 \\ 3x + 3y + 8z &= 1 \end{aligned}$	$\begin{bmatrix} 1 & 4 & -4 & 3 \\ 0 & 3 & 2 & 7 \\ 3 & 3 & 8 & 1 \end{bmatrix}$
Add -3 times the first equation to the third.	Add -3 times the first row to the third.
$\begin{aligned} x + 4y - 4z &= 3 \\ 3y + 2z &= 7 \\ -9y + 20z &= -8 \end{aligned}$	$\begin{bmatrix} 1 & 4 & -4 & 3 \\ 0 & 3 & 2 & 7 \\ 0 & -9 & 20 & -8 \end{bmatrix}$
Add 3 times the second equation to the third.	Add 3 times the second row to the third.
$\begin{aligned} x + 4y - 4z &= 3 \\ 3y + 2z &= 7 \\ 26z &= 13 \end{aligned}$	$\begin{bmatrix} 1 & 4 & -4 & 3 \\ 0 & 3 & 2 & 7 \\ 0 & 0 & 26 & 13 \end{bmatrix}$

If we were working only with augmented matrices, we would translate the last matrix on the right back to the system on the left and solve that system by backsubstitution, obtaining $(x, y, z) = (-3, 2, \frac{1}{2})$. ■

Finally, suppose that a row of an augmented matrix has all zeros except for the last entry on the right (which is nonzero). Then the associated system must be inconsistent and has no solution.

Example 7 Determine if the system associated with the given augmented matrix is consistent or inconsistent.

(a) $\begin{bmatrix} 2 & -1 & 3 & 4 \\ 0 & 4 & 1 & 8 \\ 0 & 0 & 2 & -9 \end{bmatrix}$ (b) $\begin{bmatrix} 2 & -1 & 3 & 4 \\ 0 & 4 & 1 & 8 \\ 0 & 0 & 0 & -9 \end{bmatrix}$

(c) $\begin{bmatrix} 2 & -1 & 3 & 4 \\ 0 & 4 & 1 & 8 \\ 0 & 0 & 2 & 0 \end{bmatrix}$ (d) $\begin{bmatrix} 4 & 1 & 2 & 3 & 9 & 8 & 0 \\ -5 & 18 & 13 & 1 & 2 & -3 & 0 \\ 9 & 1 & 8 & 1 & -11 & 5 & 0 \end{bmatrix}$

Solution (a) The associated system is

$$\begin{aligned} 2x - y + 3z &= 4 \\ 4y + z &= 8 \\ 2z &= -9 \end{aligned}$$

This system obviously can be solved by backsubstitution. Thus it has a solution, and hence it is consistent.

(b) This system is similar to (a) except for the last equation, which is $0 = -9$. This equation has no solution, so the system has no solutions. The system is inconsistent.

(c) This system is also similar to (a) except for the last equation, which is $2z = 0$. This has a solution, namely, $z = 0$. Thus, like (a), the system can be solved by backsubstitution (obtaining $(x, y, z) = (3, 2, 0)$), and hence the system is consistent.

(d) The associated system is

$$\begin{aligned} 4x_1 + \cdots + 8x_6 &= 0 \\ -5x_1 + \cdots - 3x_6 &= 0 \\ 9x_1 + \cdots + 5x_6 &= 0 \end{aligned}$$

This has at least one solution, which is obviously $(x_1, \ldots, x_6) = (0, \ldots, 0)$. Thus the system is consistent. ■

We shall see in the following chapters how the use of matrices not only simplifies solving linear systems, but also leads to a deeper understanding of many interesting mathematical relationships.

Exercise 1.1 In Exercises 1–6, determine if the equation is linear.

1. $3x + 4yz - 2w = 5$
2. $8x + 5\sqrt{y} + 3z = 4$
3. $5x_1 - 3x_2 + 2x_3 - \sqrt{\pi}\, x_4 = 8$
4. $15x - 2z = 3y + 5w + \sin k$, k a constant
5. $3x^2 + 4y^2 = 12$
6. $2x_1 - 3x_2^{-1} + 4x_3 = 2$

In Exercises 7–12, solve in at least two ways [as in Example 2(b)].

7. $2x + 4y = 5$
8. $3x_1 - x_2 = 6$
9. $7x + 14y - 3z = -7$
10. $2x_1 - 5x_2 + 6x_3 = 4$
11. $3x_1 - 2x_2 + 5x_3 - x_4 = 6$
12. $4x_1 - 5x_2 - 3x_3 + 4x_4 = 1$

In Exercises 13–18, graph the system and determine how many solutions it has. Do not solve.

13. $$\begin{aligned} 3x + y &= 6 \\ 6x + 2y &= 12 \end{aligned}$$
14. $$\begin{aligned} 3x + y &= 6 \\ 3x - y &= 12 \end{aligned}$$
15. $$\begin{aligned} 3x + y &= 6 \\ 6x + 2y &= 8 \end{aligned}$$
16. $$\begin{aligned} 2x - 3y &= 4 \\ 4x + 2y &= 1 \end{aligned}$$
17. $$\begin{aligned} 2x + 3y &= -1 \\ 3x - 4y &= 7 \\ 5x + y &= 4 \end{aligned}$$
18. $$\begin{aligned} 3x + 4y &= 1 \\ 5x - y &= -6 \\ x + y &= 3 \end{aligned}$$

In Exercises 19 and 20, sketch the three lines and show geometrically that there is no solution. Then find two *different* ways of changing the third equation so that there is a solution.

19. $$\begin{aligned} x + y &= 1 \\ 4x + y &= -2 \\ 2x - y &= 2 \end{aligned}$$
20. $$\begin{aligned} 2x + 3y &= 8 \\ x - y &= -1 \\ x - 2y &= 1 \end{aligned}$$

In Exercises 21–28, an augmented matrix is given. Find the associated system and solve it by backsubstitution, if possible.

21. $$\begin{bmatrix} 2 & 1 & 3 & -6 \\ 0 & -3 & -2 & -2 \\ 0 & 0 & 2 & -4 \end{bmatrix}$$
22. $$\begin{bmatrix} 3 & 2 & -2 & -2 \\ 0 & 3 & -1 & -6 \\ 0 & 0 & -2 & -6 \end{bmatrix}$$

23. $\begin{bmatrix} 4 & 1 & 3 & 2 \\ 0 & -2 & 1 & 3 \\ 0 & 0 & 6 & 0 \end{bmatrix}$

24. $\begin{bmatrix} -2 & 3 & -1 & 5 \\ 0 & -1 & 2 & 1 \\ 0 & 0 & 0 & 5 \end{bmatrix}$

25. $\begin{bmatrix} 3 & 1 & 2 & 3 & 9 \\ 0 & -2 & 1 & 2 & 2 \\ 0 & 0 & -3 & 1 & -1 \\ 0 & 0 & 0 & 2 & 4 \end{bmatrix}$

26. $\begin{bmatrix} -1 & 3 & 4 & -1 & 0 \\ 0 & 2 & 2 & 3 & 5 \\ 0 & 0 & 3 & 1 & 1 \\ 0 & 0 & 0 & -5 & -5 \end{bmatrix}$

27. $\begin{bmatrix} 4 & 1 & 2 & 3 & 1 \\ 0 & 2 & 4 & -1 & 0 \\ 0 & 0 & 3 & 1 & 2 \\ 0 & 0 & 0 & 0 & 5 \end{bmatrix}$

28. $\begin{bmatrix} 5 & 1 & 2 & 3 & 0 \\ 0 & 4 & -1 & -1 & 0 \\ 0 & 0 & 5 & 2 & 0 \\ 0 & 0 & 0 & -3 & 0 \end{bmatrix}$

In Exercises 29–32, for which values of the constant k does the system have (a) no solutions, (b) exactly one solution, (c) infinitely many solutions?

29. $\begin{aligned} 2x + 3y &= 4 \\ kx + 6y &= 8 \end{aligned}$

30. $\begin{aligned} 3x - 2y &= 1 \\ 3x + ky &= 1 \end{aligned}$

31. $\begin{aligned} -2x + y &= 3 \\ 4x - 2y &= k \end{aligned}$

32. $\begin{aligned} x - y &= 2 \\ kx - y &= k \end{aligned}$

1.2 Gaussian Elimination

In the previous section we introduced linear systems and matrices and touched lightly on Gaussian elimination. In this section we continue with a more systematic development of Gaussian elimination.

To do so we need a few definitions. First, a **pivot** of a matrix is the first nonzero entry in a row. Next we need the concept of row echelon form of a matrix. Intuitively, a matrix is in row echelon form if it has the appearance of a staircase pattern like the following matrix, where the pivots are distinctly marked (and are nonzero) and the remaining stars may or may not be zero. As we discovered in the previous section for special cases, a linear system that corresponds to a matrix in this form can be solved by backsubstitution.

$$\begin{bmatrix} 0 & * & * & * & * & * & * & * \\ 0 & 0 & * & * & * & * & * & * \\ 0 & 0 & 0 & 0 & 0 & * & * & * \\ 0 & 0 & 0 & 0 & 0 & 0 & * & * \\ 0 & 0 & 0 & 0 & 0 & 0 & 0 & 0 \end{bmatrix}$$

Now on to a formal definition.

(1.7)

DEFINITION A matrix is in **(row) echelon form** if

(a) All rows that contain only zeros are grouped at the bottom of the matrix.
(b) For each row that does not contain only zeros, the pivot appears to the right of the pivot of all rows that appear above it.

Note that a consequence of item (b) in this definition is that all the entries below and to the left of a pivot must be zero.

Example 1 The following matrices are in row echelon form. Check to see that both conditions in the definition of echelon form (1.7) are satisfied.

(a) $\begin{bmatrix} 7 & 2 & -4 & 3 & 9 & 6 \\ 0 & 0 & -5 & 2 & 3 & 1 \\ 0 & 0 & 0 & 1 & 1 & 1 \\ 0 & 0 & 0 & 0 & 0 & 0 \end{bmatrix}$ (b) $\begin{bmatrix} 0 & 0 & 0 \\ 0 & 0 & 0 \end{bmatrix}$

(c) $\begin{bmatrix} 0 & 0 & 9 & 8 & 1 \\ 0 & 0 & 0 & 0 & 2 \end{bmatrix}$ (d) $\begin{bmatrix} -1 & 2 & 3 & 8 \\ 0 & 3 & 1 & 2 \\ 0 & 0 & 5 & 9 \end{bmatrix}$ ■

Example 2 The following matrices are not in row echelon form. Determine which of the conditions in the definition of row echelon form (1.7) is not satisfied.

(a) $\begin{bmatrix} 2 & 8 & 1 & -1 \\ -3 & 4 & 2 & 2 \\ 7 & 0 & 1 & 6 \end{bmatrix}$ (b) $\begin{bmatrix} 0 & 0 & 0 \\ 0 & 0 & 1 \end{bmatrix}$

(c) $\begin{bmatrix} 8 & 1 & -2 \\ 0 & 0 & 0 \\ 0 & 4 & 1 \end{bmatrix}$ (d) $\begin{bmatrix} 4 & 0 & 3 & -2 & 1 & 0 \\ 0 & 0 & 5 & 1 & 2 & -3 \\ 0 & 0 & 0 & 0 & 0 & -5 \\ 0 & 0 & 0 & 0 & 0 & -7 \end{bmatrix}$

■

Suppose the augmented matrix for a linear system has been reduced to row echelon form. Then Example 3 illustrates how the solution can be

* *There is a more stringent form of a matrix called* ***row reduced echelon form****. For this each pivot must be one, not just nonzero, and all other entries in a column that contains a pivot must be zero. This form has a certain mathematical appeal in that it is unique (but it is sheer arithmetic torture to obtain by hand). The form we are working with requires less work, is quicker for a computer to implement, and is the basis for the way computers are programmed.*

easily found by backsubstitution, after a few simple steps. The most efficient computer programs employ this method (though there are many other methods employed for special circumstances).

Example 3 Suppose the augmented matrix for a linear system has been reduced to the given matrix in row echelon form and the variables are also as given. Solve the system.

(a) $\begin{bmatrix} 2 & 1 & -3 & 5 \\ 0 & -3 & 2 & 17 \\ 0 & 0 & 5 & -10 \end{bmatrix}$

Variables: x, y, z

(b) $\begin{bmatrix} 2 & 3 & -1 & 5 & 2 \\ 0 & 3 & 2 & -1 & 2 \\ 0 & 0 & -2 & -8 & 4 \end{bmatrix}$

Variables: x, y, z, w

(c) $\begin{bmatrix} 3 & -2 & 1 & 3 & 1 & 14 \\ 0 & 0 & 2 & 5 & -3 & 2 \\ 0 & 0 & 0 & 0 & -3 & -6 \\ 0 & 0 & 0 & 0 & 0 & 0 \end{bmatrix}$

Variables: x_1, x_2, x_3, x_4, x_5

(d) $\begin{bmatrix} 3 & -2 & 1 & 3 & 1 & 14 \\ 0 & 0 & 2 & 5 & -3 & 2 \\ 0 & 0 & 0 & 0 & -3 & -6 \\ 0 & 0 & 0 & 0 & 0 & 4 \end{bmatrix}$

Variables: x_1, x_2, x_3, x_4, x_5

Solution (a) The first step is to find the associated linear system.

$$\begin{aligned} 2x + y - 3z &= 5 \\ -3y + 2z &= 17 \\ 5z &= -10 \end{aligned}$$

The system is in triangular form and we solve it in the usual way, by backsubstitution.

$$\begin{aligned} 5z &= -10, \quad & z &= -2 \\ -3y + 2(-2) &= 17 & & \\ -3y &= 21, \quad & y &= -7 \\ 2x + (-7) - 3(-2) &= 5 & & \\ 2x &= 6, \quad & x &= 3 \end{aligned}$$

Thus the solution is $(x, y, z) = (3, -7, -2)$.

(b) The associated system is

$$\begin{aligned} 2x + 3y - z + 5w &= 2 \\ 3y + 2z - w &= 2 \\ -2z - 8w &= 4 \end{aligned}$$

We must now introduce some new terminology since there are more variables than equations. The variables x, y, and z correspond to the pivots of the augmented matrix and are called **leading variables** (or **dependent variables**). To do the second step, we must distinguish these variables from the remaining variables, which are called **free variables** (or **independent variables**). The second step is to move the free variables to the right-hand side of the equations.

$$\begin{aligned} 2x + 3y - z &= 2 - 5w \\ 3y + 2z &= 2 + w \\ -2z &= 4 + 8w \end{aligned}$$

Following the tack taken in Example 2(b) of Section 1.1, we let $w = t$

$$\begin{aligned} 2x + 3y - z &= 2 - 5t \\ 3y + 2z &= 2 + t \\ -2z &= 4 + 8t \end{aligned}$$

and observe that the system is in triangular form similar to part (a). Hence it can be solved by backsubstitution in the usual manner if we keep track of the t's.

$$\begin{aligned} -2z &= 4 + 8t, & z &= -2 - 4t \\ 3y + 2(-2 - 4t) &= 2 + t \\ 3y - 4 - 8t &= 2 + t \\ 3y &= 6 + 9t, & y &= 2 + 3t \\ 2x + 3(2 + 3t) - (-2 - 4t) &= 2 - 5t \\ 2x + 6 + 9t + 2 + 4t &= 2 - 5t \\ 2x &= -6 - 18t, & x &= -3 - 9t \end{aligned}$$

The solution is

$$(x, y, z, w) = (-3 - 9t, 2 + 3t, -2 - 4t, t), \quad t \text{ any real number}$$

(c) The associated system is

$$\begin{aligned} 3x_1 - 2x_2 + x_3 + 3x_4 + x_5 &= 14 \\ 2x_3 + 5x_4 - 3x_5 &= 2 \\ -3x_5 &= -6 \\ 0 &= 0 \end{aligned}$$

The leading variables are x_1, x_3, and x_5. We discard the last equation and move the remaining free variables to the other side.

$$\begin{aligned} 3x_1 + x_3 + x_5 &= 14 + 2x_2 - 3x_4 \\ 2x_3 - 3x_5 &= 2 - 5x_4 \\ -3x_5 &= -6 \end{aligned}$$

Referring to Example 2(c) of Section 1.1, we let $x_2 = s$ and $x_4 = t$. (If there had been many more variables that were not leading variables, we could have used subscripted independent variables such as $s_1, s_2, \ldots$.)

$$\begin{aligned} 3x_1 + x_3 + x_5 &= 14 + 2s - 3t \\ 2x_3 - 3x_5 &= 2 - 5t \\ -3x_5 &= -6 \end{aligned}$$

We now solve by backsubstitution.

$$\begin{aligned} -3x_5 &= -6, & x_5 &= 2 \\ 2x_3 - 3(2) &= 2 - 5t & & \\ 2x_3 &= 8 - 5t, & x_3 &= 4 - \tfrac{5}{2}t \\ 3x_1 + (4 - \tfrac{5}{2}t) + 2 &= 14 + 2s - 3t & & \\ 3x_1 &= 8 + 2s - \tfrac{1}{2}t, & x_1 &= \tfrac{8}{3} + \tfrac{2}{3}s - \tfrac{1}{6}t \end{aligned}$$

The solution is

$$(x_1, x_2, x_3, x_4, x_5) = (\tfrac{8}{3} + \tfrac{2}{3}s - \tfrac{1}{6}t, s, 4 - \tfrac{5}{2}t, t, 2)$$

where s and t are any real numbers.

(d) The associated system is exactly the same as in part (c) except for the last equation, which is

$$0x_1 + \cdots + 0x_5 = 4$$

Since this equation can never be satisfied (for any $x_1, \ldots, x_5$), the system has no solutions and hence is inconsistent. ■

We have just seen how easy it is to solve a system once its associated matrix is in row echelon form. The following fact is now crucial to us.

(1.8) Every matrix can be transformed to a matrix in row echelon form using (a finite number of) elementary row operations.

The method we use to do this is called **Gaussian elimination**, which is illustrated in the following examples.

Example 4 Solve the system

$$\begin{aligned} 2x_1 - x_2 - x_3 &= 3 \\ -6x_1 + 6x_2 + 5x_3 &= -3 \\ 4x_1 + 4x_2 + 7x_3 &= 3 \end{aligned}$$

Solution First find the associated augmented matrix.

$$\begin{bmatrix} 2 & -1 & -1 & 3 \\ -6 & 6 & 5 & -3 \\ 4 & 4 & 7 & 3 \end{bmatrix}$$

Next follow the four steps required to reduce this matrix to row echelon form.

Step 1 Locate the leftmost nonzero column and that column's pivot, which is the first nonzero entry in that column.

Leftmost nonzero column

Pivot $\rightarrow$

$$\begin{bmatrix} 2 & -1 & -1 & 3 \\ -6 & 6 & 5 & -3 \\ 4 & 4 & 7 & 3 \end{bmatrix}$$

Step 2 If the pivot were not in the first row, a row interchange would be necessary. When the pivot is in the first row, we skip this step.

Step 3 Make all other entries in the first column zero by adding suitable multiples of the first row to the other rows.

$$\begin{bmatrix} 2 & -1 & -1 & 3 \\ 0 & 3 & 2 & 6 \\ 4 & 4 & 7 & 3 \end{bmatrix}$$ Add 3 times row 1 to row 2.

$$\begin{bmatrix} 2 & -1 & -1 & 3 \\ 0 & 3 & 2 & 6 \\ 0 & 6 & 9 & -3 \end{bmatrix}$$ Add -2 times row 1 to row 3.

Step 4 Repeat steps 1–3 ignoring the top row. For this problem we will then have the matrix in row echelon form. (For large problems, such as in Example 5, we have to repeat step 4 several times, each time ignoring the rows where we have already found the pivots and completed steps 2 and 3.)

Pivot $\rightarrow$

$$\begin{bmatrix} 2 & -1 & -1 & 3 \\ 0 & 3 & 2 & 6 \\ 0 & 6 & 9 & -3 \end{bmatrix}$$

Leftmost

Step 1 Locate the leftmost nonzero column of what remains of the original matrix, that is, the submatrix, and its pivot.

Step 2 Interchange not necessary.

Step 3

$$\begin{bmatrix} 2 & -1 & -1 & 3 \\ 0 & 3 & 2 & 6 \\ 0 & 0 & 5 & -15 \end{bmatrix}$$ Add -2 times row 2 to row 3.

The matrix is now in row echelon form, so we find the associated system

$$\begin{aligned} 2x_1 - x_2 - x_3 &= 3 \\ 3x_2 + 2x_3 &= 6 \\ 5x_3 &= -15 \end{aligned}$$

and solve it by backsubstitution:

$$\begin{aligned} 5x_3 &= -15, & x_3 &= -3 \\ 3x_2 + 2x_3 &= 6 \\ 3x_2 + 2(-3) &= 6 \\ 3x_2 &= 12, & x_2 &= 4 \\ 2x_1 - x_2 - x_3 &= 3 \\ 2x_1 - 4 - (-3) &= 3 \\ 2x_1 &= 4, & x_1 &= 2 \end{aligned}$$

The solution is $(x_1, x_2, x_3) = (2, 4, -3)$. ■

The next example illustrates all the aspects of Gaussian elimination. After you follow this, you will be able to solve any system in this text.

Example 5 Solve the system

$$\begin{aligned} 6x_3 + 2x_4 - 4x_5 - 8x_6 &= 8 \\ 3x_3 + x_4 - 2x_5 - 4x_6 &= 4 \\ 2x_1 - 3x_2 + x_3 + 4x_4 - 7x_5 + x_6 &= 2 \\ 6x_1 - 9x_2 + 11x_4 - 19x_5 + 3x_6 &= 0 \end{aligned}$$

Solution First find the associated augmented matrix.

$$\begin{bmatrix} 0 & 0 & 6 & 2 & -4 & -8 & 8 \\ 0 & 0 & 3 & 1 & -2 & -4 & 4 \\ 2 & -3 & 1 & 4 & -7 & 1 & 2 \\ 6 & -9 & 0 & 11 & -19 & 3 & 0 \end{bmatrix}$$

Next follow, again, the four steps required to reduce this matrix to row echelon form.

Step 1 Locate the leftmost nonzero column and that column's pivot, which is the first nonzero entry in that column.

Leftmost nonzero column (first column); Pivot: 2

$$\begin{bmatrix} 0 & 0 & 6 & 2 & -4 & -8 & 8 \\ 0 & 0 & 3 & 1 & -2 & -4 & 4 \\ 2 & -3 & 1 & 4 & -7 & 1 & 2 \\ 6 & -9 & 0 & 11 & -19 & 3 & 0 \end{bmatrix}$$

Step 2 If necessary, interchange the first row with the row that contains the pivot.

$$\begin{bmatrix} 2 & -3 & 1 & 4 & -7 & 1 & 2 \\ 0 & 0 & 3 & 1 & -2 & -4 & 4 \\ 0 & 0 & 6 & 2 & -4 & -8 & 8 \\ 6 & -9 & 0 & 11 & -19 & 3 & 0 \end{bmatrix}$$ Rows 1 and 3 are interchanged.

Step 3 Make all other entries in the first column zero by adding suitable multiples of the first row to the other rows.

$$\begin{bmatrix} 2 & -3 & 1 & 4 & -7 & 1 & 2 \\ 0 & 0 & 3 & 1 & -2 & -4 & 4 \\ 0 & 0 & 6 & 2 & -4 & -8 & 8 \\ 0 & 0 & -3 & -1 & 2 & 0 & -6 \end{bmatrix}$$ Add -3 times row 1 to row 4.

Step 4 Repeat steps 1–3 ignoring the top row. Continue in this way until the entire matrix is in row echelon form. Most important, each time you return to step 1, ignore the rows already taken care of.

Step 1 Locate the leftmost nonzero column of what remains of the original matrix, i.e., the submatrix, and its pivot.

Pivot: 3 (row 2); Leftmost: third column

$$\begin{bmatrix} 2 & -3 & 1 & 4 & -7 & 1 & 2 \\ \hline 0 & 0 & 3 & 1 & -2 & -4 & 4 \\ 0 & 0 & 6 & 2 & -4 & -8 & 8 \\ 0 & 0 & -3 & -1 & 2 & 0 & -6 \end{bmatrix}$$

Step 2 Interchange not necessary.

Step 3

$$\begin{bmatrix} 2 & -3 & 1 & 4 & -7 & 1 & 2 \\ 0 & 0 & 3 & 1 & -2 & -4 & 4 \\ 0 & 0 & 0 & 0 & 0 & 0 & 0 \\ 0 & 0 & -3 & -1 & 2 & 0 & -6 \end{bmatrix}$$ Add -2 times row 2 to row 3.

$$\begin{bmatrix} 2 & -3 & 1 & 4 & -7 & 1 & 2 \\ 0 & 0 & 3 & 1 & -2 & -4 & 4 \\ 0 & 0 & 0 & 0 & 0 & 0 & 0 \\ 0 & 0 & 0 & 0 & 0 & -4 & -2 \end{bmatrix}$$ Add 1 times row 2 to row 4.

Step 4 Repeat again, ignoring rows 1 and 2 this time.

Step 1 Locate the leftmost nonzero column of the submatrix and its pivot.

$$\left[\begin{array}{rrrrrrr} 2 & -3 & 1 & 4 & -7 & 1 & 2 \\ 0 & 0 & 3 & 1 & -2 & -4 & 4 \\ \hdashline 0 & 0 & 0 & 0 & 0 & 0 & 0 \\ 0 & 0 & 0 & 0 & 0 & -4 & -2 \end{array}\right]$$

Pivot ↗ ↑ *Leftmost*

Step 2 Interchange the first row of the submatrix with the row that contains the pivot.

$$\left[\begin{array}{rrrrrrr} 2 & -3 & 1 & 4 & -7 & 1 & 2 \\ 0 & 0 & 3 & 1 & -2 & -4 & 4 \\ 0 & 0 & 0 & 0 & 0 & -4 & -2 \\ 0 & 0 & 0 & 0 & 0 & 0 & 0 \end{array}\right]$$

Rows 3 and 4 are interchanged.

Step 3 Not necessary.

The matrix is now in row echelon form, so we find the associated system and solve it as before

$$\begin{aligned} 2x_1 - 3x_2 + x_3 + 4x_4 - 7x_5 + x_6 &= 2 \\ 3x_3 + x_4 - 2x_5 - 4x_6 &= 4 \\ -4x_6 &= -2 \end{aligned}$$

The leading variables are x_1, x_3, and x_6, so move the remaining variables, that is, the free variables, to the other side.

$$\begin{aligned} 2x_1 + x_3 + x_6 &= 2 + 3x_2 - 4x_4 + 7x_5 \\ 3x_3 - 4x_6 &= 4 - x_4 + 2x_5 \\ -4x_6 &= -2 \end{aligned}$$

Finally, let $x_2 = t_1$, $x_4 = t_2$, and $x_5 = t_3$ and solve by backsubstitution.

$$\begin{aligned} -4x_6 &= -2, \quad x_6 = \tfrac{1}{2} \\ 3x_3 - 4(\tfrac{1}{2}) &= 4 - t_2 + 2t_3 \\ 3x_3 &= 6 - t_2 + 2t_3, \quad x_3 = 2 - \tfrac{1}{3}t_2 + \tfrac{2}{3}t_3 \\ 2x_1 + [2 - \tfrac{1}{3}t_2 + \tfrac{2}{3}t_3] + \tfrac{1}{2} &= 2 + 3t_1 - 4t_2 + 7t_3 \\ 2x_1 &= -\tfrac{1}{2} + 3t_1 - \tfrac{11}{3}t_2 + \tfrac{19}{3}t_3 \\ x_1 &= -\tfrac{1}{4} + \tfrac{3}{2}t_1 - \tfrac{11}{6}t_2 + \tfrac{19}{6}t_3 \end{aligned}$$

The solution is

$$(x_1, x_2, x_3, x_4, x_5, x_6) = (-\tfrac{1}{4} + \tfrac{3}{2}t_1 - \tfrac{11}{6}t_2 + \tfrac{19}{6}t_3, t_1, 2 - \tfrac{1}{3}t_2 + \tfrac{2}{3}t_3, t_2, t_3, \tfrac{1}{2})$$

where t_1, t_2, and t_3 are any real numbers. ■

Example 6 illustrates that the method works well when the number of equations and the number of variables are the same in larger systems. It also illustrates the common practice of performing the various parts of step 3 at the same time.

Example 6 Solve the system

$$\begin{aligned} 2x + 8y - z + w &= 0 \\ 4x + 16y - 3z - w &= -10 \\ -2x + 4y - z + 3w &= -6 \\ -6x + 2y + 5z + w &= 3 \end{aligned}$$

Solution The associated augmented matrix is

Pivot → 2

$$\begin{bmatrix} 2 & 8 & -1 & 1 & 0 \\ 4 & 16 & -3 & -1 & -10 \\ -2 & 4 & -1 & 3 & -6 \\ -6 & 2 & 5 & 1 & 3 \end{bmatrix}$$

The pivot is indicated and no interchanges are necessary, so we perform step 3.

$$\begin{bmatrix} 2 & 8 & -1 & 1 & 0 \\ 0 & 0 & -1 & -3 & -10 \\ 0 & 12 & -2 & 4 & -6 \\ 0 & 26 & 2 & 4 & 3 \end{bmatrix}$$

Add -2 times row 1 to row 2.
Add $+1$ times row 1 to row 3.
Add $+3$ times row 1 to row 4.

Pivot → 12

The pivot is indicated, and we must interchange rows 2 and 3.

$$\begin{bmatrix} 2 & 8 & -1 & 1 & 0 \\ 0 & 12 & -2 & 4 & -6 \\ 0 & 0 & -1 & -3 & -10 \\ 0 & 26 & 2 & 4 & 3 \end{bmatrix}$$

Now add $-\frac{13}{6}$ times row 2 to row 4.

$$\begin{bmatrix} 2 & 8 & -1 & 1 & 0 \\ 0 & 12 & -2 & 4 & -6 \\ 0 & 0 & -1 & -3 & -10 \\ 0 & 0 & \frac{19}{3} & -\frac{14}{3} & 16 \end{bmatrix}$$

Pivot → -1

The pivot is indicated, no interchanges are necessary, and we add $\frac{19}{3}$ times row 3 to row 4.

$$\begin{bmatrix} 2 & 8 & -1 & 1 & 0 \\ 0 & 12 & -2 & 4 & -6 \\ 0 & 0 & -1 & -3 & -10 \\ 0 & 0 & 0 & -\frac{71}{3} & -\frac{142}{3} \end{bmatrix}$$

The associated system is

$$\begin{aligned} 2x + 8y - z + w &= 0 \\ 12y - 2z + 4w &= -6 \\ -z - 3w &= -10 \\ -\tfrac{71}{3}w &= -\tfrac{142}{3} \end{aligned}$$

which we now solve by backsubstitution.

$$\begin{aligned} -\tfrac{71}{3}w &= -\tfrac{142}{3}, \quad & w &= 2 \\ -z - 3(2) &= -10 \\ -z &= -4, & z &= 4 \\ 12y - 2(4) + 4(2) &= -6 \\ 12y &= -6, & y &= -\tfrac{1}{2} \\ 2x + 8(-\tfrac{1}{2}) - 4 + 2 &= 0 \\ 2x &= 6, & x &= 3 \end{aligned}$$

The solution is $(x, y, z, w) = (3, -\frac{1}{2}, 4, 2)$. ■

Two Equations in Two Unknowns

Two equations in two unknowns are a little exceptional since they are so simple. However, they arise often, especially as examples in technical textbooks, and it is surprising how often they are solved in a manner more complicated than necessary. Almost always, the easiest way to solve a system of two equations in two unknowns is by Gaussian elimination, but without using the associated matrix.

Example 7 Solve the system

$$\begin{aligned} 2x + 3y &= 6 \\ 8x + 5y &= -4 \end{aligned}$$

Solution The pivot is in the first row of the first column. Go immediately to step 3 and add -4 times the first equation to the second.

$$\begin{aligned} 2x + 3y &= 6 \\ -7y &= -28 \end{aligned}$$

We solve this by backsubstitution.

$$\begin{aligned} -7y &= -28, \quad & y &= 4 \\ 2x + 3(4) &= 6 \\ 2x &= -6, & x &= -3 \end{aligned}$$

The solution is $(x, y) = (-3, 4)$. ■

Exercise 1.2 In Exercises 1–6, determine if the matrix is in row echelon form.

1. $\begin{bmatrix} 0 & 1 \\ 0 & 0 \end{bmatrix}$

2. $\begin{bmatrix} 0 & 1 \\ 2 & 0 \end{bmatrix}$

3. $\begin{bmatrix} 2 & 1 & -3 \\ 0 & 0 & 1 \end{bmatrix}$

4. $\begin{bmatrix} 3 & 0 & 1 \\ 0 & -3 & 2 \\ 0 & 4 & 1 \end{bmatrix}$

5. $\begin{bmatrix} 4 & 1 & 2 & 3 & 0 \\ 0 & 2 & 1 & -1 & 7 \\ 0 & 0 & 0 & 3 & 4 \\ 1 & 0 & 0 & 0 & 2 \end{bmatrix}$

6. $\begin{bmatrix} 0 \\ 0 \\ 0 \\ 0 \end{bmatrix}$

In Exercises 7–14, the augmented matrix for a linear system has been reduced to the given matrix in row echelon form and the variables are given. Solve the system.

7. $\begin{bmatrix} 2 & 1 & 3 \\ 0 & 2 & 6 \end{bmatrix}$
x, y

8. $\begin{bmatrix} -2 & 1 & 3 & 4 \\ 0 & 0 & 3 & -6 \end{bmatrix}$
x, y, z

9. $\begin{bmatrix} 0 & 3 & 2 & 6 \\ 0 & 0 & 4 & 2 \end{bmatrix}$
x_1, x_2, x_3

10. $\begin{bmatrix} 4 & 3 & 7 & 2 & 4 \\ 0 & 0 & 3 & 3 & -6 \end{bmatrix}$
x_1, x_2, x_3, x_4

11. $\begin{bmatrix} 2 & -1 & 1 & -2 \\ 0 & 3 & 4 & 15 \\ 0 & 0 & -2 & -6 \end{bmatrix}$
x, y, z

12. $\begin{bmatrix} -3 & 2 & 1 & 4 & 6 \\ 0 & 2 & 1 & 2 & -4 \\ 0 & 0 & 2 & 1 & 2 \end{bmatrix}$
x, y, z, w

13. $\begin{bmatrix} 2 & -1 & -3 & 4 & 2 \\ 0 & 0 & 2 & 1 & -1 \\ 0 & 0 & 0 & 3 & 9 \\ 0 & 0 & 0 & 0 & 0 \end{bmatrix}$
x_1, x_2, x_3, x_4

14. $\begin{bmatrix} 0 & 4 & -1 & 3 & 7 & 1 \\ 0 & 0 & 0 & 3 & 4 & 9 \\ 0 & 0 & 0 & 0 & 5 & 0 \end{bmatrix}$
x, y, z, u, v

In Exercises 15–28, solve the system by Gaussian elimination.

15. $$\begin{aligned} x + 2y + z &= 8 \\ -x + 3y - 2z &= 1 \\ 3x + 4y - 7z &= 10 \end{aligned}$$

16. $$\begin{aligned} x_1 + 7x_2 - 7x_3 &= 0 \\ 2x_1 + 3x_2 + x_3 &= 0 \\ x_1 - 4x_2 + 3x_3 &= 0 \end{aligned}$$

17. $$\begin{aligned} 2x_1 + 3x_2 - x_3 + x_4 &= -5 \\ 4x_1 + 5x_2 + 2x_3 - x_4 &= 4 \\ -2x_1 - x_2 - x_3 - x_4 &= 1 \\ 6x_1 + 7x_2 + x_3 - 4x_4 &= 2 \end{aligned}$$

18. $$\begin{aligned} x_1 - x_2 - x_3 &= 4 \\ x_1 + x_2 + x_3 &= 2 \end{aligned}$$

19. $$\begin{aligned} 2u - 3v + w - x + y &= 0 \\ 4u - 6v + 2w - 3x - y &= -5 \\ -2u + 3v - 2w + 2x - y &= 3 \end{aligned}$$

20. $$\begin{aligned} 3x - y &= 7 \\ 6x + 2y &= 10 \\ -3x + 4y &= -10 \end{aligned}$$

21. $$\begin{aligned} 3x_1 - x_2 + x_3 - 4x_4 &= 2 \\ 6x_1 + 3x_2 - x_3 - 4x_4 &= 3 \\ 9x_1 + 2x_2 \qquad - 8x_4 &= 6 \end{aligned}$$

22. $$\begin{aligned} 3x_1 - x_2 + x_3 - 5x_4 - x_5 &= 0 \\ 6x_1 - 2x_2 + 2x_3 - 9x_4 + x_5 &= 0 \\ -9x_1 + 3x_2 - 3x_3 + 11x_4 - x_5 &= 0 \end{aligned}$$

23. $$\begin{aligned} -x_1 + 2x_2 - x_3 &= -4 \\ 3x_1 + 4x_2 + 2x_3 &= 15 \\ -4x_1 + 6x_2 + x_3 &= -7 \end{aligned}$$

24. $$\begin{aligned} x_1 + x_2 + x_3 + x_4 + x_5 &= 5 \\ x_1 \qquad\qquad\qquad + x_5 &= -4 \\ x_1 - x_2 \qquad\qquad\qquad &= 3 \end{aligned}$$

25. $$\begin{aligned} 3x + 2y &= 0 \\ 6x + 7y &= 3 \end{aligned}$$

26. $$\begin{aligned} 4x + 5y &= 7 \\ 12x + 7y &= -3 \end{aligned}$$

27. $$\begin{aligned} 4x - 8y &= 3 \\ 3x + 2y &= 13 \end{aligned}$$

28. $$\begin{aligned} 3x - 6y &= 21 \\ 5x - 2y &= -5 \end{aligned}$$

29. If you solve the system

$$\begin{aligned} &(1) & 2x + 3y + 4z &= 5 \\ &(2) & 6x + 7y + 8z &= 9 \\ &(3) & ax + by + cz &= d \end{aligned}$$

by Gaussian elimination, are the following statements true or false?

(a) If $a = 0$, then no multiple of equation (1) would be subtracted from equation (3).

(b) If $b = 0$, then no multiple of equation (2) would be subtracted from equation (3).

(c) If $a = 0$ and $b = 0$, then no multiple of equation (1) or equation (2) would be subtracted from equation (3).

1.3 The Algebra of Matrices; Five Descriptions of the Product

In the preceding sections we have seen how matrices are helpful in Gaussian elimination. One of our goals is to do Gaussian elimination completely with matrices by having them "operate" on one another. This will help us both to understand Gaussian elimination better and to implement it on a computer more easily. In addition to computer applications, matrices arise in other contexts where knowing how matrices relate to one another is crucial. The purpose of this section is to lay the groundwork for learning these various matrix relationships.

We begin by developing a general notation for the entries of a matrix and move on to formally defining a matrix.

NOTATION We usually use capital latters for matrices and lowercase letters for entries.

Example 1

$$A = \begin{bmatrix} 2 & -1 \\ 0 & 3 \\ 4 & 1 \end{bmatrix}, \quad B = \begin{bmatrix} a & b & c \\ d & e & f \end{bmatrix}, \quad C = \begin{bmatrix} c_{11} & c_{12} & c_{13} & c_{14} \\ c_{21} & c_{22} & c_{23} & c_{24} \end{bmatrix}$$ ■

(1.9)

DEFINITION If m and n are positive integers, then an $\boldsymbol{m \times n}$ **matrix** (read "m by n matrix") is a rectangular array of m rows and n columns of the form

$$\begin{bmatrix} a_{11} & a_{12} & a_{13} & \cdots & a_{1n} \\ a_{21} & a_{22} & a_{23} & \cdots & a_{2n} \\ \vdots & \vdots & \vdots & & \vdots \\ a_{m1} & a_{m2} & a_{m3} & \cdots & a_{mn} \end{bmatrix}$$

where each a_{ij} is a number called an **entry** or **element** of the matrix. The numbers m and n are called the **dimensions** of the matrix.

This way of denoting an element of a matrix, a_{ij}, is called **double-subscript notation.** It is a convenient way to denote the location of an element in the matrix. The first subscript tells what row the element is in, and the second subscript tells what column. Sometimes, instead of writing out the matrix as was done in the definition, we write "the $m \times n$ matrix $A = (a_{ij})$." This indicates that the entries of A are denoted by a's, that we are using double-subscript notation, and that A has m rows and n columns.

In Example 1, A is a 3×2 matrix, B is 2×3, and C is 2×4. If the number of rows of a matrix $A = [a_{ij}]$ equals the number of columns, that is, if A is $n \times n$, then A is called a **square** matrix and the entries $a_{11}, a_{22}, \ldots, a_{nn}$ form the **(main) diagonal** of A.

(1.10)

DEFINITION If $A = [a_{ij}]$ and $B = [b_{ij}]$ are matrices, A is **equal** to B if their dimensions are the same and their corresponding entries are equal term by term; that is,

$$A = B \quad \text{if and only if} \quad a_{ij} = b_{ij} \text{ for every } i \text{ and } j$$

If two matrices have different dimensions, they cannot be equal.

Example 2

$$\begin{bmatrix} (\sqrt{2})^6 & 0^3 & 3^2 \\ (-2)^3 & (-1)^2 & \sqrt[3]{64} \end{bmatrix} = \begin{bmatrix} 8 & 0 & 9 \\ -8 & 1 & 4 \end{bmatrix}$$ ■

(1.11)

DEFINITION If $A = [a_{ij}]$ and $B = [b_{ij}]$ are $m \times n$ matrices, their **sum**, $A + B$, is the $m \times n$ matrix obtained by adding the corresponding entries; that is,

$$[a_{ij}] + [b_{ij}] = [a_{ij} + b_{ij}]$$

If two matrices are of different dimensions, the sum is not defined.

Example 3

$$\begin{bmatrix} 3 & -2 & -14.3 \\ 6 & -1 & 0.02 \end{bmatrix} + \begin{bmatrix} -4 & 4 & 4.5 \\ 8 & 0 & 0.25 \end{bmatrix} = \begin{bmatrix} -1 & 2 & -9.8 \\ 14 & -1 & 0.27 \end{bmatrix}$$ ■

The commutative and associative laws of addition hold (when defined) for matrices. That is, if A, B, and C are $m \times n$ matrices, then

(1.12)

$$A + B = B + A, \quad A + (B + C) = (A + B) + C$$

See Exercises 27 and 28 for an indication of the proofs. The $m \times n$ **zero matrix**, denoted by $\mathbf{0}$, is the $m \times n$ matrix whose entries are all zero. If A is an $m \times n$ matrix, then

(1.13)

$$A + \mathbf{0} = A = \mathbf{0} + A$$

so that the zero matrix is the additive identity. See Exercise 29.

If A is an $m \times n$ matrix, its **negative** is denoted by $-A$ and is the $m \times n$ matrix obtained from A by replacing each entry with its negative; that is,

(1.14)

$$-[a_{ij}] = [-a_{ij}]$$

It is easy to see that

(1.15)

$$A + (-A) = \mathbf{0} = (-A) + A$$

Traditionally, a number is called a **scalar** when discussing matrices (or vectors).

(1.16)

DEFINITION If a is a scalar and $B = [b_{ij}]$ is an $m \times n$ matrix, then the **scalar product** aB is the $m \times n$ matrix obtained by multiplying each entry of B by a, that is,

$$a[b_{ij}] = [ab_{ij}]$$

Example 4 If

$$A = \begin{bmatrix} 3 & -7 & 2 \\ -1 & 0 & 2 \end{bmatrix}$$

then

$$3A = \begin{bmatrix} 9 & -21 & 6 \\ -3 & 0 & 6 \end{bmatrix} \quad \text{and} \quad -A = (-1)A = \begin{bmatrix} -3 & 7 & -2 \\ 1 & 0 & -2 \end{bmatrix}$$ ■

The associative and distributive laws hold (when defined) for scalars and matrices. Thus if A and B are $m \times n$ matrices and a and b are scalars, then

(1.17)
$$(a + b)A = aA + bA, \quad a(A + B) = aA + aB$$
$$(ab)A = a(bA), \quad 1A = A, \quad (-1)A = -A$$

See Exercise 30.

Products

We are now ready to define the product of two matrices. The definition may seem a little strange at first, but it is justified by Theorem (1.31), by other theoretical relationships such as those in Section 1.4, and by its applications. In fact, as we shall see, the product is so rich in structure that there are several different ways of looking at it that prove useful in different contexts.

If A and B are two matrices, the only requirement restricting the definition of their product AB is that the number of columns of A must equal the number of rows of B. Their product C will then have the same number of rows as A and the same number of columns as B. Thus

(1.18)
If A is an $m \times n$ matrix and B is an $n \times p$ matrix, then

$$C = AB$$

is an $m \times p$ matrix.

To compute the element c_{ij} in the ith row and jth column of C, we first single out the ith row of A and the jth column of B:

$$\begin{bmatrix} a_{11} & a_{12} & \cdots & a_{1n} \\ \vdots & \vdots & & \vdots \\ a_{i1} & a_{i2} & \cdots & a_{in} \\ \vdots & \vdots & & \vdots \\ a_{m1} & a_{m2} & \cdots & a_{mn} \end{bmatrix} \begin{bmatrix} b_{11} & \cdots & b_{1j} & \cdots & b_{1p} \\ b_{21} & \cdots & b_{2j} & \cdots & b_{2p} \\ \vdots & & \vdots & & \vdots \\ b_{n1} & \cdots & b_{nj} & \cdots & b_{np} \end{bmatrix}$$

Next we multiply each element in the ith row of A times the corresponding element in the jth column of B, and then add the products. This gives us the formula:

(1.19)
$$c_{ij} = a_{i1}b_{1j} + a_{i2}b_{2j} + \cdots + a_{in}b_{nj}$$

NOTE This expression is called an **inner product** of the ith row of A with the jth column of B. The product of matrices usually is described by expressing the entries of $C = AB$ in this manner.

Example 5 Compute the product of the following 2×3 and 3×4 matrices (which will give us a 2×4 matrix):

$$\begin{bmatrix} 1 & 3 & -2 \\ -3 & 0 & 5 \end{bmatrix} \begin{bmatrix} 4 & 5 & -1 & 3 \\ 1 & -2 & 0 & 1 \\ 2 & -1 & 0 & 2 \end{bmatrix}$$

Solution We compute the eight entries as follows:

$$\begin{aligned} c_{11} &= 1(4) + 3(1) + (-2)(2) = 3 \\ c_{12} &= 1(5) + 3(-2) + (-2)(-1) = 1 \\ c_{13} &= 1(-1) + 3(0) + (-2)(0) = -1 \\ c_{14} &= 1(3) + 3(1) + (-2)(2) = 2 \\ c_{21} &= (-3)(4) + 0(1) + 5(2) = -2 \\ c_{22} &= (-3)(5) + 0(-2) + 5(-1) = -20 \\ c_{23} &= (-3)(-1) + 0(0) + 5(0) = 3 \\ c_{24} &= (-3)(3) + 0(1) + 5(2) = 1 \end{aligned}$$

Thus the product matrix is the 2×4 matrix

$$\begin{bmatrix} 3 & 1 & -1 & 2 \\ -2 & -20 & 3 & 1 \end{bmatrix}$$

■

An Application

Example 6 illustrates how matrices and matrix multiplication are used in inventory control and cost analysis.

Example 6 A certain store sells brand X refrigerators and freezers. The following matrix on the left gives the sales of these items for three months; the matrix on the right gives the sales price and dealer's cost of these items. Use matrix multiplication to generate a matrix that has as its entries the total dollar sales and the total retail costs of brand X items for each of the three months.

$$\begin{array}{r} \\ \text{Refrigerators} \\ \text{Freezers} \end{array} \begin{array}{c} \begin{matrix} \text{Jan.} & \text{Feb.} & \text{Mar.} \end{matrix} \\ \begin{bmatrix} 32 & 43 & 35 \\ 21 & 23 & 19 \end{bmatrix} \end{array} \qquad \begin{array}{r} \\ \text{Retail price} \\ \text{Dealer cost} \end{array} \begin{array}{c} \begin{matrix} \text{Refs.} & \text{Frzrs.} \end{matrix} \\ \begin{bmatrix} 445 & 250 \\ 310 & 180 \end{bmatrix} \end{array}$$

Solution For January the total dollar sales price of brand X items is

Retail price of refrigerator	times	No. refrigerators sold	plus	Retail price of freezer	times	No. freezers sold
445	×	32	+	250	×	21

and the total dollar cost is

Cost of refrigerator	times	No. refrigerators sold	plus	Cost of freezer	times	No. freezers sold
310	×	32	+	180	×	21

Similar computations hold for February and March, and combining the three months yields the following product.

$$\begin{bmatrix} 445 & 250 \\ 310 & 180 \end{bmatrix}\begin{bmatrix} 32 & 43 & 35 \\ 21 & 23 & 19 \end{bmatrix} = \begin{bmatrix} 19{,}490 & 24{,}885 & 20{,}325 \\ 13{,}700 & 17{,}470 & 14{,}270 \end{bmatrix}$$

Therefore, the requested matrix is

	Jan.	Feb.	Mar.
Total dollar sales	19,490	24,885	20,325
Total dealer cost	13,700	17,470	14,270

■

Properties of Matrix Multiplication

One of the most important facts about matrix multiplication is a property it does *not* have.

(1.20)

> WARNING Matrix multiplication is *not* commutative, that is, in general
>
> $$AB \neq BA$$

Example 7

$$\begin{bmatrix} 1 & 4 \\ 2 & 8 \end{bmatrix}\begin{bmatrix} 4 & -8 \\ -1 & 2 \end{bmatrix} = \begin{bmatrix} 0 & 0 \\ 0 & 0 \end{bmatrix}$$

$$\begin{bmatrix} 4 & -8 \\ -1 & 2 \end{bmatrix}\begin{bmatrix} 1 & 4 \\ 2 & 8 \end{bmatrix} = \begin{bmatrix} -12 & -48 \\ 3 & 12 \end{bmatrix}$$

■

In fact, for many pairs of matrices, AB is defined, whereas BA is not. Example 5 is such a case.

Fortunately most of the other laws of multiplication that hold for multiplication of numbers also hold for multiplication of matrices. In particular, the associative law of multiplication

(1.21)

$$A(BC) = (AB)C$$

and the two distributive laws

(1.22)
$$A(B + C) = AB + AC, \quad (A + B)C = AC + BC$$

hold (when all the products and sums are defined). See Exercises 31 and 32.

Next on the list of definitions is the identity matrix. Let I_n be the square $n \times n$ matrix with 1's along the diagonal and 0's elsewhere, so that

$$I_2 = \begin{bmatrix} 1 & 0 \\ 0 & 1 \end{bmatrix} = \begin{bmatrix} 1 & \\ & 1 \end{bmatrix}$$

$$I_3 = \begin{bmatrix} 1 & 0 & 0 \\ 0 & 1 & 0 \\ 0 & 0 & 1 \end{bmatrix} = \begin{bmatrix} 1 & & \\ & 1 & \\ & & 1 \end{bmatrix}, \quad \text{etc.*}$$

Then I_n is called the $n \times n$ **identity matrix** and is the multiplicative identity; that is,

(1.23)
If A is any $m \times n$ matrix and B is any $n \times p$ matrix,

$$AI_n = A \quad \text{and} \quad I_nB = B$$

We can now use some of these matrix properties to manipulate systems of linear equations. Consider the following system of m linear equations in n variables.

(1.24)
$$\begin{matrix} a_{11}x_1 + a_{12}x_2 + \cdots + a_{1n}x_n = b_1 \\ a_{21}x_1 + a_{22}x_2 + \cdots + a_{2n}x_n = b_2 \\ \vdots \qquad \vdots \qquad \qquad \vdots \quad \vdots \\ a_{m1}x_1 + a_{m2}x_2 + \cdots + a_{mn}x_n = b_m \end{matrix}$$

Let

$$A = \begin{bmatrix} a_{11} & a_{12} & \dots & a_{1n} \\ a_{21} & a_{22} & \dots & a_{2n} \\ \vdots & \vdots & & \vdots \\ a_{m1} & a_{m2} & \dots & a_{mn} \end{bmatrix}, \quad X = \begin{bmatrix} x_1 \\ x_2 \\ \vdots \\ x_n \end{bmatrix}, \quad B = \begin{bmatrix} b_1 \\ b_2 \\ \vdots \\ b_m \end{bmatrix}$$

Using the definition of matrix multiplication, we see that the product AX is an $m \times 1$ matrix,

(1.25)
$$AX = \begin{bmatrix} a_{11}x_1 + a_{12}x_2 + \cdots + a_{1n}x_n \\ a_{21}x_1 + a_{22}x_2 + \cdots + a_{2n}x_n \\ \vdots \qquad \vdots \qquad \qquad \vdots \\ a_{m1}x_1 + a_{m2}x_2 + \cdots + a_{mn}x_n \end{bmatrix}$$

* *We introduce here the convention that a blank entry is understood to be a zero. We shall use blanks only when this visually simplifies the matrix. Zeros should always be used when there is any question as to the location of nonzero entries.*

Using the definition of equality of matrices, the system of m equations of Equation (1.24) is equivalent to the single equation of matrices

$$AX = B \tag{1.26}$$

Equation (1.26) is called the **matrix equation associated with the system** (1.24). The matrix A is called the **coefficient matrix**, and B is called the **matrix of constants.**

Example 8 Find the matrix equation associated with the system

$$\begin{aligned} 2x - 3y + 4z &= 5 \\ 4x \qquad - z &= -2 \end{aligned}$$

Solution

$$\begin{bmatrix} 2 & -3 & 4 \\ 4 & 0 & -1 \end{bmatrix} \begin{bmatrix} x \\ y \\ z \end{bmatrix} = \begin{bmatrix} 5 \\ -2 \end{bmatrix}$$ ■

We are now ready to introduce a very important application of matrix multiplication. Suppose we have a linear system such as (1.24) and suppose in addition that each of the x's is a linear function of some other variables. For example, suppose, as in Example 8,

$$\begin{aligned} 2x - 3y + 4z &= 5 \\ 4x \qquad - z &= -2 \end{aligned} \tag{1.27}$$

and, in addition, x, y, and z are linear functions of variables s and t as follows:

$$\begin{aligned} x &= 3s - t \\ y &= 2s + 5t \\ z &= -s + 2t \end{aligned} \tag{1.28}$$

Suppose we wanted to substitute Equations (1.28) into Equations (1.27) in order to solve for s and t. We can do this directly, obtaining

$$\begin{aligned} 2(3s - t) - 3(2s + 5t) + 4(-s + 2t) &= 5 \\ 4(3s - t) \qquad\qquad - (-s + 2t) &= -2 \end{aligned}$$

This is simplified to obtain

$$\begin{aligned} -4s - 9t &= 5 \\ 13s - 6t &= -2 \end{aligned} \tag{1.29}$$

However, there is another way of doing this that usually is far easier, especially for large systems. Consider the associated matrix equations to (1.27), (1.28), and (1.29).

(1.30)
$$\begin{bmatrix} 2 & -3 & 4 \\ 4 & 0 & -1 \end{bmatrix} \begin{bmatrix} x \\ y \\ z \end{bmatrix} = \begin{bmatrix} 5 \\ -2 \end{bmatrix}, \quad \begin{bmatrix} x \\ y \\ z \end{bmatrix} = \begin{bmatrix} 3 & -1 \\ 2 & 5 \\ -1 & 2 \end{bmatrix} \begin{bmatrix} s \\ t \end{bmatrix}$$
$$\begin{bmatrix} -4 & -9 \\ 13 & -6 \end{bmatrix} \begin{bmatrix} s \\ t \end{bmatrix} = \begin{bmatrix} 5 \\ -2 \end{bmatrix}$$

We shall denote these by $AX = B$, $X = CT$, and $DT = B$ respectively. If we substitute $X = CT$ into the equation $AX = B$, we obtain $A(CT) = B$ or

$$\begin{bmatrix} 2 & -3 & 4 \\ 4 & 0 & -1 \end{bmatrix} \begin{bmatrix} 3 & -1 \\ 2 & 5 \\ -1 & 2 \end{bmatrix} \begin{bmatrix} s \\ t \end{bmatrix} = \begin{bmatrix} 5 \\ -2 \end{bmatrix}$$

If we multiply the two coefficient matrices together [we are using associativity of matrix multiplication, $A(CT) = (AC)T$], we obtain

$$\begin{bmatrix} -4 & -9 \\ 13 & -6 \end{bmatrix} \begin{bmatrix} s \\ t \end{bmatrix} = \begin{bmatrix} 5 \\ -2 \end{bmatrix}$$

Thus we obtain the matrix equation $DT = B$ simply by multiplying the coefficient matrices. This illustrates the situation in general.

(1.31)

THEOREM If one linear system with matrix equation $X = CT$

$$\begin{aligned} x_1 &= c_{11}t_1 + \cdots + c_{1p}t_p \\ &\vdots \\ x_n &= c_{n1}t_1 + \cdots + c_{np}t_p \end{aligned}$$

is substituted into a second linear system with matrix equation $AX = B$,

$$\begin{aligned} a_{11}x_1 + \cdots + a_{1n}x_n &= b_1 \\ &\vdots \\ a_{m1}x_1 + \cdots + a_{mn}x_n &= b_m \end{aligned}$$

and the results are simplified to yield a third linear system $DT = B$,

$$\begin{aligned} d_{11}t_1 + \cdots + d_{1p}t_p &= b_1 \\ &\vdots \\ d_{m1}t_1 + \cdots + d_{mp}t_p &= b_m \end{aligned}$$

then the coefficient matrix D obtained is the product

$$D = AC$$

More examples of this type of matrix manipulation are given in Exercises 21–24.

Other Descriptions of the Product*

One different way of describing a product involves the special case $AX = B$ from Equation (1.25). Thus A is $m \times n$, X is $n \times 1$, and B is $m \times 1$. Instead of multiplying out to obtain Equation (1.25), we break up A into columns, so that $A = [A_1 \vdots A_2 \vdots \cdots \vdots A_n]$, multiply the ith column by x_i, and then add them up.

(1.32) THEOREM If $A = [A_1 \vdots A_2 \vdots \cdots \vdots A_n]$ is $m \times n$ and X is $n \times 1$, then

(1.33)
$$\begin{bmatrix} a_{11}x_1 + a_{12}x_2 + \cdots + a_{1n}x_n \\ a_{21}x_1 + a_{22}x_2 + \cdots + a_{2n}x_n \\ \vdots \\ a_{m1}x_1 + a_{m2}x_2 + \cdots + a_{mn}x_n \end{bmatrix} = x_1\begin{bmatrix} a_{11} \\ a_{21} \\ \vdots \\ a_{m1} \end{bmatrix} + x_2\begin{bmatrix} a_{12} \\ a_{22} \\ \vdots \\ a_{m2} \end{bmatrix} + \cdots + x_n\begin{bmatrix} a_{1n} \\ a_{2n} \\ \vdots \\ a_{mn} \end{bmatrix}$$

so that

$$AX = x_1A_1 + x_2A_2 + \cdots + x_nA_n$$

Thus the two equations

$$AX = B \quad \text{and}$$
$$x_1A_1 + x_2A_2 + \cdots + x_nA_n = B$$

are equivalent formulations.

We introduce this concept here to give you a head start into the discussions in Section 3.4 and elsewhere in the text, where it is broadly used.

Proof Simply multiply out and add up the right-hand side of Equation (1.33) to see that we obtain the result on the left-hand side. ■

Example 9 From Example 8, we can check that

$$\begin{bmatrix} 2 & -3 & 4 \\ 4 & 0 & -1 \end{bmatrix}\begin{bmatrix} -1 \\ -5 \\ -2 \end{bmatrix} = \begin{bmatrix} 5 \\ -2 \end{bmatrix}$$

and also that

$$(-1)\begin{bmatrix} 2 \\ 4 \end{bmatrix} + (-5)\begin{bmatrix} -3 \\ 0 \end{bmatrix} + (-2)\begin{bmatrix} 4 \\ -1 \end{bmatrix} = \begin{bmatrix} 5 \\ -2 \end{bmatrix}$$ ■

* *If desired, the balance of this section may be postponed until just before Section 3.4.*

The expression $x_1A_1 + x_2A_2 + \cdots + x_nA_n$, which consists of multiplying each column by a number and adding them up, is called a **linear combination** of the columns. In writing B as $B = x_1A_1 + x_2A_2 + \cdots + x_nA_n$, we say that we **express B as a linear combination of the columns A_i**.

The remaining ways of looking at products involve matrices of all sizes. One such way is to consider each column of an $m \times n$ matrix B as an $m \times 1$ matrix and to observe that each column of a product AB is A times the corresponding column of B.

Example 10

$$AB = \begin{bmatrix} 2 & -3 & 1 \\ 4 & 1 & 2 \end{bmatrix} \begin{bmatrix} 5 & 2 \\ 0 & -7 \\ -1 & 3 \end{bmatrix}$$

$$= \left[\begin{bmatrix} 2 & -3 & 1 \\ 4 & 1 & 2 \end{bmatrix} \begin{bmatrix} 5 \\ 0 \\ -1 \end{bmatrix} \vdots \begin{bmatrix} 2 & -3 & 1 \\ 4 & 1 & 2 \end{bmatrix} \begin{bmatrix} 2 \\ -7 \\ 3 \end{bmatrix} \right]$$

$$= \left[\begin{bmatrix} 9 \\ 18 \end{bmatrix} \vdots \begin{bmatrix} 28 \\ 7 \end{bmatrix} \right] = \begin{bmatrix} 9 & 28 \\ 18 & 7 \end{bmatrix}$$

■

In general, if the jth column of B is B_j, so $B = [B_1 \vdots B_2 \vdots \cdots \vdots B_n]$, then

$$AB = A[B_1 \vdots B_2 \vdots \cdots \vdots B_n] = [AB_1 \vdots AB_2 \vdots \cdots \vdots AB_n]$$

A corresponding relationship is true for rows of the first matrix in a product.

Example 11 From Example 10,

$$[4 \quad 1 \quad 2] \begin{bmatrix} 5 & 2 \\ 0 & -7 \\ -1 & 3 \end{bmatrix} = [18 \quad 7]$$

Second row of A. *Second row of AB.*

■

This way of looking at multiplication will be very useful throughout the text. [See, for example, the proof of Theorem (5.20), Section 5.3.]

Outer Products*

The following way of looking at products has become more important with the advent of supercomputers. To make better use of the potential of parallel processors, we must rearrange the way we do mathematical computations.

* *The remainder of this section is optional. It is intended for general interest only and is not referenced elsewhere in the text.*

One method that is helpful with products is illustrated in the following example.

Example 12 Again consider the computation $AB = C$ given in Example 5,

$$\begin{bmatrix} 1 & 3 & -2 \\ -3 & 0 & 5 \end{bmatrix} \begin{bmatrix} 4 & 5 & -1 & 3 \\ 1 & -2 & 0 & 1 \\ 2 & -1 & 0 & 2 \end{bmatrix} = \begin{bmatrix} 3 & 1 & -1 & 2 \\ -2 & -20 & 3 & 1 \end{bmatrix}$$

Now A_i (each column of A) is a 2×1 matrix, B_i (each row of B) is a 1×4 matrix, and the product A_iB_i is a 2×4 matrix,

$$A_1B_1 = \begin{bmatrix} 1 \\ -3 \end{bmatrix} [4 \quad 5 \quad -1 \quad 3] = \begin{bmatrix} 4 & 5 & -1 & 3 \\ -12 & -15 & 3 & -9 \end{bmatrix}$$

$$A_2B_2 = \begin{bmatrix} 3 \\ 0 \end{bmatrix} [1 \quad -2 \quad 0 \quad 1] = \begin{bmatrix} 3 & -6 & 0 & 3 \\ 0 & 0 & 0 & 0 \end{bmatrix}$$

$$A_3B_3 = \begin{bmatrix} -2 \\ 5 \end{bmatrix} [2 \quad -1 \quad 0 \quad 2] = \begin{bmatrix} -4 & 2 & 0 & -4 \\ 10 & -5 & 0 & 10 \end{bmatrix}$$

Each of these products, an $m \times 1$ matrix multiplied by a $1 \times p$ matrix, is called an **outer product**. Amazingly, if we add up the three outer products on the right side, we get

$$A_1B_1 + A_2B_2 + A_3B_3 = C$$

■

What we just observed in Example 12 is true in general.

(1.34)

THEOREM Let A be an $m \times n$ matrix broken up in terms of its n *columns* and let B be an $n \times p$ matrix broken up in terms of its n *rows*,

$$A = [A_1 \,\vdots\, A_2 \,\vdots\, \cdots \,\vdots\, A_n] \quad \text{and} \quad B = \begin{bmatrix} B_1 \\ B_2 \\ \vdots \\ B_n \end{bmatrix}$$

Then the product $C = AB$ is the sum of n outer products, or

(1.35)

$$C = A_1B_1 + A_2B_2 + \cdots + A_nB_n$$

Proof Multiply out the right-hand side of Equation (1.35) and see that we obtain the result on the left-hand side. ■

Exercise 1.3

1. Write down the 2×3 matrices A and B such that $a_{ij} = i + j$ and $b_{ij} = (-1)^{i+j}$.
2. Write down the 2×3 matrices A and B such that $a_{ij} = ij$ and $b_{ij} = 1/(i + j)$.

In Exercises 3–6, find $A + B$, $2A$, and $-B$.

3. $A = \begin{bmatrix} 3 & 4 & 1 \\ -1 & 0 & -6 \end{bmatrix}$, $B = \begin{bmatrix} 4 & 2 & -1 \\ -2 & 5 & 3 \end{bmatrix}$

4. $A = \begin{bmatrix} 4 & 1 & 3 \\ 3 & -1 & -2 \\ 0 & 0 & 4 \end{bmatrix}$, $B = \begin{bmatrix} -3 & -2 & -3 \\ -1 & 0 & 0 \\ 8 & -2 & -4 \end{bmatrix}$

5. $A = \begin{bmatrix} 7 & 2 & 0 & -3 \\ -2 & 0 & -5 & 2 \end{bmatrix}$, $B = \begin{bmatrix} 3 & -2 & -1 & 0 \\ 5 & 4 & 0 & -8 \end{bmatrix}$

6. $A = \begin{bmatrix} 4 & 2 & 0 & 2 & 4 \end{bmatrix}$, $B = \begin{bmatrix} -3 & -2 & 1 & 0 & -4 \end{bmatrix}$

In Exercises 7–14, find AB and BA, if possible, or state "undefined."

7. $A = \begin{bmatrix} -1 & -2 \\ -3 & -4 \end{bmatrix}$, $B = \begin{bmatrix} 4 & 3 \\ -2 & -2 \end{bmatrix}$

8. $A = \begin{bmatrix} -1 & 0 & 2 \\ 2 & 1 & 3 \\ 3 & 0 & 0 \end{bmatrix}$, $B = \begin{bmatrix} 4 & 0 & -1 \\ 3 & 2 & 0 \\ 0 & 2 & -4 \end{bmatrix}$

9. $A = \begin{bmatrix} 3 & 2 & -3 \\ 0 & 1 & 0 \end{bmatrix}$, $B = \begin{bmatrix} 3 & -4 \\ 0 & 1 \\ 1 & 0 \end{bmatrix}$

10. $A = \begin{bmatrix} 1 & 3 & -3 \end{bmatrix}$, $B = \begin{bmatrix} 0 \\ 4 \\ -2 \end{bmatrix}$

11. $A = \begin{bmatrix} 3 & -1 \\ 0 & 3 \end{bmatrix}$, $B = \begin{bmatrix} 3 & -1 & 0 \\ 2 & 0 & -1 \end{bmatrix}$

12. $A = \begin{bmatrix} 5 & 3 \\ -1 & 0 \end{bmatrix}$, $B = \begin{bmatrix} 4 & -2 \\ -2 & 0 \\ 9 & 1 \end{bmatrix}$

13. $A = \begin{bmatrix} 2 & -5 \\ 1 & 0 \\ 2 & 1 \end{bmatrix}$, $B = \begin{bmatrix} 3 \\ 4 \\ 2 \end{bmatrix}$

14. $A = \begin{bmatrix} 2 \\ 1 \end{bmatrix}, \quad B = \begin{bmatrix} 3 & 0 & -4 \\ 0 & 5 & 2 \end{bmatrix}$

In Exercises 15 and 16, let

$$A = \begin{bmatrix} 1 & 4 \\ -2 & 0 \end{bmatrix}, \quad B = \begin{bmatrix} 1 & -1 \\ 0 & 1 \end{bmatrix}$$

15. Use the above matrices to show that in general $(A - B)(A + B) \neq A^2 - B^2$, where $A^2 = AA$ and $B^2 = BB$.

16. Use the above matrices to show that in general $(A + B)^2 \neq A^2 + 2AB + B^2$, where $2AB = AB + AB$.

In Exercises 17–20, for the two systems given, (a) add the systems term by term and then find the associated matrix equation and (b) find the associated matrix equations and then add the corresponding matrices.

17. $\begin{aligned} 3x - 2y &= -2 \\ 4x - y &= 5 \end{aligned} \qquad \begin{aligned} 5x + 2y &= -7 \\ 7x - 8y &= -2 \end{aligned}$

18. $\begin{aligned} 5x + y \phantom{{}- 2z} &= -5 \\ 2x + 4y - 2z &= -1 \\ 5x \phantom{{}+ 4y} + z &= 0 \end{aligned} \qquad \begin{aligned} 2y + z &= 4 \\ 3x - y - z &= 2 \\ -5x + y \phantom{{}- z} &= -2 \end{aligned}$

19. $\begin{aligned} 2x - y - z &= 1 \\ 3x + y - 4z &= -6 \end{aligned} \qquad \begin{aligned} 3x + 2y - 8z &= 9 \\ 7x + 4y + 3z &= -2 \end{aligned}$

20. $\begin{aligned} 2x - y &= 1 \\ 3x + 7y &= 2 \\ x + 6y &= 9 \end{aligned} \qquad \begin{aligned} 3x + 2y &= 7 \\ x - 3y &= -2 \\ 7x - y &= -2 \end{aligned}$

In Exercises 21–24, for the two systems given, (a) substitute the second system in the first, simplify, and then find the associated matrix equation and (b) find the associated matrix equations for the given systems and then substitute in and multiply the coefficient matrices.

21. $\begin{aligned} 3x - 4y &= 2 \\ 4x + 6y &= 4 \end{aligned} \qquad \begin{aligned} x &= 3s - t \\ y &= 4s + 2t \end{aligned}$

22. $\begin{aligned} 5x - 2y + z &= 1 \\ 4x - y + 2z &= -5 \end{aligned} \qquad \begin{aligned} x &= 3p - q \\ y &= p \phantom{{}- q} + 3r \\ z &= \phantom{3p - {}} q - r \end{aligned}$

23. $\begin{aligned} 5x - 2y &= 10 \\ 3x + 9y &= -2 \end{aligned} \qquad \begin{aligned} x &= 2p + 4q - 2r \\ y &= 3p - q + r \end{aligned}$

24. $\begin{aligned} 5x - 3y - 2z &= 4 \\ 2x + 4y - z &= 2 \\ y + 7z &= 3 \end{aligned} \qquad \begin{aligned} x &= 3p - q \\ y &= 2p + 4q \\ z &= -p + 2q \end{aligned}$

25. A store sells brand X and brand Y dishwashers. The following matrices give the sales figures and costs of these items for three months. Use matrix multiplication to determine the total dollar sales and total costs of these items for the three months.

$$\begin{array}{l} \\ \text{Brand } X \\ \text{Brand } Y \end{array} \begin{array}{c} \begin{array}{ccc} \text{Dec.} & \text{Apr.} & \text{Aug.} \end{array} \\ \begin{bmatrix} 18 & 10 & 12 \\ 19 & 12 & 14 \end{bmatrix} \end{array} \qquad \begin{array}{l} \\ \text{Retail price} \\ \text{Dealer cost} \end{array} \begin{array}{c} \begin{array}{cc} X & Y \end{array} \\ \begin{bmatrix} 350 & 260 \\ 240 & 190 \end{bmatrix} \end{array}$$

26. A cycle shop sells two grades of bicycles, Easy Roller (ER) and Super Rider (SR), manufactured by the same company. The following matrices give the sales of these items for four months and the selling price and dealer's cost of these items. Use matrix multiplication to determine the total dollar sales and total costs of this company's items for the four months.

$$\begin{array}{l} \\ \text{Easy Roller} \\ \text{Super Rider} \end{array} \begin{array}{c} \begin{array}{cccc} \text{Feb.} & \text{Mar.} & \text{Apr.} & \text{May} \end{array} \\ \begin{bmatrix} 7 & 10 & 14 & 12 \\ 5 & 7 & 7 & 7 \end{bmatrix} \end{array} \qquad \begin{array}{l} \\ \text{Retail price} \\ \text{Dealer cost} \end{array} \begin{array}{c} \begin{array}{cc} \text{ER} & \text{SR} \end{array} \\ \begin{bmatrix} 150 & 180 \\ 90 & 100 \end{bmatrix} \end{array}$$

In Exercises 27–32, use the definitions of matrix addition, multiplication, and scalar multiplication and the properties of real numbers to verify the given property. In Exercises 27–30, let $A = [a_{ij}]$, $B = [b_{ij}]$, $C = [c_{ij}]$, and the zero matrix $\mathbf{0}$ be $m \times n$ matrices.

27. Prove $A + B = B + A$. [HINT The first several steps are $A + B = [a_{ij}] + [b_{ij}] = [a_{ij} + b_{ij}]$ (by definition of addition) $= [b_{ij} + a_{ij}]$ (since addition in the real numbers is commutative).]

28. Prove $A + (B + C) = (A + B) + C$.

29. Prove $A + \mathbf{0} = A = \mathbf{0} + A$.

30. Prove the statements in (1.17).

31. Prove $A(BC) = (AB)C$ if A is $m \times n$, B is $n \times p$, and C is $p \times q$.

32. Prove $A(B + C) = AB + AC$, if A is $m \times n$ and B and C are $n \times p$.

In Exercises 33–36, show that the given A, X, and B satisfy $AX = B$. Then use the entries of X to express B as a linear combination of the columns of A.

33. $A = \begin{bmatrix} 9 & 4 \\ 0 & -2 \\ 3 & 2 \end{bmatrix}$, $X = \begin{bmatrix} -2 \\ 3 \end{bmatrix}$, $B = \begin{bmatrix} -6 \\ -6 \\ 0 \end{bmatrix}$

34. $A = \begin{bmatrix} 4 & -2 & 1 \\ -2 & 4 & -2 \\ -3 & 0 & -1 \end{bmatrix}$, $X = \begin{bmatrix} 2 \\ 4 \\ -1 \end{bmatrix}$, $B = \begin{bmatrix} -1 \\ 14 \\ -5 \end{bmatrix}$

35. $A = \begin{bmatrix} 1 & 3 & -4 \\ 8 & -1 & 2 \end{bmatrix}, \quad X = \begin{bmatrix} 3 \\ -1 \\ 2 \end{bmatrix}, \quad B = \begin{bmatrix} 8 \\ 21 \end{bmatrix}$

36. $A = \begin{bmatrix} 3 & -4 \\ -1 & 2 \end{bmatrix}, \quad X = \begin{bmatrix} 2 \\ -1 \end{bmatrix}, \quad B = \begin{bmatrix} 10 \\ -4 \end{bmatrix}$

In Exercises 37–42, find the product of the given matrices four different ways: by the definition, by expansion by columns (as in Example 10), by expansion by rows (as in Example 11), and (if covered) by outer products (as in Example 12).

37. $\begin{bmatrix} 2 & -1 \\ -1 & 3 \end{bmatrix}, \quad \begin{bmatrix} 1 & 2 \\ 4 & -1 \end{bmatrix}$

38. $\begin{bmatrix} 3 & 1 & 2 \\ 4 & -1 & -2 \\ 0 & 1 & 2 \end{bmatrix}, \quad \begin{bmatrix} 3 & -1 & 0 \\ 0 & 2 & 1 \\ 1 & 0 & 1 \end{bmatrix}$

39. $\begin{bmatrix} 2 & 3 \\ 4 & 1 \end{bmatrix}, \quad \begin{bmatrix} 5 & 0 & 1 \\ -1 & 2 & 1 \end{bmatrix}$

40. $\begin{bmatrix} 1 & 2 \\ 2 & 1 \\ -1 & 3 \end{bmatrix}, \quad \begin{bmatrix} 2 & 0 \\ -1 & 1 \end{bmatrix}$

41. $\begin{bmatrix} -1 & 4 & 1 \\ 3 & 0 & 1 \end{bmatrix}, \quad \begin{bmatrix} 3 & 1 \\ 2 & -1 \\ 0 & 1 \end{bmatrix}$

42. $\begin{bmatrix} -1 & 2 & 1 \\ 4 & 0 & -2 \end{bmatrix}, \quad \begin{bmatrix} 0 & 3 & 1 \\ 1 & -1 & 2 \\ 2 & 5 & 1 \end{bmatrix}$

1.4 Inverses and Elementary Matrices

In this section we continue to lay the groundwork for the matrix description of Gaussian elimination by covering inverses and elementary matrices. As a byproduct we also obtain a simple scheme for finding inverses.

(1.36)

DEFINITION If A is an $n \times n$ matrix, an **inverse** of A is an $n \times n$ matrix A^{-1} with the property

$$AA^{-1} = I = A^{-1}A$$

where $I = I_n$, the $n \times n$ identity matrix.

NOTE In this definition we said "an" inverse and not "the" inverse. At this point we must allow for the possibility that a given matrix might have more than one inverse (if it has any). However, later we shall show that a given matrix can have at most one inverse. If a matrix A does have an inverse, A is said to be **invertible**.

It is important first to note that not all $n \times n$ matrices are invertible.

Example 1 Show the following matrices have no inverses.

(a) $\begin{bmatrix} 0 & 0 \\ 0 & 0 \end{bmatrix}$ (b) $\begin{bmatrix} 1 & 2 \\ 0 & 0 \end{bmatrix}$

Solution (a) The fact that the zero matrix has no inverse should not be surprising since the number 0 has no inverse either. In fact the two proofs are identical:

There is no matrix B such that $\mathbf{0}B = I$ because $\mathbf{0}B = \mathbf{0}$ for all B (and $\mathbf{0} \neq I$).

(b) This is very similar to (a). Let A be the given matrix and let B be any 2×2 matrix. Thus

$$A = \begin{bmatrix} 1 & 2 \\ 0 & 0 \end{bmatrix} \quad \text{and} \quad B = \begin{bmatrix} a & b \\ c & d \end{bmatrix}$$

Then

$$AB = \begin{bmatrix} 1 & 2 \\ 0 & 0 \end{bmatrix}\begin{bmatrix} a & b \\ c & d \end{bmatrix} = \begin{bmatrix} a + 2c & b + 2d \\ 0 & 0 \end{bmatrix}$$

Thus we can see no matter what B is, the (2, 2) entry in AB cannot equal 1 so

$$AB \neq \begin{bmatrix} 1 & 0 \\ 0 & 1 \end{bmatrix}$$

■

The following examples show that many $n \times n$ matrices do have inverses.

Example 2 Show

$$\begin{bmatrix} 5 & 2 \\ 3 & 1 \end{bmatrix}^{-1} = \begin{bmatrix} -1 & 2 \\ 3 & -5 \end{bmatrix}$$

Solution Multiply to determine that the product yields the identity matrix:

$$\begin{bmatrix} 5 & 2 \\ 3 & 1 \end{bmatrix}\begin{bmatrix} -1 & 2 \\ 3 & -5 \end{bmatrix} = \begin{bmatrix} -5 + 6 & 10 - 10 \\ -3 + 3 & 6 - 5 \end{bmatrix} = \begin{bmatrix} 1 & 0 \\ 0 & 1 \end{bmatrix}$$

and

$$\begin{bmatrix} -1 & 2 \\ 3 & -5 \end{bmatrix}\begin{bmatrix} 5 & 2 \\ 3 & 1 \end{bmatrix} = \begin{bmatrix} -5+6 & -2+2 \\ 15-15 & 6-5 \end{bmatrix} = \begin{bmatrix} 1 & 0 \\ 0 & 1 \end{bmatrix}$$ ■

Example 3 Show that if $ad - bc \neq 0$,

$$\begin{bmatrix} a & b \\ c & d \end{bmatrix}^{-1} = \frac{1}{ad - bc}\begin{bmatrix} d & -b \\ -c & a \end{bmatrix}$$

Solution Multiply as in Example 2. This is a very useful formula and should be memorized. ■

Note that by Exercise 43 if $ad - bc = 0$, then $\begin{bmatrix} a & b \\ c & d \end{bmatrix}$ has no inverse.

Unfortunately formulas for inverses of higher dimensional square matrices are fairly complicated. However, fortunately we have the following.

(1.37)

> IMPORTANT FACT Inverses are very important for theoretical purposes, such as deriving formulas, but it is very seldom necessary actually to find inverses to solve computational problems.

For example, suppose A is an invertible $n \times n$ matrix and we wish to solve the equation $AX = B$ for X. Then theoretically $X = A^{-1}B$, so it appears we should compute A^{-1} and then multiply $A^{-1}B$ to find X. However, for this type of problem, Gaussian elimination is almost always easier, takes fewer steps, and leads to less round-off error when programmed on a computer.

Properties of Inverses

We now turn to some properties of matrix inverses.

(1.38)

> THEOREM If an $n \times n$ matrix has an inverse, that inverse is unique.

Proof Suppose A has an inverse A^{-1} and B is a (possibly different) matrix such that $BA = I$. Then

$$B = BI = B(AA^{-1}) = (BA)A^{-1} = IA^{-1} = A^{-1}$$ ■

Theorem (1.38) will be used repeatedly, often as follows: If A is an $n \times n$ matrix, and if we can find another matrix B such that $BA = I$ and $AB = I$, then A^{-1} exists [by Definition (1.36)] and equals B [by Theorem (1.38)]. The first use of this will be to prove Theorem (1.39).

(1.39) **THEOREM** If A and B are invertible $n \times n$ matrices, then AB is invertible and

$$(AB)^{-1} = B^{-1}A^{-1}$$

Proof

$$(B^{-1}A^{-1})(AB) = B^{-1}(A^{-1}A)B = B^{-1}IB = I$$
$$(AB)(B^{-1}A^{-1}) = A(BB^{-1})A^{-1} = AIA^{-1} = I$$

By the discussion preceding Theorem (1.39), we are done. ■

WARNING Note the reversed order of the inverses and how the proof requires this order to be reversed.

It follows from Theorem (1.39) that if $A_1, \ldots, A_k$ are each $n \times n$ invertible matrices, then

(1.40)
$$(A_1 \cdots A_k)^{-1} = A_k^{-1} \cdots A_1^{-1}$$

Elementary Matrices

We now turn to elementary matrices. They are invertible, and their inverses are easy to determine. We shall see that elementary matrices are "building blocks" for all invertible matrices, a feature that makes them very important.

Recall that we defined elementary row operations in Definition (1.6), Section 1.1. This definition is built upon here to define elementary matrices.

(1.41) **DEFINITION** Let e be an elementary row operation. Then the $n \times n$ **elementary matrix** E associated with e is the matrix obtained by applying e to the $n \times n$ identity matrix. Thus

$$E = e(I)^*$$

Example 4 illustrates elementary matrices and the notation we use to denote them.

Example 4 (a)

$$M_3 = \begin{bmatrix} 1 & & & \\ & 1 & & \\ & & -5 & \\ & & & 1 \end{bmatrix}$$

Multiply the third row by -5. The M is for "multiply" and the subscript 3 is for the row acted upon.

* *Here $e(I)$ is a notational device representing the elementary operation e applied to the matrix I.*

(b)
$$P_{12} = \begin{bmatrix} 0 & 1 & \\ 1 & 0 & \\ & & 1 \end{bmatrix}$$

Permute (or interchange) the first and second rows. The P is for "permute" and the subscripts indicate the rows interchanged.

In fact the matrices P_{ij} are called **elementary permutation matrices**, and we shall see they are very important to both theoretical and computational mathematics.

(c)
$$E_{12} = \begin{bmatrix} 1 & & \\ -3 & 1 & \\ & & 1 \end{bmatrix}$$

Add -3 times row 1 to row 2. The E stands for "elementary" and the subscripts 1 and 2 represent the rows acted upon. ■

When we wish to refer to an elementary matrix in general (and we do not know which type it is), we shall use the symbol E or E_i (for elementary or ith elementary matrix). The number of subscripts and the context should help avoid confusion.

NOTE The identity matrix itself is an elementary matrix. There are several different ways you can see this. See Exercise 18.

One of the reasons that elementary matrices are so important is that they have the following property.

(1.42)

THEOREM Let e be an elementary operation and let E be the corresponding $m \times m$ elementary matrix $E = e(I)$. Then for every $m \times n$ matrix A,

$$e(A) = EA$$

That is, an elementary row operation can be performed on A by multiplying A on the left by the corresponding elementary matrix.

The proof is omitted; Example 5 illustrates the idea.

Example 5 Let

$$A = \begin{bmatrix} 2 & -2 & 3 & 1 \\ 1 & 0 & 2 & 4 \\ -6 & 2 & 1 & 5 \end{bmatrix}, \quad M_1 = \begin{bmatrix} 2 & & \\ & 1 & \\ & & 1 \end{bmatrix}$$

$$P_{23} = \begin{bmatrix} 1 & & \\ & 0 & 1 \\ & 1 & 0 \end{bmatrix}, \quad E_{13} = \begin{bmatrix} 1 & & \\ 0 & 1 & \\ 3 & 0 & 1 \end{bmatrix}$$

(a) Observe that M_1 was obtained from I_3 by multiplying the first row by 2. Now compute

$$M_1A = \begin{bmatrix} 2 & & \\ & 1 & \\ & & 1 \end{bmatrix} \begin{bmatrix} 2 & -2 & 3 & 1 \\ 1 & 0 & 2 & 4 \\ -6 & 2 & 1 & 5 \end{bmatrix} = \begin{bmatrix} 4 & -4 & 6 & 2 \\ 1 & 0 & 2 & 4 \\ -6 & 2 & 1 & 5 \end{bmatrix}$$

and observe that the last matrix was obtained from A by multiplying the first row by 2.

(b) Observe that P_{23} was obtained from I_3 by interchanging rows 2 and 3. Now compute

$$P_{23}A = \begin{bmatrix} 1 & & \\ & 0 & 1 \\ & 1 & 0 \end{bmatrix} \begin{bmatrix} 2 & -2 & 3 & 1 \\ 1 & 0 & 2 & 4 \\ -6 & 2 & 1 & 5 \end{bmatrix} = \begin{bmatrix} 2 & -2 & 3 & 1 \\ -6 & 2 & 1 & 5 \\ 1 & 0 & 2 & 4 \end{bmatrix}$$

and observe that the last matrix was obtained from A by interchanging rows 2 and 3.

(c) Observe that E_{13} was obtained from I_3 by adding 3 times row 1 to row 3. Now compute

$$E_{13}A = \begin{bmatrix} 1 & & \\ 0 & 1 & \\ 3 & 0 & 1 \end{bmatrix} \begin{bmatrix} 2 & -2 & 3 & 1 \\ 1 & 0 & 2 & 4 \\ -6 & 2 & 1 & 5 \end{bmatrix} = \begin{bmatrix} 2 & -2 & 3 & 1 \\ 1 & 0 & 2 & 4 \\ 0 & -4 & 10 & 8 \end{bmatrix}$$

and observe that the last matrix was obtained from A by adding 3 times row 1 to row 3. ■

We shall see that this theorem is important for both theoretical and computational reasons in that it leads to a deeper understanding of Gaussian elimination and its computer implementation.

An important property of elementary matrices is that they are invertible.

(1.43) **THEOREM** Each elementary matrix is invertible and its inverse is an elementary matrix of the same type.

The proof is given after Example 6.

Example 6 (a) Let

$$M_2 = \begin{bmatrix} 1 & & \\ & 7 & \\ & & 1 \end{bmatrix}.$$

Then
$$M_2^{-1} = \begin{bmatrix} 1 & & \\ & \frac{1}{7} & \\ & & 1 \end{bmatrix}$$
because
$$\begin{bmatrix} 1 & & \\ & 7 & \\ & & 1 \end{bmatrix}\begin{bmatrix} 1 & & \\ & \frac{1}{7} & \\ & & 1 \end{bmatrix} = \begin{bmatrix} 1 & & \\ & 1 & \\ & & 1 \end{bmatrix} = \begin{bmatrix} 1 & & \\ & \frac{1}{7} & \\ & & 1 \end{bmatrix}\begin{bmatrix} 1 & & \\ & 7 & \\ & & 1 \end{bmatrix}$$

(b) Let
$$P_{12} = \begin{bmatrix} 0 & 1 & \\ 1 & 0 & \\ & & 1 \end{bmatrix}.$$
Then $P_{12}^{-1} = P_{12}$ because
$$\begin{bmatrix} 0 & 1 & \\ 1 & 0 & \\ & & 1 \end{bmatrix}\begin{bmatrix} 0 & 1 & \\ 1 & 0 & \\ & & 1 \end{bmatrix} = \begin{bmatrix} 1 & & \\ & 1 & \\ & & 1 \end{bmatrix}$$

(c) Let
$$E_{24} = \begin{bmatrix} 1 & & & \\ & 1 & & \\ & 0 & 1 & \\ & 5 & 0 & 1 \end{bmatrix}.$$
Then
$$E_{24}^{-1} = \begin{bmatrix} 1 & & & \\ & 1 & & \\ & 0 & 1 & \\ & -5 & 0 & 1 \end{bmatrix}$$
because
$$\begin{bmatrix} 1 & & & \\ & 1 & & \\ & 0 & 1 & \\ & 5 & 0 & 1 \end{bmatrix}\begin{bmatrix} 1 & & & \\ & 1 & & \\ & 0 & 1 & \\ & -5 & 0 & 1 \end{bmatrix} = \begin{bmatrix} 1 & & & \\ & 1 & & \\ & & 1 & \\ & & & 1 \end{bmatrix}$$
$$= \begin{bmatrix} 1 & & & \\ & 1 & & \\ & 0 & 1 & \\ & -5 & 0 & 1 \end{bmatrix}\begin{bmatrix} 1 & & & \\ & 1 & & \\ & 0 & 1 & \\ & 5 & 0 & 1 \end{bmatrix}$$
■

Example 6 illustrates the following proof of Theorem (1.43).

Proof of Theorem (1.43) First each elementary operation e can be reversed or undone by a similar elementary operation e^{-1}, called the **inverse operation**.

- Multiplying a row by $c \neq 0$ is undone by multiplying that same row by c^{-1}.
- Interchanging two rows is undone by interchanging those same two rows.
- Adding a multiple m of one row to another is undone by adding $-m$ times the first row to the second.

Next suppose $E = e(I)$. Then if we perform the inverse operation on E, we obtain I back. That is,

(1.44)
$$e^{-1}(E) = e^{-1}(e(I)) = I$$

But by Theorem (1.37) $e^{-1}(E) = e^{-1}(I)E$, so that

$$e^{-1}(I)E = e^{-1}(E) = I$$

Thus $e^{-1}(I) = E^{-1}$ by Theorem (1.38) ■

The last line is important enough to set off.

(1.45) COROLLARY Let e be an elementary operation and $E = e(I)$ a corresponding elementary matrix. Then

$$e^{-1}(I) = E^{-1}$$

Also, if e is an elementary permutation, then $e^{-1} = e$, so we have

(1.46) COROLLARY An elementary permutation matrix P is its own inverse; that is,

$$P = P^{-1}$$

Diagonal Matrices

Elementary matrices of the form of M_3 in Example 4(a) are special cases of diagonal matrices. A **diagonal matrix** is a square matrix in which all the *non*diagonal entries are zero. Note that the diagonal entries may or may not be zero.

Example 7 Each of the following is a diagonal matrix:

$$\begin{bmatrix} 4 & 0 \\ 0 & 3 \end{bmatrix}, \quad \begin{bmatrix} -5 & 0 & 0 \\ 0 & 0 & 0 \\ 0 & 0 & \frac{1}{2} \end{bmatrix}, \quad \begin{bmatrix} 10 & 0 & 0 & 0 \\ 0 & -2 & 0 & 0 \\ 0 & 0 & 3 & 0 \\ 0 & 0 & 0 & 0 \end{bmatrix}, \quad \mathbf{0} = \begin{bmatrix} 0 & 0 & 0 \\ 0 & 0 & 0 \\ 0 & 0 & 0 \end{bmatrix}$$

■

One reason why diagonal matrices are important is the effect they produce when they multiply other matrices.

(1.47) THEOREM Let

$$D = \begin{bmatrix} d_1 & & & \\ & d_2 & & \\ & & \ddots & \\ & & & d_n \end{bmatrix}$$

be a $n \times n$ diagonal matrix.

(a) Let A be an $n \times p$ matrix and B be an $m \times n$ matrix. The result of the product DA is to multiply the ith row of A by d_i; the result of the product BD is to multiply the ith column by d_i.

(b) The diagonal matrix D is invertible exactly when every d_i is nonzero. In this case

$$D^{-1} = \begin{bmatrix} d_1^{-1} & & & \\ & d_2^{-1} & & \\ & & \ddots & \\ & & & d_n^{-1} \end{bmatrix}$$

Proof Simply multiply out and see what happens. ■

Example 8 It is straightforward to see

$$\begin{bmatrix} 4 & 0 \\ 0 & 3 \end{bmatrix}\begin{bmatrix} 4 & 0 & -7 \\ -2 & 3 & 5 \end{bmatrix} = \begin{bmatrix} 16 & 0 & -28 \\ -6 & 9 & 15 \end{bmatrix}$$

$$\begin{bmatrix} -5 & -2 \\ 4 & 11 \\ -1 & 3 \end{bmatrix}\begin{bmatrix} 4 & 0 \\ 0 & 3 \end{bmatrix} = \begin{bmatrix} -20 & -6 \\ 16 & 33 \\ -4 & 9 \end{bmatrix}$$

$$\begin{bmatrix} 4 & 0 \\ 0 & 3 \end{bmatrix}\begin{bmatrix} \frac{1}{4} & 0 \\ 0 & \frac{1}{3} \end{bmatrix} = \begin{bmatrix} 1 & 0 \\ 0 & 1 \end{bmatrix}, \text{ so that } \begin{bmatrix} 4 & 0 \\ 0 & 3 \end{bmatrix}^{-1} = \begin{bmatrix} \frac{1}{4} & 0 \\ 0 & \frac{1}{3} \end{bmatrix}$$ ■

Row Equivalence and the Main Theorem

Suppose you can obtain a matrix B from a matrix A by performing elementary row operations. That is, suppose

$$e_k \cdots e_2 e_1 A = B \tag{1.48}$$

Then you can get from B back to A (by $A = e_1^{-1}e_2^{-1} \cdots e_k^{-1}B$), and we say A and B are **row equivalent**. For example, we can restate (1.8) in Section 1.2 as

(1.49) **THEOREM** Every $m \times n$ matrix A is row equivalent to a matrix U in row echelon form.

We are now ready to state our main theorem.

(1.50) **THEOREM** Let A be an $n \times n$ matrix. Then the following are equivalent.

(a) A is invertible.
(b) $AX = B$ has a unique solution for any B.
(c) A is **nonsingular**, which means that the equation $AX = \mathbf{0}$ has only the trivial solution $X = \mathbf{0}$.
(d) A is row equivalent to I_n.
(e) A is a product of elementary matrices.

Many people who read this theorem for the first time will view this as a bizarre collection of statements to be equivalent. This view is not unreasonable, but in some sense a good part of linear algebra is proving as many statements are equivalent to (a) as possible. Statement (e) implies a very useful fact, that elementary matrices are the building blocks of invertible matrices. Part (c) gives the definition of nonsingular and technically it means something different from invertible. But "invertible" and "nonsingular" are equivalent, by Theorem (1.50), and it is common practice to use the two words interchangeably. Also, we say A is **singular** if it is not invertible. For example, A is singular if a row of zeros occurs in the matrix, and a 2×2 matrix $\begin{bmatrix} a & b \\ c & d \end{bmatrix}$ is singular if $ad - bc = 0$.

A method for finding the inverse of an arbitrary invertible matrix is hidden in the proof of Theorem (1.50). Before proving the theorem itself, let us cover that method explicitly. The idea of the construction is that if $E_1, \ldots, E_k$ is a sequence of elementary matrices such that $E_k \cdots E_1 A = I$, then the product $E_k \cdots E_1$ must be A^{-1} by Theorem (1.38). Thus $E_k \cdots E_1 I = A^{-1}$ or $e_k \cdots e_1 I = A^{-1}$, where e_i is the elementary operation associated with E_i. The method, illustrated in Example 9, chooses the operations in an order that is easiest for most matrices.

Example 9 Find the inverse of

$$A = \begin{bmatrix} 2 & 1 & 0 \\ -4 & -1 & -3 \\ 3 & 1 & 2 \end{bmatrix}$$

Solution First augment the matrix with the identity, forming $[A \,\vdots\, I]$, an $n \times 2n$ matrix.

$$\left[\begin{array}{rrr:rrr} 2 & 1 & 0 & 1 & 0 & 0 \\ -4 & -1 & -3 & 0 & 1 & 0 \\ 3 & 1 & 2 & 0 & 0 & 1 \end{array}\right]$$

Doing a row operation e on this matrix corresponds to doing e on A and e on I simultaneously. Our goal is to transform this matrix into $[I \,\vdots\, ?]$, using elementary row operations. According to the paragraph directly preceding this example, we must have $? = A^{-1}$. Thus we shall proceed as in Gaussian elimination, reducing A to echelon form by performing elementary operations, but by performing the operations *on the whole $n \times 2n$ matrix.*

$$\left[\begin{array}{rrr:rrr} 2 & 1 & 0 & 1 & 0 & 0 \\ 0 & 1 & -3 & 2 & 1 & 0 \\ 0 & -\frac{1}{2} & 2 & -\frac{3}{2} & 0 & 1 \end{array}\right]$$

$$\left[\begin{array}{rrr:rrr} 2 & 1 & 0 & 1 & 0 & 0 \\ 0 & 1 & -3 & 2 & 1 & 0 \\ 0 & 0 & \frac{1}{2} & -\frac{1}{2} & \frac{1}{2} & 1 \end{array}\right]$$

Next, like Gaussian elimination backward, we reduce the left-hand side to diagonal form, *working with the last column first*, then the next to last, and so on (and still performing the operations on the whole matrix).

$$\left[\begin{array}{rrr:rrr} 2 & 1 & 0 & 1 & 0 & 0 \\ 0 & 1 & 0 & -1 & 4 & 6 \\ 0 & 0 & \frac{1}{2} & -\frac{1}{2} & \frac{1}{2} & 1 \end{array}\right]$$

$$\left[\begin{array}{rrr:rrr} 2 & 0 & 0 & 2 & -4 & -6 \\ 0 & 1 & 0 & -1 & 4 & 6 \\ 0 & 0 & \frac{1}{2} & -\frac{1}{2} & \frac{1}{2} & 1 \end{array}\right]$$

Finally, divide each row by the diagonal element (so in this case, divide row 1 by 2, row 2 by 1, and row 3 by $\frac{1}{2}$, obtaining $[I \,\vdots\, A^{-1}]$,

$$\left[\begin{array}{rrr:rrr} 1 & 0 & 0 & 1 & -2 & -3 \\ 0 & 1 & 0 & -1 & 4 & 6 \\ 0 & 0 & 1 & -1 & 1 & 2 \end{array}\right]$$

so

$$A^{-1} = \begin{bmatrix} 1 & -2 & -3 \\ -1 & 4 & 6 \\ -1 & 1 & 2 \end{bmatrix}$$

You can verify that we have found A^{-1} by showing directly $AA^{-1} = I = A^{-1}A$. ■

If you wish to determine if A^{-1} exists, you can start the process. It will become apparent by the end of the first set of reductions whether A is singular or nonsingular.

Example 10 Determine if A^{-1} exists if

$$A = \begin{bmatrix} 2 & 1 & 0 \\ -4 & -1 & -3 \\ 3 & 1 & \frac{3}{2} \end{bmatrix}$$

Solution Proceed as in Example 9, manipulating $[A \,\vdots\, I]$ to try to reach $[I \,\vdots\, A^{-1}]$

$$\left[\begin{array}{ccc:ccc} 2 & 1 & 0 & 1 & 0 & 0 \\ -4 & -1 & -3 & 0 & 1 & 0 \\ 3 & 1 & \frac{3}{2} & 0 & 0 & 1 \end{array}\right]$$
$$\left[\begin{array}{ccc:ccc} 2 & 1 & 0 & 1 & 0 & 0 \\ 0 & 1 & -3 & 2 & 1 & 0 \\ 0 & -\frac{1}{2} & \frac{3}{2} & -\frac{3}{2} & 0 & 1 \end{array}\right]$$
$$\left[\begin{array}{ccc:ccc} 2 & 1 & 0 & 1 & 0 & 0 \\ 0 & 1 & -3 & 2 & 1 & 0 \\ 0 & 0 & 0 & -\frac{1}{2} & \frac{1}{2} & 1 \end{array}\right]$$

Since we have obtained a row of zeros on the left-hand side, A is singular. ■

OPTIONAL* Now comes the proof of Theorem (1.50). However, we first prove two lemmas that will allow the proof of the theorem to go more smoothly.

(1.51) **LEMMA** Let U be an $n \times n$ matrix in row echelon form. If $u_{ii} \neq 0$, $1 \leq i \leq n$, then U is row equivalent to I.

Proof Essentially we do Gaussian elimination in reverse. Since $u_{nn} \neq 0$, we can find multiples of the last row to add to the higher rows to make $u_{in} = 0$, $i < n$.

$$\begin{bmatrix} u_{11} & u_{12} & \cdots & u_{1n} \\ & u_{22} & \cdots & u_{2n} \\ & & \cdots & \\ & & & u_{nn} \end{bmatrix} \rightarrow \begin{bmatrix} u_{11} & u_{12} & \cdots & u_{1n-1} & 0 \\ & u_{22} & \cdots & u_{2n-1} & 0 \\ & & \cdots & & \\ & & & u_{n-1n-1} & 0 \\ & & & & u_{nn} \end{bmatrix}$$

* *It is possible to skip the remainder of this section without losing the continuity of the subject matter. However, the proofs are not difficult and understanding them certainly enhances the overall understanding of the material.*

Now since $u_{n-1n-1} \neq 0$, we can make $u_{in-1} = 0$, $i < n - 1$, and keep going like this until we get to a diagonal matrix.

$$\begin{bmatrix} u_{11} & & \\ & \ddots & \\ & & u_{nn} \end{bmatrix}$$

Finally, since each $u_{ii} \neq 0$, we can multiply each row by u_{ii}^{-1} to obtain I. ■

(1.52)

LEMMA If A and B are row equivalent $m \times n$ matrices, then the equations $AX = \mathbf{0}$ and $BX = \mathbf{0}$ have exactly the same solutions.

Proof Suppose A is row equivalent to B, so that $e_k \cdots e_1 A = B$. Letting $e_i(I) = E_i$, we see by Theorem (1.42) that

$$E_k \cdots E_1 A = B$$

Then $AX = \mathbf{0} \Rightarrow E_k \cdots E_1 AX = E_k \cdots E_1 \mathbf{0} = \mathbf{0} \Rightarrow BX = \mathbf{0}$. Thus any solution to $AX = \mathbf{0}$ is a solution to $BX = \mathbf{0}$. By reversing the process, we can go the other way, so we are done. ■

Proof of Theorem (1.50) We shall establish the equivalences by proving the sequence of implications (a) $\Rightarrow$ (b) $\Rightarrow$ (c) $\Rightarrow$ (d) $\Rightarrow$ (e) $\Rightarrow$ (a).*

(a) $\Rightarrow$ (b). (NOTE This part of the proof is really quite formal. It is included mainly to illustrate how to go about such formalities.) We assume the matrix A is invertible. Pick B. We wish to show the equation $AX = B$ has a solution X and that the solution is unique.

EXISTENCE To show there is a solution to $AX = B$, knowing that A^{-1} exists, let $X = A^{-1}B$. Then

$$AX = A(A^{-1}B) = (AA^{-1})B = IB = B$$

Thus this X works, so a solution exists.

UNIQUENESS Let X_1 be a possibly different solution. Then

$$\begin{aligned} AX_1 &= B \\ A^{-1}(AX_1) &= A^{-1}B \\ (A^{-1}A)X_1 &= A^{-1}B \\ IX_1 &= A^{-1}B \\ X_1 &= A^{-1}B \end{aligned}$$

* *The symbol "$\Rightarrow$" is an abbreviation for the word "implies."*

Thus X_1 is in fact the same as X, so X is unique.

(b) $\Rightarrow$ (c). To show $AX = B$ has a unique solution implies A is nonsingular. In the equation $AX = B$, set $B = \mathbf{0}$. Observe $X = \mathbf{0}$ is a solution. By (b), this is the only solution; that is, $AX = \mathbf{0}$ has only the trivial solution, so we are done.

(c) $\Rightarrow$ (d). To show A is nonsingular implies A is row equivalent to I_n. By Theorem (1.49), A is row equivalent to a matrix U in row echelon form. Thus U looks like

$$\begin{bmatrix} u_{11} & u_{12} & \cdots & u_{1n} \\ & u_{22} & \cdots & u_{2n} \\ & & & \vdots \\ & & & u_{nn} \end{bmatrix}$$

where the u_{ij}s, $j \geq i$, may or may not be zero. If we show this U is row equivalent to I, we are done.

Since $AX = \mathbf{0}$ has only the trivial solution, $UX = \mathbf{0}$ has only the trivial solution by Lemma (1.52). If $u_{ii} \neq 0$, all i, then we are done by Lemma (1.51). Suppose some $u_{ii} = 0$. Let j be the smallest index such that $u_{jj} = 0$. Then the system

$$\begin{bmatrix} u_{11} & \cdots & u_{1j-1} \\ & \ddots & \vdots \\ & & u_{j-1j-1} \end{bmatrix} \begin{bmatrix} x_1 \\ \vdots \\ x_{j-1} \end{bmatrix} = -\begin{bmatrix} u_{1j} \\ \vdots \\ u_{j-1j} \end{bmatrix}$$

can easily be solved for $x_1, \ldots, x_{j-1}$ by backsubstitution (since each $u_{ii} \neq 0$, $i < j$). Then

$$X = \begin{bmatrix} x_1 \\ \vdots \\ x_{j-1} \\ 1 \\ 0 \\ \vdots \\ 0 \end{bmatrix}$$

can be seen to be a nontrivial solution to $AX = \mathbf{0}$. This is a contradiction.

(d) $\Rightarrow$ (e). To show A is row equivalent to I_n implies A is a product of elementary matrices. Since A is row equivalent to I, there is a sequence of row operations $e_1, \ldots, e_k$ such that

$$e_k \cdots e_2 e_1 A = I$$

Letting $E_i = e_i(I)$, we see by Theorem (1.42) that

$$E_k \cdots E_2 E_1 A = I \qquad \Rightarrow \qquad A = E_1^{-1} \cdots E_k^{-1}$$

We are now done by Theorem (1.43).

(e) ⇒ (a). To show that A is a product of elementary matrices implies A is invertible. This follows immediately since elementary matrices are invertible and products of invertible matrices are invertible. ■

Exercise 1.4 In Exercises 1–4, for the given elementary row operation e, find its inverse operation, e^{-1}, and the elementary matrices associated with e and e^{-1}.

1. Interchange the first and third row of 4×4 matrices.
2. Add -2 times the first row to the second row of 3×3 matrices.
3. Multiply the third row of a 3×5 matrix by $\frac{2}{3}$.
4. Add 7 times the fourth row to the second row of 4×8 matrices.

In Exercises 5–8, an elementary row operation e and a matrix A are given. Find the associated elementary matrix E, compute eA and EA, and compare.

5. Interchange the second and third rows of

$$A = \begin{bmatrix} 1 & -2 & 3 & 0 \\ 4 & 1 & 1 & 7 \\ 3 & -1 & 1 & -1 \end{bmatrix}$$

6. Add 3 times the second row to the first of

$$\begin{bmatrix} 5 & -1 & 5 \\ 7 & 3 & -2 \\ 8 & 1 & 2 \\ 6 & 0 & -1 \end{bmatrix}$$

7. Add -2 times the third row to the first of

$$\begin{bmatrix} 2 & 0 & 3 \\ 4 & -1 & 1 \\ 3 & 3 & 1 \end{bmatrix}$$

8. Multiply the second row of

$$\begin{bmatrix} 1 & -2 & 3 & -1 \\ 4 & 0 & 8 & 2 \end{bmatrix}$$

by $\frac{1}{4}$.

In Exercises 9–14, determine if the given matrix is an elementary matrix. If it is, find the corresponding elementary operation.

9. $\begin{bmatrix} 1 & 0 & 2 \\ 0 & 1 & 0 \\ 0 & 0 & 1 \end{bmatrix}$

10. $\begin{bmatrix} 0 & 0 & 1 \\ 0 & 1 & 0 \\ 1 & 0 & 0 \end{bmatrix}$

11. $\begin{bmatrix} 0 & 1 & 0 & 0 \\ 1 & 0 & 0 & 0 \\ 0 & 0 & 0 & 1 \\ 0 & 0 & 1 & 0 \end{bmatrix}$

12. $\begin{bmatrix} 1 & 0 & 0 \\ 2 & 1 & 0 \\ 0 & 2 & 1 \end{bmatrix}$

13. $\begin{bmatrix} 1 & 0 & 0 & 0 \\ 3 & 1 & 0 & 0 \\ 0 & 0 & 1 & 0 \end{bmatrix}$

14. $\begin{bmatrix} 0 & 1 & 0 \\ 1 & 0 & 0 \\ 0 & 0 & 0 \end{bmatrix}$

In Exercises 15–17, let

$$A = \begin{bmatrix} 0 & 1 & 2 & 3 \\ 4 & 5 & 6 & 7 \\ 8 & 9 & 10 & 11 \end{bmatrix}, \quad B = \begin{bmatrix} 4 & 5 & 6 & 7 \\ 0 & 1 & 2 & 3 \\ 8 & 9 & 10 & 11 \end{bmatrix}, \quad C = \begin{bmatrix} 4 & 5 & 6 & 7 \\ 8 & 9 & 10 & 11 \\ 0 & 1 & 2 & 3 \end{bmatrix}$$

15. Find elementary matrices E_1 and E_2 such that $E_1A = B$ and $E_2B = A$. How are E_1 and E_2 related?

16. Find elementary matrices E_3 and E_4 such that $E_3B = C$ and $E_4C = B$. How are E_3 and E_4 related?

17. Is it possible to find elementary matrices E_5 and E_6 such that $E_5A = C$ and $E_6C = A$? Justify your answer.

18. Explain why the identity matrix I is an elementary matrix in two *different* ways.

19. If E_1 and E_2 are elementary matrices, explain under what circumstances E_1E_2 is an elementary matrix.

20. Consider the matrix

$$A = \begin{bmatrix} 1 & 0 \\ 2 & 3 \end{bmatrix}$$

(a) Find elementary matrices E_1 and E_2 such that $E_2E_1A = I$.
(b) Write A^{-1} as a product of elementary matrices.
(c) Write A as a product of elementary matrices.

In Exercises 21–24, compute the products DA and BD. Compare the results with Theorem (1.47).

21. $D = \begin{bmatrix} -5 & \\ & 2 \end{bmatrix}, \quad A = \begin{bmatrix} -3 & 2 & -6 \\ 13 & 4 & 10 \end{bmatrix}, \quad B = \begin{bmatrix} -6 & 1 \\ 12 & 15 \\ -7 & -2 \end{bmatrix}$

22. $D = \begin{bmatrix} 2 & & \\ & 0 & \\ & & -\frac{1}{2} \end{bmatrix}, \quad A = \begin{bmatrix} 10 & 3 \\ -2 & 15 \\ --6 & 18 \end{bmatrix}, \quad B = \begin{bmatrix} -7 & 3 & -8 \\ 11 & 5 & 16 \end{bmatrix}$

23. $D = \begin{bmatrix} 0 & & \\ & -2 & \\ & & 1 \end{bmatrix}, \quad A = B = \begin{bmatrix} 10 & 8 & 2 \\ -2 & 2 & -3 \\ 5 & 3 & -12 \end{bmatrix}$

24. $D = \begin{bmatrix} -2 & & & \\ & -5 & & \\ & & 3 & \\ & & & -1 \end{bmatrix}, \quad A = \begin{bmatrix} 10 & 1 \\ 6 & -2 \\ -2 & 7 \\ -1 & -5 \end{bmatrix},$

$B = \begin{bmatrix} -3 & 7 & -5 & 0 \\ 12 & 4 & 11 & 2 \end{bmatrix}$

In Exercises 25–36, determine if the given matrix has an inverse, and find the inverse if it exists. Check your answer by multiplying $A \cdot A^{-1}$ to get I.

25. $\begin{bmatrix} -1 & 2 \\ -3 & 5 \end{bmatrix}$

26. $\begin{bmatrix} 2 & 3 \\ -3 & -5 \end{bmatrix}$

27. $\begin{bmatrix} 3 & -2 \\ 6 & -4 \end{bmatrix}$

28. $\begin{bmatrix} 1 & 0 & 1 \\ 0 & 1 & 1 \\ 2 & 3 & 5 \end{bmatrix}$

29. $\begin{bmatrix} 1 & 0 & 1 \\ 1 & 1 & -1 \\ 0 & 1 & 0 \end{bmatrix}$

30. $\begin{bmatrix} 2 & 1 & 3 \\ 0 & 0 & 1 \\ 3 & 1 & 2 \end{bmatrix}$

31. $\begin{bmatrix} 1 & 2 & 3 & 1 \\ 1 & 3 & 3 & 2 \\ 2 & 4 & 3 & 3 \\ 1 & 1 & 1 & 1 \end{bmatrix}$

32. $\begin{bmatrix} 1 & 0 & 0 & 0 \\ -2 & 1 & 0 & 0 \\ 3 & 2 & 1 & 0 \\ 1 & -2 & 0 & 1 \end{bmatrix}$

33. $\begin{bmatrix} \frac{1}{2} & 0 & \frac{1}{2} \\ 0 & 1 & 0 \\ \frac{1}{2} & 0 & -\frac{1}{2} \end{bmatrix}$

34. $\begin{bmatrix} 1 & 1 \\ 3 & 2 \end{bmatrix}, \begin{bmatrix} 2 & 5 \\ 1 & 2 \end{bmatrix},$ and $\begin{bmatrix} 1 & 1 & 0 & 0 \\ 3 & 2 & 0 & 0 \\ 0 & 0 & 2 & 5 \\ 0 & 0 & 1 & 2 \end{bmatrix}$

35. $\begin{bmatrix} 3 & & \\ & \frac{1}{4} & \\ & & 5 \end{bmatrix}$

36. $\begin{bmatrix} -8 & & & \\ & \frac{1}{5} & & \\ & & \frac{2}{3} & \\ & & & -\frac{2}{5} \end{bmatrix}$

37. Suppose A is $m \times m$, B is $n \times n$, and both are invertible. Let C be the $(m + n) \times (m + n)$ matrix

$$C = \begin{bmatrix} A & 0 \\ 0 & B \end{bmatrix}$$

Can you find C^{-1}? Use Theorem (1.38) to justify your answer.

In Exercises 38–41, assume a, b, c, d are all nonzero. Find the inverse of the given matrix.

38. $\begin{bmatrix} a & & & \\ & b & & \\ & & c & \\ & & & d \end{bmatrix}$

39. $\begin{bmatrix} & & & a \\ & & b & \\ & c & & \\ d & & & \end{bmatrix}$

40. $\begin{bmatrix} 0 & a & & \\ b & 0 & & \\ & & 0 & c \\ & & d & 0 \end{bmatrix}$

41. $\begin{bmatrix} 1 & & & \\ a & 1 & & \\ & a & 1 & \\ & & a & 1 \end{bmatrix}$

42. Show the inverse of $\begin{bmatrix} \cos\theta & \sin\theta \\ -\sin\theta & \cos\theta \end{bmatrix}$ is $\begin{bmatrix} \cos(-\theta) & \sin(-\theta) \\ -\sin(-\theta) & \cos(-\theta) \end{bmatrix}$.

43. Show that if $ad - bc = 0$, then $A = \begin{bmatrix} a & b \\ c & d \end{bmatrix}$ has no inverse. [HINT If $b = 0$, then either $a = 0$ or $d = 0$, and use an argument similar to Example 1(b). If $b \neq 0$, $c = \dfrac{ad}{b}$, and proceed in a similar way.]

44. Find five 2×2 matrices A such that $A^2 = I$ (and discover the surprising fact that there are matrices A other than I and $-I$ having the property $A^{-1} = A$).

45. There are three types of elementary column operations on a matrix that are analogous to the three types of elementary row operations listed in Definition (1.6), Section 1.1.

    (a) List the three types of elementary column operations.
    (b) State the definition corresponding to Definition (1.41) that would apply to elementary column operations.
    (c) Suppose we modify Theorem (1.42) to apply to elementary column operations e. Explain why it would now read: if $E = e(I)$, then $e(A) = AE$. (Notice the reverse order!)
    (d) Suppose E is an elementary matrix obtained by an elementary *row* operation. For each of the three types of row operations, determine if that same E could also be obtained by applying an elementary *column* operation to I.

1.5 Gaussian Elimination as a Matrix Factorization

We are now ready to describe Gaussian elimination in terms of matrices multiplied by one another. This will include the famous "LU decomposition" as well as other forms that will simplify programming Gaussian elimination on a computer. This first example illustrates the factorization.

Example 1 Let

$$A = \begin{bmatrix} 2 & 8 & 1 & 1 \\ 4 & 13 & 3 & -1 \\ -2 & -5 & -3 & 3 \\ -6 & -18 & -1 & 1 \end{bmatrix}$$

Then, without augmenting A, Gaussian elimination could proceed as follows:

$$\begin{bmatrix} 2 & 8 & 1 & 1 \\ 4 & 13 & 3 & -1 \\ -2 & -5 & -3 & 3 \\ -6 & -18 & -1 & 1 \end{bmatrix} \to \begin{bmatrix} 2 & 8 & 1 & 1 \\ 0 & -3 & 1 & -3 \\ 0 & 3 & -2 & 4 \\ 0 & 6 & 2 & 4 \end{bmatrix}$$

$$\to \begin{bmatrix} 2 & 8 & 1 & 1 \\ 0 & -3 & 1 & -3 \\ 0 & 0 & -1 & 1 \\ 0 & 0 & 4 & -2 \end{bmatrix} \to \begin{bmatrix} 2 & 8 & 1 & 1 \\ 0 & -3 & 1 & -3 \\ 0 & 0 & -1 & 1 \\ 0 & 0 & 0 & 2 \end{bmatrix}$$

Denote this sequence by

$$A \to A_1 \to A_2 \to A_3$$

Earlier we said that A_3 was in echelon form and denoted it by U. When such a U is square, it is also said to be **upper triangular**; that is, U is square and all the entries below the diagonal are zero.

$$U = A_3 = \begin{bmatrix} 2 & 8 & 1 & 1 \\ 0 & -3 & 1 & -3 \\ 0 & 0 & -1 & 1 \\ 0 & 0 & 0 & 2 \end{bmatrix}$$

A matrix L is called **lower triangular** if it is square and the entries above the diagonal are 0. It is called **unit lower triangular** if it is lower triangular and all the diagonal entries are 1,

$$L = \begin{bmatrix} 1 & 0 & 0 & \cdots & 0 \\ * & 1 & 0 & \cdots & 0 \\ * & * & 1 & \cdots & 0 \\ \vdots & \vdots & \vdots & & \vdots \\ * & * & * & \cdots & 1 \end{bmatrix}$$

Unit upper triangular is defined analogously.

Now we shall see how to go through this same Gaussian elimination procedure using matrices. To go from A to A_1, we have

1. added -2 times row 1 to row 2.
2. added 1 times row 1 to row 3.
3. added 3 times row 1 to row 4.

Referring to Section 1.4, we see

$$A_1 = E_{14}E_{13}E_{12}A$$

where

$$E_{12} = \begin{bmatrix} 1 & & & \\ -2 & 1 & & \\ & & 1 & \\ & & & 1 \end{bmatrix}, \quad E_{13} = \begin{bmatrix} 1 & & & \\ & 1 & & \\ 1 & & 1 & \\ & & & 1 \end{bmatrix},$$

$$E_{14} = \begin{bmatrix} 1 & & & \\ & 1 & & \\ & & 1 & \\ 3 & & & 1 \end{bmatrix}$$

Note that each E_{1j} is unit lower triangular. To go from A_1 to A_2, we have

1. added 1 times row 2 to row 3.
2. added 2 times row 2 to row 4.

As previously, we see

$$A_2 = E_{24}E_{23}A_1$$

where

$$E_{23} = \begin{bmatrix} 1 & & & \\ & 1 & & \\ & 1 & 1 & \\ & & & 1 \end{bmatrix}, \quad E_{24} = \begin{bmatrix} 1 & & & \\ & 1 & & \\ & 0 & 1 & \\ & 2 & 0 & 1 \end{bmatrix}$$

Note that each E_{2j} is unit lower triangular.

Finally, to go from A_2 to $A_3 = U$, we have added 4 times row 3 to row 4. Then $U = A_3 = E_{34}A_2$, where

$$E_{34} = \begin{bmatrix} 1 & & & \\ & 1 & & \\ & & 1 & \\ & & 4 & 1 \end{bmatrix}$$

and again we have a unit lower triangular matrix.

Putting this altogether, we have

$$U = A_3 = E_{34}A_2 = E_{34}E_{24}E_{23}A_1 = E_{34}E_{24}E_{23}E_{14}E_{13}E_{12}A \qquad \blacksquare$$

You can see that, in general, if we start with an arbitrary square matrix A and can proceed through Gaussian elimination without any interchanges, the process will yield

(1.53)
$$U = E_k \cdots E_1 A$$

If we solve Equation (1.53) for A, we obtain

(1.54)
$$A = E_1^{-1} \cdots E_k^{-1} U$$

By Theorem (1.43), Section 1.4, we know exactly what each E_i^{-1} looks like:

(1.55)
$$\text{If} \quad E_i = \begin{bmatrix} 1 & & \\ & \ddots & \\ m_i & & 1 \end{bmatrix}, \quad \text{then} \quad E_i^{-1} = \begin{bmatrix} 1 & & \\ & \ddots & \\ -m_i & & 1 \end{bmatrix}$$

Thus $E_1^{-1} \cdots E_k^{-1}$ is a product of unit lower triangular matrices. Now it happens that:

(1.56)

> THEOREM (a) The product of two lower triangular matrices is a lower triangular matrix.
> (b) The product of two unit lower triangular matrices is a unit lower triangular matrix.
> (c) Corresponding statements regarding upper triangular matrices are also true.

The proof is left to the exercises. (See Exercises 37 and 38.) Hence if we let

(1.57)
$$L = E_1^{-1} \cdots E_k^{-1}$$

then L is unit lower triangular. Substituting L into Equation (1.54), we have now demonstrated one of our main theorems.

(1.58)

> THEOREM If A is square and Gaussian elimination can be performed without any interchanges on A, yielding nonzero pivots, then
>
> $$A = LU$$
>
> where U is upper triangular and L is unit lower triangular. This is called the ***LU* decomposition** of A, and it is unique.

The proof of uniqueness is left as an exercise. (See Exercise 43.)

Computing LU Decompositions

We shall now see how easy it is to compute an LU decomposition (when a square matrix has one; that is when Gaussian elimination can be performed with no interchanges). We already know that U is simply the end result of Gaussian elimination and L is the product of the E_i^{-1}s. Example 2 illustrates the fortunate fact that the E_i^{-1}s are always nicely ordered, so that finding L requires no additional computation (once we have found the multiples that Gaussian elimination requires).*

Example 2 Find the LU decomposition for the matrix A of Example 1,

$$A = \begin{bmatrix} 2 & 8 & 1 & 1 \\ 4 & 13 & 3 & -1 \\ -2 & -5 & -3 & 3 \\ -6 & -18 & -1 & 1 \end{bmatrix}$$

Solution Our goal is to factor A as $A = LU$, where

$$L = \begin{bmatrix} 1 & & & \\ * & 1 & & \\ * & * & 1 & \\ * & * & * & 1 \end{bmatrix}, \qquad U = \begin{bmatrix} * & * & * & * \\ & * & * & * \\ & & * & * \\ & & & * \end{bmatrix}$$

Our method will be to use the sequence from in Example 1

$$A \to A_1 \to A_2 \to A_3 = U \tag{1.59}$$

to obtain the multiples m_i that form L. The important thing to remember is

$$L = E_1^{-1} \cdots E_k^{-1} \qquad \text{where} \qquad E_i^{-1} = \begin{bmatrix} 1 & & \\ & \ddots & \\ & -m_i & 1 \end{bmatrix} \tag{1.60}$$

Thus from the multiples m_i we obtain in Gaussian elimination, we use the *negatives*, $-m_i$, to form L. We start with

$$A = \begin{bmatrix} 2 & 8 & 1 & 1 \\ 4 & 13 & 3 & -1 \\ -2 & -5 & -3 & 3 \\ -6 & -18 & -1 & 1 \end{bmatrix} \qquad \text{and} \qquad L = \begin{bmatrix} 1 & & & \\ * & 1 & & \\ * & * & 1 & \\ * & * & * & 1 \end{bmatrix}$$

---

* *At this point, some people may wish to examine this fortuitous ordering in detail, while others may wish simply to accept it and go on. For more details, see Exercises 44 and 45.*

We see that equation $UX = Y$ is easily solved by backsubstitution, once we know Y, and $LY = B$ is easily solved by **forward substitution**, which is just like backsubstitution, except in the opposite order. Example 4 illustrates this.

Example 4 For

$$A = \begin{bmatrix} 2 & 4 & -4 \\ 1 & -4 & 3 \\ -6 & -9 & 10 \end{bmatrix} \quad \text{and} \quad B = \begin{bmatrix} 12 \\ -21 \\ -24 \end{bmatrix},$$

solve the equation $AX = B$ using the LU decomposition found in Example 3.

Solution By Example 3 the LU decomposition for A is

$$\begin{bmatrix} 2 & 4 & -4 \\ 1 & -4 & 3 \\ -6 & -9 & 10 \end{bmatrix} = \begin{bmatrix} 1 & & \\ \frac{1}{2} & 1 & \\ -3 & -\frac{1}{2} & 1 \end{bmatrix} \begin{bmatrix} 2 & 4 & -4 \\ & -6 & 5 \\ & & \frac{1}{2} \end{bmatrix}$$

To solve $AX = B$ or $LUX = B$, first we let $Y = UX$. We next solve $LY = B$ and finally solve $UX = Y$. Now $LY = B$ is

$$\begin{bmatrix} 1 & & \\ \frac{1}{2} & 1 & \\ -3 & -\frac{1}{2} & 1 \end{bmatrix} \begin{bmatrix} y_1 \\ y_2 \\ y_3 \end{bmatrix} = \begin{bmatrix} 12 \\ -21 \\ -24 \end{bmatrix} \quad \text{or} \quad \begin{aligned} y_1 \qquad\qquad &= 12 \\ \tfrac{1}{2}y_1 + y_2 \qquad &= -21 \\ -3y_1 - \tfrac{1}{2}y_2 + y_3 &= -24 \end{aligned}$$

Solving this by forward substitution, we obtain

$$\begin{aligned} y_1 &= 12 \\ \tfrac{1}{2}(12) + y_2 &= -21 \\ y_2 &= -21 - 6, \quad y_2 = -27 \\ -3(12) - \tfrac{1}{2}(-27) + y_3 &= -24 \\ y_3 &= -24 + 36 - \tfrac{27}{2} \\ y_3 &= 12 - \tfrac{27}{2}, \quad y_3 = -\tfrac{3}{2} \end{aligned}$$

Thus $(y_1, y_2, y_3) = (12, -27, -\frac{3}{2})$. Now we solve $UX = Y$ for X.

$$\begin{bmatrix} 2 & 4 & -4 \\ & -6 & 5 \\ & & \frac{1}{2} \end{bmatrix} \begin{bmatrix} x_1 \\ x_2 \\ x_3 \end{bmatrix} = \begin{bmatrix} 12 \\ -27 \\ -\frac{3}{2} \end{bmatrix} \quad \text{or} \quad \begin{aligned} 2x_1 + 4x_2 - 4x_3 &= 12 \\ -6x_2 + 5x_3 &= -27 \\ \tfrac{1}{2}x_3 &= -\tfrac{3}{2} \end{aligned}$$

$$\begin{aligned} \tfrac{1}{2}x_3 &= -\tfrac{3}{2}, \quad x_3 = -3 \\ -6x_2 + 5(-3) &= -27 \\ -6x_2 &= -12, \quad x_2 = 2 \\ 2x_1 + 4(2) - 4(-3) &= 12 \\ 2x_1 &= -8, \quad x_1 = -4 \end{aligned}$$

Therefore the solution is $(x_1, x_2, x_3) = (-4, 2, -3)$. ■

Why Use LU Decompositions?

Now that we see how to solve systems using LU decompositions, we have a very important question to address. Since we can already solve the system $AX = B$ by Gaussian elimination, why should we do it this way? To answer this, we first observe that to solve a system $AX = B$ one time, it takes the same amount of work by either method. In fact in Appendix B we show that if A is $n \times n$, it takes roughly $n^3/3$ operations* to solve $AX = B$ by Gaussian elimination. If we were to examine the procedures carefully, we would see that the operations used to solve $AX = B$ are exactly the same operations used to find the LU decomposition (in a slightly different order).

However, there are many real-world situations in which we need to solve the system $AX = B$ for many different B's but the *same* A. (See Exercises 21–24 here and Inverse Iteration in Section 5.8 for examples.) Using LU decompositions, we have to find L and U only for the first B, and this takes most of the work. Thereafter we can use the same L and U for all the other B's, just solve $LY = B$ and $UX = Y$ as above, by forward and backsubstitution. This takes only n^2 operations, which we show in Appendix B. This leads to considerable savings in work (and hence in time and money). For example, suppose A is 100×100 (not really all that big in terms of realworld problems). Then to solve $AX = B$, Gaussian elimination takes about $n^3/3 = 100^3/3 \approx 330{,}000$ operations. But if we already have the LU decomposition for A (from a previous problem), then solving $LY = B$ and $UX = Y$ takes only $n^2 = 100^2 = 10{,}000$ operations. Obviously 10,000 is a considerable savings over 330,000!

LU Decompositions and Row Interchanges

We have seen that A has an LU decomposition if we can perform Gaussian elimination on A without interchanging any rows. However, we know there are matrices A, such as

$$A = \begin{bmatrix} 0 & 1 & 2 \\ 3 & 4 & 5 \\ 6 & 7 & 8 \end{bmatrix}$$

for which we would have to interchange rows in order to get nonzero pivots. Moreover, we see in Section 1.7 that there are matrices for which, during Gaussian elimination, we would want to interchange rows for numerical reasons. We shall now discuss what is meant by an LU decomposition in such a situation, and we shall also see an easy, but inefficient, method for obtaining them. An alternative approach and rigorous explanation of an efficient method is given in Appendix A.†

* *See the first paragraph of Appendix B for a definition of "operations."*

† *This is the simplest thing to do for an introductory course. If an efficient method is desired, Appendix A may be substituted for the balance of this section.*

Suppose now that row interchanges are performed on a matrix A during Gaussian elimination. If we know what the row interchanges are ahead of time, we can do them first and then perform Gaussian elimination on the new A without interchanges. Each interchange of two rows can be performed by multiplying on the left by an elementary permutation matrix P_i. Thus if we do all the interchanges at first, we have $P_n \cdots P_1 A$ or simply PA where $P = P_n \cdots P_1$ is a permutation matrix. (A **permutation matrix** is a matrix that is a product of elementary permutation matrices.)* The main result is then

(1.63)

> THEOREM If A is nonsingular, then there is a permutation matrix P such that
>
> $$PA = LU$$
>
> For a fixed A, different P's may yield different LU decompositions. Once P is fixed, then L and U are unique.

Thus our method is to run through Gaussian elimination twice. The first time is to determine any row interchanges and the permutation matrix P. The second is to determine L and U so that $PA = LU$. Of course, efficient methods do this in one pass, but for matrices of small dimension, this method presents sufficient insight into the underlying mathematics of $PA = LU$ decompositions. For 2×2 matrices, the process is particularly simple.

Example 5 Find the $PA = LU$ decomposition for $A = \begin{bmatrix} 0 & 2 \\ 3 & 4 \end{bmatrix}$.

Solution To obtain a nonzero entry in the first position, we must interchange the two rows. But then the resulting matrix is a U in upper triangular form. So we take L as the identity matrix and the $PA = LU$ decomposition is

$$\begin{bmatrix} 0 & 1 \\ 1 & 0 \end{bmatrix}\begin{bmatrix} 0 & 2 \\ 3 & 4 \end{bmatrix} = \begin{bmatrix} 1 & \\ 0 & 1 \end{bmatrix}\begin{bmatrix} 3 & 4 \\ & 2 \end{bmatrix}$$ ■

Example 6 illustrates that 3×3 matrices are not difficult either.

Example 6 Find the $PA = LU$ decomposition for $A = \begin{bmatrix} 0 & 0 & 3 \\ 2 & 4 & -1 \\ 6 & 5 & 5 \end{bmatrix}$.

* *See also Review Exercises 25, 26.*

Solution To obtain a nonzero entry in the (1, 1) position, we interchange the first two rows. We next add -3 times row 1 to row 3.

$$\begin{bmatrix} 0 & 0 & 3 \\ 2 & 4 & -1 \\ 6 & 5 & 5 \end{bmatrix} \rightarrow \begin{bmatrix} 2 & 4 & -1 \\ 0 & 0 & 3 \\ 6 & 5 & 5 \end{bmatrix} \rightarrow \begin{bmatrix} 2 & 4 & -1 \\ 0 & 0 & 3 \\ 0 & -7 & 8 \end{bmatrix}$$

To obtain a nonzero entry in the (2, 2) position, we now interchange rows 2 and 3

$$\begin{bmatrix} 2 & 4 & -1 \\ 0 & 0 & 3 \\ 0 & -7 & 8 \end{bmatrix} \rightarrow \begin{bmatrix} 2 & 4 & -1 \\ 0 & -7 & 8 \\ 0 & 0 & 3 \end{bmatrix}$$

and obtain a matrix U in upper triangular form. We next set out to find a P and an L so that PA is L times this U. We first obtain P by applying the permutations we just used, in the *same* order, to I.

$$I = \begin{bmatrix} 1 & & \\ & 1 & \\ & & 1 \end{bmatrix} \xrightarrow[\text{rows 1 and 2.}]{\text{Interchange}} \begin{bmatrix} 0 & 1 & \\ 1 & 0 & \\ & & 1 \end{bmatrix}$$

$$\xrightarrow[\text{rows 2 and 3.}]{\text{Interchange}} \begin{bmatrix} 0 & 1 & 0 \\ 0 & 0 & 1 \\ 1 & 0 & 0 \end{bmatrix} = P$$

Then

$$PA = \begin{bmatrix} 0 & 1 & 0 \\ 0 & 0 & 1 \\ 1 & 0 & 0 \end{bmatrix}\begin{bmatrix} 0 & 0 & 3 \\ 2 & 4 & -1 \\ 6 & 5 & 5 \end{bmatrix} = \begin{bmatrix} 2 & 4 & -1 \\ 6 & 5 & 5 \\ 0 & 0 & 3 \end{bmatrix} = A_1$$

If we have done our work correctly, we can find an LU decomposition for A_1. We do this exactly as we did earlier in this section. We start with

$$A_1 = \begin{bmatrix} 2 & 4 & -1 \\ 6 & 5 & 5 \\ 0 & 0 & 3 \end{bmatrix}, \quad L = \begin{bmatrix} 1 & & \\ * & 1 & \\ * & * & 1 \end{bmatrix}$$

Adding -3 times row 1 to row 2
Adding 0 times row 1 to row 3

Place the negatives of these multiples in the corresponding positions of L.

$$A_2 = \begin{bmatrix} 2 & 4 & -1 \\ 0 & -7 & 8 \\ 0 & 0 & 3 \end{bmatrix}, \quad L = \begin{bmatrix} 1 & & \\ 3 & 1 & \\ 0 & * & 1 \end{bmatrix}$$

This is the U that we obtained earlier. Hence there is no more to do, the remaining unknown entry of L is zero, and the $PA = LU$ decomposition for A is

$$\begin{bmatrix}0 & 1 & 0\\0 & 0 & 1\\1 & 0 & 0\end{bmatrix}\begin{bmatrix}0 & 0 & 3\\2 & 4 & -1\\6 & 5 & 5\end{bmatrix}=\begin{bmatrix}1 & & \\3 & 1 & \\0 & 0 & 1\end{bmatrix}\begin{bmatrix}2 & 4 & -1\\ & -7 & 8\\ & & 3\end{bmatrix} \tag{1.64}$$

■

Solving Systems Using $PA = LU$

Suppose we have a decomposition $PA = LU$, and we want to solve the equation $AX = B$. We first multiply through by P, obtaining

$$PAX = PB \tag{1.65}$$

If we let $PB = B'$ and substitute $PA = LU$, we obtain

$$LUX = B' \tag{1.66}$$

This is exactly the same form as we obtained before, so we proceed as we did then. We let $Y = UX$, solve $LY = B'$ by forward substitution, and then solve $UX = Y$ by backsubstitution.

Example 7 Solve $AX = B$, where

$$A=\begin{bmatrix}0 & 0 & 3\\2 & 4 & -1\\6 & 5 & 5\end{bmatrix},\qquad B=\begin{bmatrix}6\\-10\\-7\end{bmatrix}$$

using the $PA = LU$ decomposition found in Example 6.

Solution By Equation (1.64) in Example 6, the $PA = LU$ decomposition is

$$\begin{bmatrix}0 & 1 & 0\\0 & 0 & 1\\1 & 0 & 0\end{bmatrix}\begin{bmatrix}0 & 0 & 3\\2 & 4 & -1\\6 & 5 & 5\end{bmatrix}=\begin{bmatrix}1 & & \\3 & 1 & \\0 & 0 & 1\end{bmatrix}\begin{bmatrix}2 & 4 & -1\\ & -7 & 8\\ & & 3\end{bmatrix}$$

We first compute

$$B' = PB = \begin{bmatrix}0 & 1 & 0\\0 & 0 & 1\\1 & 0 & 0\end{bmatrix}\begin{bmatrix}6\\-10\\-7\end{bmatrix}=\begin{bmatrix}-10\\-7\\6\end{bmatrix}$$

To solve $PAX = B'$, we substitute LU for PA and proceed to solve $LUX = B'$. We first solve $LY = B'$.

$$\begin{bmatrix}1 & & \\3 & 1 & \\0 & 0 & 1\end{bmatrix}\begin{bmatrix}y_1\\y_2\\y_3\end{bmatrix}=\begin{bmatrix}-10\\-7\\6\end{bmatrix}\quad\text{or}\quad\begin{aligned}y_1 \qquad &= -10\\3y_1 + y_2 \qquad &= -7\\y_3 &= 6\end{aligned}$$

We obtain $(y_1, y_2, y_3) = (-10, 23, 6)$ and finish by solving $UX = Y$.

$$\begin{bmatrix} 2 & 4 & -1 \\ & -7 & 8 \\ & & 3 \end{bmatrix}\begin{bmatrix} x_1 \\ x_2 \\ x_3 \end{bmatrix} = \begin{bmatrix} -10 \\ 23 \\ 6 \end{bmatrix} \quad \text{or} \quad \begin{aligned} 2x_1 + 4x_2 - x_3 &= -10 \\ -7x_2 + 8x_3 &= 23 \\ 3x_3 &= 6 \end{aligned}$$

The answer, which you can easily check, is $(x_1, x_2, x_3) = (-2, -1, 2)$. ■

Exercise 1.5

In Exercises 1–8, find the *LU* decomposition for the given matrix. (No interchanges should be required.)

1. $\begin{bmatrix} 2 & 4 \\ 6 & 3 \end{bmatrix}$

2. $\begin{bmatrix} 4 & -6 \\ -3 & 5 \end{bmatrix}$

3. $\begin{bmatrix} 2 & 1 & 3 \\ -2 & 5 & 1 \\ 4 & 2 & 4 \end{bmatrix}$

4. $\begin{bmatrix} 5 & 1 & 2 \\ 10 & 3 & 8 \\ -10 & 4 & -7 \end{bmatrix}$

5. $\begin{bmatrix} 3 & -2 & -1 \\ -6 & -4 & 2 \\ -3 & -2 & 4 \end{bmatrix}$

6. $\begin{bmatrix} 4 & -2 & 4 \\ 2 & 3 & -1 \\ -8 & 2 & 5 \end{bmatrix}$

7. $\begin{bmatrix} 1 & 3 & 2 & -1 \\ 2 & 5 & 3 & 2 \\ -3 & 2 & -1 & 2 \\ 1 & 1 & 3 & 1 \end{bmatrix}$

8. $\begin{bmatrix} 2 & -2 & 6 & -4 \\ 2 & -5 & 2 & 2 \\ -4 & 1 & 3 & 2 \\ 1 & 5 & 1 & -2 \end{bmatrix}$

In Exercises 9–16, use the *LU* decomposition given for *A* to solve the equation $AX = B$, for the given *B*. Do *not* find *A*.

9. $A = \begin{bmatrix} 1 & \\ 3 & 1 \end{bmatrix}\begin{bmatrix} 2 & -3 \\ & 4 \end{bmatrix}, \quad B = \begin{bmatrix} 23 \\ 25 \end{bmatrix}$

10. $A = \begin{bmatrix} 1 & \\ -\frac{1}{2} & 1 \end{bmatrix}\begin{bmatrix} 2 & 2 \\ & 3 \end{bmatrix}, \quad B = \begin{bmatrix} -24 \\ -6 \end{bmatrix}$

11. $A = \begin{bmatrix} 1 & & \\ 2 & 1 & \\ -3 & 1 & 1 \end{bmatrix}\begin{bmatrix} 2 & -1 & 0 \\ & 3 & 1 \\ & & 4 \end{bmatrix}, \quad B = \begin{bmatrix} 4 \\ 6 \\ -22 \end{bmatrix}$

12. $A = \begin{bmatrix} 1 & & \\ -4 & 1 & \\ -2 & 3 & 1 \end{bmatrix}\begin{bmatrix} 3 & -2 & 2 \\ & -2 & -1 \\ & & -3 \end{bmatrix}, \quad B = \begin{bmatrix} -8 \\ 42 \\ 64 \end{bmatrix}$

13. $A = \begin{bmatrix} 1 & & \\ 5 & 1 & \\ -3 & 2 & 1 \end{bmatrix}\begin{bmatrix} 4 & -1 & 2 \\ & 2 & 1 \\ & & -7 \end{bmatrix}, \quad B = \begin{bmatrix} -7 \\ -27 \\ 23 \end{bmatrix}$

14. $A = \begin{bmatrix} 1 & & \\ 0 & 1 & \\ -5 & -4 & 1 \end{bmatrix}\begin{bmatrix} -5 & 2 & 6 \\ & 1 & -2 \\ & & 4 \end{bmatrix}, \quad B = \begin{bmatrix} 6 \\ -5 \\ -12 \end{bmatrix}$

15. $A = \begin{bmatrix} 1 & & & \\ 2 & 1 & & \\ -3 & 9 & 1 & \\ -1 & 4 & -2 & 1 \end{bmatrix} \begin{bmatrix} 2 & 1 & 0 & -2 \\ & 3 & 1 & -1 \\ & & -2 & 1 \\ & & & 2 \end{bmatrix}, \quad B = \begin{bmatrix} 7 \\ 16 \\ -11 \\ 13 \end{bmatrix}$

16. $A = \begin{bmatrix} 1 & & & \\ 3 & 1 & & \\ 0 & 2 & 1 & \\ -2 & -1 & 3 & 1 \end{bmatrix} \begin{bmatrix} -3 & -1 & 2 & 1 \\ & 4 & -1 & 2 \\ & & -2 & -1 \\ & & & 2 \end{bmatrix}, \quad B = \begin{bmatrix} 4 \\ 13 \\ 6 \\ -7 \end{bmatrix}$

In Exercises 17–20, rewrite the system in matrix form and then solve using LU decompositions.

17. $\begin{aligned} 2x_1 + 3x_2 &= -1 \\ 6x_1 + 13x_2 &= -7 \end{aligned}$

18. $\begin{aligned} 2x_1 - 2x_2 &= 0 \\ -x_1 + 4x_2 &= 2 \end{aligned}$

19. $\begin{aligned} -2x_1 + 4x_2 - x_3 &= -15 \\ -4x_1 + 7x_2 - 2x_3 &= -27 \\ 6x_1 - 11x_2 + 6x_3 &= 39 \end{aligned}$

20. $\begin{aligned} 2x_1 + x_2 - 4x_3 &= 6 \\ -6x_1 - 5x_2 + 11x_3 &= -9 \\ 2x_1 - 3x_2 - 9x_3 &= 7 \end{aligned}$

The solution to $A^3X = B$, when A^3 is difficult to compute, can be found by solving three problems: $AY = B$, $AZ = Y$, and finally $AX = Z$. (A similar pattern of problems holds for $A^nX = B$, where n is any positive integer.) This illustrates the economy of using LU decompositions (see "Why Use LU Decompositions?"). In Exercises 21–25, solve $A^3X = B$.

21. Use A and B in Exercise 9.

22. Use A and B in Exercise 10.

23. Use A and B in Exercise 11.

24. Use A and B in Exercise 12.

In Exercises 25–32, find a $PA = LU$ decomposition for the given matrix. Check your answer by multiplying PA and LU.

25. $\begin{bmatrix} 0 & 2 \\ -3 & 4 \end{bmatrix}$

26. $\begin{bmatrix} 0 & -2 \\ 5 & 3 \end{bmatrix}$

27. $\begin{bmatrix} 0 & 2 & 3 \\ -2 & 1 & 4 \\ 6 & -7 & 2 \end{bmatrix}$

28. $\begin{bmatrix} 0 & 4 & -1 \\ 3 & 2 & 4 \\ 6 & -6 & 1 \end{bmatrix}$

29. $\begin{bmatrix} 2 & 3 & -1 \\ -4 & -6 & 5 \\ 4 & 1 & 2 \end{bmatrix}$

30. $\begin{bmatrix} 3 & -2 & -4 \\ -6 & 4 & 3 \\ 6 & 1 & 2 \end{bmatrix}$

31. $\begin{bmatrix} -2 & 3 & -1 & 2 \\ 4 & -6 & 2 & 3 \\ 6 & -1 & -2 & 3 \\ -8 & -4 & -3 & -1 \end{bmatrix}$

32. $\begin{bmatrix} 0 & 2 & 0 & -3 \\ 1 & 2 & -1 & 3 \\ 3 & 4 & -3 & -4 \\ 5 & -2 & -3 & -1 \end{bmatrix}$

In Exercises 33–36, use the $PA = LU$ decomposition to solve the equation $AX = B$, for the given B. Do *not* find A.

33. $\begin{bmatrix} & 1 \\ 1 & \end{bmatrix} A = \begin{bmatrix} 1 & \\ 2 & 1 \end{bmatrix} \begin{bmatrix} 2 & -1 \\ & -3 \end{bmatrix}, \quad B = \begin{bmatrix} 22 \\ 8 \end{bmatrix}$

34. $\begin{bmatrix} 0 & 1 & 0 \\ 0 & 0 & 1 \\ 1 & 0 & 0 \end{bmatrix} A = \begin{bmatrix} 1 & & \\ -2 & 1 & \\ 3 & 2 & 1 \end{bmatrix} \begin{bmatrix} 2 & -3 & -1 \\ & 3 & -2 \\ & & -3 \end{bmatrix}, \quad B = \begin{bmatrix} 29 \\ 7 \\ -13 \end{bmatrix}$

35. $\begin{bmatrix} 0 & 0 & 1 \\ 1 & 0 & 0 \\ 0 & 1 & 0 \end{bmatrix} A = \begin{bmatrix} 1 & & \\ 2 & 1 & \\ -3 & -2 & 1 \end{bmatrix} \begin{bmatrix} -3 & -1 & 0 \\ & -2 & -2 \\ & & -1 \end{bmatrix}, \quad B = \begin{bmatrix} -8 \\ 10 \\ -7 \end{bmatrix}$

36. $\begin{bmatrix} 0 & 1 & 0 & 0 \\ 0 & 0 & 1 & 0 \\ 0 & 0 & 0 & 1 \\ 1 & 0 & 0 & 0 \end{bmatrix} A = \begin{bmatrix} 1 & & & \\ -2 & 1 & & \\ 0 & -3 & 1 & \\ -1 & -1 & 2 & 1 \end{bmatrix} \begin{bmatrix} 2 & -1 & 0 & 2 \\ & -3 & 1 & -2 \\ & & -2 & 1 \\ & & & -1 \end{bmatrix}, \quad B = \begin{bmatrix} -5 \\ 4 \\ -1 \\ -17 \end{bmatrix}$

37. Suppose $A = [a_{ij}]$ and $B = [b_{ij}]$ are two $n \times n$ lower triangular matrices. Thus $a_{ij} = 0$ and $b_{ij} = 0$ if $j > i$. Let $AB = C = [c_{ij}]$. Show that C is lower triangular. (HINT $c_{ij} = a_{i1}b_{1j} + a_{i2}b_{2j} + \cdots + a_{in}b_{nj}$. If $j > i$, then one of the factors of $a_{ik}b_{kj}$ must be zero; hence c_{ij} is a sum of zeros.)

38. In Exercise 37 show that if A and B are unit lower triangular, then C is also unit lower triangular. (HINT Show $c_{ii} = 0 + 1 + 0$.)

39. Suppose L is unit lower triangular and U is upper triangular. Show that if $L = U$, then $L = U = I$.

40. Suppose L is unit lower triangular. Show that L is invertible and L^{-1} is also unit lower triangular. (HINT First show we can write L as a product of elementary matrices E of the form 1's on the diagonal and a single entry m below the diagonal.)

41. Suppose L is lower triangular with nonzero entries on the diagonal. Show that L is invertible and L^{-1} is also lower triangular. (HINT First write L as $L = DL'$; L' is unit lower triangular, D is diagonal and invertible.)

42. Using arguments similar to Exercises 40 and 41, show that if U is unit upper triangular (upper triangular with nonzero entries on the diagonal), then U^{-1} exists and has the same form.

43. Show that LU decompositions are unique, that is, if $A = LU$ and $A = L'U'$, then $L = L'$ and $U = U'$. (HINT From $LU = L'U'$, use Exercises 40–42 to obtain $L^{-1}L' = U'U^{-1}$ and show that this is a unit lower triangular matrix equal to an upper triangular matrix. Apply Exercise 39.)

44. Let

$$E_1 = \begin{bmatrix} 1 & & \\ a & 1 & \\ & & 1 \end{bmatrix}, \quad E_2 = \begin{bmatrix} 1 & & \\ & 1 & \\ & b & 1 \end{bmatrix}, \quad \text{and} \quad A = \begin{bmatrix} 1 & & \\ a & 1 & \\ & b & 1 \end{bmatrix}$$

where $a, b \neq 0$. Show that $E_1E_2 = A$ but $E_2E_1 \neq A$. (Do not just multiply out blindly; use the fact that E_1 and E_2 are elementary matrices.)

45. Suppose

$$E = \begin{bmatrix} 1 & & & \\ & 1 & & \\ & & \ddots & \\ & m & & \ddots & \\ & & & & 1 \end{bmatrix}$$

with m in the (a, b)th entry, $a > b$. Suppose L is unit lower triangular with $l_{ij} = 0$ for $i < a$ and $l_{ab} = 0$. Show that EL is L with m inserted in the (a, b)th entry. Note that Exercise 44 shows this may not be true if $l_{ij} \neq 0$ for $i < a$.

1.6 Transposes, Symmetry, and Band Matrices; An Application

It is usually the case that matrices that occur in applications, especially large matrices, have special properties. Many of these properties are immediately obvious to even the casual observer. For example, the matrix might be symmetrical, **sparse** (which means most of the entries are zero), or a **band matrix** (which means the nonzero entries are concentrated near the diagonal). Such special properties usually considerably reduce the work needed to solve the associated systems. The purpose of this section is to discuss these special properties and then give an application. A second application is given in Appendix C.

Transposes and Symmetry

(1.67) DEFINITION A matrix $A = [a_{ij}]$ is called **symmetric** if it is square and $a_{ij} = a_{ji}$, for all i and j.

Example 1 The matrices

$$\begin{bmatrix} 1 & 2 \\ 2 & 3 \end{bmatrix}, \quad \begin{bmatrix} 2 & -1 & 3 \\ -1 & 5 & 0 \\ 3 & 0 & -7 \end{bmatrix}$$

are both symmetric. ■

To discuss symmetric matrices fully, it is convenient to introduce the *transpose* A^T of an $m \times n$ matrix A.

(1.68)

DEFINITION If A is an $m \times n$ matrix, its **transpose**, A^T, is the $n \times m$ matrix obtained from A by making each row of A the corresponding column of A^T.

Example 2 If

$$A = \begin{bmatrix} 2 & -1 & 3 \\ 0 & 5 & -2 \end{bmatrix}, \quad B = \begin{bmatrix} 2 & 5 \\ -3 & 7 \end{bmatrix}, \quad C = \begin{bmatrix} 5 \\ -1 \\ -6 \end{bmatrix}$$

then

$$A^T = \begin{bmatrix} 2 & 0 \\ -1 & 5 \\ 3 & -2 \end{bmatrix}, \quad B^T = \begin{bmatrix} 2 & -3 \\ 5 & 7 \end{bmatrix}, \quad C^T = [5 \quad -1 \quad -6]$$ ■

One obvious feature is the following theorem.

(1.69)

THEOREM An $n \times n$ matrix is symmetric if and only if $A = A^T$.

Transposes are very important in their own right, and following are a few of their important properties.

(1.70)

THEOREM

(a) $(A^T)^T = A$.
(b) $(A + B)^T = A^T + B^T$, if A and B are both $m \times n$.
(c) $(AB)^T = B^T A^T$, if A is $m \times n$ and B is $n \times p$.
(d) If A is $n \times n$ and invertible, then A^T is invertible and

$$(A^T)^{-1} = (A^{-1})^T$$

Proof of (d) Parts (a) and (b) are easy to see. Part (c) is left to the exercises, but *notice the reverse order*. To prove (d), if A is invertible, then

$$\begin{aligned} AA^{-1} &= I & &\text{and} & A^{-1}A &= I \\ (AA^{-1})^T &= I^T & &\text{and} & (A^{-1}A)^T &= I^T \\ (A^{-1})^T A^T &= I & &\text{and} & A^T(A^{-1})^T &= I \end{aligned}$$

by part (c) and since $I^T = I$. Thus $(A^{-1})^T$ plays the role of $(A^T)^{-1}$, so the two are the same by Theorem (1.38), Section 1.4. ■

One of the very important uses of A^T will be given when we discuss dot products. Another use can be seen in the exercises when we define a general permutation matrix P and show that $P^T = P^{-1}$. (See Exercises 23, 24.)

Symmetric matrices have important properties, and we shall see a few of them at different places in this text. One easy to state but very important property is that they can be stored in a computer in only half as much space. But the most important property of symmetric matrices will be discussed in Section 5.4, where we show that eigenvalues and eigenvectors of symmetric matrices are quite special.

Band Matrices

Applications often give rise to matrices with special form. One such common special form is a *band matrix*, which is a square matrix with all of its nonzero entries on or near its diagonal. This is illustrated in Example 3.

Example 3 Suppose a supporting truss for a bridge is as indicated in Figure 1.3a. There are nine pivots labeled 1–9.

Figure 1.3

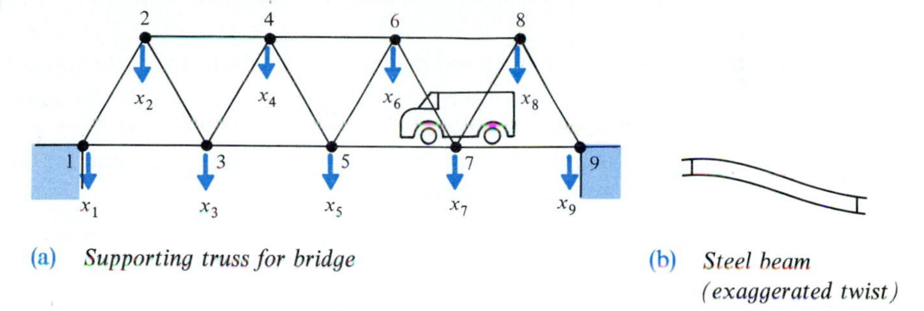

(a) *Supporting truss for bridge* (b) *Steel beam (exaggerated twist)*

No material is completely stiff, so the beams will bend somewhat, both from the weight of the structure and from the weight of vehicles on the bridge. Let x_i be the amount of vertical displacement at the ith pivot. If the seventh pivot is pushed down somewhat, say by a truck, then the fifth, sixth, eighth, and ninth pivots will also be *directly* pushed down, because they are connected to the seventh pivot by a beam. The fourth pivot will be pushed down by the fifth and sixth pivots, but the fourth will not be directly affected by the seventh pivot, as there is no beam connecting the fourth and seventh pivots. It turns out that, under certain hypotheses, the relationship between x_7 and the other displacements is linear and gives rise to an equation of the form

$$a_{75}x_5 + a_{76}x_6 + a_{77}x_7 + a_{78}x_8 + a_{79}x_9 = b_7$$

Each displacement x_i gives rise to a similar equation, and altogether we obtain a system

$$\begin{aligned}
&a_{11}x_1 + a_{12}x_2 + a_{13}x_3 &&= b_1\\
&a_{21}x_1 + a_{22}x_2 + a_{23}x_3 + a_{24}x_4 &&= b_2\\
&a_{31}x_1 + a_{32}x_2 + a_{33}x_3 + a_{34}x_4 + a_{35}x_5 &&= b_3\\
&\qquad a_{42}x_2 + a_{43}x_3 + a_{44}x_4 + a_{45}x_5 + a_{46}x_6 &&= b_4\\
&&&\vdots\\
&\qquad\qquad a_{97}x_7 + a_{98}x_8 + a_{99}x_9 &&= b_9
\end{aligned} \tag{1.71}$$

The associated matrix equation is $AX = B$, where

$$A = \begin{bmatrix} a_{11} & a_{12} & a_{13} & & & & & & \\ a_{21} & a_{22} & a_{23} & a_{24} & & & & & \\ a_{31} & a_{32} & a_{33} & a_{34} & a_{35} & & & & \\ & a_{42} & a_{43} & a_{44} & a_{45} & a_{46} & & & \\ & & \ddots & \ddots & \ddots & \ddots & \ddots & & \\ & & & & & a_{86} & a_{87} & a_{88} & a_{89} \\ & & & & & & a_{97} & a_{98} & a_{99} \end{bmatrix}$$

It is clear that the matrix A is a band matrix. In fact it turns out that $a_{ij} = a_{ji}$ (since the relationship between x_i and x_j is the same as the relationship between x_j and x_i), so that A is in fact a symmetric band matrix. ■

Example 4 It really should be pointed out that the mathematical description in the previous example is too simple for accurate predictions. Beams twist as well as bend, as indicated in Figure 1.3b. When all such things are considered, the resulting relationships are still linear, but much larger. The resulting matrix looks something like

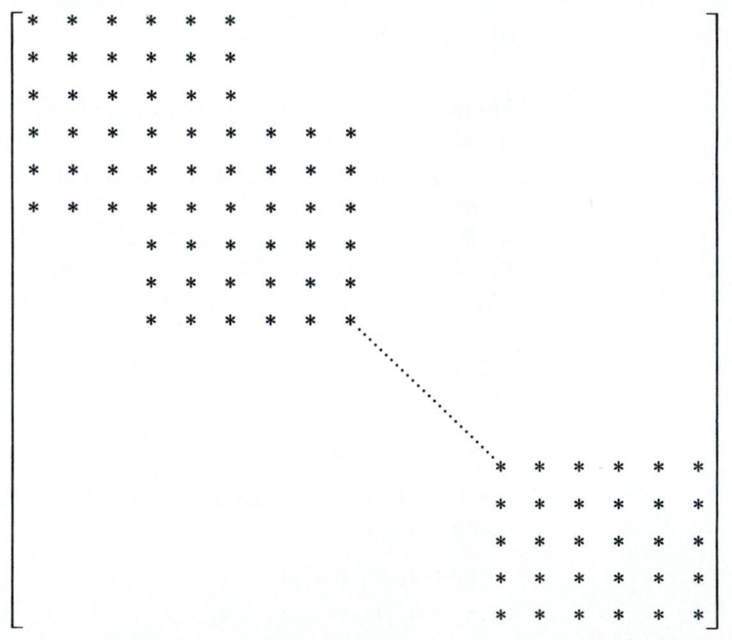

This matrix is still a symmetric band matrix, but it is *large*. Details as to how such systems are derived can be found in textbooks such as W. Weaver and J. Gere, *Matrix Analysis of Framed Structures*, D. Van Nostrand. ■

Matrices that arise in realistic descriptions of applied problems are usually large, even ridiculously large. (In some applications $10{,}000 \times 10{,}000$ are not uncommon!) Fortunately the techniques we learn to handle small matrices efficiently greatly help us with large matrices. For instance, Example 5 illustrates how important LU decompositions are for band matrices.

Example 5 For

$$A = \begin{bmatrix} 1 & 1 & & \\ 2 & 4 & 1 & \\ & 6 & 7 & 1 \\ & & 12 & 2 \end{bmatrix}$$

you can check that the LU decomposition A is

$$\begin{bmatrix} 1 & 1 & & \\ 2 & 4 & 1 & \\ & 6 & 7 & 1 \\ & & 12 & 2 \end{bmatrix} = \begin{bmatrix} 1 & & & \\ 2 & 1 & & \\ & 3 & 1 & \\ & & 3 & 1 \end{bmatrix} \begin{bmatrix} 1 & 1 & & \\ & 2 & 1 & \\ & & 4 & 1 \\ & & & -1 \end{bmatrix}$$

and that

$$A^{-1} = \begin{bmatrix} \frac{1}{2} & \frac{1}{4} & -\frac{1}{4} & \frac{1}{8} \\ \frac{1}{2} & -\frac{1}{4} & \frac{1}{4} & -\frac{1}{8} \\ -3 & \frac{3}{2} & -\frac{1}{2} & \frac{1}{4} \\ 18 & -9 & 3 & -1 \end{bmatrix}$$

■

This illustrates the fact that in the LU decomposition of a band matrix the factors are always band matrices (of the same bandwidth), but the inverse of a band matrix is usually full. If you picture the matrix in Example 5 as being 100×100 (not very large for real applications), you can appreciate the considerable savings of using LU decompositions rather than inverses.

Exercise 1.6

In Exercises 1–6 find the transpose of the given matrix.

1. $\begin{bmatrix} 2 & -3 \\ 4 & 1 \end{bmatrix}$

2. $\begin{bmatrix} 4 & -1 & 3 & -1 \end{bmatrix}$

3. $\begin{bmatrix} 5 & 7 & 2 \\ 8 & -1 & -3 \\ -4 & 2 & 4 \end{bmatrix}$

4. $\begin{bmatrix} 8 & -1 \\ 3 & 5 \\ -2 & 5 \\ 1 & 2 \\ -3 & -5 \end{bmatrix}$

5. $\begin{bmatrix} 2 & 4 & 9 \\ -1 & 7 & 1 \\ 3 & -3 & 3 \\ 0 & -2 & -1 \end{bmatrix}$

6. $\begin{bmatrix} 2 & -1 & 3 & -1 \\ 4 & 0 & -1 & 2 \end{bmatrix}$

In Exercises 7–12, for the given matrix A, compute A^T, $(A^T)^{-1}$, A^{-1}, and $(A^{-1})^T$. Compare $(A^T)^{-1}$ and $(A^{-1})^T$.

7. $\begin{bmatrix} 2 & -1 \\ -5 & 3 \end{bmatrix}$

8. $\begin{bmatrix} 1 & & & \\ 2 & 1 & & \\ & 3 & 1 & \\ & & 4 & 1 \end{bmatrix}$

9. $\begin{bmatrix} 1 & & & \\ 2 & 1 & & \\ -3 & & 1 & \\ 4 & & & 1 \end{bmatrix}$

10. $\begin{bmatrix} 1 & & \\ 2 & 1 & \\ 3 & 2 & 1 \end{bmatrix}$

11. $\begin{bmatrix} 1 & 3 & 2 \\ & 1 & \\ & & 1 \end{bmatrix}$

12. $\begin{bmatrix} 1 & & & \\ & -2 & & \\ & & 3 & \\ 2 & & & -4 \end{bmatrix}$

13. What is A^T if $A = BC^2D^T$?

14. What is A^T if $A = 5B^{-1}C^TD^3$?

In Exercises 15–18, for the given tridiagonal matrix A,

(a) Find the LU decomposition for A.
(b) Find A^{-1}.
(c) Solve the equation $AX = B$, for the given B, in two ways:
   (i) Using the LU decomposition found in part (a).
   (ii) Computing $X = A^{-1}B$.

15. $\begin{bmatrix} 1 & 1 & \\ 1 & 2 & 1 \\ & 1 & 3 \end{bmatrix}$, $\begin{bmatrix} 1 \\ -3 \\ -10 \end{bmatrix}$

16. $\begin{bmatrix} 2 & 3 & \\ -2 & -6 & 1 \\ & 6 & -4 \end{bmatrix}$, $\begin{bmatrix} 7 \\ -18 \\ 26 \end{bmatrix}$

17. $\begin{bmatrix} 2 & \frac{1}{2} & & \\ 4 & 2 & 1 & \\ & 4 & 2 & 1 \\ & & 4 & 2 \end{bmatrix}$, $\begin{bmatrix} 1 \\ 3 \\ -6 \\ 4 \end{bmatrix}$

18. $\begin{bmatrix} 1 & 1 & & \\ 4 & 2 & -1 & \\ & 4 & 3 & -1 \\ & & 4 & 4 \end{bmatrix}$, $\begin{bmatrix} 4 \\ 14 \\ 6 \\ 8 \end{bmatrix}$

19. Explain why A is symmetric if and only if $A^T = A$.

DEFINITION A matrix is **skew symmetric** if and only if $A^T = -A$.

20. If B is any square matrix, show

(a) $B + B^T$ is symmetric
(b) $B - B^T$ is skew symmetric

21. Show every diagonal element of a skew-symmetric matrix is zero.

22. If B is any $m \times n$ matrix then show BB^T and B^TB are both symmetric.

A **permutation matrix** is a square matrix with exactly a single 1 in each row and column; all other entries are zero. (Also see Review Exercises 25, 26.)

23. Write down the six 3×3 permutation matrices P, pair up each with its inverse (possibly itself), and observe $P^{-1} = P^T$.

24. Show $P^{-1} = P^T$ for any $n \times n$ permutation matrix (as defined above) by showing that the 1's and 0's are in exactly the right place to yield $PP^T = I$.

1.7 Numerical and Programming Considerations: Partial Pivoting, Overwriting Matrices, and Ill-Conditioned Systems*

The objective of this section is to discuss a few of the most important remaining considerations when programming a computer to solve systems of linear equations.

Partial Pivoting

One consideration to keep in mind when programming a computer is round-off error. Round-off error in a computer can cause considerable difficulty, as illustrated in the following example. However, some of this difficulty can be avoided if care is used.

Example 1 Solve the system

$$\begin{aligned} 10^{-8}x + y &= 1 \\ x + y &= 2 \end{aligned}$$

by Gaussian elimination, but round off all computations as you go along to eight significant figures (as some computers do).

Solution First observe that the solution is $(x, y) = (1, 1)$ to eight significant figures.

To solve, add -10^8 times the first equation to the second equation, obtaining (to eight significant figures)

$$\begin{aligned} 10^{-8}x + y &= 1 \\ -10^8 y &= -10^8 \end{aligned}$$

* *This section is optional. However, all students are encouraged at least to skim through the subsection on ill-conditioned systems so that they can become acquainted with this potentially disastrous pitfall.*

Solving by backsubstitution, we obtain

$$y = 1$$
$$10^{-8}x + 1 = 1, \quad x = 0$$

Thus we obtain $(x, y) = (0, 1)$, an answer in considerable error! ■

The difficulty comes because the coefficient a_{11} is very small relative to a_{21}, and this illustrates one of the maxims of numerical mathematics.

(1.72)

> If a number's being equal to zero causes difficulty mathematically, then its being close to zero will cause difficulty numerically (i.e., on a computer).

Thus if a_{11} had been equal to zero, we would have been forced **to pivot** (which means, in this context, to interchange rows in order to obtain a new pivot). Since it was close to zero, we should have pivoted anyway (and we ran into difficulty because we did not).

Let us see what happens when we do pivot:

$$x + y = 2$$
$$10^{-8}x + y = 1$$

Add -10^{-8} times the first equation to the second equation (and round off to eight significant figures).

$$x + y = 2$$
$$y = 1$$

Solving by backsubstitution, we get the "correct" answer $(x, y) = (1, 1)$. This suggests the following strategy.

(1.73)

> When proceeding through Gaussian elimination, a computer should compare each pivot with all the other possible pivots in the same column. The computer should choose the largest (in absolute value) of these candidates and make it the pivot (by interchanging corresponding rows). This process is called **partial pivoting**.

We illustrate this procedure of partial pivoting in Example 2.

Example 2 Use partial pivoting and Gaussian elimination to solve the system,

$$\begin{aligned} x_1 + x_2 + x_3 &= -2 \\ 4x_1 + 16x_2 + 64x_3 &= 100 \\ 2x_1 + 4x_2 + 8x_3 &= 6 \end{aligned}$$

Solution We proceed as described in Section 1.2 and first form the augmented matrix.

$$\begin{bmatrix} 1 & 1 & 1 & -2 \\ 4 & 16 & 64 & 100 \\ 2 & 4 & 8 & 6 \end{bmatrix}$$

If there were no other considerations, we would probably not interchange any rows because $a_{11} = 1$. However, partial pivoting requires that we examine the first column and use the entry with largest absolute value. This is $a_{21} = 4$, so we interchange rows 1 and 2, obtaining

$$\begin{bmatrix} 4 & 16 & 64 & 100 \\ 1 & 1 & 1 & -2 \\ 2 & 4 & 8 & 6 \end{bmatrix}$$

We now make $a_{i1} = 0$, $i \geq 2$ by adding $-\frac{1}{4}$ times row 1 to row 2 and by adding $-\frac{1}{2}$ times row 1 to row 3.

$$\begin{bmatrix} 4 & 16 & 64 & 100 \\ 0 & -3 & -15 & -27 \\ 0 & -4 & -24 & -44 \end{bmatrix}$$

We next examine a_{i2}, $i \geq 2$, for the entry of largest absolute value. This is $a_{32} = -4$, and we interchange rows 2 and 3:

$$\begin{bmatrix} 4 & 16 & 64 & 100 \\ 0 & -4 & -24 & -44 \\ 0 & -3 & -15 & -27 \end{bmatrix}$$

We now make $a_{32} = 0$ by adding $-\frac{3}{4}$ times row 2 to row 3.

$$\begin{bmatrix} 4 & 16 & 64 & 100 \\ 0 & -4 & -24 & -44 \\ 0 & 0 & 3 & 6 \end{bmatrix}$$

Next, we find the associated linear system.

$$\begin{aligned} 4x_1 + 16x_2 + 64x_3 &= 100 \\ -4x_2 - 24x_3 &= -44 \\ 3x_3 &= 6 \end{aligned}$$

We solve this by backsubstitution.

$$\begin{aligned} 3x_3 &= 6, & x_3 &= 2 \\ -4x_2 - 24(2) &= -44 & & \\ -4x_2 &= 4, & x_3 &= -1 \\ 4x_1 + 16(-1) + 64(2) &= 100 & & \\ 4x_1 &= -12, & x_1 &= -3 \end{aligned}$$

Thus the solution is $(x_1, x_2, x_3) = (-3, -1, 2)$. ■

Suppose we also want the $PA = LU$ decomposition induced by partial pivoting. Then we continue with what we have calculated so far and proceed as described in Section 1.5 to obtain

$$\begin{bmatrix} & 1 & \\ & & 1 \\ 1 & & \end{bmatrix}\begin{bmatrix} 1 & 1 & 1 \\ 4 & 16 & 64 \\ 2 & 4 & 8 \end{bmatrix} = \begin{bmatrix} 1 & & \\ \frac{1}{2} & 1 & \\ \frac{1}{4} & \frac{3}{4} & 1 \end{bmatrix}\begin{bmatrix} 4 & 16 & 64 \\ & -4 & -24 \\ & & 3 \end{bmatrix}$$

as you can easily check.

Overwriting Matrices

For most situations we do not actually need the matrix A once an LU decomposition or $PA = LU$ decomposition has been found. Thus we are free to overwrite the entries of A and, consequently, use less storage space.* In addition U is upper triangular and L is unit lower triangular. We certainly do not need to store the 0's below the diagonal in U or above the diagonal in L. We even do not have to store the 1's on the diagonal of L. Thus when we are done, we can store the u_{ij}, $j \geq i$, in the upper triangle of A and the l_{ij}, $j < i$, from L in the lower triangle of A. This is illustrated in the next example.

Example 3 Starting with the A from Example 2,

$$A = \begin{bmatrix} 1 & 1 & 1 \\ 4 & 16 & 64 \\ 2 & 4 & 8 \end{bmatrix}$$

we can overwrite A with the interesting parts of L and U as indicated.

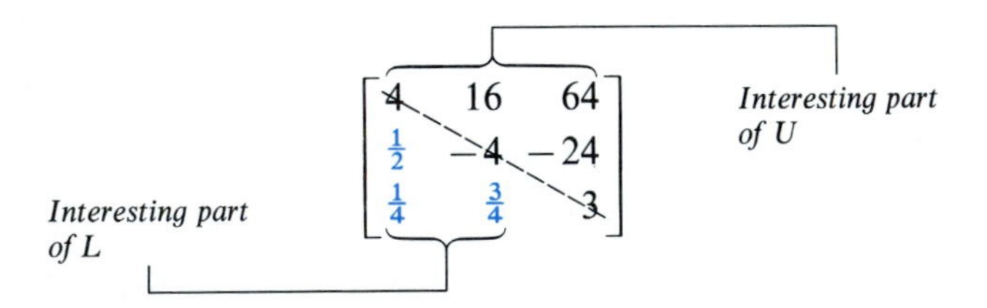

■

We must now take care of P. We only need to store the interesting information it conveys. Because P only tells us with which row the ith row is interchanged (if any), its interesting information can be conveyed with a simple "vector."

---

* *If we do need both A and the factorization, we can simply make a copy of A and work with that.*

Example 4 Using the factorization from Example 2, all the information in the P can be stored in the ordered pair

$$(2, 3)$$

where:

- The 2 in the first position indicates row 1 is interchanged with row 2.
- The 3 in the second position indicates row 2 is next interchanged with row 3. ■

Ill-Conditioned Systems

We begin this subsection with two easy but illustrative examples.

Example 5 Solve the systems

(a) $\begin{aligned} x + 2y &= 3 \\ 3x - 2y &= 1 \end{aligned}$ (b) $\begin{aligned} x + 2y &= 3 \\ 3x - 2y &= 1.008 \end{aligned}$

Solution You can easily check that the answer to (a) is (1, 1) and the answer to (b) is (1.002, 0.999). As you probably would have guessed, a very small change in this system produces a very small change in the answer. ■

Example 6 Solve the systems

(a) $\begin{aligned} x + 2y &= 3 \\ 1.001x + 2y &= 3.001 \end{aligned}$ (b) $\begin{aligned} x + 2y &= 3 \\ 1.001x + 2y &= 3.003 \end{aligned}$

Solution You can easily check that the answer to (a) is (1, 1) and the answer to (b) is (3, 0). Here a very small change in this system produces a tremendous change in the answer! ■

Herein lies the rub: In some systems very small changes in the data produce comparatively large changes in the answer. Systems with this characteristic are called **ill conditioned**. Ill-conditioned problems are very difficult to handle because if we wish a prescribed number of significant figures in the solution, we must determine accurately many more significant figures in the constants we start with. This is undesirable at best, and may even be impossible if the constants are obtained from physical data.

It is surprising and unfortunate how many approaches to real-world problems turn out to lead to ill-conditioned systems. When this happens, alternative approaches that lead to less ill-conditioned systems must be found. For example, the normal equations for least-squares problems are ill conditioned. These are discussed in Chapter 4, and one of the alternative methods for solving least-squares problems, the QR decomposition, is presented there.

This subsection has two purposes: to make you aware that systems can be ill conditioned and to give you some intuition as to what causes ill-conditioning in linear systems. If you have an actual problem that you suspect might be ill conditioned, you may have to approach an expert for help and an alternative method of solution. Different methods work effectively in different situations, and dealing with special ill-conditioned systems is a topic of current research.

The Geometry of Ill-Conditioning

To gain some geometric insight, we turn to the graphs of the systems in Examples 5 and 6. See Figure 1.4.

Figure 1.4

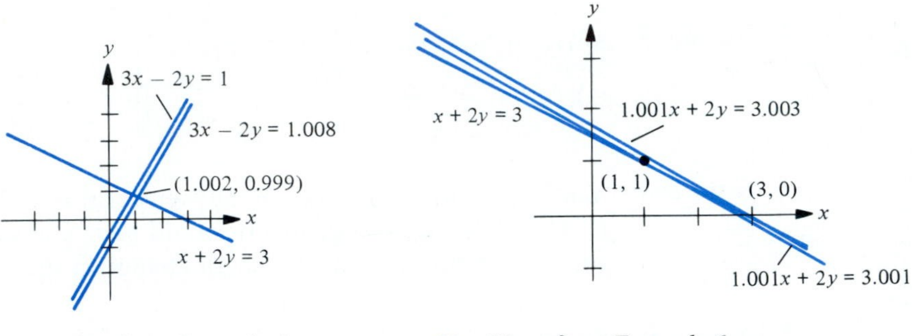

(a) *Lines from Example 5* (b) *Lines from Example 6*

In Figure 1.4a, look at the two lines that correspond to Example 5(a). These two lines intersect at a fairly large angle. A small change in the coefficients (leading to Example 5(b)) changes one line by only a small amount. Since the angle between the lines is large, a small change in one line produces only a small change in the intersection and correspondingly a small change in the solution.

Now examine Figure 1.4b. The two lines, which correspond to Example 6(a), are nearly parallel, so they intersect at a very small angle. A small change in a coefficient [leading to Example 6(b)] again changes one line by only a small amount. Now, however, since the lines are nearly parallel, a small change in one line produces a huge change in the intersection and correspondingly a huge change in the solution.

This geometry generalizes to three dimensions. An equation like $2x - 3y + 4z = 5$ represents a plane in $\mathbb{R}^3$. See Example 6, Section 3.3. A system of three such equations

$$\begin{aligned} 2x - 3y + 4z &= 5 \\ 4x - 4y + 2z &= 3 \\ -2x + 3y - 5z &= 8 \end{aligned}$$

represents three planes in 3-space. If the three planes intersect in one point, the system has a unique solution. If, in addition, they intersect at large angles (though the geometry is a little more complicated here), then the system is well conditioned, and small changes in the constants produce small changes in the solution. If, however, they intersect at small angles, then the system is ill conditioned, and small changes in the constants produce large changes in the solution.

It is fun, and sometimes even instructive, to generalize this to four and higher dimensions, but we shall not do this here.

A Numerical Example

We conclude our discussion of ill-conditioning with a numerical example. We shall use it to approach ill-conditioning of linear systems from a slightly different viewpoint and to give a brief introduction on how to detect this problem. Consider the matrix

$$A = \begin{bmatrix} 1 & 1 & 1 & 1 \\ 1 & 2 & 3 & 4 \\ 1 & 2 & 1 & 2 \\ 4+f & 4 & 5 & 5 \end{bmatrix} \tag{1.74}$$

This letter "f" is for "fudge factor." If $f = 0$, then A is singular. (Row 2 is row 3 plus twice row 4 minus 8 times row 1.) You can check that if $f \neq 0$, then A is nonsingular.

Suppose we start with $f = 1$ and let f become closer and closer to zero. Then, in some sense, A becomes closer and closer to being singular. We shall see in the numerical example that the system $AX = B$ also becomes more and more ill conditioned.

To see this, we start with an X we know, say $X = [1 \quad 2 \quad 3 \quad 4]^T$, and let $B = AX$. Thus

$$B = AX = \begin{bmatrix} 1 & 1 & 1 & 1 \\ 1 & 2 & 3 & 4 \\ 1 & 2 & 1 & 2 \\ 4+f & 4 & 5 & 5 \end{bmatrix} \begin{bmatrix} 1 \\ 2 \\ 3 \\ 4 \end{bmatrix} = \begin{bmatrix} 10 \\ 30 \\ 16 \\ 47+f \end{bmatrix} \tag{1.75}$$

For this B we do know the solution to the system $AX = B$, but our purpose is to see how close a computer comes to the answer. In successive tries we let $f = 1, 10^{-2}, 10^{-4}, \ldots, 10^{-14}$. The system $AX = B$ was solved on a CDC Cyber 170 model 750 computer for each value of f. This computer works to about 14 significant figures (in single precision). The results are given in Table 1.1.

Table 1.1 The solution to $AX = B$ as computed on a CDC Cyber, where

$$A = \begin{bmatrix} 1 & 1 & 1 & 1 \\ 1 & 2 & 3 & 4 \\ 1 & 2 & 1 & 2 \\ 4+f & 4 & 5 & 5 \end{bmatrix}, \qquad B = \begin{bmatrix} 10 \\ 30 \\ 16 \\ 47+f \end{bmatrix}$$

Note that the solution is exactly $(x_1, x_2, x_3, x_4) = (1, 2, 3, 4)$ for all values of $f \neq 0$.

f	x_1	x_2	x_3	x_4	*Condition Number*
1	0.99999999999979	2.0000000000002	3.0000000000004	3.9999999999999	9.12×10^1
10^{-2}	0.99999999997586	2.0000000000242	3.0000000000240	3.9999999999760	6.13×10^3
10^{-4}	0.99999999971571	2.0000000002843	3.0000000002842	3.9999999997158	6.16×10^5
10^{-6}	1.00000000426324	1.9999999573676	2.9999999573675	4.0000000426325	6.16×10^7
10^{-8}	0.99998010481911	2.0000198951809	3.0000198951808	3.9999801048192	6.16×10^9
10^{-10}	0.99886315191112	2.0011368480889	3.0011368480886	3.9988631519113	6.16×10^{11}
10^{-12}	0.79856115107924	2.2014388489208	3.2014388489209	3.7985611510791	6.24×10^{13}
10^{-14}		Answers are completely absurd.			8.66×10^{15}

You can see from Table 1.1 that as $f \to 0$, the computer loses more and more significant figures. In fact we can count the number of significant digits lost as follows. We know mathematically the answers are exactly $(x_1, x_2, x_3, x_4) = (1., 2., 3., 4.)$, by the way we constructed the problem. Since the computer works to 14 significant figures, an idealist would hope for answers such as, for x_2,

$$x_2 = 1.99999999999996 \qquad \text{or} \qquad x_2 = 2.00000000000002$$

since such answers *when rounded to 14 significant digits* give the correct answer. However, the computer almost never does that well and, when the problem is ill conditioned, does much worse. For example, examine x_1 when $f = 10^{-8}$. We see

f	x_1	x_1 rounded to 5 significant figures
10^{-8}	0.99998010481911	1.0000

To get the correct answer, the best we can do is to round x_1 to five significant figures. Thus instead of the hoped for 14 significant figures, we obtain only five significant figures, so *we have lost nine significant figures.* Altogether the number of significant figures lost is compiled in Table 1.2. The last column in both Tables 1.1 and 1.2 is headed "condition number," which is our next topic of discussion.

Table 1.2

f	*Number of Significant Figures Lost*	*Condition Number*
1	1	9.12×10^1
10^{-2}	3	6.13×10^3
10^{-4}	4	6.16×10^5
10^{-6}	6	6.16×10^7
10^{-8}	9	6.16×10^9
10^{-10}	11	6.16×10^{11}
10^{-12}	13	6.24×10^{13}
10^{-14}	all	8.66×10^{15}

Condition Number

To every nonsingular matrix, A, there is associated a special number called the **condition number** of A, $c(A)$. Unfortunately the definition of condition number requires matrix norms, which puts it in the realm of a higher course in linear algebra. However, we can still make some use of the condition number of a matrix. The condition number is greater than or equal to 1, and the larger $c(A)$ is, the more ill conditioned A is. In particular, if you examine Table 1.2, you can see the following rough rule of thumb.

(1.76)

IMPORTANT RELATIONSHIP If A is a nonsingular matrix and $c(A) \approx 10^k$, then you cannot trust the last k significant digits when the solution to a problem $AX = B$ is calculated on a computer.

You can see how important this relationship is. In fact all that you need is a rough estimate of the condition number in order to make use of the relationship given in (1.76). Some subroutine packages, such as LINPACK, have subroutines that estimate $c(A)$. (In fact the estimates for $c(A)$ in Table 1.1 were obtained using LINPACK.) Such routines make estimates of $c(A)$ readily available, and with these estimates you can determine how much of a computed answer you can trust.

Exercise 1.7 In Exercises 1–2, solve the system in two ways: (a) Using no pivoting and (b) Using partial pivoting. Perform all computations to four significant figures and check your answers.

1. $\begin{aligned} 0.0001x + y &= 2 \\ x + y &= 3 \end{aligned}$

2. $\begin{aligned} 0.0002x - y &= 1 \\ 4x + 3y &= 5 \end{aligned}$

In Exercises 3–10, solve the system using partial pivoting. Keep track of the interchanges as you may need them in Exercises 11–18.

3. $\begin{aligned} 2x + y &= 1 \\ 4x - 3y &= 17 \end{aligned}$

4. $\begin{aligned} 2x + 2y &= 2 \\ 6x + 5y &= 9 \end{aligned}$

5. $\begin{aligned} -2x + 3y + z &= -12 \\ 4x - 8y + 3z &= 28 \\ x + y + z &= 1 \end{aligned}$

6. $\begin{aligned} x + 2y - 3z &= 9 \\ 2x - 3y + 3z &= -2 \\ 4x + 8y - z &= 3 \end{aligned}$

7. $\begin{aligned} 4x_1 - 2x_2 + x_3 &= 9 \\ -2x_1 + 3x_2 - x_3 &= -10 \\ 6x_1 + 12x_2 + 6x_3 &= -3 \end{aligned}$

8. $\begin{aligned} 2x + 3y + 4z &= 1 \\ 3x - 3y + 2z &= -3 \\ -4x - 6y - 4z &= 0 \end{aligned}$

9. $$\begin{aligned} 3x_1 - 2x_2 + 3x_3 + x_4 &= -11 \\ 2x_1 + x_2 + x_3 - 4x_4 &= -18 \\ 4x_1 + 8x_2 + 4x_3 - x_4 &= 8 \\ x_1 + x_2 + x_3 + x_4 &= 4 \end{aligned}$$

10. $$\begin{aligned} x_1 - 2x_2 + x_3 - x_4 &= 5 \\ 2x_1 + 2x_2 + 3x_3 + x_4 &= 5 \\ -2x_1 + 4x_2 - 2x_3 + 2x_4 &= -10 \\ 4x_1 + 2x_2 + 2x_3 - 4x_4 &= 13 \end{aligned}$$

In Exercises 11–18, the matrix, A, is the coefficient matrix from the corresponding Exercises 3–10. Find the $PA = LU$ decomposition for A using partial pivoting.

11. $\begin{bmatrix} 2 & 1 \\ 4 & -3 \end{bmatrix}$

12. $\begin{bmatrix} 2 & 2 \\ 6 & 5 \end{bmatrix}$

13. $\begin{bmatrix} -2 & 3 & 1 \\ 4 & -8 & 3 \\ 1 & 1 & 1 \end{bmatrix}$

14. $\begin{bmatrix} 1 & 2 & -3 \\ 2 & -3 & 3 \\ 4 & 8 & -1 \end{bmatrix}$

15. $\begin{bmatrix} 4 & -2 & 1 \\ -2 & 3 & -1 \\ 6 & 12 & 6 \end{bmatrix}$

16. $\begin{bmatrix} 2 & 3 & 4 \\ 3 & -3 & 2 \\ -4 & -6 & -4 \end{bmatrix}$

17. $\begin{bmatrix} 3 & -2 & 3 & 1 \\ 2 & 1 & 1 & -4 \\ 4 & 8 & 4 & -1 \\ 1 & 1 & 1 & 1 \end{bmatrix}$

18. $\begin{bmatrix} 1 & -2 & 1 & -1 \\ 2 & 2 & 3 & 1 \\ -2 & 4 & -2 & 2 \\ 4 & 2 & 2 & -4 \end{bmatrix}$

In Exercises 19–24, solve and graph the system. Decide from the graph whether or not the system is ill conditioned. Finally, perform the indicated change to the system and see the effect on the solution.

19. $$\begin{aligned} 2x + y &= 3 \\ 2.001x + y &= 3.002 \end{aligned}$$
Change the second equation to: $= 3.001$

20. $$\begin{aligned} 3x + 2y &= -1 \\ 3x + 2.001y &= -1.002 \end{aligned}$$
Change the second equation to: -1.005

21. $$\begin{aligned} 2x + y &= 3 \\ 2x - y &= 5 \end{aligned}$$
Change the second equation to: $= 5.004$

22. $$\begin{aligned} 3x + 2y &= 0 \\ 3x - y &= 9 \end{aligned}$$
Change the second equation to: $= 9.009$

23. $$\begin{aligned} 4x - 3y &= 18 \\ 4x - 3.001y &= 18.002 \end{aligned}$$
Change the second equation to: $= 17.998$

24. $$\begin{aligned} 2x - 5y &= -3 \\ 2x - 5.002y &= -3.002 \end{aligned}$$
Change the second equation to: $= -3.006$

Review Exercises

In Exercises 1–4: (a) find the associated matrix equations and (b) solve the system by Gaussian elimination.

1. $$\begin{aligned} 2x + 3y - z &= -5 \\ 4x - y - 2z &= 4 \\ -2x - y + z &= 1 \end{aligned}$$

2. $$\begin{aligned} 3x + 4y - z &= -3 \\ -9x - 5y + 10z &= 2 \\ 6x + 5y - 6z &= -1 \end{aligned}$$

3. $$\begin{aligned} x + w &= 4 \\ x - y &= 3 \\ x + z &= 4 \\ y - z &= -3 \end{aligned}$$

4. $$\begin{aligned} x + y + z + w &= 2 \\ 2x + 2y - z - 4w &= -2 \end{aligned}$$

In Exercises 5–7, let

$$A = \begin{bmatrix} 1 & 2 \\ 0 & 1 \end{bmatrix}, \qquad B = \begin{bmatrix} 1 & 0 \\ -3 & 1 \end{bmatrix}$$

5. Compute AB and $A + B$.
6. Compute A^T, B^T, and $(AB)^T$.
7. Compute A^{-1}, B^{-1}, $(AB)^{-1}$.

In Exercises 8–10, A is an arbitrary $3 \times n$ matrix.

8. For what 3×3 matrix E will EA subtract the third row of A from the first?
9. For what 3×3 matrix P will PA interchange the second and third rows of A?
10. For what 3×3 matrix M will MA multiply the first row of A by -5 and leave the remaining rows unchanged?
11. Suppose E is a 2×2 matrix such that, for any $2 \times k$ matrix A, EA is A with 2 times the first row added to the second row.
    (a) What is $E^{10}A$?
    (b) Find E, E^{10}, $10E$.
12. Suppose P is a 2×2 matrix such that, for any $2 \times k$ matrix A, PA is A with its two rows interchanged.
    (a) What is P^2A? $P^{25}A$?
    (b) Find P, P^2, P^{25}.
13. Sketch the system

$$\begin{aligned} 2x + 3y &= 5 \\ 4x + ky &= 10 \end{aligned}$$

for various values of k.

(a) For which values of k (if any) does the system have no solution, one solution, or infinitely many solutions?
(b) For which values of k (roughly) would you guess the system is ill conditioned?

14. Same question as Exercise 13 for the system

$$\begin{aligned} x + ky &= 1 \\ kx + y &= 1 \end{aligned}$$

15. By trial and error, find examples of 2×2 matrices A, B, C, and D with the properties:

(a) $A^2 = -I$ (A has only real entries).
(b) $B \neq \mathbf{0}$ but $B^2 = \mathbf{0}$.
(c) $CD \neq \mathbf{0}$ and $CD = -DC$.

16. If A and B are square matrices, which of the following equals $(A + B)^2$?

(a) $(B + A)^2$
(b) $A^2 + 2AB + B^2$
(c) $A^2 + AB + BA + B^2$
(d) $(A + B)(B + A)$
(e) $A(A + B) + B(A + B)$

17. Under what conditions on a, b, and c are the following products invertible?

(a) $\begin{bmatrix} 1 & & \\ 4 & 2 & \\ 6 & 5 & 3 \end{bmatrix}\begin{bmatrix} a & 1 & 1 \\ & b & 1 \\ & & c \end{bmatrix}$ (b) $\begin{bmatrix} 1 & & \\ 4 & 2 & \\ 6 & 5 & 3 \end{bmatrix}\begin{bmatrix} a & & \\ & b & \\ & & c \end{bmatrix}\begin{bmatrix} 3 & 2 & 1 \\ & 5 & 4 \\ & & 6 \end{bmatrix}$

18. Show that if the inverse of A^2 is B, then the inverse of A is AB.* (Thus A is invertible $\Leftrightarrow$ A^2 is invertible.)

19. Find the inverse and LU decomposition for

$$\begin{bmatrix} 2 & 4 \\ 4 & 7 \end{bmatrix}$$

20. Find the LU decomposition of

$$\begin{bmatrix} 1 & 1 & & \\ 1 & 2 & 1 & \\ & 1 & 3 & 1 \\ & & 1 & 4 \end{bmatrix}$$

21. Suppose A and B are 5×3 matrices, $X = \begin{bmatrix} 1 \\ 0 \\ 0 \end{bmatrix}$, and $AX = BX$. What can you say about A and B?

* *Technically, at this point we can show only that AB is a right inverse of A. (To complete the story, see Exercise 44, Section 3.7.)*

22. Use Exercise 21 as a hint to show: If A and B are $m \times n$ matrices, $AX = BX$ for all $n \times 1$ matrices X, then $A = B$.

23. Using partial pivoting, find the $PA = LU$ decomposition for

$$A = \begin{bmatrix} 0 & 2 & 3 \\ 2 & 1 & 2 \\ 0 & 3 & 4 \end{bmatrix}$$

and use it to solve $AX = B$ for

$$B = \begin{bmatrix} -8 \\ -1 \\ -11 \end{bmatrix}$$

24. If A is $n \times n$ and invertible and A^{-1} is symmetric, is A necessarily symmetric? Explain.

25. Let P be an arbitrary $n \times n$ permutation matrix (that is, a matrix with a single 1, in each row and column and a zero for each remaining entry).

    (a) Argue that there are at most $n - 1$ elementary permutation matrices P_i such that $P_{n-1} \cdots P_1 P = I$.
    (b) Using the fact that $P_i^{-1} = P_i$, conclude that P is a product of elementary matrices.

26. Let $P_1, \cdots, P_k$ be $n \times n$ elementary permutation matrices. Let X be an $n \times n$ matrix with the property defined in Exercise 25.

    (a) Show that $P_i X$ retains this property.
    (b) Show that $P_k \cdots P_2 P_1$ has this property.

    NOTE Exercises 25 and 26 show that the two definitions of a permutation matrix (as given on pages 63, 75) are equivalent.

27. If $P = P_k \cdots P_1$, where each P_i is an elementary permutation matrix, then $P^{-1} = P^T$. Compare this result with Exercise 24, Section 1.6. (HINT $P_i = P_i^{-1} = P_i^T$, and do not forget to reverse the order.)

28. Find examples of $n \times n$ matrices A and B such that:

    (a) A and B are invertible but $A + B$ is not.
    (b) A and B are not invertible but $A + B$ is.
    (c) A, B, and $A + B$ are all invertible.

29. Assume that A, B, and $A + B$ are all invertible. Show that $A^{-1} + B^{-1}$ is also invertible, and find a formula for it. (HINT Simplify $A^{-1}(B + A)B^{-1}$, and use Theorem 1.39, Section 1.4.)

30. If A is invertible, which of the following properties are preserved by its inverse?

    (a) A is symmetric

(b) A is triangular.
(c) A is unit triangular.
(d) A is tridiagonal.
(e) All entries are whole numbers.
(f) All entries are fractions, including whole numbers like $\frac{2}{1}$.

31. Answer true or false. Give a reason if true or a counterexample if false.

(a) If A and B are invertible, then BA is invertible.
(b) If A and B are symmetric, then AB is symmetric.
(c) If all diagonal entries of A are zero, then A is singular.
(d) If $A^2 + A = I$, then A is invertible and $A^{-1} = A + I$.

Chapter 2

DETERMINANTS

Determinants were first suggested by Leibniz in 1693, but their fundamental properties were discovered over the ensuing years by several different mathematicians. For example, in 1750 Gabriel Cramer published *Cramer's rule*, but it was probably known to Maclaurin in 1729: Cramer had the better notation, however, and this is probably why the method was associated with him.

For two centuries people continued to study determinants extensively because of the many fascinating mathematical relationships and applications involving determinants. By 1900 enough was known about determinants to fill a four-volume tome, published by Thomas Muir in 1923.

Today, however, there is far less emphasis on determinants than at the turn of the century. In modern mathematics determinants play an important but narrow role in theory and a small role in computations (mostly in low-dimensional problems). In particular be sure to note the comments made at the end of Section 2.1 and after Cramer's rule in Section 2.4. However, determinants are involved in many fascinating relationships (Cramer's rule being one of them), and you can find more than a few mathematicians who have stumbled across "new" interconnections in their work, become enthralled, and spent considerable time investigating various consequences—only to learn later it was all published by someone in 1903. Section 2.5 provides an overview of applications.

2.1 The Determinant Function*

The determinant is a function that associates with every *square* matrix, A, a number called the *determinant* of A, which is denoted by $\det(A)$. This function

* *This chapter provides a full coverage of determinants. Section 5.1 is a one-section alternative coverage for instructors who prefer a quick overview of determinants. Both Chapter 2 and Section 5.1 are located where most instructors who opt them will use them, but either can be covered anytime after Chapter 1 and before Section 5.2.*

has the very important property that $\det(A) \neq 0$ if and only if A is nonsingular. We begin this section discussing permutations, since they play a central role in the definition.

(2.1)

DEFINITION A **permutation** of a set of integers is a reordering of those integers without omissions or repetitions.

Example 1 There are six permutations of the set $\{1, 2, 3\}$, namely,

$$\begin{array}{lll} (1, 2, 3), & (2, 1, 3), & (3, 1, 2) \\ (1, 3, 2), & (2, 3, 1), & (3, 2, 1) \end{array}$$

■

An **elementary permutation** is a permutation that interchanges exactly two numbers. In Example 1, (2, 1, 3), (3, 2, 1), and (1, 3, 2) are elementary permutations, and the other three are not.

Example 2 There are 24 permutations of $\{1, 2, 3, 4\}$, and you should see a systematic procedure for listing them all. One way is to use a **permutation tree**, which is illustrated in Figure 2.1a. In any case all 24 are listed in Figure 2.1b. As you can see, there are four choices for the first entry, and each choice has a column. Once a first choice is made, there are three choices left for the second entry, so there are $4 \cdot 3$ ways of filling the first two entries. Once these two choices are made, there are two choices left for the third entry, so there are $4 \cdot 3 \cdot 2$ ways of filling the first three entries. Finally there is only one choice left for the last position, so altogether there are $4 \cdot 3 \cdot 2 \cdot 1 = 24$ ways of filling all four entries. You can extend this reasoning and reach the conclusion that the set $\{1, 2, \ldots, n\}$ has $n(n-1)(n-2) \cdots 2 \cdot 1 = n!$ different permutations.

Figure 2.1(a)

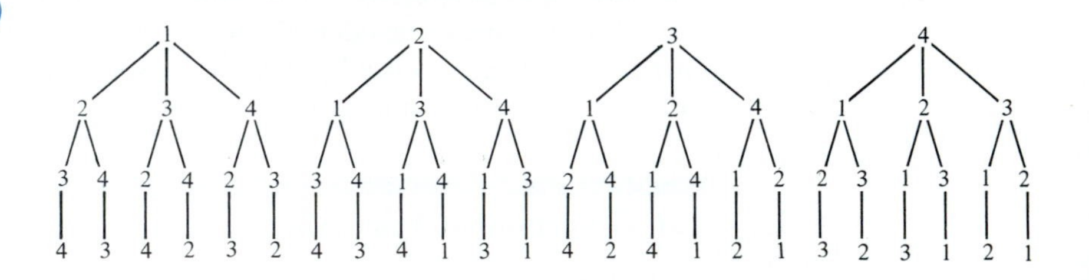

Figure 2.1(b)

(1, 2, 3, 4)	(2, 1, 3, 4)	(3, 1, 2, 4)	(4, 1, 2, 3)
(1, 2, 4, 3)	(2, 1, 4, 3)	(3, 1, 4, 2)	(4, 1, 3, 2)
(1, 3, 2, 4)	(2, 3, 1, 4)	(3, 2, 1, 4)	(4, 2, 1, 3)
(1, 3, 4, 2)	(2, 3, 4, 1)	(3, 2, 4, 1)	(4, 2, 3, 1)
(1, 4, 2, 3)	(2, 4, 1, 3)	(3, 4, 1, 2)	(4, 3, 1, 2)
(1, 4, 3, 2)	(2, 4, 3, 1)	(3, 4, 2, 1)	(4, 3, 2, 1)

You can also check that there are six elementary permutations of $\{1, 2, 3, 4\}$. See Exercises 34 and 35. ■

We shall denote a general permutation of the set $\{1, 2, \ldots, n\}$ by $(j_1, j_2, \ldots, j_n)$. This means j_1 is the first integer, j_2 is the second integer, and so forth.

(2.2)

DEFINITION An **inversion** occurs in a permutation $(j_1, \ldots, j_n)$ whenever a larger number precedes a smaller one. To find the number of inversions for a given permutation $(j_1, \ldots, j_n)$,

Count the number of integers that both follow j_1 and are less than j_1.

Count the number of integers that both follow j_2 and are less than j_2.

Continue in the same manner through j_{n-1}. The sum of all these numbers is called the **number of inversions** in the permutation.

Example 3 (a) The number of inversions in (6, 1, 4, 2, 5, 3) is $5 + 0 + 2 + 0 + 1 = 8$.
(b) The number of inversions in (3, 4, 2, 1) is $2 + 2 + 1 = 5$.
(c) The number of inversions in (1, 2, 3, 4, 5) is $0 + 0 + 0 + 0 = 0$. ■

(2.3)

DEFINITION A permutation is called **even** if it has an even number of inversions; it is called **odd** if it has an odd number of inversions.

Example 4 In Example 3, (a) and (c) are even and (b) is odd. ■

Example 5 An elementary permutation is odd. First consider the elementary permutation (1, 6, 3, 4, 5, 2). You can see that

The number 6 precedes	3, 4, 5, 2	which make four inversions
In addition to 6, the numbers	3, 4, 5	precede 2, which make three more inversions

By alignment you can see that the numbers between 2 and 6 lead to two times three inversions and the 2 leads to one more, so altogether the number of inversions is odd. In general, for any two numbers i and j, if $i < j$ and $(1, \ldots, j, \ldots, i, \ldots, n)$ is the elementary permutation of $\{1, \ldots, n\}$ that interchanges i and j, then

The number j precedes	$i+1, \ldots, j-1, i$	which make $j-i$ inversions
In addition to j, the numbers	$i+1, \ldots, j-1$	precede i, which make $j-i-1$ inversions

Again you can see that the numbers between i and j lead to $2(j-i-1)$ inversions and the i leads to one more, so altogether the number of inversions is odd. ■

We now know enough about permutations to take the next important step towards defining the determinant. Let A be an $n \times n$ matrix

$$A = \begin{bmatrix} a_{11} & \cdots & a_{1n} \\ \vdots & & \vdots \\ a_{n1} & \cdots & a_{nn} \end{bmatrix}$$

(2.4) DEFINITION An **elementary product** from A is a product of n entries of A, exactly one from each row and each column.

Example 6 (a) The matrix

$$\begin{bmatrix} a_{11} & a_{12} \\ a_{21} & a_{22} \end{bmatrix}$$

has two elementary products, namely,

$$a_{11}a_{22}, \qquad a_{21}a_{12}$$

(b) The matrix

$$\begin{bmatrix} a_{11} & a_{12} & a_{13} \\ a_{21} & a_{22} & a_{23} \\ a_{31} & a_{32} & a_{33} \end{bmatrix}$$

has six elementary products, namely,

$$\begin{array}{lll} a_{11}a_{22}a_{33}, & a_{11}a_{23}a_{32}, & a_{12}a_{21}a_{33} \\ a_{12}a_{23}a_{31}, & a_{13}a_{21}a_{32}, & a_{13}a_{22}a_{31} \end{array}$$

■

Example 6 illustrates that an elementary product from the $n \times n$ matrix A is a product of n entries of A. Since there must be exactly one entry from each row, an elementary product can be rewritten in the form

$$a_{1j_1}a_{2j_2}\cdots a_{nj_n}$$

Since there must be exactly one entry from each column, the numbers $(j_1, j_2, \ldots, j_n)$ must be a permutation of $\{1, 2, \ldots, n\}$. Thus there is one distinct elementary product for each permutation of $\{1, 2, \ldots, n\}$. Altogether, an $n \times n$ matrix has $n!$ elementary products.

(2.5)

DEFINITION A **signed elementary product** is an elementary product multiplied by $+1$ or -1,

$$\pm a_{1j_1} a_{2j_2} \cdots a_{nj_n}$$

This sign is $+$ or $-$ depending or whether the permutation $(j_1, j_2, \ldots, j_n)$ is even or odd.

Example 7 Using the matrices and elementary products from Example 6:

(a) $A = \begin{bmatrix} a_{11} & a_{12} \\ a_{21} & a_{22} \end{bmatrix}$

Elementary Product $a_{1j_1}a_{2j_2}$	*Associated Permutation* (j_1, j_2)	*Even or Odd*	*Signed Elementary Product*
$a_{11}a_{22}$	(1, 2)	Even	$a_{11}a_{22}$
$a_{12}a_{21}$	(2, 1)	Odd	$-a_{12}a_{21}$

(b)

$A = \begin{bmatrix} a_{11} & a_{12} & a_{13} \\ a_{21} & a_{22} & a_{23} \\ a_{31} & a_{32} & a_{33} \end{bmatrix}$

Elementary Product $a_{1j_1}a_{2j_2}a_{3j_3}$	*Associated Permutation* (j_1, j_2, j_3)	*Even or Odd*	*Signed Elementary Product*
$a_{11}a_{22}a_{33}$	(1, 2, 3)	Even	$a_{11}a_{22}a_{33}$
$a_{11}a_{23}a_{32}$	(1, 3, 2)	Odd	$-a_{11}a_{23}a_{32}$
$a_{12}a_{21}a_{33}$	(2, 1, 3)	Odd	$-a_{12}a_{21}a_{33}$
$a_{12}a_{23}a_{31}$	(2, 3, 1)	Even	$a_{12}a_{23}a_{31}$
$a_{13}a_{21}a_{32}$	(3, 1, 2)	Even	$a_{13}a_{21}a_{32}$
$a_{13}a_{22}a_{31}$	(3, 2, 1)	Odd	$-a_{13}a_{22}a_{31}$

■

We are now ready to take the final step and define the determinant function.

(2.6)

DEFINITION The **determinant function** is a function that associates with every square matrix, A, a number, denoted by $\det(A)$ or $\det A$, such that $\det(A)$ is the sum of all signed elementary products from A.

Using the notation of Section 4.1, we could write

$$\det : \{\text{square matrices}\} \to \mathbb{R}$$

to emphasize that the domain of the function det is the collection of all square matrices and the range is the set of real numbers.

Example 8 From Example 7, the definition of the determinant function (2.6) gives

(a) $\det \begin{bmatrix} a_{11} & a_{12} \\ a_{21} & a_{22} \end{bmatrix} = a_{11}a_{22} - a_{21}a_{12}$

(b) $\det \begin{bmatrix} a_{11} & a_{12} & a_{13} \\ a_{21} & a_{22} & a_{23} \\ a_{31} & a_{32} & a_{33} \end{bmatrix} = a_{11}a_{22}a_{33} + a_{12}a_{23}a_{31} + a_{13}a_{21}a_{32} - a_{11}a_{23}a_{32} - a_{12}a_{21}a_{33} - a_{13}a_{22}a_{31}$ ■

The formulas in Example 8 are simple and can be easily memorized. A simple diagram to help memorize them is given in Figure 2.2.

Figure 2.2

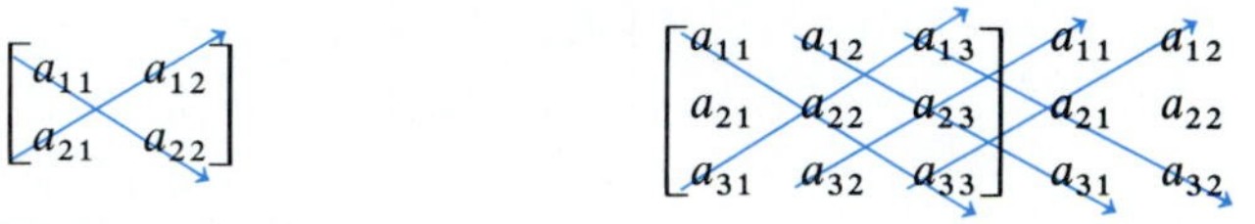

(a) Down arrow means $+a_{11}a_{22}$; up arrow means $-a_{12}a_{21}$.

(b) Down arrows mean $+a_{11}a_{22}a_{33}$, $+a_{12}a_{23}a_{31}$, $+a_{13}a_{21}a_{32}$; up arrows mean $-a_{31}a_{22}a_{13}$, $-a_{32}a_{23}a_{11}$, $-a_{33}a_{21}a_{12}$

Although the diagrams in Figure 2.2 work for 2×2 and 3×3 matrices, there are no corresponding diagrams for $n \times n$ matrices when $n > 3$. This is illustrated in the next example, where our main purpose is to emphasize the definition of the determinant function. We shall see easier ways to compute determinants in the following sections.

Example 9 Evaluate the determinants of the given matrices.

(a) $A = \begin{bmatrix} 2 & -3 \\ 4 & -5 \end{bmatrix}$ (b) $B = \begin{bmatrix} 0 & 2 & 4 \\ 3 & 0 & 0 \\ 0 & 1 & -1 \end{bmatrix}$

(c) $C = \begin{bmatrix} 0 & 0 & 0 & 0 & 1 \\ 0 & 1 & 0 & 0 & 0 \\ 0 & 0 & 0 & 1 & 0 \\ 1 & 0 & 0 & 0 & 0 \\ 0 & 0 & 1 & 0 & 0 \end{bmatrix}$

Solution (a) By Example 8(a), $\det(A) = 2(-5) - (4)(-3) = -10 + 12 = 2$.

(b) *Using the definition.* B has exactly two nonzero elementary products, $b_{12}b_{21}b_{33}$ and $b_{13}b_{21}b_{32}$, which are odd and even respectively. Thus by the definition of determinant or Example 6,

$$\begin{aligned}\det B &= -b_{12}b_{21}b_{33} + b_{13}b_{21}b_{32}\\ &= -2(3)(-1) + 4(3)(1) = 6 + 12 = 18\end{aligned}$$

Using the method in Figure 2.2. We see

$$\det B = \begin{bmatrix} 0 & 2 & 4 \\ 3 & 0 & 0 \\ 0 & 1 & -1 \end{bmatrix} \begin{matrix} 0 & 2 \\ 3 & 0 \\ 0 & 1 \end{matrix}$$

$$\begin{aligned}&= 0(0)(-1) + 2(0)(0) + 4(3)(1) - 0(0)(4) - 1(0)(0) - (-1)(3)(2)\\ &= 0 + 0 + 12 - 0 - 0 + 6 = 18\end{aligned}$$

(c) The matrix C has exactly one nonzero elementary product, $c_{15}c_{22}c_{34}c_{41}c_{53}$. The permutation (5, 2, 4, 1, 3) is odd, so the signed permutation is

$$-c_{15}c_{22}c_{34}c_{41}c_{53} = -1(1)(1)(1)(1) = -1$$

Thus $\det C = -1$ since this is the sum of all the signed elementary products.

Note that if we were to try to generalize the diagrams in Figure 2.2 to this case, we would obtain $\det C = 0$, whereas we now know $\det C = -1$. ■

As mentioned in the introduction, outside of a classroom it is rare that a determinant is actually computed when $n > 3$. But even when a determinant is computed, it is seldom that the definition is used directly. The reason is that an $n \times n$ matrix has $n!$ elementary products. So directly computing a determinant of a 25×25 matrix would require computing the sum of 25!, which is approximately 1.55×10^{25} products of 25 numbers each, a hard task for even our largest computers. The methods presented in the ensuing sections enable us to compute the determinant of an $n \times n$ matrix with approximately $\frac{2}{3}n^3$ multiplications and additions. Hence a 25×25 determinant requires only about $\frac{2}{3}25^3 \approx 6400$ operations, a far more reasonable task.

NOTATION An alternate notation for the determinant of a matrix A is $|A|$, rather than $\det(A)$. Using this notation, the determinants in Examples 9(a) and (b) can be written

$$\begin{vmatrix} 2 & -3 \\ 4 & -5 \end{vmatrix}^{*} = 2, \quad \begin{vmatrix} 0 & 2 & 4 \\ 3 & 0 & 0 \\ 0 & 1 & -1 \end{vmatrix} = 18$$ ■

* *Note that technically we should write* $\left|\begin{bmatrix} 2 & -3 \\ 4 & -5 \end{bmatrix}\right|$, *but this is always condensed as illustrated.*

Exercise 2.1 In Exercises 1–6, determine the number of inversions in the given permutation and whether it is even or odd.

1. (3, 2, 1)
2. (4, 2, 1, 3)
3. (3, 5, 1, 2, 4)
4. (6, 2, 5, 1, 3, 4)
5. (2, 1, 4, 3, 6, 5)
6. (7, 6, 5, 4, 3, 2, 1)

In Exercises 7–12, you are given a product of entries from a 4×4 matrix. Determine if it is an elementary product. If it is, find its sign.

7. $a_{13}a_{22}a_{34}a_{41}$
8. $a_{11}a_{24}a_{33}a_{42}$
9. $a_{14}a_{23}a_{33}a_{41}$
10. $a_{13}a_{22}a_{34}a_{42}$
11. $a_{12}a_{24}a_{31}a_{43}$
12. $a_{11}a_{22}a_{33}$

In Exercises 13–32, use the definition to evaluate the determinant of the given matrix.

13. $\begin{bmatrix} 1 & 2 \\ 3 & 6 \end{bmatrix}$
14. $\begin{bmatrix} 2 & -1 \\ 4 & 3 \end{bmatrix}$
15. $\begin{bmatrix} 3 & -8 \\ 2 & -7 \end{bmatrix}$
16. $\begin{bmatrix} 4 & -1 \\ -8 & 2 \end{bmatrix}$
17. $\begin{bmatrix} 1 & 2 & -1 \\ 2 & 3 & 1 \\ 4 & 1 & -4 \end{bmatrix}$
18. $\begin{bmatrix} 2 & 4 & 1 \\ 1 & -3 & 1 \\ -2 & -1 & 2 \end{bmatrix}$
19. $\begin{bmatrix} 2 & 0 & -1 \\ 3 & 1 & 2 \\ 4 & -3 & 1 \end{bmatrix}$
20. $\begin{bmatrix} 3 & 1 & 2 \\ -2 & 1 & 0 \\ 1 & -2 & 3 \end{bmatrix}$
21. $\begin{bmatrix} 0 & 2 & 0 \\ 1 & 0 & 0 \\ 0 & 0 & 3 \end{bmatrix}$
22. $\begin{bmatrix} 0 & 0 & 1 \\ 0 & -2 & 0 \\ -3 & 0 & 0 \end{bmatrix}$
23. $\begin{bmatrix} x & x-1 \\ 2 & 3 \end{bmatrix}$
24. $\begin{bmatrix} x-2 & 3 \\ x & x+1 \end{bmatrix}$
25. $\begin{bmatrix} x+5 & 0 & x \\ 0 & x-1 & 0 \\ 2 & 0 & x \end{bmatrix}$
26. $\begin{bmatrix} 0 & 1 & x-1 \\ x & 0 & 2 \\ 0 & 3 & x+1 \end{bmatrix}$
27. $\begin{bmatrix} 0 & 0 & 1 & 0 \\ 0 & 2 & 0 & 0 \\ 0 & 0 & 0 & 3 \\ -2 & 0 & 0 & 0 \end{bmatrix}$
28. $\begin{bmatrix} 0 & 2 & 0 & 0 \\ 0 & 0 & 1 & 0 \\ 3 & 0 & 0 & 0 \\ 0 & 0 & 0 & -1 \end{bmatrix}$
29. $\begin{bmatrix} 0 & 1 & 0 & 0 & 0 \\ 0 & 0 & 0 & 0 & 1 \\ 1 & 0 & 0 & 0 & 0 \\ 0 & 0 & 1 & 0 & 0 \\ 0 & 0 & 0 & 1 & 0 \end{bmatrix}$
30. $\begin{bmatrix} 0 & 0 & 0 & 0 & 1 & 0 \\ 0 & 1 & 0 & 0 & 0 & 0 \\ 0 & 0 & 0 & 1 & 0 & 0 \\ 1 & 0 & 0 & 0 & 0 & 0 \\ 0 & 0 & 0 & 0 & 0 & 1 \\ 0 & 0 & 1 & 0 & 0 & 0 \end{bmatrix}$

31. $\begin{bmatrix} 2 & -1 & 0 & 3 \\ 3 & 4 & 0 & 1 \\ 2 & -1 & 0 & 5 \\ 1 & -2 & 0 & 5 \end{bmatrix}$

32. $\begin{bmatrix} -1 & 2 & 3 & 4 \\ 5 & 1 & 2 & 8 \\ 9 & 8 & 7 & 2 \\ 0 & 0 & 0 & 0 \end{bmatrix}$

33. Suppose A is an $n \times n$ matrix with a column or row of all zeros (such as Exercise 31 or 32). Explain why each elementary product must be zero and hence $\det(A) = 0$.

34. List the elementary permutations of $\{1, 2, 3, 4\}$.

35. Determine how many elementary permutations of $\{1, 2, \ldots, n\}$ there are.

2.2 Evaluating Determinants

In this section we uncover several important facts about determinants that lead the way to a more efficient way of evaluating determinants. This method involves reducing the matrix to upper triangular form (much as in Graussian elimination), and it requires considerably fewer operations than using the definition directly.

Two Easy Special Cases

We first consider two types of matrices whose determinants are extremely easy to evaluate.

(2.7) **THEOREM** If a square matrix A has a row or column of all zeros, then $\det A = 0$.

Proof By the definition of an elementary product (2.4), every elementary product contains a factor from each column and each row. Hence every signed elementary product contains zero as a factor, so the product is zero. Thus det A is a sum of zeros and hence is zero. ■

Example 1 If

$$A = \begin{bmatrix} 2 & 3 & -1 & -2 \\ 4 & 8 & 1 & 3 \\ 0 & 0 & 0 & 0 \\ 5 & 1 & 2 & 7 \end{bmatrix}, \quad B = \begin{bmatrix} 3 & 0 & -1 & -2 \\ -2 & 0 & 1 & 5 \\ 4 & 0 & 4 & -1 \\ 3 & 0 & 1 & 3 \end{bmatrix}$$

then $\det A = 0$ and $\det B = 0$ by Theorem (2.7). ■

(2.8)

THEOREM Let A be a triangular matrix, either upper or lower. Then $\det A$ is the product of the diagonal entries,

$$\det A = a_{11}a_{22}\cdots a_{nn}$$

Proof By the definition of a signed elementary product (2.5), $a_{11}a_{22}\cdots a_{nn}$ is a signed elementary product. We shall show the other elementary products of A are zero. Let $a_{1j_1}a_{2j_2}\cdots a_{nj_n}$ be any other elementary product, so that $(j_1, j_2, \ldots, j_n) \neq (1, 2, \ldots, n)$. Then some $j_i < i$ and some $j_k > k$. (In fact the first $j_k \neq k$ is greater than k and the last $j_i \neq i$ is less than i.)

If A is upper triangular, then $a_{pq} = 0$ if $p > q$, so $a_{ij_i} = 0$.

If A is lower triangular, then $a_{pq} = 0$ if $p < q$, so $a_{kj_k} = 0$.

In either case the product $a_{1j_1}a_{2j_2}\cdots a_{nj_n}$ contains zero as a factor, so the product is zero. ■

Example 2 If

$$A = \begin{bmatrix} 2 & 4 & -2 & 3 \\ & -1 & 1 & -8 \\ & & -3 & 1 \\ & & & -2 \end{bmatrix}, \quad B = \begin{bmatrix} 2 & & & \\ 3 & -1 & & \\ 4 & -2 & -3 & \\ 5 & 1 & 2 & 5 \end{bmatrix}$$

then $\det A = 2(-1)(-3)(-2) = -12$, $\det B = 2(-1)(-3)(5) = 30$ by Theorem (2.8). ■

Determinants of Elementary Matrices

Elementary matrices are building blocks of invertible matrices [see Theorem (1.50), Section 1.4], so knowing their determinants will be useful.

(2.9)

THEOREM The determinants of the elementary matrices are as follows:

(a) If $E = \begin{bmatrix} 1 & & & \\ & 1 & m_{ij} & \\ & & \ddots & \\ & & & 1 \end{bmatrix}$ or

$E = \begin{bmatrix} 1 & & & \\ & 1 & & \\ & & \ddots & \\ m_{ij} & & & 1 \end{bmatrix}$, then $\det(E) = 1$.

(b) If $E = \begin{bmatrix} 1 & & & & \\ & \ddots & & & \\ & & m & & \\ & & & \ddots & \\ & & & & 1 \end{bmatrix}$, then $\det E = m$.

(c) If P is an elementary permutation matrix,

$$P = \begin{bmatrix} 1 & & & & & & \\ & \ddots & & & & & \\ & & 0 & \cdots & 1 & & \\ & & \vdots & \ddots & \vdots & & \\ & & 1 & \cdots & 0 & & \\ & & & & & \ddots & \\ & & & & & & 1 \end{bmatrix}$$

then $\det P = -1$.

Proof Parts (a) and (b) follow directly from Theorem (2.8).

(c) Suppose P is obtained from the identity matrix by interchanging rows i and j, $i < j$. Then P has exactly one nonzero elementary product, namely,

$$a_{11} \cdots a_{ij} \cdots a_{ji} \cdots a_{nn}, \quad \text{each } a_{qp} = 1$$

The corresponding permutation is $(1, \ldots, j, \ldots, i, \ldots, n)$. This is an elementary permutation and hence is odd, which we saw in Example 5, Section 2.1. By the definition of det P [Definition (2.6), Section 2.1],

$$\det P = -a_{11} \cdots a_{ij} \cdots a_{ji} \cdots a_{nn} = -1(1) \cdots (1) = -1$$ ■

Example 3

If

$$E = \begin{bmatrix} 1 & & \\ 2 & 1 & \\ 0 & 0 & 1 \end{bmatrix}, \quad M = \begin{bmatrix} 1 & & \\ & 1 & \\ & & 3 \end{bmatrix}, \quad P = \begin{bmatrix} & & 1 \\ & 1 & \\ 1 & & \end{bmatrix}$$

then $\det E = 1$, $\det M = 3$, and $\det P = -1$ by Theorem (2.9). ■

Theorem (2.10) will help us greatly simplify our computations.

(2.10) THEOREM If E is an elementary $n \times n$ matrix and A is any $n \times n$ matrix, then

$$\det(EA) = (\det E)(\det A)$$

A quick look at Theorem (2.9) gives us the determinants of the three types of elementary matrices. This in conjunction with Theorem (2.10) gives us the following corollary.

(2.11)

COROLLARY (a) If B is obtained from A by adding a multiple of one row of A to another row, then $\det B = \det A$.

(b) If B is obtained from A by multiplying a row of A by the number m, then $\det B = m \det A$.

(c) If B is obtained from A by interchanging two rows of A, then $\det B = -\det A$.

The proof of Theorem (2.10) is somewhat fussy in some of the details, so we shall not give it. In the exercises the proofs of a few special cases are suggested, and from these you can get a good idea how the proof goes in general.

Example 4 Let

$$A = \begin{bmatrix} 2 & -1 & 3 \\ 0 & 3 & 1 \\ 3 & -2 & -1 \end{bmatrix}, \qquad B_1 = \begin{bmatrix} 8 & -4 & 12 \\ 0 & 3 & 1 \\ 3 & -2 & -1 \end{bmatrix}$$

$$B_2 = \begin{bmatrix} 0 & 3 & 1 \\ 2 & -1 & 3 \\ 3 & -2 & -1 \end{bmatrix}, \qquad B_3 = \begin{bmatrix} 2 & -1 & 3 \\ -4 & 5 & -5 \\ 3 & -2 & -1 \end{bmatrix}$$

You can show directly from the definition that $\det A = -32$. Observe that B_1 is obtained from A by multiplying row 1 by 4. By Corollary (2.11b),

$$\det B_1 = 4 \det A = -128$$

Observe that B_2 is obtained from A by interchanging rows 1 and 2. By Corollary (2.11c),

$$\det B_2 = -\det A = 32$$

Observe that B_3 is obtained from A by adding -2 times row 1 to row 2. By Corollary (2.11a),

$$\det B_3 = \det A = -32$$

Also, if

$$E_1 = \begin{bmatrix} 4 & & \\ & 1 & \\ & & 1 \end{bmatrix}, \qquad E_2 = \begin{bmatrix} 0 & 1 & \\ 1 & 0 & \\ & & 1 \end{bmatrix}, \qquad E_3 = \begin{bmatrix} 1 & & \\ -2 & 1 & \\ & & 1 \end{bmatrix}$$

you can see that $B_i = E_iA$, $i = 1, 2, 3$. By Theorem (2.9), $\det E_1 = 4$, $\det E_2 = -1$, and $\det E_3 = 1$. Hence you can see that $\det B_i = \det E_i \det A$, $i = 1, 2, 3$. ■

Computing Determinants by Row Reduction

We are now ready to describe a simplified process for evaluating determinants. The idea is to use row operations to reduce a matrix to triangular form (much as we did with Gaussian elimination). For each step we use Corollary (2.11) to tell us how the determinant is changed (if it is). Once in triangular form, the determinant is easily evaluated using Theorem (2.8). This process is illustrated in the next example.

Example 5 Evaluate $\det A$ where

$$A = \begin{bmatrix} 0 & 4 & 1 \\ 6 & -9 & 3 \\ 4 & -2 & -3 \end{bmatrix}$$

Solution For this procedure using $|A|$ for $\det A$ simplifies the description.

$$\det A = \begin{vmatrix} 0 & 4 & 1 \\ 6 & -9 & 3 \\ 4 & -2 & -3 \end{vmatrix}$$

$$= -\begin{vmatrix} 6 & -9 & 3 \\ 0 & 4 & 1 \\ 4 & -2 & -3 \end{vmatrix} \quad \text{by (2.11c) (interchanged rows 1 and 2)}$$

$$= -3\begin{vmatrix} 2 & -3 & 1 \\ 0 & 4 & 1 \\ 4 & -2 & -3 \end{vmatrix} \quad \text{by (2.11b) (factored a 3 from row 1)}$$

$$= -3\begin{vmatrix} 2 & -3 & 1 \\ 0 & 4 & 1 \\ 0 & 4 & -5 \end{vmatrix} \quad \text{by (2.11a) (added } -2 \text{ times row 1 to row 3)}$$

$$= -3\begin{vmatrix} 2 & -3 & 1 \\ 0 & 4 & 1 \\ 0 & 0 & -6 \end{vmatrix} \quad \text{by (2.11a) (added } -1 \text{ times row 2 to row 3, yielding a matrix in triangular form)}$$

$$= -3(2)(4)(-6) \quad \text{by Theorem (2.8)}$$

$$= 144$$

■

We conclude the discussion of evaluating determinants with one last important observation.

(2.12) **THEOREM** If A is a square matrix with some row a multiple of another row, then $\det A = 0$.

Proof If one row of A is m times a second row, then add $-m$ times the second row to the first, obtaining a matrix B. This means B has a zero row, so $\det B = 0$ by Theorem (2.7). Because $\det B = \det A$ by Corollary (2.11a), $\det A = 0$. ■

Example 6 Each of the following matrices has one row a multiple of another. Hence the determinant of each matrix is zero, by Theorem (2.12).

$$\begin{bmatrix} 3 & -4 \\ -6 & 8 \end{bmatrix}, \quad \begin{bmatrix} 2 & -1 & 3 \\ 3 & 2 & 1 \\ 4 & -2 & 6 \end{bmatrix}, \quad \begin{bmatrix} 2 & -1 & -3 & 2 \\ 4 & -2 & -6 & 1 \\ 3 & 5 & 1 & 4 \\ -8 & 4 & 12 & -2 \end{bmatrix}$$ ■

Exercise 2.2

In Exercises 1–12, evaluate the determinant of the given matrix by inspection, using Theorems (2.7), (2.8), (2.9), and (2.11).

1. $\begin{bmatrix} 2 & 1 \\ -4 & -2 \end{bmatrix}$

2. $\begin{bmatrix} 2 & 3 \\ 0 & -3 \end{bmatrix}$

3. $\begin{bmatrix} 4 & 0 \\ 2 & 0 \end{bmatrix}$

4. $\begin{bmatrix} 0 & 1 \\ 1 & 0 \end{bmatrix}$

5. $\begin{bmatrix} -2 & 3 & -1 \\ 0 & 1 & -2 \\ 0 & 0 & 4 \end{bmatrix}$

6. $\begin{bmatrix} 1 & & \\ & 1 & 2 \\ & & 1 \end{bmatrix}$

7. $\begin{bmatrix} 7 & -6 & 0 \\ 4 & 3 & 0 \\ 5 & -1 & 0 \end{bmatrix}$

8. $\begin{bmatrix} 1 & & \\ & 1 & \\ & & -3 \end{bmatrix}$

9. $\begin{bmatrix} 1 & & \\ & 0 & 1 \\ & 1 & 0 \end{bmatrix}$

10. $\begin{bmatrix} -2 & 3 & 1 & 2 \\ -4 & 3 & 2 & 2 \\ 6 & -9 & -3 & -6 \\ 4 & -6 & 1 & 1 \end{bmatrix}$

11. $\begin{bmatrix} 1 & 3 & 2 & -1 \\ 4 & 3 & -1 & 3 \\ 0 & 0 & 0 & 0 \\ 5 & 1 & 3 & -2 \end{bmatrix}$

12. $\begin{bmatrix} 2 & & & \\ 1 & -1 & & \\ 3 & 1 & -2 & \\ 1 & 0 & 3 & -3 \end{bmatrix}$

In Exercises 13–20, evaluate the determinant of the given matrix by first reducing it to triangular form.

13. $\begin{bmatrix} 2 & 1 \\ -4 & 1 \end{bmatrix}$

14. $\begin{bmatrix} 3 & 2 \\ 4 & 3 \end{bmatrix}$

15. $\begin{bmatrix} 2 & -1 & 3 \\ 5 & 10 & -5 \\ 1 & 2 & 3 \end{bmatrix}$

16. $\begin{bmatrix} -3 & 1 & 2 \\ 6 & 2 & 1 \\ -9 & 1 & 2 \end{bmatrix}$

17. $\begin{bmatrix} 0 & 0 & 4 & 3 \\ 2 & 4 & 6 & -4 \\ -3 & 9 & 3 & 6 \\ -2 & -6 & 4 & 2 \end{bmatrix}$

18. $\begin{bmatrix} 2 & 1 & 0 & 1 \\ -4 & 3 & 1 & -2 \\ 2 & 2 & -1 & 0 \\ 6 & 1 & 3 & 1 \end{bmatrix}$

19. $\begin{bmatrix} 3 & -1 & 2 & 4 \\ 6 & 8 & 3 & 1 \\ 0 & 5 & 1 & -3 \\ -9 & -2 & 1 & 3 \end{bmatrix}$

20. $\begin{bmatrix} 0 & 1 & 2 & 0 \\ 4 & 0 & 1 & 0 \\ 0 & 3 & 2 & 1 \\ -2 & 0 & 1 & 3 \end{bmatrix}$

21. Evaluate the following determinant by first reducing it to triangular form, and show that:

$$\begin{vmatrix} 1 & 1 & 1 \\ a & b & c \\ a^2 & b^2 & c^2 \end{vmatrix} = (b-a)(c-a)(c-b)$$

In Exercises 22 and 23, evaluate the given determinants by modifying the proof of Theorem (2.8).

22. $\begin{vmatrix} 0 & 0 & a_{13} \\ 0 & a_{22} & a_{23} \\ a_{31} & a_{32} & a_{33} \end{vmatrix} = -a_{31}a_{22}a_{13}$

23. $\begin{vmatrix} 0 & 0 & 0 & a_{14} \\ 0 & 0 & a_{23} & a_{24} \\ 0 & a_{32} & a_{33} & a_{34} \\ a_{41} & a_{42} & a_{43} & a_{44} \end{vmatrix} = a_{41}a_{32}a_{23}a_{14}$

In Exercises 24–27, assume

$$\begin{vmatrix} a & b & c \\ d & e & f \\ g & h & i \end{vmatrix} = 3$$

Use Theorem (2.11) to evaluate the given determinant.

24. $\begin{vmatrix} g & h & i \\ d & e & f \\ 2a & 2b & 2c \end{vmatrix}$

25. $\begin{vmatrix} g & h & i \\ a & b & c \\ d & e & f \end{vmatrix}$

26. $\begin{vmatrix} a & b & c \\ a+d & b+e & c+f \\ 3g & 3h & 3i \end{vmatrix}$

27. $\begin{vmatrix} 2g & 2h & 2i \\ -a & -b & -c \\ -5d & -5e & -5f \end{vmatrix}$

In Exercises 28–30, prove the given case of Theorem (2.11) from the definition, Definition (2.6), Section 2.1.

28. $$\begin{vmatrix} ma_{11} & ma_{12} & ma_{13} \\ a_{21} & a_{22} & a_{23} \\ a_{31} & a_{32} & a_{33} \end{vmatrix} = m \begin{vmatrix} a_{11} & a_{12} & a_{13} \\ a_{21} & a_{22} & a_{23} \\ a_{31} & a_{32} & a_{33} \end{vmatrix}$$

29. $$\begin{vmatrix} a_{21} & a_{22} & a_{23} \\ a_{11} & a_{12} & a_{13} \\ a_{31} & a_{32} & a_{33} \end{vmatrix} = - \begin{vmatrix} a_{11} & a_{12} & a_{13} \\ a_{21} & a_{22} & a_{23} \\ a_{31} & a_{32} & a_{33} \end{vmatrix}$$

30. $$\begin{vmatrix} a_{11} & a_{12} & a_{13} \\ a_{21} + ka_{11} & a_{22} + ka_{12} & a_{23} + ka_{13} \\ a_{31} & a_{32} & a_{33} \end{vmatrix} = \begin{vmatrix} a_{11} & a_{12} & a_{13} \\ a_{21} & a_{22} & a_{23} \\ a_{31} & a_{32} & a_{33} \end{vmatrix}$$

2.3 Properties of Determinants

In this section we shall prove several important facts about determinants. We begin with what is for us the most important property of inverses.

(2.13) **THEOREM** A square matrix A is invertible if and only if $\det A \neq 0$.

Proof First consider the special case that U is a matrix that is upper triangular. Then we know

$$U \text{ is invertible} \Leftrightarrow \text{all the diagonal entries } u_{ii} \neq 0^*$$
$$\det U = u_{11}u_{22} \cdots u_{nn} \qquad \text{by Theorem (2.8), Section 2.2}$$

Putting these two statements together yields

(2.14) $$U \text{ is invertible} \Leftrightarrow \det U \neq 0$$

Now let A be any square matrix. Then there are elementary matrices $E_1, \ldots, E_k$ such that

(2.15) $$E_1 \cdots E_k A = U$$

Since the E_i's are invertible, we have

(2.16) $$A \text{ is invertible} \Leftrightarrow U \text{ is invertible}$$

* *See Theorem (1.50) and the proof that (c) ⇒ (d), Section 1.4.*

Now take the determinant of both sides of (2.15) and apply the fact that $\det(EB) = (\det E)(\det B)$ [Theorem (2.10), Section 2.2] over and over.

(2.17)
$$\begin{aligned} \det(U) &= \det(E_1 \cdots E_k A) = (\det E_1)\det(E_2 \cdots E_k A) \\ &= (\det E_1)(\det E_2)\det(E_3 \cdots E_k A) \\ &\ \vdots \\ &= (\det E_1)(\det E_2)\cdots(\det E_k)\det A \end{aligned}$$

Since $\det E_i \neq 0$ [by Theorem (2.9), Section 2.2], we see from (2.17).

(2.18)
$$\det U \neq 0 \quad \Leftrightarrow \quad \det A \neq 0$$

Putting (2.16) and (2.14) and (2.18) together (in that order), we have

(2.19)
$$A \text{ is invertible} \Leftrightarrow U \text{ is invertible} \Leftrightarrow \det U \neq 0 \Leftrightarrow \det A \neq 0$$

This completes the proof. ■

Example 1 Suppose $A = \begin{bmatrix} 1 & 2 \\ 3 & -4 \end{bmatrix}$ and $B = \begin{bmatrix} 2 & -1 \\ -6 & 3 \end{bmatrix}$. Then

$$|A| = \begin{vmatrix} 1 & 2 \\ 3 & -4 \end{vmatrix} = 1(-4) - 3(2) = -10$$

$$|B| = \begin{vmatrix} 2 & -1 \\ -6 & 3 \end{vmatrix} = 2(3) - (-6)(-1) = 0$$

Thus by Theorem (2.13), A is invertible and B is not. ■

WARNING As we have mentioned before, computing determinants is fraught with numerical dangers and hardships. Although Theorem (2.13) appears extremely simple, if you need to determine if a matrix is invertible, especially if it is larger than 3×3, it is computationally easier and wiser to use the methods discussed in Chapter 1.

The principal used in Equation (2.16) is central in the proof of the next important property.

(2.20) **THEOREM** If A and B are $n \times n$ matrices, then

(2.21)
$$\det(AB) = (\det A)(\det B)$$

Proof Suppose A is invertible. Then A is a product of elementary matrices

(2.22)
$$A = E_1 E_2 \cdots E_k$$

by Theorem (1.50), Section 1.4. Then, as in the previous proof, applying the fact that $\det(EC) = (\det E)(\det C)$ [Theorem (2.10), Section 2.2] over and over, we have

(2.23)
$$
\begin{aligned}
\det(A) &= \det(E_1E_2\cdots E_k)\\
&= (\det E_1)\det(E_2\cdots E_k)\\
&\;\;\vdots\\
&= (\det E_1)(\det E_2)\cdots(\det E_k)
\end{aligned}
$$

Applying the same principle again,

(2.24)
$$
\begin{aligned}
\det(AB) &= \det(E_1E_2\cdots E_kB)\\
&= (\det E_1)\det(E_2\cdots E_kB)\\
&\;\;\vdots\\
&= (\det E_1)(\det E_2)\cdots(\det E_k)(\det B)
\end{aligned}
$$

Substituting (2.23) into (2.24) yields $\det(AB) = (\det A)(\det B)$.

Suppose A is not invertible. Then AB is also not invertible. (See Exercise 15.) Thus both $\det(AB) = 0$ and $\det(A) = 0$ [so $(\det A)(\det B) = 0$], and we have $\det(AB) = (\det A)(\det B)$ in this case also. ■

Example 2 Suppose $A = \begin{bmatrix} 2 & 3 \\ -4 & 1 \end{bmatrix}$ and $B = \begin{bmatrix} 4 & 2 \\ 3 & 1 \end{bmatrix}$ Then

$$|A| = \begin{vmatrix} 2 & 3 \\ -4 & 1 \end{vmatrix} = 2 - (-12) = 14, \qquad |B| = \begin{vmatrix} 4 & 2 \\ 3 & 1 \end{vmatrix} = 4 - 6 = -2$$

$$AB = \begin{bmatrix} 2 & 3 \\ -4 & 1 \end{bmatrix}\begin{bmatrix} 4 & 2 \\ 3 & 1 \end{bmatrix} = \begin{bmatrix} 17 & 7 \\ -13 & -7 \end{bmatrix}, \quad |AB| = -119 - (-91) = -28$$

You can see that $|AB| = |A||B|$. ■

Transposes and Determinants

The relationship between the determinants of a matrix and its transpose is extremely simple.

(2.25)

> THEOREM If A is an $n \times n$ matrix and A^T is its transpose, then
>
> $$\det(A) = \det(A^T)$$

The proof of this theorem is trivial for 1×1 or 2×2 matrices. (See Exercise 27.) If $n > 2$, the proof is not so simple. It goes back to the definition of determinant [Definition (2.6), Section 2.1] and shows that the *signed* products of A and A^T are identical. This requires some fussy, detailed arguments regarding permutations, so we shall not give the proof. However, see Exercise 28 for the 3×3 case.

Example 3 If

$$A = \begin{bmatrix} 2 & 2 & 1 \\ 3 & 1 & -2 \\ -1 & 2 & 1 \end{bmatrix}, \quad \text{then} \quad A^T = \begin{bmatrix} 2 & 3 & -1 \\ 2 & 1 & 2 \\ 1 & -2 & 1 \end{bmatrix} \quad \text{and}$$

$$\begin{aligned} |A| &= \begin{vmatrix} 2 & 2 & 1 \\ 3 & 1 & -2 \\ -1 & 2 & 1 \end{vmatrix} \\ &= 2(1)(1) + 2(-2)(-1) + 1(3)(2) - (-1)(1)(1) - 2(-2)(2) - 1(3)(2) \\ &= 2 + 4 + 6 + 1 + 8 - 6 = 15 \end{aligned}$$

$$\begin{aligned} |A^T| &= \begin{vmatrix} 2 & 3 & -1 \\ 2 & 1 & 2 \\ 1 & -2 & 1 \end{vmatrix} \\ &= 2(1)(1) + 3(2)(1) + (-1)(2)(-2) - 1(1)(-1) - (-2)(2)(2) - 1(2)(3) \\ &= 2 + 6 + 4 + 1 + 8 - 6 = 15 \end{aligned}$$

You can see that $|A| = |A^T|$. ■

Column Operations

All that we know about row operations can now be applied to column operations. For example, suppose we multiply the third column of a matrix A by 5. Then we can do this by multiplying the third row of A^T by 5, calling the resulting matrix B and taking the transpose of this, B^T. We then see

$$\begin{aligned} \det(A\text{—with its third column multiplied by 5}) &= \det(B^T) \\ &= \det(B) \qquad \text{by Theorem (2.25)} \\ &= \det(A^T\text{—with its third row multiplied by 5}) \\ &= 5\det(A^T) \quad \text{by Corollary (2.11b), Section 2.2} \\ &= 5\det(A) \quad \text{by Theorem (2.25)} \end{aligned}$$

In this way you can see that the effect of any column operation on the determinant of a matrix is the same effect obtained by using the corresponding row operation. Combining this with Corollary (2.11), Section 2.2, gives us

(2.26)

THEOREM Let A be a square matrix.

(a) If B is obtained from A by adding a multiple of one column of A to another column, then $\det B = \det A$.

(b) If B is obtained from A by multiplying a column of A by a number m, then $\det B = m \det A$.

(c) If B is obtained from A by interchanging two columns of A, then $\det B = -\det A$.

Example 4 Let

$$A = \begin{bmatrix} 2 & -1 & 3 \\ 0 & 3 & 1 \\ 3 & -2 & -1 \end{bmatrix}, \qquad B_1 = \begin{bmatrix} 2 & -3 & 3 \\ 0 & 9 & 3 \\ 3 & -6 & -1 \end{bmatrix}$$

$$B_2 = \begin{bmatrix} -1 & 2 & 3 \\ 3 & 0 & -1 \\ -2 & 3 & -1 \end{bmatrix}, \qquad B_3 = \begin{bmatrix} 2 & -1 & 5 \\ 0 & 3 & -5 \\ 3 & -2 & 3 \end{bmatrix}$$

You can show directly that $\det A = -32$. Observe that B_1 is obtained from A by multiplying column 2 by 3. So, by Theorem (2.26b),

$$\det B_1 = 3 \det A = -96$$

Observe that B_2 is obtained from A by interchanging columns 1 and 2. So, by Theorem (2.26c),

$$\det B_2 = -\det A = 32$$

Finally, observe that B_3 is obtained from A by adding -2 times column 2 to column 3. Then, by Theorem (2.26a),

$$\det B_3 = \det A = -32$$

Of course, you can check these three facts directly. ■

Two Elementary Facts

Next on our list of properties of determinants to cover are two elementary facts. The first elementary fact is an elementary relationship.

(2.27) THEOREM If A is an $n \times n$ matrix and k is a number, then

$$\det(kA) = k^n \det A$$

Proof We know a common factor of any row can be "moved through" the det sign [Corollary (2.11b), Section 2.2]. Since each of the n rows of kA has k as a common factor, we can move all n of the k's out to obtain

$$\det(kA) = k \cdots k \det A = k^n \det A$$ ■

Example 5 Let

$$A = \begin{bmatrix} 2 & -1 \\ 4 & 5 \end{bmatrix} \quad \text{so that} \quad 3A = \begin{bmatrix} 6 & -3 \\ 12 & 15 \end{bmatrix}$$

Then

$$|A| = 2(5) - (-1)(4) = 14, \qquad |3A| = 6(15) - (-3)12 = 126$$

so that

$$|3A| = 126 = 9(14) = 3^2(14) = 3^2|A|$$ ■

The second elementary fact is an elementary nonrelationship.

(2.28)

IMPORTANT FACT The two quantities

$$\det(A + B) \qquad \text{and} \qquad \det A + \det B$$

are in general *not* the same.

Example 6 Let

$$A = \begin{bmatrix} 2 & -1 \\ 4 & 5 \end{bmatrix}, \qquad B = \begin{bmatrix} 1 & 3 \\ 1 & 2 \end{bmatrix} \qquad \text{so that} \qquad A + B = \begin{bmatrix} 3 & 2 \\ 5 & 7 \end{bmatrix}$$

Then

$$|A| = 2(5) - 4(-1) = 14, \qquad |B| = 1(2) - 1(3) = -1, \qquad \text{and}$$
$$|A + B| = 3(7) - 5(2) = 11$$

You can see that $|A| + |B|$ and $|A + B|$ are different. ■

Exercise 2.3 In Exercises 1–6, evaluate $\det(E)$, $\det(A)$, and $\det(EA)$. Then verify that $\det(E)\det(A) = \det(EA)$.

1. $E = \begin{bmatrix} 0 & 1 \\ 1 & 0 \end{bmatrix}, \quad A = \begin{bmatrix} 2 & 3 \\ 5 & 1 \end{bmatrix}$

2. $E = \begin{bmatrix} 1 & 0 \\ 1 & 1 \end{bmatrix}, \quad A = \begin{bmatrix} 1 & -1 \\ -2 & 1 \end{bmatrix}$

3. $E = \begin{bmatrix} 2 & & \\ & 1 & \\ & & 1 \end{bmatrix}, \quad A = \begin{bmatrix} 3 & 1 & 2 \\ 0 & 2 & 0 \\ 5 & 0 & 0 \end{bmatrix}$

4. $E = \begin{bmatrix} 1 & & \\ & -2 & \\ & & 1 \end{bmatrix}, \quad A = \begin{bmatrix} 0 & 0 & 2 \\ 3 & -1 & -2 \\ 4 & 1 & 0 \end{bmatrix}$

5. $E = \begin{bmatrix} 1 & -2 \\ 0 & 1 \end{bmatrix}, \quad A = \begin{bmatrix} -3 & 1 \\ 2 & 1 \end{bmatrix}$

6. $E = \begin{bmatrix} 0 & 1 \\ 1 & 0 \end{bmatrix}, \quad A = \begin{bmatrix} 4 & 1 \\ 1 & 1 \end{bmatrix}$

In Exercises 7 and 8, evaluate $\det(A)$, $\det(B)$, and $\det(AB)$. Then verify that $\det(A)\det(B) = \det(AB)$.

7. $A = \begin{bmatrix} 3 & 2 & 1 & -1 \\ 0 & -2 & 4 & 1 \\ 0 & 0 & 5 & 1 \\ 0 & 0 & 0 & -1 \end{bmatrix}$, $B = \begin{bmatrix} 4 & 1 & -1 & 2 \\ 0 & 3 & 1 & -1 \\ 0 & 0 & -3 & 4 \\ 0 & 0 & 0 & 2 \end{bmatrix}$

8. $A = \begin{bmatrix} 2 & 1 & 3 \\ 1 & 0 & 0 \\ 2 & -1 & 1 \end{bmatrix}$, $B = \begin{bmatrix} 4 & 1 & 2 \\ 2 & 3 & 0 \\ 4 & 1 & 0 \end{bmatrix}$

In Exercises 9–14, use determinants to determine if the given matrix is invertible.

9. $\begin{bmatrix} 2 & -3 \\ -4 & 6 \end{bmatrix}$

10. $\begin{bmatrix} 8 & -2 \\ -4 & 1 \end{bmatrix}$

11. $\begin{bmatrix} 1 & 3 & -2 \\ 4 & 1 & 3 \\ 2 & -5 & 7 \end{bmatrix}$

12. $\begin{bmatrix} 4 & -1 & 3 \\ 2 & 8 & 1 \\ 0 & -17 & 1 \end{bmatrix}$

13. $\begin{bmatrix} 3 & 1 & 2 \\ 1 & -1 & 0 \\ 2 & 2 & 3 \end{bmatrix}$

14. $\begin{bmatrix} 3 & 1 & -2 \\ -1 & 2 & 1 \\ 4 & -1 & 3 \end{bmatrix}$

15. Suppose A is not invertible. Show that AB is not invertible as follows:

(a) If B is invertible, suppose AB is also, and consider the product $(AB)B^{-1}$.
(b) If B is not invertible, find $\mathbf{x} \neq \mathbf{0}$ such that $B\mathbf{x} = \mathbf{0}$ and consider $(AB)\mathbf{x}$.

In Exercises 16–18, compute $\det(A)$ and $\det(A^T)$.

16. $A = \begin{bmatrix} 1 & 2 \\ 3 & -4 \end{bmatrix}$

17. $\begin{bmatrix} 2 & 3 & 1 \\ 0 & -3 & 2 \\ 0 & 0 & 4 \end{bmatrix}$

18. $\begin{bmatrix} 3 & 1 & -1 \\ 2 & -1 & 2 \\ 4 & 1 & 1 \end{bmatrix}$

In Exercises 19–24, evaluate $\det(A)$, $\det(E)$ and $\det(AE)$. Then verify that $\det(A)\det(E) = \det(AE)$. Note that if E is an elementary matrix obtained by performing a column operation on I, then AE is the matrix obtained from A by performing that same column operation on A.

19. $A = \begin{bmatrix} 2 & 1 \\ -3 & 0 \end{bmatrix}$, $E = \begin{bmatrix} 0 & 1 \\ 1 & 0 \end{bmatrix}$

20. $A = \begin{bmatrix} 3 & -2 \\ 1 & -4 \end{bmatrix}$, $E = \begin{bmatrix} 1 & 0 \\ 1 & 1 \end{bmatrix}$

21. $A = \begin{bmatrix} 4 & 0 & 2 \\ 1 & 1 & 0 \\ -3 & 0 & 0 \end{bmatrix}$, $E = \begin{bmatrix} -2 & & \\ & 1 & \\ & & 1 \end{bmatrix}$

22. $A = \begin{bmatrix} 4 & 1 & 3 \\ 0 & -2 & 0 \\ 3 & 1 & 5 \end{bmatrix}, \quad E = \begin{bmatrix} 1 & & \\ & 3 & \\ & & 1 \end{bmatrix}$

23. $A = \begin{bmatrix} 4 & 5 \\ 7 & 12 \end{bmatrix}, \quad E = \begin{bmatrix} 1 & -2 \\ & 1 \end{bmatrix}$

24. $A = \begin{bmatrix} 3 & -2 \\ -5 & 3 \end{bmatrix}, \quad E = \begin{bmatrix} 0 & 1 \\ 1 & 0 \end{bmatrix}$

25. If A is $n \times n$ and $\det(-A) = \det(A)$, what can you say about n?

26. If $A = \begin{bmatrix} 2 & 3 \\ & -4 \end{bmatrix}$, so that $\det A = -8$, find a matrix B such that $\det(A + B) = 0$ and $\det B \neq 8$.

27. Let $A = [a]$ or $A = \begin{bmatrix} a & b \\ c & d \end{bmatrix}$. Find $\det(A)$ and $\det(A^T)$ and see that they are equal.

28. Let

$$A = \begin{bmatrix} a_{11} & a_{12} & a_{13} \\ a_{21} & a_{22} & a_{23} \\ a_{31} & a_{32} & a_{33} \end{bmatrix}$$

List the six signed elementary products of A [you might review Example 5(b), Section 2.1] and then of A^T. See that the two lists are the same so, by the definition of determinant, $\det(A) = \det(A^T)$.

2.4 Cofactor Expansion; Cramer's Rule

In this section we devise a method for evaluating determinants that is useful both in theory and in computing determinants of low-dimensional matrices by hand. We also use this method to derive Cramer's rule, a formula for the solution to $AX = B$, A being a square and nonsingular matrix. This formula is simple and elegant and consequently is traditionally quite popular. However, as we shall explain, it is really only of theoretical importance as it is a computational nightmare for a computer.

Cofactor Expansion

We first define two of the key terms used with this method and then describe the process of evaluating a determinant by "expansion by cofactors."

(2.29)

DEFINITION Suppose in an $n \times n$ matrix A we delete the ith row and jth column to obtain an $(n-1) \times (n-1)$ matrix. The determinant of this submatrix is called the (i, j)th **minor** of A and is denoted by M_{ij}. The number $(-1)^{i+j}M_{ij}$ is called the (i, j)th **cofactor** of A and is denoted by C_{ij}.

Example 1 Let

$$A = \begin{bmatrix} 2 & -3 & 1 \\ 4 & 0 & -2 \\ 3 & -1 & -3 \end{bmatrix}$$

Then

(a) $$M_{11} = \begin{vmatrix} 2 & -3 & 1 \\ 4 & 0 & -2 \\ 3 & -1 & -3 \end{vmatrix} = \begin{vmatrix} 0 & -2 \\ -1 & -3 \end{vmatrix} = 0(-3) - (-1)(-2) = -2$$

$$C_{11} = (-1)^{1+1}M_{11} = (+1)M_{11} = -2$$

(b) $$M_{12} = \begin{vmatrix} 2 & -3 & 1 \\ 4 & 0 & -2 \\ 3 & -1 & -3 \end{vmatrix} = \begin{vmatrix} 4 & -2 \\ 3 & -3 \end{vmatrix} = 4(-3) - 3(-2) = -6$$

$$C_{12} = (-1)^{1+2}M_{12} = (-1)M_{12} = 6$$

(c) $$M_{23} = \begin{vmatrix} 2 & -3 & 1 \\ 4 & 0 & -2 \\ 3 & -1 & -3 \end{vmatrix} = \begin{vmatrix} 2 & -3 \\ 3 & -1 \end{vmatrix} = 2(-1) - 3(-3) = 7$$

$$C_{23} = (-1)^{2+3}M_{23} = (-1)M_{23} = -7$$

■

The following way of evaluating a determinant is called **expansion by cofactors**.

(2.30)

THEOREM Let A be an $n \times n$ matrix. Then det A can be evaluated by expanding by cofactors along any row or along any column in the following way.

$$\det A = a_{i1}C_{i1} + a_{i2}C_{i2} + \cdots + a_{in}C_{in}, \quad 1 \le i \le n$$

or

$$\det A = a_{1j}C_{1j} + a_{2j}C_{2j} + \cdots + a_{nj}C_{nj}, \quad 1 \le j \le n$$

Idea of Proof Let A be a general 3×3 matrix,

$$A = \begin{bmatrix} a_{11} & a_{12} & a_{13} \\ a_{21} & a_{22} & a_{23} \\ a_{31} & a_{32} & a_{33} \end{bmatrix}$$

We know [see Example 8(b), Section 2.1]

$$\begin{aligned} \det(A) = a_{11}a_{22}a_{33} + a_{12}a_{23}a_{31} + a_{13}a_{21}a_{32} \\ - a_{11}a_{23}a_{32} - a_{12}a_{21}a_{33} - a_{13}a_{22}a_{31} \end{aligned}$$

This can be rewritten as

$$\det(A) = a_{11}(a_{22}a_{33} - a_{23}a_{32}) + a_{12}(a_{23}a_{31} - a_{21}a_{33}) + a_{13}(a_{21}a_{32} - a_{22}a_{31})$$

You can readily verify that the expressions in parentheses are exactly the cofactors C_{11}, C_{12}, and C_{13}. Thus we have shown that

$$\det(A) = a_{11}C_{11} + a_{12}C_{12} + a_{13}C_{13}$$

which means that $\det(A)$ can be evaluated by expansion of cofactors along the first row. The other rows and the columns can be handled similarly. The situation in general is similar also, but the details are quite tedious. ■

Example 2

Let

$$A = \begin{bmatrix} 2 & -3 & 1 \\ 4 & 0 & -2 \\ 3 & -1 & -3 \end{bmatrix}$$

as in Example 1. Then by Example 8, Section 2.1,

$$\begin{aligned} |A| &= 2(0)(-3) + (-3)(-2)(3) + 1(4)(-1) - 3(0)(1) \\ &\quad -(-1)(-2)(2) - (-3)(4)(-3) \\ &= 0 + 18 - 4 - 0 - 4 - 36 = -26 \end{aligned}$$

If we expand along the first row, we obtain

$$\begin{aligned} &2C_{11} + (-3)C_{12} + 1C_{13} \\ &\quad = 2(-1)^{1+1}\begin{vmatrix} 0 & -2 \\ -1 & -3 \end{vmatrix} - 3(-1)^{1+2}\begin{vmatrix} 4 & -2 \\ 3 & -3 \end{vmatrix} + 1(-1)^{1+3}\begin{vmatrix} 4 & 0 \\ 3 & -1 \end{vmatrix} \\ &\quad = 2[0(-3) - (-1)(-2)] + 3[4(-3) - 3(-2)] + [4(-1) - 3(0)] \\ &\quad = 2(-2) + 3(-6) + (-4) = -26 = |A| \end{aligned}$$

If we expand along the second row, we obtain

$$\begin{aligned} 4C_{21} + 0C_{22} - 2C_{23} &= 4(-1)^{2+1}\begin{vmatrix} -3 & 1 \\ -1 & -3 \end{vmatrix} + 0 + (-2)(-1)^{2+3}\begin{vmatrix} 2 & -3 \\ 3 & -1 \end{vmatrix} \\ &= -4[-3(-3) - (-1)(1)] + 2[(2(-1) - 3(-3)] \\ &= -4(10) + 2(7) = -26 = |A| \end{aligned}$$

If we expand along the third column, we obtain

$$\begin{aligned}
&1C_{13} + (-2)C_{23} + (-3)C_{33} \\
&\quad = 1(-1)^{1+3}\begin{vmatrix} 4 & 0 \\ 3 & -1 \end{vmatrix} - 2(-1)^{2+3}\begin{vmatrix} 2 & -3 \\ 3 & -1 \end{vmatrix} - 3(-1)^{3+3}\begin{vmatrix} 2 & -3 \\ 4 & 0 \end{vmatrix} \\
&\quad = 1[4(-1) - 3(0)] + 2[2(-1) - 3(-3)] - 3[2(0) - 4(-3)] \\
&\quad = -4 + 2(7) - 3(12) = -26 = |A|
\end{aligned}$$

These last three calculations illustrate Theorem (2.29). ■

Example 3 If

$$A = \begin{bmatrix} 3 & 4 & -1 & 0 \\ 4 & -1 & 0 & 3 \\ -6 & 4 & 8 & -2 \\ -1 & 1 & 2 & 7 \end{bmatrix},$$

then

$$\begin{aligned}
|A| = 3(-1)^{1+1}\begin{vmatrix} -1 & 0 & 3 \\ 4 & 8 & -2 \\ 1 & 2 & 7 \end{vmatrix} + 4(-1)^{1+2}\begin{vmatrix} 4 & 0 & 3 \\ -6 & 8 & -2 \\ -1 & 2 & 7 \end{vmatrix} \\
+ (-1)(-1)^{1+3}\begin{vmatrix} 4 & -1 & 3 \\ -6 & 4 & -2 \\ -1 & 1 & 7 \end{vmatrix} + 0(-1)^{1+4}\begin{vmatrix} 4 & -1 & 0 \\ -6 & 4 & 8 \\ -1 & 1 & 2 \end{vmatrix}
\end{aligned}$$

We next compute the four 3×3 determinants as we have done before and eventually calculate that $|A| = -1162$, as you can check. ■

It was just suggested to complete Example 3 by evaluating four 3×3 determinants. Although it is possible to do this, it would require considerable work. In practice, when working by hand, cofactor expansion is used in conjunction with row or column operations to provide a computationally reasonable procedure to evaluate a determinant. This is illustrated in the next example.

Example 4 Let

$$A = \begin{bmatrix} 3 & 4 & -1 & 0 \\ 4 & -1 & 0 & 3 \\ -6 & 4 & 8 & -2 \\ -1 & 1 & 2 & 7 \end{bmatrix}$$

as in Example 3. Then

$$|A| = \begin{vmatrix} 3 & 4 & -1 & 0 \\ 4 & -1 & 0 & 3 \\ -6 & 4 & 8 & -2 \\ -1 & 1 & 2 & 7 \end{vmatrix}$$

$$= \begin{vmatrix} 0 & 7 & 5 & 21 \\ 0 & 3 & 8 & 31 \\ 0 & -2 & -4 & -44 \\ -1 & 1 & 2 & 7 \end{vmatrix} \quad \text{By adding appropriate multiples of row 4 to the other rows}$$

$$= 0 - 0 + 0 - (-1)\begin{vmatrix} 7 & 5 & 21 \\ 3 & 8 & 31 \\ -2 & -4 & -44 \end{vmatrix} \quad \text{By cofactor expansion along column 1}$$

$$= -2\begin{vmatrix} 7 & 5 & 21 \\ 3 & 8 & 31 \\ 1 & 2 & 22 \end{vmatrix} \quad \text{By factoring out a } -2 \text{ from row 3}$$

$$= -2\begin{vmatrix} 7 & -9 & -133 \\ 3 & 2 & -35 \\ 1 & 0 & 0 \end{vmatrix} \quad \text{By adding appropriate multiples of column 1 to the other columns}$$

$$= -2(1)\begin{vmatrix} -9 & -133 \\ 2 & -35 \end{vmatrix} - 0 + 0 \quad \text{By cofactor expansion along row 3}$$

$$= -2[-9(-35) - 2(-133)] = -2[581] = -1162$$

■

The Adjoint and a Theoretical Formula for A^{-1}

We now develop a wonderful formula for A^{-1}. Alas, this formula is a computational nightmare (except for 2×2 and 3×3 matrices). But it is fascinating, it is extremely useful theoretically, and it is very useful in the derivation of practical formulas (see Section 2.5).

For a $n \times n$ matrix A, the (i, j)th cofactor of A is described in Definition (2.29), is denoted by C_{ij}, and is a *number*. The **matrix of cofactors** of A has the form

$$\begin{bmatrix} C_{11} & C_{12} & \cdots & C_{1n} \\ C_{21} & C_{22} & \cdots & C_{2n} \\ & \cdots\cdots\cdots & & \\ C_{n1} & C_{n2} & \cdots & C_{nn} \end{bmatrix} \tag{2.31}$$

The *transpose* of this matrix of cofactors is the *adjoint* of A.

(2.32)

DEFINITION If A is an $n \times n$ matrix, the **adjoint** of A, denoted by adj(A), is the transpose of the matrix of cofactors,

$$\operatorname{adj}(A) = \begin{bmatrix} C_{11} & C_{21} & \cdots & C_{n1} \\ C_{12} & C_{22} & \cdots & C_{n2} \\ & \cdots\cdots\cdots & & \\ C_{1n} & C_{2n} & \cdots & C_{nn} \end{bmatrix}$$

The adjoint of A plays the following extremely important role.

(2.33) THEOREM If A is an $n \times n$ matrix, then

(2.34)
$$A \operatorname{adj}(A) = \det(A)I$$

Thus if $\det(A) \neq 0$ so that A^{-1} exists, then

(2.35)
$$A^{-1} = \frac{1}{\det(A)} \operatorname{adj}(A)$$

The proof follows Example 5.

Example 5 Let

$$A = \begin{bmatrix} 2 & -3 & 1 \\ 4 & 0 & -2 \\ 3 & -1 & -3 \end{bmatrix}$$

from Example 1. Then the matrix of cofactors is

$$\begin{bmatrix} \begin{vmatrix} 0 & -2 \\ -1 & -3 \end{vmatrix} & -\begin{vmatrix} 4 & -2 \\ 3 & -3 \end{vmatrix} & \begin{vmatrix} 4 & 0 \\ 3 & -1 \end{vmatrix} \\ -\begin{vmatrix} -3 & 1 \\ -1 & -3 \end{vmatrix} & \begin{vmatrix} 2 & 1 \\ 3 & -3 \end{vmatrix} & -\begin{vmatrix} 2 & -3 \\ 3 & -1 \end{vmatrix} \\ \begin{vmatrix} -3 & 1 \\ 0 & -2 \end{vmatrix} & -\begin{vmatrix} 2 & 1 \\ 4 & -2 \end{vmatrix} & \begin{vmatrix} 2 & -3 \\ 4 & 0 \end{vmatrix} \end{bmatrix} = \begin{bmatrix} -2 & 6 & -4 \\ -10 & -9 & -7 \\ 6 & 8 & 12 \end{bmatrix}$$

Taking the transpose, we obtain

$$\operatorname{adj}(A) = \begin{bmatrix} -2 & -10 & 6 \\ 6 & -9 & 8 \\ -4 & -7 & 12 \end{bmatrix}$$

From Example 2 we know that $\det(A) = -26$. To check Equation (2.34),

$$A\,\text{adj}(A) = \begin{bmatrix} 2 & -3 & 1 \\ 4 & 0 & -2 \\ 3 & -1 & -3 \end{bmatrix} \begin{bmatrix} -2 & -10 & 6 \\ 6 & -9 & 8 \\ -4 & -7 & 12 \end{bmatrix} = \begin{bmatrix} -26 & 0 & 0 \\ 0 & -26 & 0 \\ 0 & 0 & -26 \end{bmatrix}$$

$$= -26 \begin{bmatrix} 1 & & \\ & 1 & \\ & & 1 \end{bmatrix} = \det(A)I$$

Thus we see that $A\left[\dfrac{1}{-26}\,\text{adj}(A)\right] = I$ or

$$A^{-1} = \frac{1}{-26}\,\text{adj}(A) = \frac{1}{-26} \begin{bmatrix} -2 & -10 & 6 \\ 6 & -9 & 8 \\ -4 & -7 & 12 \end{bmatrix}$$ ■

Proof of Theorem (2.33) First consider the product of A with its adjoint,

$$A\,\text{adj}(A) = \begin{bmatrix} a_{11} & a_{12} & \cdots & a_{1n} \\ & \cdots\cdots & & \\ a_{i1} & a_{i2} & \cdots & a_{in} \\ & \cdots\cdots & & \\ a_{n1} & a_{n2} & \cdots & a_{nn} \end{bmatrix} \begin{bmatrix} C_{11} & \cdots & C_{j1} & \cdots & C_{n1} \\ C_{12} & \cdots & C_{j2} & \cdots & C_{n2} \\ \vdots & & & & \vdots \\ C_{1n} & \cdots & C_{jn} & \cdots & C_{nn} \end{bmatrix}$$

As indicated by the tinted row and column, the (i, j)th entry of this product is given by

$$a_{i1}C_{j1} + a_{i2}C_{j2} + \cdots + a_{in}C_{jn} \tag{2.36}$$

If $i = j$, then this sum is simply the cofactor expansion of A along its ith row. By Theorem (2.30), this sum is det A.

If $i \neq j$, then this sum is zero. To see this, form the matrix B by taking A and replacing the jth row of A with the ith row of A (keeping everything else in A the same):

$$B = \begin{bmatrix} a_{11} & a_{12} & \cdots & a_{1n} \\ & \cdots\cdots & & \\ a_{i1} & a_{i2} & \cdots & a_{in} \\ & \cdots\cdots & & \\ a_{i1} & a_{i2} & \cdots & a_{in} \\ & \cdots\cdots & & \\ a_{n1} & a_{n2} & \cdots & a_{nn} \end{bmatrix} \begin{matrix} \\ \\ \longleftarrow \textit{ith row of B} \\ \\ \longleftarrow \textit{jth row of B} \\ \\ \\ \end{matrix}$$

Now $\det B = 0$ [by Theorem (2.12), Section 2.2], since two rows of B are the same. If we expand $\det B$ by cofactors along the jth row, then we obtain the expression in (2.36). (Since A and B are identical off the jth row, the corresponding cofactors are the same.) Thus, for $i \neq j$,

$$\text{Expression (2.36)} = \det B = 0$$

This proves Equation (2.35). For the expression in (2.36), if $\det A \neq 0$, we divide Equation (2.35) by $\det A$ to obtain

$$A\left[\frac{1}{\det A}\operatorname{adj}(A)\right] = I \qquad \text{or} \qquad A^{-1} = \frac{1}{\det A}\operatorname{adj}(A)$$

by uniqueness of inverses [see Theorem (1.38), Section 1.4]. ■

Cramer's Rule

Cramer's rule is a traditional topic in linear algebra. Certainly one of the reasons is the intriguing, almost magical, way in which it works. It is fun, the rule is simple to state, and the outcome is extremely powerful. For these reasons Cramer's rule is very important theoretically. It is also purported to be a powerful computational tool. However, this is *not* the case, as we shall explain after stating the rule and giving a few examples.

Consider the system of n equations in n unknowns, $AX = B$. Suppose $\det(A) \neq 0$, so that A is invertible and the system $AX = B$ has a unique solution. For each j, $1 \leq j \leq n$, the matrix obtained from A by replacing the jth column of A with B is denoted by A_j:

(2.37)
$$A_j = \begin{bmatrix} a_{11} & \cdots & b_1 & \cdots & a_{1n} \\ a_{21} & \cdots & b_2 & \cdots & a_{2n} \\ \vdots & & \vdots & & \vdots \\ a_{n1} & \cdots & b_n & \cdots & a_{nn} \end{bmatrix} \quad (\text{b column is the } j\text{th column})$$

With this terminology in place, we can now state Cramer's rule.

(2.38) **THEOREM (CRAMER'S RULE)** Let $AX = B$ be a system of n equations in n unknowns, where $\det A \neq 0$. If A_j is a matrix given in Equation (2.37), $1 \leq j \leq n$, then the unique solution to the system $AX = B$ is given by

$$x_1 = \frac{\det A_1}{\det A}, \quad x_2 = \frac{\det A_2}{\det A}, \quad \ldots, \quad x_n = \frac{\det A_n}{\det A}$$

The proof follows from Theorem (2.33) and is given after Example 7.

Example 6 Use Cramer's rule to solve the system

$$\begin{aligned} 2x - y &= 7 \\ 4x + 3y &= -1 \end{aligned}$$

Solution Let $A = \begin{bmatrix} 2 & -1 \\ 4 & 3 \end{bmatrix}$, so that $\det(A) = 6 - (-4) = 10$. Then

$$A_1 = \begin{bmatrix} 7 & -1 \\ -1 & 3 \end{bmatrix} \quad \text{and} \quad \det(A_1) = 21 - (1) = 20$$

$$A_2 = \begin{bmatrix} 2 & 7 \\ 4 & -1 \end{bmatrix} \quad \text{and} \quad \det(A_2) = -2 - 28 = -30$$

Then

$$x = \frac{\det A_1}{\det A} = \frac{20}{10} = 2, \quad y = \frac{\det A_2}{\det A} = \frac{-30}{10} = -3$$

Thus the solution is $(x, y) = (2, -3)$, as you can easily check. ■

Example 7 Use Cramer's rule to solve the system

$$\begin{aligned} 3x + 2y + 3z &= 4 \\ -2x - 4y + 2z &= -12 \\ 2x + 3z &= 0 \end{aligned}$$

Solution We first construct A, A_1, A_2, and A_3.

$$A = \begin{bmatrix} 3 & 2 & 3 \\ -2 & -4 & 2 \\ 2 & 0 & 3 \end{bmatrix}, \quad A_1 = \begin{bmatrix} 4 & 2 & 3 \\ -12 & -4 & 2 \\ 0 & 0 & 3 \end{bmatrix}$$

$$A_2 = \begin{bmatrix} 3 & 4 & 3 \\ -2 & -12 & 2 \\ 2 & 0 & 3 \end{bmatrix}, \quad A_3 = \begin{bmatrix} 3 & 2 & 4 \\ -2 & -4 & -12 \\ 2 & 0 & 0 \end{bmatrix}$$

Following Cramer's rule, we evaluate the various determinants by expanding along the bottom row

$$|A| = \begin{vmatrix} 3 & 2 & 3 \\ -2 & -4 & 2 \\ 2 & 0 & 3 \end{vmatrix} = 2\begin{vmatrix} 2 & 3 \\ -4 & 2 \end{vmatrix} - 0 + 3\begin{vmatrix} 3 & 2 \\ -2 & -4 \end{vmatrix}$$

$$= 2(4 + 12) + 3(-12 + 4) = 2(16) + 3(-8) = 32 - 24 = 8$$

$$|A_1| = \begin{vmatrix} 4 & 2 & 3 \\ -12 & -4 & 2 \\ 0 & 0 & 3 \end{vmatrix} = 0 + 0 + 3\begin{vmatrix} 4 & 2 \\ -12 & -4 \end{vmatrix} = 3(-16 + 24) = 24$$

$$|A_2| = \begin{vmatrix} 3 & 4 & 3 \\ -2 & -12 & 2 \\ 2 & 0 & 3 \end{vmatrix} = 2\begin{vmatrix} 4 & 3 \\ -12 & 2 \end{vmatrix} + 3\begin{vmatrix} 3 & 4 \\ -2 & -12 \end{vmatrix}$$

$$= 2(8 + 36) + 3(-36 + 8) = 2(44) + 3(-28) = 4$$

$$|A_3| = \begin{vmatrix} 3 & 2 & 4 \\ -2 & -4 & -12 \\ 2 & 0 & 0 \end{vmatrix} = 2\begin{vmatrix} 2 & 4 \\ -4 & -12 \end{vmatrix} = 2(-24 + 16) = -16$$

Thus the solution is

$$x_1 = \frac{|A_1|}{|A|} = \frac{24}{8} = 3, \qquad x_2 = \frac{|A_2|}{|A|} = \frac{4}{8} = \frac{1}{2}, \qquad x_3 = \frac{|A_3|}{|A|} = \frac{-16}{8} = -2$$

You can check that $(x_1, x_2, x_3) = (3, \frac{1}{2}, -2)$ works. ■

Proof of Cramer's Rule, Theorem (2.38) Since $\det A \neq 0$, A is invertible. Using Equation (2.35), the unique solution to $AX = B$ is

$$X = A^{-1}B = \frac{1}{\det A}\operatorname{adj}(A)\,B = \frac{1}{\det A}\begin{bmatrix} C_{11} & C_{21} & \cdots & C_{n1} \\ C_{12} & C_{22} & \cdots & C_{n2} \\ & \cdots\cdots\cdot & & \\ C_{1n} & C_{2n} & \cdots & C_{nn} \end{bmatrix}\begin{bmatrix} b_1 \\ b_2 \\ \vdots \\ b_n \end{bmatrix}$$

By multiplying out, we see that

$$x_i = \frac{1}{\det A}(b_1C_{1i} + b_2C_{2i} + \cdots + b_nC_{ni})$$

But the sum in parentheses is exactly the cofactor expansion of $\det A_i$ along the ith column. Thus $x_i = |A_i|/|A|$. ■

Numerical Considerations

Suppose we were to count carefully the number of operations it takes to solve a system of two equations in two unknowns, such as Example 6. We would find it takes eight multiplications/divisions using Cramer's rule and six using Gaussian elimination. This minor difference in no way begins to reflect the tremendous difference in the amount of computation in larger systems, which is caused by the following fact.

(2.39)

> IMPORTANT FACT Suppose A is an $n \times n$ matrix. If n is "large," then it takes roughly the same number of operations to evaluate $\det(A)$, *using the most efficient method possible*, as it does to solve the system $AX = B$ by Gaussian elimination.

Here "large" means n^3 is large compared with n^2, which is a value judgment. For example, if $n = 10$, then for some people $n^3 = 1000$ is large as compared with $n^2 = 100$. If $n = 100$, then for almost any purpose, $n^3 = 1{,}000{,}000$ is large compared with $n^2 = 10{,}000$.

Suppose A is $n \times n$ with n large, and you wish to solve $AX = B$. Then to use Cramer's rule you must compute $(n + 1)$ determinants, $|A|, |A_1|, \ldots, |A_n|$, each one of which takes essentially as much work as solving the problem by Gaussian elimination. (You still have n more divisions, $x_1 = |A_1|/|A|$, and so on, but this is minor.) Thus altogether we have

(2.40)

> Suppose A is $n \times n$ and n is large. Then solving the system $AX = B$ by Cramer's rule takes roughly $(n + 1)$ times as much work as solving by Gaussian elimination.

Of course, more work means it takes longer (and hence costs more), and there is more chance for computer error to build up. This is why the methods discussed in Chapter 1 are preferable for computations.

However, Cramer's rule remains a fond tool for mathematicians, engineers, and scientists working on a small-dimensional problem with messy numbers. For example, sitting at a desk with a hand-held calculator and faced with a problem like

$$\begin{aligned} 3.17984x + 5.99812y &= 12.3561 \\ 7.42495x - 9.11874y &= -8.4515 \end{aligned}$$

many people would probably use Cramer's rule. Perhaps they would also for a similar 3×3 problem. For a 4×4 or larger problem, however, it is time to turn to a programmable calculator, PCMatlab, or the like, and to revert to Gaussian elimination.

Exercise 2.4

In Exercises 1–4, (a) find all the minors and (b) find all the cofactors of the given matrix.

1. $\begin{bmatrix} -2 & 1 \\ 3 & 4 \end{bmatrix}$

2. $\begin{bmatrix} 3 & 5 \\ -1 & -2 \end{bmatrix}$

3. $\begin{bmatrix} 2 & 1 & 0 \\ 1 & -1 & 2 \\ 0 & 1 & 0 \end{bmatrix}$

4. $\begin{bmatrix} 5 & -1 & -2 \\ 0 & 2 & 1 \\ 0 & 3 & 0 \end{bmatrix}$

5. Evaluate the determinant of the matrix in Exercise 1 by cofactor expansion along the second row.

6. Evaluate the determinant of the matrix in Exercise 2 by cofactor expansion along the first column.

7. Evaluate the determinant of the matrix in Exercise 3 by cofactor expansion along the second column.

8. Evaluate the determinant of the matrix in Exercise 4 by cofactor expansion along the third row.

In Exercises 9–14, evaluate the determinant of the given matrix by cofactor expansion along the row or column of your choice.

9. $\begin{bmatrix} 2 & -1 & 3 \\ 4 & 8 & 2 \\ 0 & 0 & -1 \end{bmatrix}$

10. $\begin{bmatrix} 3 & 8 & 1 \\ 2 & 0 & -2 \\ 5 & 0 & 2 \end{bmatrix}$

11. $\begin{bmatrix} -4 & 3 & 5 \\ 0 & -1 & 1 \\ 0 & 2 & 5 \end{bmatrix}$

12. $\begin{bmatrix} 8 & 0 & 0 \\ 3 & -1 & 2 \\ 4 & 1 & 4 \end{bmatrix}$

13. $\begin{bmatrix} 4 & 0 & 1 & 0 \\ 0 & -1 & 0 & 0 \\ 5 & 2 & -1 & 0 \\ 8 & 0 & 0 & 3 \end{bmatrix}$

14. $\begin{bmatrix} 0 & 2 & 1 & 0 \\ -1 & 0 & 0 & 6 \\ 0 & 3 & 2 & 0 \\ 0 & 0 & 0 & 4 \end{bmatrix}$

In Exercises 15–22, evaluate the determinant using a combination of cofactor expansions and elementary operations, as in Example 4.

15. $\begin{bmatrix} 2 & -1 \\ 4 & 1 \end{bmatrix}$

16. $\begin{bmatrix} 3 & 6 \\ 4 & -8 \end{bmatrix}$

17. $\begin{bmatrix} 1 & 2 & 3 \\ 2 & -1 & -2 \\ -3 & 1 & 2 \end{bmatrix}$

18. $\begin{bmatrix} 3 & 1 & 2 \\ 2 & -2 & 4 \\ 5 & 1 & -3 \end{bmatrix}$

19. $\begin{bmatrix} 2 & -1 & 1 & 2 \\ 4 & 1 & 2 & 1 \\ -3 & 1 & 1 & 2 \\ 1 & -1 & 2 & 1 \end{bmatrix}$

20. $\begin{bmatrix} 3 & -1 & 2 & 1 \\ 1 & 4 & 8 & -1 \\ -1 & 1 & 1 & 1 \\ 2 & -1 & 1 & -2 \end{bmatrix}$

21. $\begin{bmatrix} 1 & 1 & 1 \\ a & b & c \\ a^2 & b^2 & c^2 \end{bmatrix}$

22. $\begin{bmatrix} 1 & w & w^2 & w^3 \\ 1 & x & x^2 & x^3 \\ 1 & y & y^2 & y^3 \\ 1 & z & z^2 & z^3 \end{bmatrix}$

In Exercises 23–28, find the matrix of cofactors, the adjoint, and the inverse [using Equation (2.35)] of the given matrix. Verify that I is the product of the matrix with the inverse you found.

23. The matrix given in Exercise 15.

24. The matrix given in Exercise 16.

25. The matrix given in Exercise 17.

26. The matrix given in Exercise 18.

27. The matrix given in Exercise 9.

28. The matrix given in Exercise 10.

In Exercises 29–34, solve the system using Cramer's rule.

29. $\begin{aligned} 2x - y &= 8 \\ 3x + y &= 7 \end{aligned}$

30. $\begin{aligned} 2x + 3y &= -4 \\ 5x + 2y &= 2 \end{aligned}$

31. $\begin{aligned} 3x + 2y + z &= -2 \\ 2x - 4y + 3z &= 7 \\ 4x - 2y + 3z &= 0 \end{aligned}$

32. $\begin{aligned} 2x - y + z &= -1 \\ 3x + 4y - z &= -1 \\ 4x - y + 2z &= -1 \end{aligned}$

33. $\begin{aligned} 2x - y + z &= 8 \\ 3x + 3y - 2z &= 3 \\ -x - 2y - z &= 1 \end{aligned}$

34. $\begin{aligned} 2x - 3y - z &= 2 \\ x + y &= -3 \\ -y + z &= -2 \end{aligned}$

2.5 Applications

Applications of many mathematical topics often require lengthy explanations and advanced material. This is unfortunately true in the case of determinants. For this reason, *this section contains general information only*. Detailed explanations of the applications discussed here are beyond the scope and level of this text. The brief introductions provided should help you appreciate some of the applications that you may encounter later.

We begin with the most elementary facts.

det $A \neq 0$

As we have already noted

$$\text{A matrix is invertible} \Leftrightarrow \text{its determinant is nonzero}$$

This is the most widespread application of determinants. Sometimes it is used simply for notational purposes; an author or instructor may find it easier to say "det $A \neq 0$" instead of "A is invertible" or "A is nonsingular." In other applications, however, the use of det A is not superficial; describing (and sometimes finding) eigenvalues, as we shall see in Chapter 5, is one such case. The next two subsections essentially discuss other applications of det $A \neq 0$.

Jacobians

You may have already seen functions like

$$\mathbf{f}(x, y, z) = (y \sin 2x, 3x^2z - 4yz, e^{4xyz})$$

Think of it as $\mathbf{f} = (f_1, f_2, f_3)$, so $f_1(x, y, z) = y \sin 2x$, and so on. If you have seen partial derivatives, you may have seen the matrix

$$\begin{bmatrix} \partial_1 f_1 & \partial_2 f_1 & \partial_3 f_1 \\ \partial_1 f_2 & \partial_2 f_2 & \partial_3 f_2 \\ \partial_1 f_3 & \partial_2 f_3 & \partial_3 f_3 \end{bmatrix} = \begin{bmatrix} 2y \cos 2x & \sin 2x & 0 \\ 6xz & -4z & 3x^2 - 4y \\ 4yze^{4xyz} & 4xze^{4xyz} & 4xye^{4xyz} \end{bmatrix} \tag{2.41}$$

The determinant of this matrix is called the **Jacobian** of **f**; in later courses you will think of the matrix itself as the derivative of **f**. (The Jacobian is named after Carl Gustav Jacob Jacobi (1804–1851). Born into a wealthy banking family in Potsdam, Germany, Jacobi contributed very original work in many fields; his work is especially noteworthy in the theory of elliptic functions.) If you take advanced calculus or certain engineering, physics, chemistry, or economics courses, you will learn to use Jacobians. Where the Jacobian is nonzero, the matrix in Equation (2.41) is nonsingular, which means that the function **f** has certain nice properties; where the Jacobian is zero, the graph of **f** may have severe links in it that make the function very difficult to analyze. Details will have to wait for later courses.

Wronskians

You may have already seen some differential equations. The equation

$$x^2y'' + xy' - 4y = 0, \qquad x > 0 \tag{2.42}$$

is a second-order differential equation called *Euler's equation*, which has two special solutions: $f_1(x) = x^2$, and $f_2(x) = x^{-2}$. Both of these solutions have the property that any solution $y = f(x)$ of Equation (2.42) can be expressed as $f(x) = a_1f_1(x) + a_2f_2(x)$. One way to determine that the pair of solutions, f_1 and f_2, have this special property is to verify that their **Wronskian**

$$W(f_1, f_2) = \det\begin{bmatrix} f_1(x) & f_2(x) \\ f'_1(x) & f'_2(x) \end{bmatrix} = \det\begin{bmatrix} x^2 & x^{-2} \\ 2x & -2x^{-3} \end{bmatrix} = -4x^{-1}$$

is nonzero for all $x > 0$. This condition holds in general. For a nth order linear differential equation, we try to find n solutions $f_1, \ldots, f_n$ such that their Wronskian

$$W(f_1, \ldots, f_2) = \det\begin{bmatrix} f_1 \cdots f_n \\ \cdots\cdots \\ f_1^{(n-1)} \cdots f_n^{(n-1)} \end{bmatrix} \neq 0$$

The Wronskian was named after its inventor Josef Wronski (1778–1853). Born in Poland, Wronski was really a philosopher who did mathematics as a hobby.

Cross Products

Some mathematical relationships are extremely complex, and any notational gimmicks that can simplify them are to be cherished. Describing cross products in terms of determinants is one such case.

First, what are cross products? If you do not know, do not *believe* what you are about to read; go *do* it. Remove the front wheel from a bicycle. Using Figure 2.3 as a guide, place the end of the axle on your fist and, being careful of your foot, let go. The wheel will surely fall. Next, pick up the wheel, hold onto the axle, and turn the wheel until it is spinning reasonably quickly.

Figure 2.3

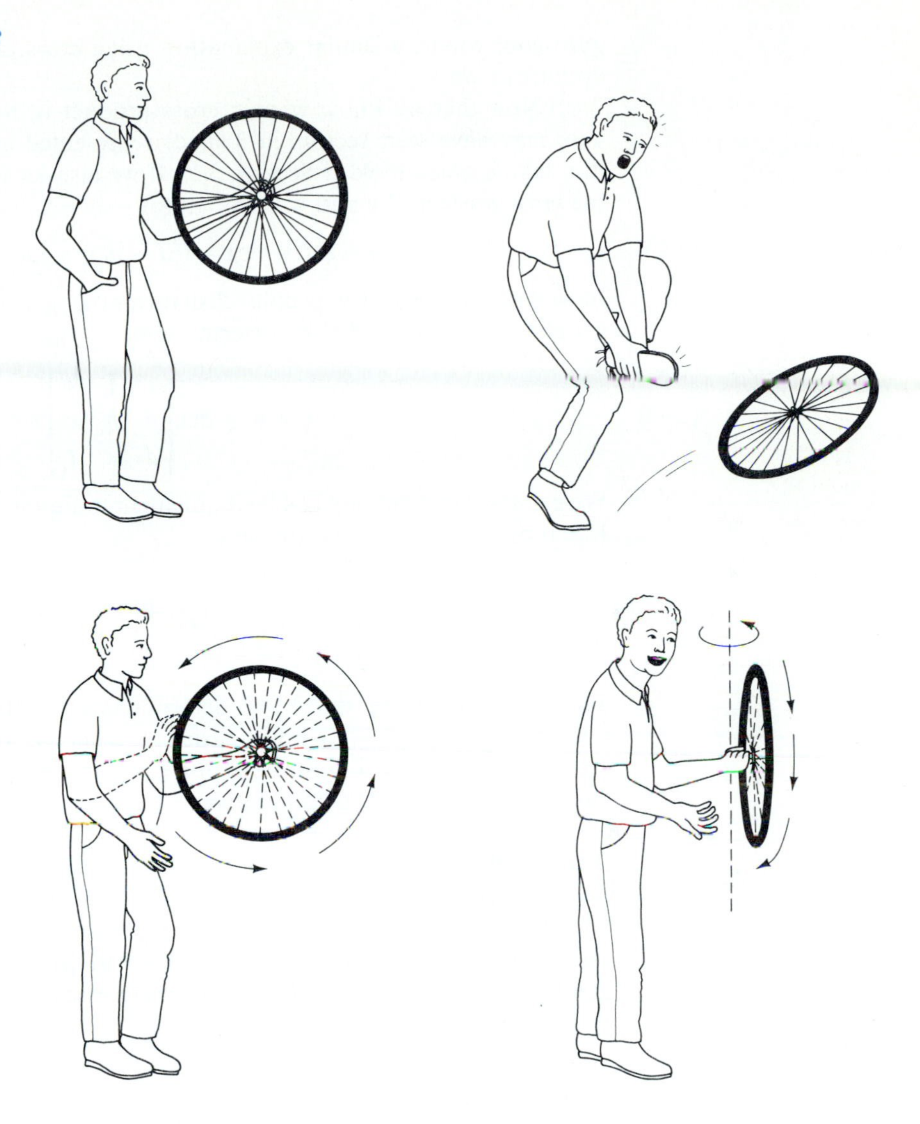

Then put the axle on your fist and let go again. There is no need to worry about your foot this time; until the wheel slows down, it will *not* fall. The axle will rotate on your fist, either clockwise or counterclockwise, depending on which way you spun the wheel. Gravity exerts a force downward, and the *spinning* wheel creates a force along the wheel's axle. These two forces combine to form a single force perpendicular to them both. This resulting force is called their *cross product* and is what keeps the axel rotating. You must see this to believe it!

Actually, whenever a cyclist leans into a turn, the cross-product force makes the turn easier. The force resulting from a cross product also makes a gyroscope rotate; a similar explanation using cross products describes why cyclotrons work.

Now that we know what a cross product is, how do we describe it? You may have seen vectors in 3-space represented as $\mathbf{v} = a\mathbf{i} + b\mathbf{j} + c\mathbf{k}$. (If not, take a quick look at Section 3.1.) If we also let $\mathbf{w} = d\mathbf{i} + e\mathbf{j} + f\mathbf{k}$, then the **cross product** of $\mathbf{v}$ and $\mathbf{w}$ is the vector

$$\mathbf{v} \times \mathbf{w} = (bf - ce)\mathbf{i} + (cd - af)\mathbf{j} + (ae - bd)\mathbf{k} \tag{2.43}$$

Now this is a formula few people relish memorizing. Fortunately, a notational simplification is provided by determinants:

$$\mathbf{v} \times \mathbf{w} = \det \begin{bmatrix} \mathbf{i} & \mathbf{j} & \mathbf{k} \\ a & b & c \\ d & e & f \end{bmatrix} \tag{2.44}$$

We can see that Equation (2.44) is equivalent to Equation (2.43) by expanding Equation (2.44) across the top row

$$\begin{vmatrix} \mathbf{i} & \mathbf{j} & \mathbf{k} \\ a & b & c \\ d & e & f \end{vmatrix} = \begin{vmatrix} b & c \\ e & f \end{vmatrix}\mathbf{i} - \begin{vmatrix} a & c \\ d & f \end{vmatrix}\mathbf{j} + \begin{vmatrix} a & b \\ d & e \end{vmatrix}\mathbf{k}$$

and then simplifying the 2×2 determinants to obtain Equation (2.43). Clearly, Equation (2.44) is much easier to memorize than Equation (2.43)!

You might notice from Equation (2.44) that $\mathbf{v} \times \mathbf{w} = -(\mathbf{w} \times \mathbf{v})$, so this "multiplication" is not commutative.

Area and Volume

As indicated in Figure 2.4, two points in a plane determine a parallelogram and three points in 3-space determine a parallelepiped. You can see that if the points are represented by 2×1 or 3×1 matrices, then the other vertices can be expressed in terms of sums. In these cases, the area and volume can

Figure 2.4

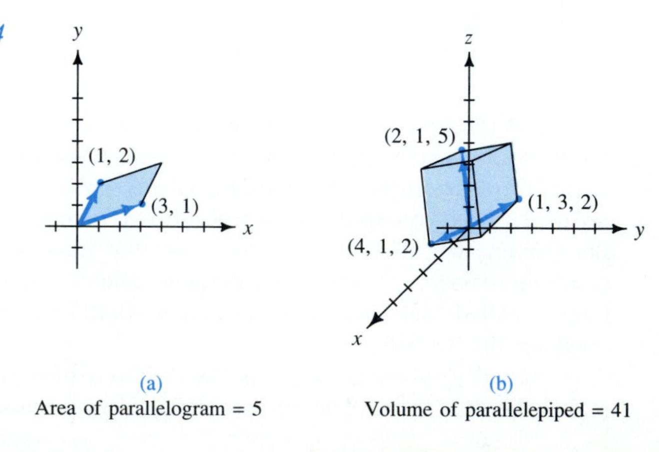

(a)
Area of parallelogram = 5

(b)
Volume of parallelepiped = 41

be expressed in terms of determinants. For example, for the figures given in Figure 2.4, the area A and the volume V are

$$A = \det[X \quad Y] = \det\begin{bmatrix} 3 & 1 \\ 1 & 2 \end{bmatrix} = 5$$

$$V = \det[X \quad Y \quad Z] = \det\begin{bmatrix} 4 & 1 & 2 \\ 1 & 3 & 2 \\ 2 & 1 & 5 \end{bmatrix} = 41$$

The proof of this can be found in most calculus textbooks. This application is a foundation for the next application discussed here.

Multiple Integrals

If you are majoring in a physical science, engineering, or mathematics, eventually you will use multiple integrals. Sometimes with multiple integrals you will need to change coordinate systems and then describe the effects. If, for example, you have seen polar coordinates (in, say, trigonometry), you know that $x = r \cos \theta$ and that $y = r \sin \theta$. The relationship between double integrals in rectangular and polar coordinates is

$$\iint f \, dx \, dy = \iint f \det\begin{bmatrix} \dfrac{\partial x}{\partial r} & \dfrac{\partial x}{\partial \theta} \\ \dfrac{\partial y}{\partial r} & \dfrac{\partial y}{\partial \theta} \end{bmatrix} dr \, d\theta$$

$$= \iint f \det\begin{bmatrix} \cos \theta & -r \sin \theta \\ \sin \theta & r \cos \theta \end{bmatrix} dr \, d\theta = \iint fr \, dr \, d\theta$$

The determinant plays a corresponding rôle in three-dimensional (and higher-dimensional) integrals.

Higher-Level Applications

The formula $A \operatorname{adj}(A) = \det(A)I$ and Cramer's rule, presented in Section 2.4, are useful tools in abstract algebra, algebraic geometry, and elsewhere. These abstract tools also can be used to explain relationships that tell us how to solve problems on a computer. One such example can be found in Section 7-9 of Parlett, *The Symmetric Eigenvalue Problem*, where the formula $A \operatorname{adj}(A) = \det(A)I$ is used to derive an interesting relationship among eigenvalues and eigenvectors:

$$\operatorname{adj}[\lambda_i I - A] = (\lambda_i - \lambda_1) \cdots (\lambda_i - \lambda_{i-1})(\lambda_i - \lambda_{i+1}) \cdots (\lambda_i - \lambda_n) V_i V_i^T$$

This relationship in turn is used to explain aspects of the Lanczos Algorithm, which is currently the only computationally effective way of finding eigenvalues and eigenvectors of large, sparse matrices.

In summary, this section provides a brief indication of the diverse applications of determinants. We can see that determinants have both historical interest and contemporary usefulness.

Review Exercises

In Exercises 1–4, let

$$A = \begin{bmatrix} 0 & 0 & 0 & 0 & 1 \\ 0 & 0 & 1 & 0 & 0 \\ 0 & 1 & 0 & 0 & 0 \\ 1 & 0 & 0 & 0 & 0 \\ 0 & 0 & 0 & 1 & 0 \end{bmatrix}$$

1. Find det A using the definition.
2. Find det A by cofactor expansion.
3. Find det A by first reducing A to triangular form using row operations.
4. If

$$B = \begin{bmatrix} 2 & 0 & 0 & 0 & 1 \\ 0 & 0 & 1 & 0 & 0 \\ 0 & 1 & 0 & 0 & 0 \\ 1 & 0 & 0 & 0 & 0 \\ 0 & 0 & 0 & 1 & 0 \end{bmatrix}$$

explain why det B = det A.

5. If an $n \times n$ matrix A has all integers for entries, will det A necessarily be an integer? Explain.
6. If A and A^{-1} both have all integers for entries, explain why det $A = \pm 1$.
7. Use Cramer's rule to show that if det $A = \pm 1$ and A has all integers for entries, then A^{-1} also has all integers for entries.
8. Suppose A is 3×3, all entries are either 0 or 1, and there are exactly four 1's.
   (a) Find an example of such an A with all elementary products equal to zero.
   (b) Prove A cannot have more than one elementary product equal to 1.
   (c) What are the possibilities for det(A)?
9. Suppose n is odd, A and B are $n \times n$ matrices, and $AB = -BA$. Show that $[\det(A + B)]^2 = \det(A^2 + B^2)$.
10. In a 3×3 matrix, where could you place the fewest number of zeros to assure that the determinant is zero?

11. If A is 3×3, $\det A = 3$, B is obtained from A by multiplying the second row by 4, C is obtained from B by adding 5 times row 2 to row 3, and D is obtained from C by interchanging rows 1 and 3, what is $\det D$?

12. Use Cramer's rule to solve the system

$$\begin{aligned} u &= x \cos\theta - y \sin\theta \\ v &= x \sin\theta + y \cos\theta \end{aligned}$$

for x and y in terms of u and v.

13. Suppose a matrix A has all entries of 0's and 1's. Must $\det A = 0$ or ± 1?

Chapter 3
VECTOR SPACES

Once we write vectors in 3-space as (x_1, x_2, x_3), it is not hard to generalize and write "vectors" in "27-space" as $(x_1, x_2, \ldots, x_{27})$. The first question is: Why would you want to do this? The historical answer is quite straightforward. It helps us find answers to problems we need to solve. Most real-life problems have *many* more variables than just two or three. If there happen to be 27 variables, then it is useful to work in "27-space." Fortunately much of the geometry of 2- and 3-space generalizes to "n-space," and this helps us with problem solving considerably.

After a review of some of the properties of 2- and 3-space that are important to linear algebra, this chapter formalizes the generalization of those properties to n-space. In so doing, this chapter lays the groundwork for the remainder of this text and, indeed, the whole body of linear algebra.

3.1 Vectors in 2- and 3-Spaces*

Many quantities, such as area, amount of money invested, and weight, are described using a single number. Other quantities, such as force, relativity, and change in the stock market, are described using a number and a direction. This type of quantity can be represented by vectors. To describe vectors intuitively, we first give an analogy.

We all know what rational numbers are and that the ratios

$$\frac{2}{3}, \frac{4}{6}, \frac{6}{9}, \frac{-2}{-3}, \frac{10}{15} \cdots$$

* *Readers familiar with this material may go directly to Section 3.2 with no loss of continuity. However, Theorems (3.5), (3.7), and (3.11) and their proofs might be quickly reviewed, as there will be brief references to them in later sections.*

are all *different* ratios that represent the *same* rational number. We usually use the ratio $\frac{2}{3}$ to represent that rational number, as it is the "simplest" of all the ratios. There are occasions, however, when we need a rational number in terms of a different ratio, for example, if we wanted the sum $\frac{2}{3} + \frac{1}{5}$, we would replace $\frac{2}{3}$ with $\frac{10}{15}$, $\frac{1}{5}$ with $\frac{3}{15}$, and so on.

A **vector** is used to represent a number and a direction. To do this, we use an *arrow* or *directed line segment*. For example, the force exerted by the wind on an airplane in Figure 3.1a is represented by the vector **F**.

Figure 3.1

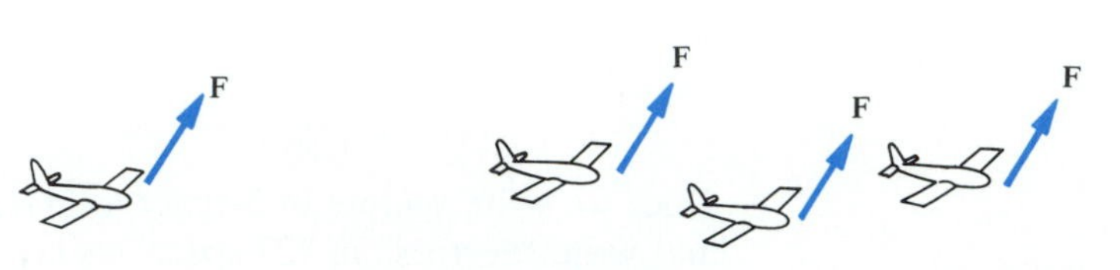

(a) *Force of the wind on an airplane* (b) *Same wind exerts the same force on different airplanes at different places*

However the same wind will exert the same force, that is, push with the same magnitude in the same direction, on different airplanes that are in different (nearby) places. See Figure 3.1b. Thus we want *different arrows* to represent the *same vector* if they point in the same direction and have the same length. In such a situation we say these different arrows **represent** the same vector.

If two (or more) forces act on the same object, then that object is affected in a way illustrated in Figure 3.2a.

Figure 3.2

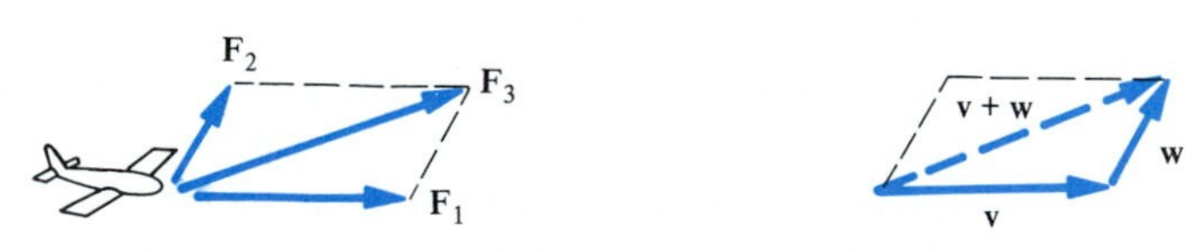

(a) $\mathbf{F}_1$ *is the force due to the motor,* $\mathbf{F}_2$ *is the force due to the wind* (b) *The sum of* **v** *and* **w**

Here $\mathbf{F}_1$ and $\mathbf{F}_2$ are described, but the airplane actually moves as if the force $\mathbf{F}_3$ were acting on it. We think of $\mathbf{F}_3$ as the *resultant* or *sum* of $\mathbf{F}_1$ and $\mathbf{F}_2$. This leads to the definition of vector sum, illustrated in Figure 3.2b. Suppose **v** and **w** are two vectors. Visualize the starting point of **w** moved to the head of **v**. Then the **sum** $\mathbf{v} + \mathbf{w}$ is the vector which is (represented by) the arrow starting at the base of **v** and ending at the tip of **w**.

Example 1 Suppose a man were rowing a boat due north at 2 mph across a stream flowing due east at 2 mph. Then the actual direction and speed of the boat would be northeast at $2\sqrt{2} \approx 2.8$ mph, as illustrated in Figure 3.3.

Figure 3.3

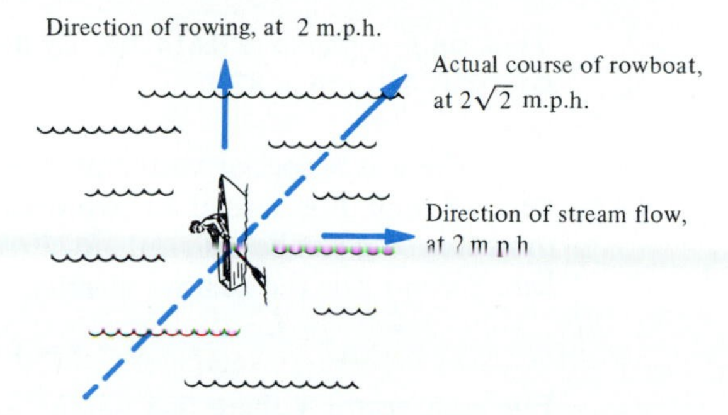

■

Example 2 Suppose a small plane is flying on a heading of 90°. (The **heading** is the direction the plane is pointed and is measured clockwise from due north. Thus the plane is pointed due east.) Suppose there is a 25-mph wind in the direction of 45° (northeast). Assuming the airspeed of the plane is 125 mph (the **airspeed** is the speed relative to the air, *not* the ground), find the ground speed and actual course of the plane.

Solution

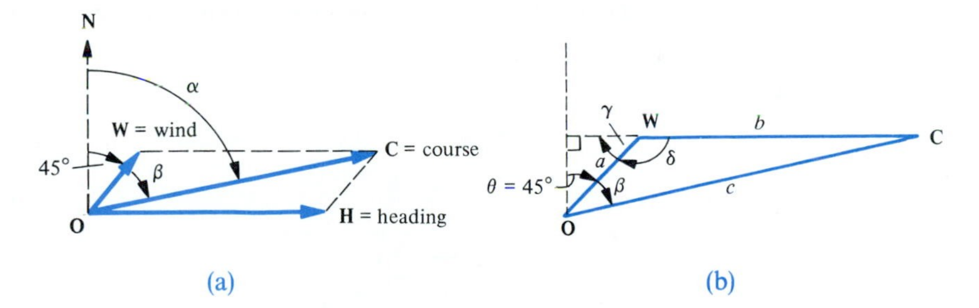

Figure 3.4

Referring to Figure 3.4a, the heading and speed of the plane are represented by **H** and the direction and speed of the wind are represented by **w**. We wish to find the length of the vector **c** and the angle $\alpha = 45° + \beta$, where β is the angle **WOC**. From Figure 3.4a we extract the triangle indicated in Figure 3.4b. Given that angle θ is 45°, $a = 25$, and $b = 125$, we can see that $\gamma = 90° - \theta = 45°$ and $\delta = 180° - \gamma = 135°$.

By the Law of Cosines,

$$\begin{aligned} c^2 &= a^2 + b^2 - 2ab \cos \delta \\ &= 25^2 + 125^2 - 2(25)(125) \cos(135°) \\ &\approx 20669.417 \end{aligned}$$

Thus $c \approx 143.769$, so the ground speed of the plane is approximately 144 mph. To find β, we use the Law of Sines,

$$\frac{\sin \beta}{b} = \frac{\sin \delta}{c}$$

Thus $\sin \beta = \frac{b}{c} \sin \delta \approx 0.6147942$ so $\beta \approx 37.9°$. Therefore, the plane's actual course is $45° + \beta \approx 82.9°$. ■

There is a special vector of zero length called the **zero vector** and is denoted by $\mathbf{0}$. It is special for two reasons. First it has no particular natural direction, so it can be assigned any direction that is convenient for any problem. Second it is the **additive identity**, which means that for any vector $\mathbf{v}$

$$\mathbf{0} + \mathbf{v} = \mathbf{v} = \mathbf{v} + \mathbf{0}$$

For each vector $\mathbf{v}$ there is a unique vector called the **negative** (or **additive inverse**) of $\mathbf{v}$ and denoted by $-\mathbf{v}$. The negative of $\mathbf{v}$ has the same length as $\mathbf{v}$ but points in the opposite direction. It also has the special relationship

$$\mathbf{v} + (-\mathbf{v}) = \mathbf{0} = -\mathbf{v} + \mathbf{v}$$

The negative is used to define **subtraction** by

$$\mathbf{v} - \mathbf{w} = \mathbf{v} + (-\mathbf{w})$$

Thus one way to find $\mathbf{v} - \mathbf{w}$ is to take $\mathbf{w}$, find its negative $-\mathbf{w}$, and then construct the sum $\mathbf{v} + (-\mathbf{w})$, as illustrated in Figure 3.5a.

Figure 3.5

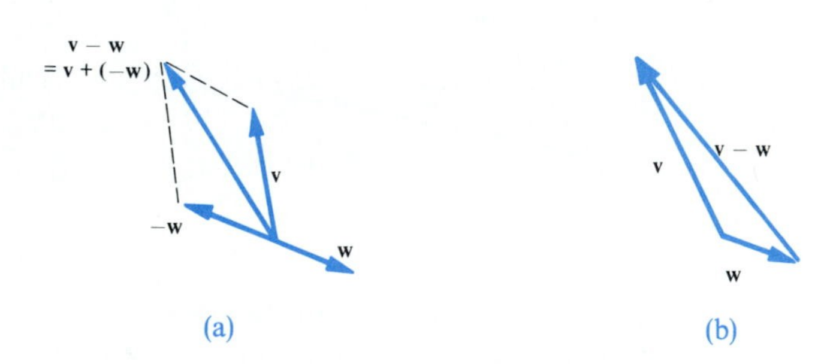

However, it is often easier to construct $\mathbf{v} - \mathbf{w}$ by observing it has the property

$$\mathbf{w} + (\mathbf{v} - \mathbf{w}) = \mathbf{v}$$

Thus to find $\mathbf{v} - \mathbf{w}$, just find the vector that when added to $\mathbf{w}$ gives $\mathbf{v}$. See Figure 3.5b.

If the vectors are in a plane or in 3-space, then we usually take the representative of each vector that starts at the origin. Then the vector is uniquely determined by the point at its arrow tip, and we often identify the vector with that point at its tip, as indicated in Figure 3.6.

Figure 3.6

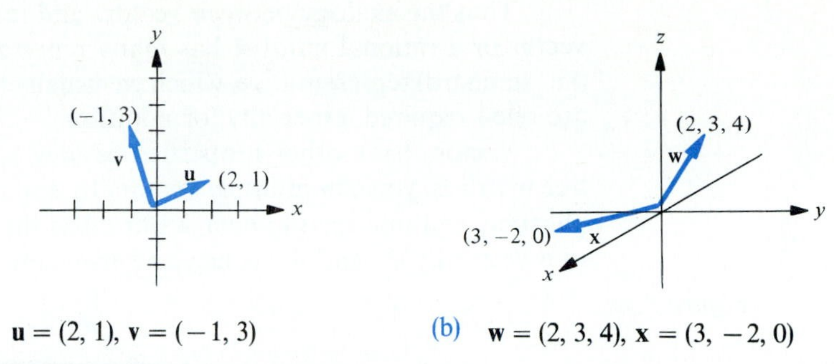

(a) $\mathbf{u} = (2, 1), \mathbf{v} = (-1, 3)$ (b) $\mathbf{w} = (2, 3, 4), \mathbf{x} = (3, -2, 0)$

When this is done, then the coordinates of the point at the tip are called the **coordinates** of the vector. For example, the coordinates of $\mathbf{v}$ in Figure 3.6a are -1 and 3, and the coordinates of $\mathbf{w}$ in Figure 3.6b are 2, 3, and 4. However, while identifying a vector with the point at its tip simplifies descriptions nicely, this identification is frought with danger for the beginning student. It works *only* when the vector starts at the origin. See Figure 3.7.

Figure 3.7

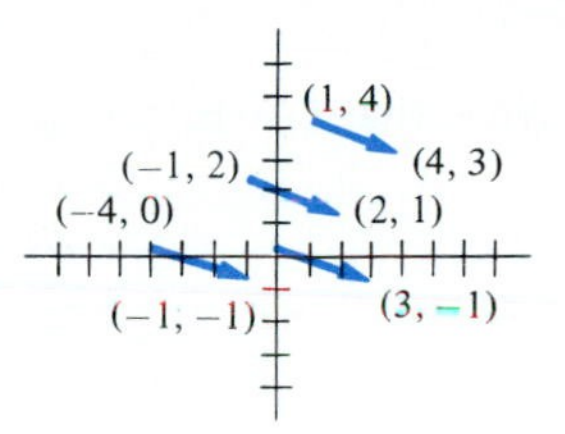

Several representatives of $\mathbf{u} = (3, -1)$

The description of addition is nicely simplified by coordinates. If $\mathbf{v} = (v_1, v_2)$ and $\mathbf{w} = (w_1, w_2)$ then

$$\mathbf{v} + \mathbf{w} = (v_1 + w_1, v_2 + w_2)$$

This is illustrated in Figure 3.8.

Figure 3.8

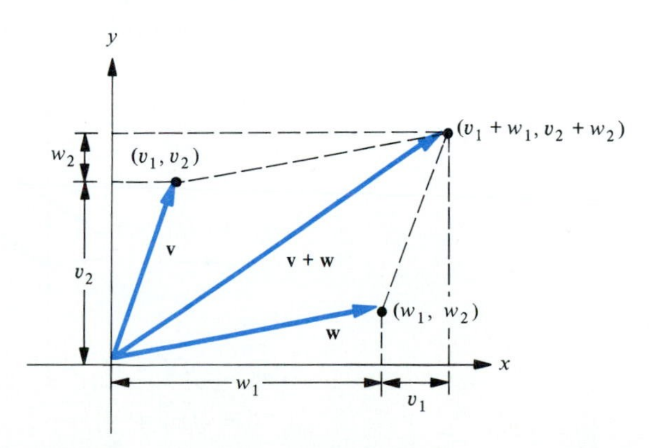

Thus the analogy between vectors and rational numbers is complete: a vector or a rational number has many representatives; there is a canonical (i.e., standard) representative, which we usually take; but other representatives are often required, especially for addition.

Vectors have other properties besides addition and subtraction. Just like matrices, you can multiply vectors by scalars. This is called **scalar multiplication**, and multiplying by a scalar c has the effect of changing the length of a vector by $|c|$, and if c is negative reversing the direction. See Figure 3.9.

Figure 3.9

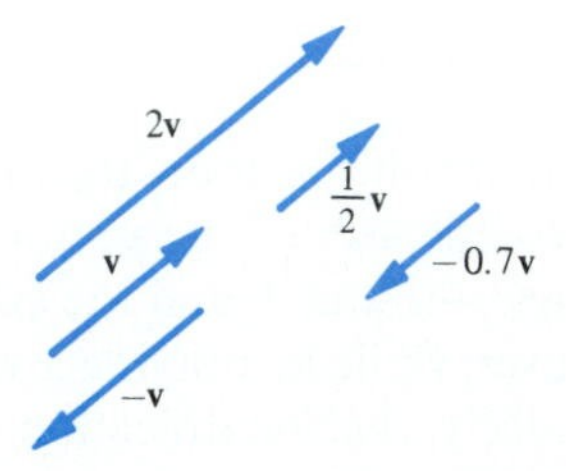

For the coordinate description, if $\mathbf{v} = (a, b, c)$ and r is a scalar, then

$$r\mathbf{v} = r(a, b, c) = (ra, rb, rc)$$

Note that $(-1)\mathbf{v} = -\mathbf{v}$.

DEFINITION Suppose $\mathbf{v}$ goes from the point $\mathbf{Q} = (y_1, y_2, y_3)$ to the point $\mathbf{R} = (z_1, z_2, z_3)$. The **(Euclidean) distance**, d, between these two points is

$$d = \sqrt{(y_1 - z_1)^2 + (y_2 - z_2)^2 + (y_3 - z_3)^2}$$

Thus the **length** or **norm** of $\mathbf{v}$, which we will denote by $\|\mathbf{v}\|$, is given by

$$\|\mathbf{v}\| = \sqrt{(y_1 - z_1)^2 + (y_2 - z_2)^2 + (y_3 - z_3)^2}$$

If, as a special case, we pick the canonical representative of $\mathbf{v}$ that starts at the origin, then, as illustrated in Figure 3.10, $\mathbf{v} = (x_1, x_2, x_3)$ where $x_i =$

Figure 3.10

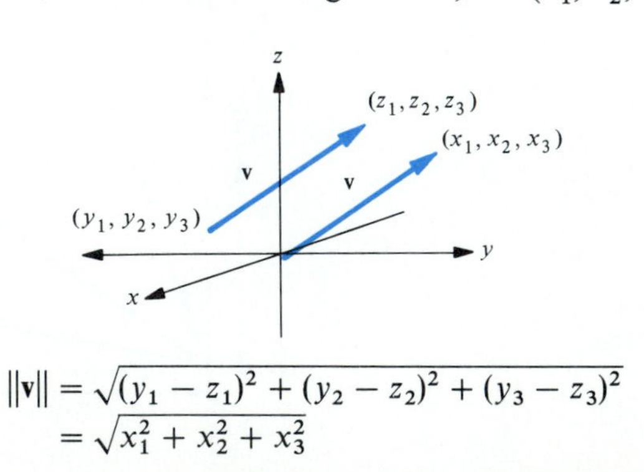

$$\|\mathbf{v}\| = \sqrt{(y_1 - z_1)^2 + (y_2 - z_2)^2 + (y_3 - z_3)^2}$$
$$= \sqrt{x_1^2 + x_2^2 + x_3^2}$$

$y_i - z_i$, for $i = 1, 2, 3$, and

$$\|\mathbf{v}\| = \|(x_1, x_2, x_3)\| = \sqrt{x_1^2 + x_2^2 + x_3^2}$$

Suppose that **v** and **w** are vectors (in 2- or 3-space) that are positioned so that their initial points coincide. The **angle between v and w** is the angle θ determined by **v** and **w** that satisfies $0 \le \theta \le \pi$, as illustrated in Figure 3.11.

Figure 3.11

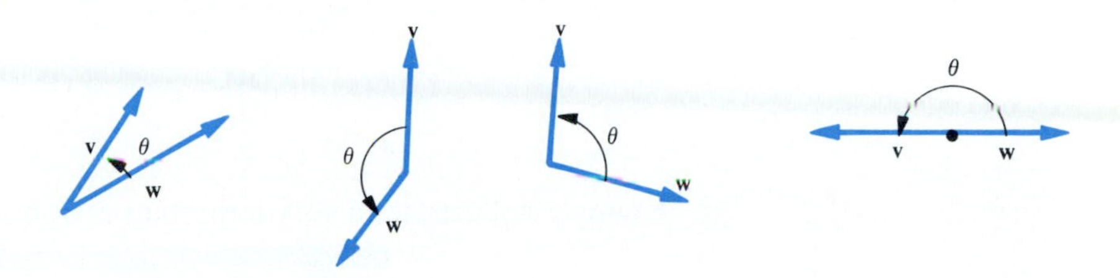

NOTE If **v** or **w** = **0**, we usually take $\theta = 0$.

We are now ready to define a type of product that associates a *number* or *scalar* (and *not* another vector) to pairs of vectors. This product will be *very* useful in helping us with problems that require perpendicular vectors, in finding least-squares fit to data, and in a variety of other problems.

(3.1)

DEFINITION If **u** and **v** are vectors and θ is the angle between them, then their **dot product** (or **scalar product** or **(Euclidean) inner product**) is denoted by $\mathbf{u} \cdot \mathbf{v}$ and is defined by

$$\mathbf{u} \cdot \mathbf{v} = \|\mathbf{u}\| \, \|\mathbf{v}\| \cos \theta$$

Example 3 If $\mathbf{u} = (0, 2, 0)$ and $\mathbf{v} = (0, 1, 1)$, then $\theta = 45°$ and

$$\mathbf{u} \cdot \mathbf{v} = \|\mathbf{u}\| \, \|\mathbf{v}\| \cos \theta = \sqrt{0^2 + 2^2 + 0^2}\sqrt{0^2 + 1^2 + 1^2}\,\frac{1}{\sqrt{2}} = 2$$

See Figure 3.12a. ■

Figure 3.12a

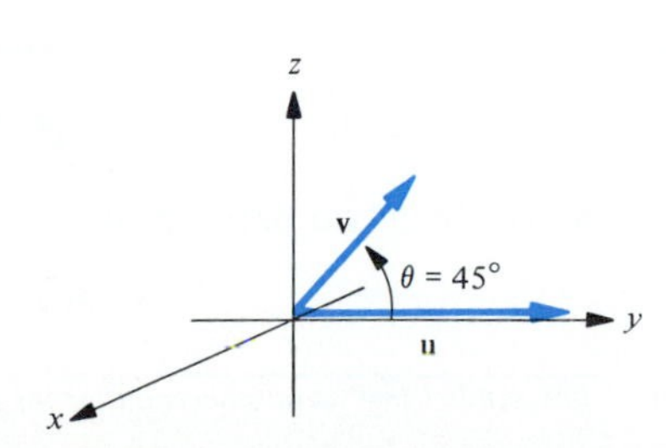

Example 4 If $\mathbf{u} = (1, 0, 1)$ and $\mathbf{v} = (-1, 0, 1)$, then $\theta = 90°$ and

$$\mathbf{u} \cdot \mathbf{v} = \|\mathbf{u}\|\,\|\mathbf{v}\| \cos \theta = \sqrt{2}\sqrt{2}0 = 0$$

See Figure 3.12b. ■

Figure 3.12b

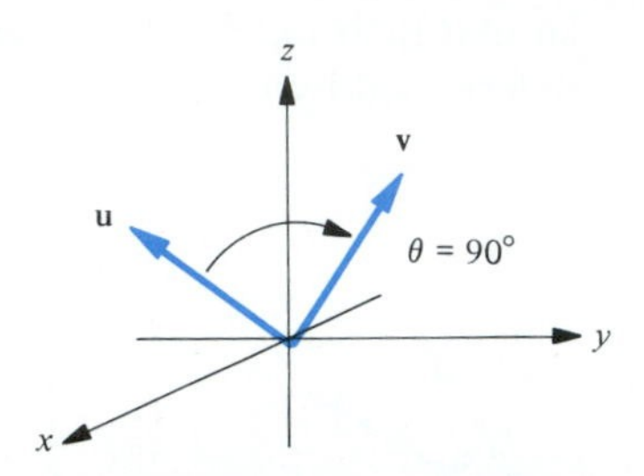

Example 4 illustrates a very important relationship:

(3.2)

> Let $\mathbf{u}$ and $\mathbf{v}$ be two nonzero vectors and let θ be the angle between them.
>
> DEFINITION We say $\mathbf{u}$ and $\mathbf{v}$ are **perpendicular** if θ is $90°$.
>
> THEOREM Nonzero vectors $\mathbf{u}$ and $\mathbf{v}$ are perpendicular
>
> $$\Leftrightarrow \quad \cos \theta = 0$$
> $$\Leftrightarrow \quad \mathbf{u} \cdot \mathbf{v} = 0*$$

Proof The last step follows since the only way $\mathbf{u} \cdot \mathbf{v} = \|\mathbf{u}\|\,\|\mathbf{v}\| \cos \theta$ can be zero is for $\cos \theta = 0$, *if* $\mathbf{u}$ and $\mathbf{v}$ are both nonzero. ■

It is natural to ask what we do if $\mathbf{u}$ or $\mathbf{v}$ is the zero vector; it is the usual convention to say the zero vector is perpendicular to every vector. Thus altogether,

(3.3)

> THEOREM For all vectors $\mathbf{u}$ and $\mathbf{v}$,
>
> $$\mathbf{u} \cdot \mathbf{v} = 0 \quad \Leftrightarrow \quad \mathbf{u} \text{ and } \mathbf{v} \text{ are perpendicular}$$

Another important result is

(3.4)

> THEOREM For any vector $\mathbf{u}$,
>
> $$\mathbf{u} \cdot \mathbf{u} = \|\mathbf{u}\|^2 \quad \text{or equivalently} \quad \|\mathbf{u}\| = \sqrt{\mathbf{u} \cdot \mathbf{u}}$$

Proof Since the angle between $\mathbf{u}$ and itself is $0°$ and $\cos 0° = 1$, we have

$$\mathbf{u} \cdot \mathbf{u} = \|\mathbf{u}\|\,\|\mathbf{u}\| \cos 0° = \|\mathbf{u}\|^2$$ ■

* *The symbol "⇔" is an abbreviation for the phrase "if and only if."*

In most cases, however, it is difficult to find the dot product of two vectors directly from the definition (3.1), since θ, and hence $\cos\theta$, is often hard to compute. Fortunately there is a very easy formula.

(3.5) **THEOREM** If $\mathbf{u} = (u_1, u_2, u_3)$ and $\mathbf{v} = (v_1, v_2, v_3)$, then

$$\mathbf{u} \cdot \mathbf{v} = u_1v_1 + u_2v_2 + u_3v_3$$

Of course, if $\mathbf{u} = (u_1, u_2)$, $\mathbf{v} = (v_1, v_2)$, then $\mathbf{u} \cdot \mathbf{v} = u_1v_1 + u_2v_2$. We first give two brief examples and then prove the theorem.

Example 5 From Example 3, if $\mathbf{u} = (0, 2, 0)$ and $\mathbf{v} = (0, 1, 1)$, then

$$\mathbf{u} \cdot \mathbf{v} = 0(0) + 2(1) + 0(1) = 2$$

■

Example 6 From Example 4, if $\mathbf{u} = (1, 0, 1)$ and $\mathbf{v} = (-1, 0, 1)$ then

$$\mathbf{u} \cdot \mathbf{v} = 1(-1) + 0(0) + 1(1) = -1 + 0 + 1 = 0$$

■

Proof of Theorem (3.5)

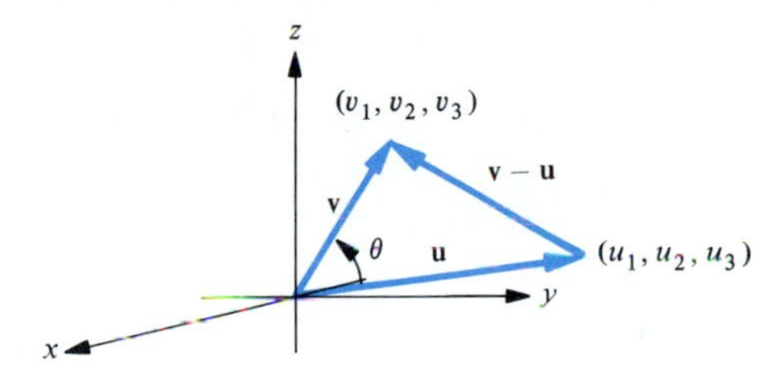

Figure 3.13

Let θ be the angle between $\mathbf{u}$ and $\mathbf{v}$. Draw in the vector $\mathbf{v} - \mathbf{u}$ (see Figure 3.13). By the Law of Cosines

$$\|\mathbf{v} - \mathbf{u}\|^2 = \|\mathbf{v}\|^2 + \|\mathbf{u}\|^2 - 2\|\mathbf{u}\|\,\|\mathbf{v}\| \cos\theta$$

We rewrite this as (3.1),

$$\|\mathbf{u}\|\,\|\mathbf{v}\| \cos\theta = \tfrac{1}{2}(\|\mathbf{v}\|^2 + \|\mathbf{u}\|^2 - \|\mathbf{v} - \mathbf{u}\|^2)$$

or, by Definition (3.1),

$$\mathbf{u} \cdot \mathbf{v} = \tfrac{1}{2}(\|\mathbf{v}\|^2 + \|\mathbf{u}\|^2 - \|\mathbf{v} - \mathbf{u}\|^2) \tag{3.6}$$

Using $\mathbf{v} - \mathbf{u} = (v_1 - u_1, v_2 - u_2, v_3 - u_3)$, we see

$$\begin{aligned}\|\mathbf{v} - \mathbf{u}\|^2 &= (v_1 - u_1)^2 + (v_2 - u_2)^2 + (v_3 - u_3)^2 \\ &= v_1^2 - 2v_1u_1 + u_1^2 + v_2^2 - 2v_2u_2 + u_2^2 + v_3^2 - 2v_3u_3 + u_3^2 \\ &= v_1^2 + v_2^2 + v_3^2 + u_1^2 + u_2^2 + u_3^2 - 2(u_1v_1 + u_2v_2 + u_3v_3)\end{aligned}$$

or

$$\|\mathbf{v} - \mathbf{u}\|^2 = \|\mathbf{v}\|^2 + \|\mathbf{u}\|^2 - 2(u_1v_1 + u_2v_2 + u_3v_3)$$

Substituting the last equation into (3.6) and simplifying, we obtain

$$\mathbf{u} \cdot \mathbf{v} = u_1v_1 + u_2v_2 + u_3v_3$$ ■

Example 7 A rectangular solid has height 1 and a square base where each edge has length 2. Find the angle between a vertical edge and a diagonal.

Solution

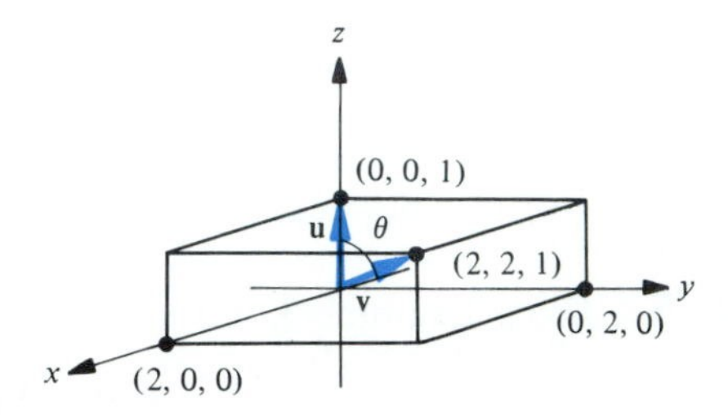

Figure 3.14

Set up the problem in 3-space as indicated in Figure 3.14. Then the vector $\mathbf{u} = (0, 0, 1)$ represents a vertical edge and $\mathbf{v} = (2, 2, 1)$ represents a diagonal. We wish to find θ. Now

$$\mathbf{u} \cdot \mathbf{v} = 0(2) + 0(2) + 1(1) = 1, \qquad \|\mathbf{u}\| = 1, \qquad \|\mathbf{v}\| = 3$$

Using

$$\mathbf{u} \cdot \mathbf{v} = \|\mathbf{u}\| \, \|\mathbf{v}\| \cos \theta$$

we find

$$\cos \theta = \frac{1}{1(3)} = \frac{1}{3}$$

Thus

$$\theta = \cos^{-1}\left(\frac{1}{3}\right) \approx 70.53^\circ$$ ■

Using the dot product formula given in (3.5), it is easy to verify the following properties.

(3.7)

> THEOREM If $\mathbf{u}$, $\mathbf{v}$, and $\mathbf{w}$ are vectors and r is a scalar, then
>
> (a) $\mathbf{u} \cdot \mathbf{v} = \mathbf{v} \cdot \mathbf{u}$
> (b) $\mathbf{u} \cdot (\mathbf{v} + \mathbf{w}) = \mathbf{u} \cdot \mathbf{v} + \mathbf{u} \cdot \mathbf{w}$
> (c) $r(\mathbf{u} \cdot \mathbf{v}) = (r\mathbf{u}) \cdot \mathbf{v} = \mathbf{u} \cdot (r\mathbf{v})$

The verification is left to the exercises. (See Exercises 51–53.)

Before continuing, we note that a vector of length 1 is called a **unit vector**. In particular, if $\mathbf{v} \neq \mathbf{0}$, then $\dfrac{\mathbf{v}}{\|\mathbf{v}\|}$ is a unit vector in the same direction as $\mathbf{v}$.

Projections

Vector projections arise in many everyday situations. For example, suppose you are pushing a lawnmower through the grass with the handle at a 40° angle with the ground. Whatever force **F** you push along the handle, the projection of **F** parallel to the ground $\mathbf{F}_P$ will make the lawnmower move (see Figure 3.15a). If you push the lawnmower uphill, then you must also push against a projection $\mathbf{W}_P$ of the weight **W** of the lawnmower. (See Figure 3.15b.) Of course, if the handle is perpendicular to the ground, then $\mathbf{F}_P = \mathbf{0}$ (the zero vector) and the lawnmower does not move (although the handle could break).

Figure 3.15

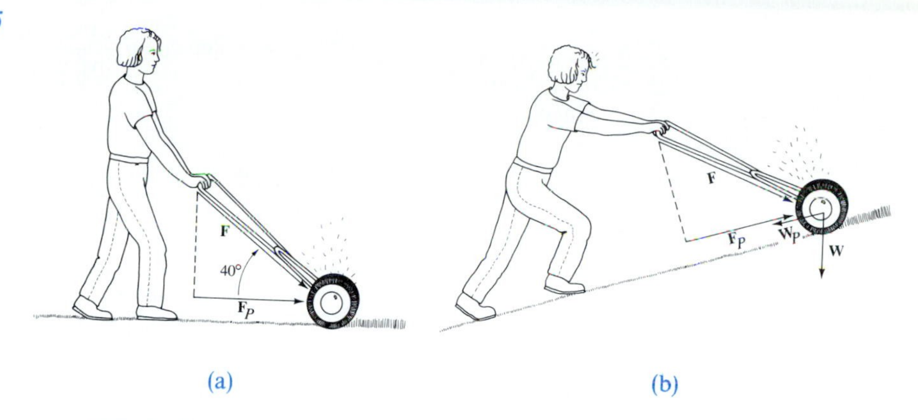

(a) (b)

We now describe the projection of one vector **u** on another vector **v**. If **v** is the zero vector, then the projection of **u** on **v** is just the zero vector. So suppose $\mathbf{v} \neq \mathbf{0}$, **u** is another vector, θ is the angle between **u** and **v**, and we wish to find the projection, **w**, of **u** on **v**. Two possibilities for **u** are pictured in Figure 3.16.

Figure 3.16

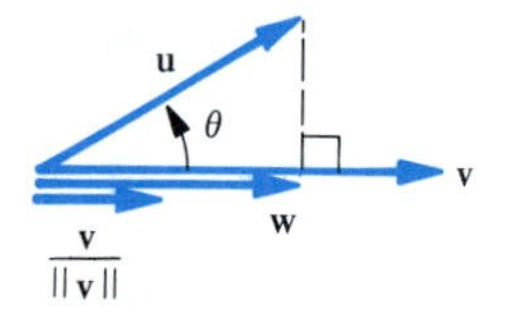

(a) *The projection is in the same direction as* **v**, *so* $\cos\theta > 0$.

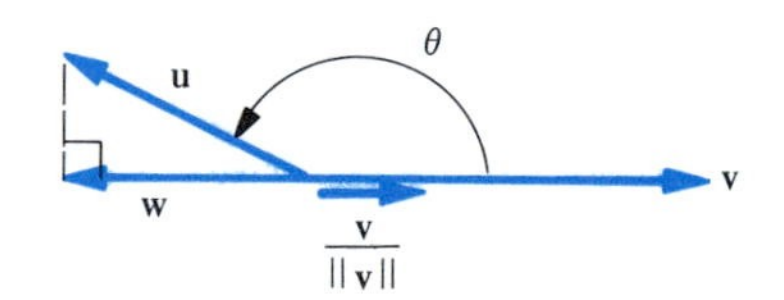

(b) *The projection is in the opposite direction of* **v**, *so* $\cos\theta < 0$

Observe first that $\left(\text{since } \dfrac{\mathbf{v}}{\|\mathbf{v}\|} \text{ is a } \textit{unit} \text{ vector}\right)$

(3.8a) $$\mathbf{w} = \|\mathbf{w}\| \frac{\mathbf{v}}{\|\mathbf{v}\|}, \qquad \text{if } \mathbf{w} \text{ points in the same direction as } \mathbf{v}$$

(3.8b) $$\mathbf{w} = -\|\mathbf{w}\| \frac{\mathbf{v}}{\|\mathbf{v}\|}, \qquad \text{if } \mathbf{w} \text{ points in the opposite direction of } \mathbf{v}$$

Next, as can be seen from the right triangle in Figure 3.16,

(3.9) $$\|\mathbf{w}\| = \|\mathbf{u}\|\,|\cos\theta|$$

However,

(3.10) **w** points in the same or opposite direction as **v** depending on whether

$$\cos\theta > 0 \qquad \text{or} \qquad \cos\theta < 0$$

Thus, putting Equations (3.8)–(3.10) together, we have

$$\mathbf{w} = \|\mathbf{u}\| \cos\theta \frac{\mathbf{v}}{\|\mathbf{v}\|}$$

Since

$$\mathbf{u}\cdot\mathbf{v} = \|\mathbf{u}\|\,\|\mathbf{v}\| \cos\theta \qquad \text{or} \qquad \|\mathbf{u}\| \cos\theta = \frac{\mathbf{u}\cdot\mathbf{v}}{\|\mathbf{v}\|}$$

we have

$$\mathbf{w} = \frac{\mathbf{u}\cdot\mathbf{v}}{\|\mathbf{v}\|}\frac{\mathbf{v}}{\|\mathbf{v}\|} = \frac{\mathbf{u}\cdot\mathbf{v}}{\|\mathbf{v}\|^2}\mathbf{v} = \frac{\mathbf{u}\cdot\mathbf{v}}{\mathbf{v}\cdot\mathbf{v}}\mathbf{v}$$

If **u** is perpendicular to **v**, the last formula also works. Altogether we have

(3.11) THEOREM If **u** is any vector and $\mathbf{v} \neq \mathbf{0}$, then the projection of **u** on **v** is given by

$$\frac{\mathbf{u}\cdot\mathbf{v}}{\|\mathbf{v}\|^2}\mathbf{v} = \frac{\mathbf{u}\cdot\mathbf{v}}{\mathbf{v}\cdot\mathbf{v}}\mathbf{v}$$

Example 8 Find the projection, **w**, of $\mathbf{u} = (2, 1, -3)$ on $\mathbf{v} = (-4, 1, -3)$.

Solution We find

$$\mathbf{u}\cdot\mathbf{v} = 2(-4) + 1(1) + (-3)(-3) = 2$$

and

$$\|\mathbf{v}\|^2 = (-4)^2 + 1^2 + (-3)^2 = 26$$

Then

$$\mathbf{w} = \frac{\mathbf{u}\cdot\mathbf{v}}{\|\mathbf{v}\|^2}\mathbf{v} = \frac{2}{26}\mathbf{v} = \left(\frac{-4}{13}, \frac{1}{13}, \frac{-3}{13}\right)$$

■

Exercise 3.1

1. A man can row his boat at 3 mph. He is in a stream flowing from north to south at 4 mph. How fast and in what direction will he travel if he rows:

   (a) Directly downstream?
   (b) Directly upstream?
   (c) Due east (across the stream)?
   (d) Northeast (45° on a navigator's compass)?
   (e) Southeast (135°)?

2. A duck is flying directly north at 24 km/h (kilometers per hour). How fast and in what direction will the bird actually travel if the wind is blowing at 30 km/h

   (a) Directly north?
   (b) Directly south?
   (c) From the west?
   (d) Toward the southeast (135°)?
   (e) Toward the northwest (315°)?

3. Assume that a portion of the Mississippi River is one mile wide and that the current is 6 mph. Points A and B are directly opposite each other on the banks. If a man starts from A rowing at 4 mph and keeps his boat parallel to AB,

   (a) What is his actual velocity?
   (b) How far downriver from B will he land?
   (c) How long will it take him to land?
   (d) How long would it take him to land if there were no current?

4. An airplane flies over the northwest corner of Colorado heading due south at 200 mph with a crosswind of 25 mph blowing directly east. If it is 275 miles between the northern and southern borders of Colorado,

   (a) What is the actual velocity of the plane?
   (b) How far east of the southwestern corner of Colorado will the plane cross the border?
   (c) How long will it take the plane to cross the southern border of Colorado?
   (d) How long would it take the plane to reach the southern border if there were no wind?

5. A plane is flying at an airspeed of 400 mph with a heading of 328°. There is a wind of 40 mph in the direction of 243°. Find the course and ground speed of the plane.

6. A ship is sailing at 21 km/h on a heading of 107° through a crosscurrent of 3 km/h in the direction of 21°. Find the actual speed and direction of the ship.

7. There is a 50 km/h wind in the direction of 36°. A pilot wishes to fly on a course of 121° with a ground speed of 400 km/h. Determine her heading and airspeed.

8. Two people want to take their motorboat to an island 20 miles away on a heading of 349°. If there is a 2-mph crosscurrent in the direction of 40°, in what direction and at what speed should they head if they want to take two hours for the trip?

In Exercises 9–14, express the given vector in terms of its coordinates.

9. The vector from (2, 3) to (1, 4).

10. The vector from $(4, 7, -1)$ to $(8, -1, -8)$.

11. The vector from $(5, -11, 2)$ to $(0, 0, 0)$.

12. The vector from $(-6, 1, 2)$ to $(-6, 1, 2)$.

13. The vector from the origin to the end point of the vector from $(-3, 7, 2)$ in the direction and with the length of $\mathbf{u} = (2, -3, 4)$.

14. The vector from the origin to the initial point of the vector to $(6, 4, -3)$ in the opposite direction and with twice the length of $(-2, 3, 4)$.

In Exercises 15–18, $\mathbf{u} = (3, -1, 2)$, $\mathbf{v} = (4, -1, 5)$, $\mathbf{w} = (8, -7, -6)$. Find:

15. $\mathbf{u} - \mathbf{v}$

16. $2\mathbf{v} + 3\mathbf{w}$

17. $4(2\mathbf{u} - \mathbf{w})$

18. $3\mathbf{v} - 4(5\mathbf{u} - 6\mathbf{w})$

In Exercises 19–22, find the distance between the given points.

19. (2, 3), (4, 1)

20. $(-3, -4)$, $(5, 2)$

21. $(3, 1, -2)$, $(4, -5, 4)$

22. $(8, 1, 3)$, $(-4, -2, -6)$

In Exercises 23–26, find the norm.

23. $\|(3, -4)\|$

24. $\|2\mathbf{v} - \mathbf{w}\|$, $\mathbf{v} = (-3, 1, 2)$, $\mathbf{w} = (4, 1, 4)$

25. $\|-3\mathbf{u}\| + 3\|\mathbf{u}\|$, for $\mathbf{u} = (-2, 1, 3)$

26. $\|\mathbf{u}\| + \|\mathbf{v}\|$ and $\|\mathbf{u} + \mathbf{v}\|$ for $\mathbf{u} = (3, -1, 2)$, $\mathbf{v} = (1, 1, 1)$

27. If $\mathbf{w} \neq 0$, what is $\left\| \frac{1}{\|\mathbf{w}\|} \mathbf{w} \right\|$? Why?

28. Find all numbers r such that $\|r(2, -1, -2\| = 6$.

In Exercises 29–34, find $\mathbf{u} \cdot \mathbf{v}$ and then the angle, θ, between $\mathbf{u}$ and $\mathbf{v}$.

29. $\mathbf{u} = \mathbf{v} = (2, -2, 1)$

30. $\mathbf{u} = (-2, 1)$, $\mathbf{v} = -\mathbf{u}$

31. $\mathbf{u} = (1, 1, 1)$, $\mathbf{v} = (2, -1, -1)$

32. $\mathbf{u} = (3, -1, 2)$, $\mathbf{v} = (1, 1, -1)$

33. $\mathbf{u} = (2, 4, -1)$, $\mathbf{v} = (3, 1, 3)$

34. $\mathbf{u} = (-1, -2, 3)$, $\mathbf{v} = (4, 1, 1)$

In Exercises 35–40, compute the projection of **u** on **v**.

35. $\mathbf{u} = \mathbf{0}$, $\mathbf{v} = (1, 2)$

36. $\mathbf{u} = (1, 1)$, $\mathbf{v} = (1, -1)$

37. $\mathbf{u} = (3, 8)$, $\mathbf{v} = (-2, 4)$

38. $\mathbf{u} = (4, 1, 2)$, $\mathbf{v} = (-1, 2, 3)$

39. $\mathbf{u} = (8, -1, -2)$, $\mathbf{v} = (3, 4, -1)$

40. $\mathbf{u} = (3, 1, 3)$, $\mathbf{v} = (4, 1, -2)$

In Exercises 41–44, $\mathbf{u} = (3, -2)$, $\mathbf{v} = (2, 5)$, $\mathbf{w} = (4, -1)$. Find:

41. $2\mathbf{u} \cdot (3\mathbf{v} - 4\mathbf{w})$.

42. $\|(\mathbf{u} \cdot \mathbf{v})\mathbf{w}\|$.

43. $\|\mathbf{u}\|(\mathbf{v} \cdot \mathbf{w})$

44. $(\|\mathbf{u}\|\mathbf{v}) \cdot \mathbf{w}$

In Exercises 45–48, explain why the given expression makes no sense. Here **u**, **v**, and **w** are vectors in $\mathbb{R}^3$ and r is a scalar.

45. $(\mathbf{u} \cdot \mathbf{v}) \cdot \mathbf{w}$

46. $\mathbf{u} + (\mathbf{v} \cdot \mathbf{w})$

47. $\|\mathbf{u} \cdot \mathbf{v}\|$

48. $r \cdot (\mathbf{u} + \mathbf{v})$

49. Verify the identity

$$\|\mathbf{u} + \mathbf{v}\|^2 + \|\mathbf{u} - \mathbf{v}\|^2 = 2\|\mathbf{u}\|^2 + 2\|\mathbf{v}\|^2$$

[HINT Use Theorem (3.4).]

50. Verify the identity

$$\mathbf{u} \cdot \mathbf{v} = \tfrac{1}{4}\|\mathbf{u} + \mathbf{v}\|^2 - \tfrac{1}{4}\|\mathbf{u} - \mathbf{v}\|^2$$

Same hint as in Exercise 49.

In Exercises 51–53, let $\mathbf{u} = (u_1, u_2, u_3)$, $\mathbf{v} = (v_1, v_2, v_3)$, and $\mathbf{w} = (w_1, w_2, w_3)$. Let r be a number, and use Theorem (3.5).

51. Verify Theorem (3.7a) $\mathbf{u} \cdot \mathbf{v} = \mathbf{v} \cdot \mathbf{u}$.

52. Verify Theorem (3.7b) $\mathbf{u} \cdot (\mathbf{v} + \mathbf{w}) = \mathbf{u} \cdot \mathbf{v} + \mathbf{u} \cdot \mathbf{w}$.

53. Verify Theorem (3.7c) $r(\mathbf{u} \cdot \mathbf{v}) = (r\mathbf{u}) \cdot \mathbf{v} = \mathbf{u} \cdot (r\mathbf{v})$.

3.2 Euclidean *n*-Space

The introduction to this chapter suggests that it is easy to generalize from 2- and 3-space to *n*-space. Historically, however, it was not that easy. Using ordered pairs and triples to locate points in 2- and 3-space was well understood in the 17th century, but it was not until the 19th century that mathematicians and physicists readily used larger tuples. Now, standing on their shoulders with a century of evolution, we can describe that transition easily.

(3.12)

DEFINITION Suppose n is a positive integer. An **ordered n-tuple** of real numbers is an ordered set of n real numbers and is denoted by $(x_1, x_2, \ldots, x_n)$. The set of all ordered n-tuples of real numbers is called (real) **n-space** and is denoted by $\mathbb{R}^n$.

If $\mathbf{x} = (x_1, \ldots, x_n)$ and $\mathbf{y} = (y_1, \ldots, y_n)$ are in $\mathbb{R}^n$, then $\mathbf{x}$ is **equal** to $\mathbf{y}$ if and only if $x_1 = y_1, \ldots, x_n = y_n$. (This follows by what we mean when we say two ordered sets are equal.) As in 2- and 3-space, the numbers $x_1, \ldots, x_n$ are called the **coordinates** of $\mathbf{x}$.

As before, we will call ordered 2-tuples "ordered pairs" and ordered 3-tuples "ordered triples." If $n = 1$, we identify the 1-tuple (a) with the real number a and we usually write $\mathbb{R}$ instead of $\mathbb{R}^1$.

We will often call a point $\mathbf{x} = (x_1, x_2, \ldots, x_n)$ a **vector** in $\mathbb{R}^n$. The **origin** is the **zero vector**, $\mathbf{0} = (0, 0, \ldots, 0)$. Even when $n > 3$, we pretend we visualize directed line segments or arrows in $\mathbb{R}^n$. As in the previous sections, we identify the point $\mathbf{x} = (x_1, \ldots, x_n)$ with the arrow starting at the origin and ending at the point $\mathbf{x} = (x_1, \ldots, x_n)$. In this way we can define the **angle** θ between $\mathbf{x} = (x_1, \ldots, x_n)$ and any $\mathbf{y} = (y_1, \ldots, y_n)$ in the usual way, using the "plane" in $\mathbb{R}^n$ containing those two vectors.

The algebra of 2- and 3-space extends to n-space quite naturally.

(3.13)

DEFINITION If $\mathbf{x} = (x_1, \ldots, x_n)$ and $\mathbf{y} = (y_1, \ldots, y_n)$ are vectors in $\mathbb{R}^n$ and r is a real number (or scalar), then

The **sum** $\mathbf{x} + \mathbf{y}$ is defined by

$$\mathbf{x} + \mathbf{y} = (x_1 + y_1, \ldots, x_n + y_n)$$

The **scalar product** $r\mathbf{x}$ is defined by

$$r\mathbf{x} = (rx_1, \ldots, rx_n)$$

The **negative** (or **additive inverse**) of $\mathbf{x}$, $-\mathbf{x}$, is

$$-\mathbf{x} = (-x_1, \ldots, -x_n)$$

subtraction, $\mathbf{x} - \mathbf{y}$, is defined by

$$\mathbf{x} - \mathbf{y} = \mathbf{x} + (-\mathbf{y})$$

Example 1 Let $\mathbf{x} = (2, 1, -3, 0, 4)$ and $\mathbf{y} = (-2, -5, 7, 4, 11)$ be in $\mathbb{R}^5$. Then

(a) $\mathbf{x} + \mathbf{y} = (0, -4, 4, 4, 15)$.
(b) $3\mathbf{x} = (6, 3, -9, 0, 12)$.
(c) $-\mathbf{x} = (-2, -1, 3, 0, -4)$.
(d) $\mathbf{x} - \mathbf{y} = (4, 6, -10, -4, -7)$. ■

Altogether, addition and scalar multiplication satisfy the following very important properties.

(3.14)

THEOREM If $\mathbf{u}$, $\mathbf{v}$, and $\mathbf{w}$ are vectors in $\mathbb{R}^n$ and r and s are numbers (scalars) then:

(a) $\mathbf{u} + \mathbf{v} = \mathbf{v} + \mathbf{u}$ (commutative law of addition)
(b) $(\mathbf{u} + \mathbf{v}) + \mathbf{w} = \mathbf{u} + (\mathbf{v} + \mathbf{w})$ (associative law of addition)
(c) $\mathbf{u} + \mathbf{0} = \mathbf{u}$ (additive identity)
(d) $\mathbf{u} + (-\mathbf{u}) = \mathbf{0}$ (additive inverse)
(e) $r(\mathbf{u} + \mathbf{v}) = r\mathbf{u} + r\mathbf{v}$ (distributive law)
(f) $(r + s)\mathbf{u} = r\mathbf{u} + s\mathbf{u}$ (distributive law)
(g) $(rs)\mathbf{u} = r(s\mathbf{u})$ (associative law)
(h) $1\mathbf{u} = \mathbf{u}$ (multiplicative identity)

The proof is left to the exercises. See Exercises 38 and 39.

In the next section, we shall define a vector space to be set on which addition and scalar multiplication are defined so that all properties of Theorem (3.14) are satisfied. Vector spaces arise very naturally in a variety of situations with linear problems we wish to solve. Hence we shall be referring to these properties in several different contexts. In fact we have already shown in Section 1.3 that addition and scalar multiplication of $m \times n$ matrix satisfy these properties.

Dot Product, Norm, and Length

The concepts of dot product, norm or length, and distance also extend to n-space quite naturally.

(3.15)

DEFINITION If $\mathbf{x} = (x_1, \ldots, x_n)$ and $\mathbf{y} = (y_1, \ldots, y_n)$ are vectors in $\mathbb{R}^n$, then the **dot product** (or **(Euclidean) inner product**) of $\mathbf{x}$ and $\mathbf{y}$, $\mathbf{x} \cdot \mathbf{y}$, is defined by

$$\mathbf{x} \cdot \mathbf{y} = x_1y_1 + x_2y_2 + \cdots + x_ny_n$$

The **(Euclidean) norm** or **length** of $\mathbf{x}$, $\|\mathbf{x}\|$, is defined by

$$\|\mathbf{x}\| = \sqrt{x_1^2 + x_2^2 + \cdots + x_n^2}$$

The **(Euclidean) distance** between $\mathbf{x}$ and $\mathbf{y}$, $d = d(\mathbf{x}, \mathbf{y})$ is defined by

$$d(\mathbf{x}, \mathbf{y}) = \sqrt{(x_1 - y_1)^2 + (x_2 - y_2)^2 + \cdots + (x_n - y_n)^2}$$

By **Euclidean *n*-space** we mean $\mathbb{R}^n$, together with these definitions of distance and inner product.

We can see immediately that

$$\|\mathbf{x}\| = \sqrt{\mathbf{x} \cdot \mathbf{x}} = d(\mathbf{x}, \mathbf{0}) \quad \text{and} \quad d(\mathbf{x}, \mathbf{y}) = \|\mathbf{x} - \mathbf{y}\| \tag{3.16}$$

are true. See Exercises 32 and 33.

Example 2 Let $\mathbf{u} = (2, -1, 8, 7, 3)$ and $\mathbf{v} = (-1, -8, -5, 1, 2)$, vectors in $\mathbb{R}^5$. Compute

(a) $\mathbf{u} \cdot \mathbf{v}$, (b) $\|\mathbf{u}\|$ and $\|\mathbf{v}\|$, (c) $d(\mathbf{u}, \mathbf{v})$

Solution (a) $\mathbf{u} \cdot \mathbf{v} = 2(-1) + (-1)(-8) + 8(-5) + 7(1) + 3(2) = -21$

(b) $\|\mathbf{u}\| = \sqrt{2^2 + (-1)^2 + 8^2 + 7^2 + 3^2} = \sqrt{127} \approx 11.2694$

$\|\mathbf{v}\| = \sqrt{(-1)^2 + (-8)^2 + (-5)^2 + 1^2 + 2^2} = \sqrt{95} \approx 9.74679$

(c) $d(\mathbf{u}, \mathbf{v}) = \sqrt{(2-(-1))^2 + (-1-(-8))^2 + (8-(-5))^2 + (7-1)^2 + (3-2)^2}$

$= \sqrt{3^2 + 7^2 + 13^2 + 6^2 + 1^2} = \sqrt{264} \approx 16.2481$

■

As mentioned earlier, if $\mathbf{x}$ and $\mathbf{y}$ are vectors in $\mathbb{R}^n$, then we can define the angle between them in the usual way using the plane in $\mathbb{R}^n$ determined by $\mathbf{x}$ and $\mathbf{y}$. We then get

(3.17)

THEOREM If $\mathbf{x}$ and $\mathbf{y}$ are vectors in $\mathbb{R}^n$ and θ is the angle between them, then

$$\mathbf{x} \cdot \mathbf{y} = \|\mathbf{x}\| \, \|\mathbf{y}\| \cos \theta$$

The proof is essentially the same as the proof of Theorem (3.5), Section 3.1, just appropriately modified by taking vectors in $\mathbb{R}^n$, not $\mathbb{R}^3$.

Example 3 Find the angle between the vectors **x** and **y** of Example 2.

Solution Using the computations of Example 2, parts (a) and (b),

$$\cos\theta = \frac{\mathbf{u}\cdot\mathbf{v}}{\|\mathbf{u}\|\,\|\mathbf{v}\|} = \frac{-21}{\sqrt{127}\sqrt{95}}$$

Thus

$$\theta = \cos^{-1}\left(\frac{-21}{\sqrt{127}\sqrt{95}}\right) \approx 101.02^\circ$$ ■

It is often very useful to represent a point in n-space, $\mathbf{x} = (x_1, \ldots, x_n)$, as an $n \times 1$ matrix or a $1 \times n$ matrix:

$$\mathbf{x} = \begin{bmatrix} x_1 \\ \vdots \\ x_n \end{bmatrix} \qquad \text{or} \qquad \mathbf{x} = [x_1 \quad \cdots \quad x_n]$$

This is very natural since addition of such matrices

$$\mathbf{x} + \mathbf{y} = \begin{bmatrix} x_1 \\ \vdots \\ x_n \end{bmatrix} + \begin{bmatrix} y_1 \\ \vdots \\ y_n \end{bmatrix} = \begin{bmatrix} x_1 + y_1 \\ \vdots \\ x_n + y_n \end{bmatrix}$$

$$\mathbf{x} + \mathbf{y} = [x_1 \quad \cdots \quad x_n] + [y_1 \quad \cdots \quad y_n] = [x_1 + y_1 \quad \cdots \quad x_n + y_n]$$

corresponds to addition of tuples

$$\mathbf{x} + \mathbf{y} = (x_1, \ldots, x_n) + (y_1, \ldots, y_n) = (x_1 + y_1, \ldots, x_n + y_n)$$

The corresponding statement for scalar multiplication is also true. The only real difference with these three representatives is how the components are displayed. Hence hereafter we shall use these representations interchangeably.

One noticeable difference is that from now on $n \times 1$ matrices and $1 \times n$ matrices will often be considered as vectors in $\mathbb{R}^n$ and hence be denoted by lowercase boldface letters. For example, a system of linear equations will be denoted by

$$A\mathbf{x} = \mathbf{b} \qquad \text{instead of} \qquad AX = B$$

as before.

Projections

We illustrate this new notation with projections in $\mathbb{R}^n$. Suppose **u** and **v** are two vectors in $\mathbb{R}^n$, $\mathbf{v} \neq \mathbf{0}$, and **u** is not a multiple of **v**. Then the whole discussion on projections in the previous section applies here. Working in the plane (in $\mathbb{R}^n$) determined by $\mathbf{u}$ and $\mathbf{v}$, we see all the formulas hold that precede Theorem (3.11), Section 3.1, so we have

(3.18)

THEOREM If $\mathbf{u}$ and $\mathbf{v}$ are any vectors in $\mathbb{R}^n$ with $\mathbf{v} \neq 0$, then the projection of $\mathbf{u}$ on $\mathbf{v}$ is given by

$$\frac{\mathbf{u} \cdot \mathbf{v}}{\|\mathbf{v}\|^2}\mathbf{v} = \frac{\mathbf{u} \cdot \mathbf{v}}{\mathbf{v} \cdot \mathbf{v}}\mathbf{v}$$

Example 4 If $\mathbf{x} = (-3, 4, 7, 1, -2, 0)$ and $\mathbf{y} = (1, -2, 0, 2, -3, 4)$ in $\mathbb{R}^6$, compute the projection of $\mathbf{y}$ on $\mathbf{x}$.

Solution By Theorem (3.18) the projection is

$$\begin{aligned}\frac{\mathbf{y} \cdot \mathbf{x}}{\mathbf{x} \cdot \mathbf{x}}\mathbf{x} &= \frac{-3 - 8 + 0 + 2 + 6 + 0}{9 + 16 + 49 + 1 + 4 + 0}\mathbf{x} \\ &= \frac{-3}{79}\mathbf{x} \\ &= \left(\tfrac{9}{79}, -\tfrac{12}{79}, -\tfrac{21}{79}, -\tfrac{3}{79}, \tfrac{6}{79}, 0\right)\end{aligned}$$ ■

Exercise 3.2 In Exercises 1–10, express the given vector in terms of its coordinates.

1. The vector from $(3, -1, 8, 2, 4)$ to $(1, 5, 0, 8, -1)$.
2. The vector from $(8, -1, -2, -3)$ to $(9, 4, 7, 3)$.
3. The vector from $(9, 8, 7, 5, -3, -9)$ to $(0, 0, 0, 0, 0, 0)$.
4. The vector from $(5, -1, 8, -2, 3)$ to $(5, -1, 8, -2, 3)$.
5. The vector from the origin to the end point of $\mathbf{w}$, where $\mathbf{w}$ has the direction and length of $(2, -1, 8, 4, 1)$ and $\mathbf{w}$ starts from $(3, -2, 8, -1, 4)$.
6. The vector from the origin to the initial point of the vector to $(-3, 8, 6, 1)$ in the opposite direction and with half the length of $(-2, 4, 2, 6)$.

In Exercises 7–10,

$$\mathbf{u} = \begin{bmatrix} 3 \\ -1 \\ 2 \\ 4 \end{bmatrix}, \quad \mathbf{v} = \begin{bmatrix} 4 \\ -1 \\ 5 \\ 8 \end{bmatrix}, \quad \mathbf{w} = \begin{bmatrix} 8 \\ -7 \\ -6 \\ 1 \end{bmatrix}$$

7. $\mathbf{u} - \mathbf{v}$

8. $2\mathbf{v} + 3\mathbf{w}$

9. $4(2\mathbf{u} - \mathbf{w})$

10. $3\mathbf{v} - 4(5\mathbf{u} - 6\mathbf{w})$

In Exercises 11–14, find the distance between the given points.

11. $(2, -3, 4, 1, -1)$, $(4, 1, 4, 2, -1)$

12. $[5 \quad 1 \quad 8 \quad -1 \quad 2 \quad 9]$, $[4 \quad 1 \quad 4 \quad 3 \quad 2 \quad 8]$

13. $\begin{bmatrix} 3 \\ -1 \\ 2 \\ -1 \\ 4 \end{bmatrix}, \begin{bmatrix} 4 \\ -5 \\ 4 \\ 0 \\ 4 \end{bmatrix}$

14. $\begin{bmatrix} 8 \\ 1 \\ 3 \\ 1 \end{bmatrix}, \begin{bmatrix} -4 \\ -3 \\ 2 \\ 3 \end{bmatrix}$

In Exercises 15–18, find the norm

15. $\|(1, -2, 0, 2)\|$

16. $\left\|\begin{bmatrix} 8 \\ -1 \\ 2 \\ 5 \end{bmatrix}\right\|$

17. $\|2\mathbf{u} - 3\mathbf{w}\|$ for $\mathbf{u} = [5 \quad 1 \quad 2 \quad -4]$, $\mathbf{w} = [2 \quad 6 \quad -2 \quad 1]$

18. $\|\mathbf{u}\| + \|\mathbf{v}\|$, $\|\mathbf{u} + \mathbf{v}\|$ for $\mathbf{u} = (3, -1, -2, 1, 4)$, $\mathbf{v} = (1, 1, 1, 1, 1)$

19. If $\mathbf{w} \neq 0$, what is $\left\|\dfrac{1}{\|\mathbf{w}\|}\mathbf{w}\right\|$? Why?

20. Find all numbers r such that $\|r(1, 0, -3, -1, 4, 1)\| = 1$.

In Exercises 21–26, find $\mathbf{u} \cdot \mathbf{v}$ and then the angle θ between $\mathbf{u}$ and $\mathbf{v}$.

21. $\mathbf{u} = \mathbf{v} = (2, -3, -1, 4, 1)$

22. $\mathbf{u} = [5 \quad 1 \quad 2 \quad 0] = -\mathbf{v}$

23. $\mathbf{u} = (4, 1, 8, 2)$, $\mathbf{v} = (-2, -6, 1, 3)$

24. $\mathbf{u} = \begin{bmatrix} 3 \\ -1 \\ 2 \\ 1 \end{bmatrix}, \quad \mathbf{v} = \begin{bmatrix} 1 \\ 0 \\ -1 \\ -1 \end{bmatrix}$

25. $\mathbf{u} = \begin{bmatrix} 2 \\ 1 \\ 3 \\ -2 \\ 1 \end{bmatrix}, \quad \mathbf{v} = \begin{bmatrix} -1 \\ -2 \\ 3 \\ 2 \\ 1 \end{bmatrix}$

26. $\mathbf{u} = \begin{bmatrix} 1 \\ 0 \\ -1 \\ 0 \\ 1 \\ 0 \\ -1 \end{bmatrix}, \quad \mathbf{v} = \begin{bmatrix} 1 \\ 2 \\ 1 \\ 1 \\ 1 \\ 0 \\ 1 \end{bmatrix}$

In Exercises 27–31, compute the projection of $\mathbf{u}$ on $\mathbf{v}$.

27. $\mathbf{u} = 0, \quad \mathbf{v} = (1, 2, 1, 3, -1)$

28. $\mathbf{u} = \begin{bmatrix} 1 \\ 2 \\ 1 \\ -1 \\ 3 \end{bmatrix}, \quad \mathbf{v} = \begin{bmatrix} -3 \\ 1 \\ -1 \\ 1 \\ 1 \end{bmatrix}$

29. $\mathbf{u} = [2 \quad -1 \quad 3 \quad 0 \quad 1], \quad \mathbf{v} = [4 \quad 2 \quad 1 \quad 0 \quad 1]$

30. $\mathbf{u} = (1, 2, -3, 1), \quad \mathbf{v} = (4, -1, 2, -1)$

31. $\mathbf{u} = (5, 0, 1, 0, -1), \quad \mathbf{v} = (3, 2, -1, 1, 2)$

In Exercises 32–39, let $\mathbf{u} = (u_1, \ldots, u_n)$, $\mathbf{v} = (v_1, \ldots, v_n)$, $\mathbf{w} = (w_1, \ldots, w_n)$, and use Definition (3.15).

32. Prove $\|\mathbf{u}\| = \sqrt{\mathbf{u} \cdot \mathbf{u}} = d(\mathbf{u}, \mathbf{0})$.

33. Prove $\|\mathbf{u} - \mathbf{v}\| = d(\mathbf{u}, \mathbf{v})$.

34. Prove $\|r\mathbf{u}\| = |r|\,\|\mathbf{u}\|$.

35. Prove $\mathbf{u} \cdot \mathbf{v} = \mathbf{v} \cdot \mathbf{u}$.

36. Prove $\mathbf{u} \cdot (\mathbf{v} + \mathbf{w}) = \mathbf{u} \cdot \mathbf{v} + \mathbf{u} \cdot \mathbf{w}$

37. Prove $r(\mathbf{u} \cdot \mathbf{v}) = (r\mathbf{u}) \cdot \mathbf{v} = \mathbf{u} \cdot (r\mathbf{v})$.

38. Verify parts 1, 3, 5, and 7 of Theorem (3.14).

39. Verify parts 2, 4, 6, and 8 of Theorem (3.14).

3.3 General Vector Spaces

In the previous section, Theorem (3.14) lists eight properties that hold in $\mathbb{R}^n$. However, these same properties hold in many other contexts. It is usually useful to point out when they hold because knowing this can help us understand what is going on mathematically, which in turn can help us solve certain kinds of problems. Hence we will define a "vector space" to be a set of objects in which those eight properties hold and a "vector" will simply mean an object in such a space. In addition to being useful, vector spaces turn up in surprising and interesting places.

(3.19)

DEFINITION Let V be a set on which addition and scalar multiplication are defined. (For emphasis, this means for every $\mathbf{u}$ and $\mathbf{v}$ in V and for every number r, the sum $\mathbf{u} + \mathbf{v}$ is in V and the scalar product $r\mathbf{u}$ is in V. This is abbreviated by saying "V is **closed** under addition and scalar multiplication.") Suppose the following axioms are satisfied for every $\mathbf{u}$, $\mathbf{v}$, and $\mathbf{w}$ in V and for all numbers r and s.

(a) $\mathbf{u} + \mathbf{v} = \mathbf{v} + \mathbf{u}$.
(b) $(\mathbf{u} + \mathbf{v}) + \mathbf{w} = \mathbf{u} + (\mathbf{v} + \mathbf{w})$.
(c) There is a special member $\mathbf{0}$ (called the **zero vector**) of V such that $\mathbf{u} + \mathbf{0} = \mathbf{u}$, for all $\mathbf{u}$ in V.
(d) For every member $\mathbf{u}$ of V, there is a **negative**, $-\mathbf{u}$, in V such that $\mathbf{u} + (-\mathbf{u}) = \mathbf{0}$.
(e) $r(\mathbf{u} + \mathbf{v}) = r\mathbf{u} + r\mathbf{v}$.
(f) $(r + s)\mathbf{u} = r\mathbf{u} + s\mathbf{u}$.
(g) $(rs)\mathbf{u} = r(s\mathbf{u})$.
(h) $1\mathbf{u} = \mathbf{u}$.

Then V is called a **vector space**, and the members of V are called **vectors**.

Thus there are times that by a "vector" we simply mean a member of some vector space V, and the "vector" might bear little resemblance to a "directed line segment."

NOTE By a "number" in Definition (3.19) is meant a "real number," so what has been defined is a "real vector space." In a few places in this text we use complex numbers (and then technically we are working in a "complex vector space") but such instances will be rare and carefully noted. In fact even numbers other than real or complex numbers are used. You will get a slight introduction and flavor of such applications in coding theory, Section 3.9. Vector spaces over arbitrary fields are discussed in Section 6.5.

Several examples follow, starting with objects you will find quite familiar and working toward some that are more abstract. The more abstract examples will be referred to occasionally throughout this and the following chapters; they will be used to illustrate certain concepts as we go along. However, the thrust of the next three chapters is mainly in $\mathbb{R}^n$; the abstract vector spaces will be more fully developed in Chapter 6.

Example 1 $\mathbb{R}^n$ with the usual addition and scalar multiplication. By Theorem (3.14), Section 3.2, this is a vector space. ■

Example 2 $\mathbb{R}^2$ with the usual addition, but scalar multiplication is given by

$$r(x_1, x_2) = (rx_1, x_2)$$

This is *not* a vector space because

$$(r + s)(x_1, x_2) = ((r + s)x_1, x_2) = (rx_1 + sx_1, x_2)$$
$$r(x_1, x_2) + s(x_1, x_2) = (rx_1, x_2) + (sx_1, x_2) = (rx_1 + sx_1, 2x_2)$$

Therefore,

$$(r + s)(x_1, x_2) \neq r(x_1, x_2) + s(x_1, x_2)$$

so that axiom (f) of the definition of a vector space (3.19) fails. ■

Example 3 Consider a straight line ℓ through the origin in $\mathbb{R}^3$ with the usual addition and scalar multiplication. This is a vector space. Pick any vector $\mathbf{v} \neq \mathbf{0}$ on the line, say $\mathbf{v} = (a, b, c)$. Then the line ℓ is equal to all scalar multiples of $\mathbf{v}$,

$$\ell = \{s\mathbf{v} \mid s \text{ is any real number}\}$$

or

$$\ell = \{(sa, sb, sc) \mid s \text{ is any real number}\} \tag{3.20}$$

It is straightforward to check that all the axioms are satisfied, and this is left as an exercise. (See Exercises 8–10.) ■

You should note that (3.20) is equivalent to saying a point (x, y, z) is on the line ℓ if and only if

$$\left.\begin{aligned} x &= at \\ y &= bt \\ z &= ct \end{aligned}\right\} \quad \text{for some real number } t \tag{3.21}$$

Equations (3.21) are called the **parametric equations** for a straight line (through the origin).

Example 4 Consider a straight line ℓ through the origin in $\mathbb{R}^n$ with the usual definition of addition and scalar multiplication. As in Example 3, this is a vector space. If we pick any $\mathbf{v} = (a_1, \ldots, a_n) \neq \mathbf{0}$ on the line ℓ, then

$$\ell = \{s\mathbf{v} \mid s \text{ any real number}\} = \{(sa_1, \ldots, sa_n) \mid s \text{ any real number}\}$$ ■

Example 5 A straight line in $\mathbb{R}^3$ that does *not* go through the origin with the usual definitions of addition and scalar multiplication. This is not a vector space. The easiest reason is that the zero vector $\mathbf{0}$ is not in the set ℓ, but there are other reasons as well. See Exercises 2–4. ■

Example 6 Consider a plane P through the origin in $\mathbb{R}^3$ with the usual definitions of addition and scalar multiplication. This is a vector space. In fact P is described by an equation of the form

$$ax + by + cz = 0 \tag{3.22}$$

where at least one of a, b, or $c \neq 0$. We can use this equation to see that P is closed under addition. Suppose $\mathbf{v} = (x_1, x_2, x_3)$ and $\mathbf{w} = (y_1, y_2, y_3)$ are in the plane P. Then $\mathbf{v}$ and $\mathbf{w}$ satisfy Equation (3.22) so

$$ax_1 + bx_2 + cx_3 = 0 \qquad \text{and} \qquad ay_1 + by_2 + cy_3 = 0$$

Adding these two equations yields

$$ax_1 + bx_2 + cx_3 + ay_1 + by_2 + cy_3 = 0 + 0$$

or

$$a(x_1 + y_1) + b(x_2 + y_2) + c(x_3 + y_3) = 0$$

Thus the vector $\mathbf{v} + \mathbf{w}$ satisfies Equation (3.22) and so $\mathbf{v} + \mathbf{w}$ is in P. See Exercises 5–7 for the remaining properties. ■

We now begin a discussion of some more abstract vector spaces.

Example 7 Consider the set of all $m \times n$ matrices (where m and n are fixed). The definitions of addition and scalar multiplication are those given in Section 1.3. By those definitions, addition and scalar multiplication are closed. The zero "vector" $\mathbf{0}$ is the zero matrix. If the "vector" $\mathbf{u}$ is the $m \times n$ matrix A, then its negative $-\mathbf{u}$ is the matrix $-A$. The various associative, commutative, and distributive properties are all discussed in Section 1.3 and all hold. Thus this is a vector space, and, in fact, it will be quite important to us. In this context, a "vector" is an "$m \times n$ matrix." ■

(3.23) **NOTATION** We shall denote the vector space of all $m \times n$ matrices by M_{mn}.

Example 8 Let P_n denote the set of all polynomials (with real coefficients) of degree less than or equal to n. If $p = a_0 + a_1x + \cdots + a_nx^n$ and $q = b_0 + b_1x + \cdots + b_nx^n$ are two polynomials in P_n, then addition and scalar multiplication are defined by

$$p + q = (a_0 + b_0) + (a_1 + b_1)x + \cdots + (a_n + b_n)x^n$$
$$rp = (ra_0) + (ra_1)x + \cdots + (ra_n)x^n$$

It is immediately clear that P_n is closed under addition and scalar multiplication, and it is straightforward to check all the axioms. See Exercises 11 and 12. Thus P_n is a vector space, and in this context a "vector" is simply a polynomial. The zero vector $\mathbf{0}$ is the zero polynomial $\mathbf{0} = 0 + 0x + \cdots + 0x^n$ and if the vector $\mathbf{u}$ is the polynomial $p = a_0 + a_1x + \cdots + a_nx^n$, then its negative $-\mathbf{u}$ is the polynomial $-p = -a_0 + (-a_1)x + \cdots + (-a_n)x^n$. P_n is a very important vector space in approximation theory. ■

Example 9 Let $C[a, b]$ be the set of all continuous real-valued functions defined on the interval $[a, b]$. For example, if f is the function with rule $f(x) = 1/x$, then f is in $C[1, 2]$ but f is not in $C[-1, 1]$. (Why?) If $f = f(x)$ and $g = g(x)$ are in $C[a, b]$ and r is a real number, then the sum $f + g$ and scalar multiple rf are the functions defined by

$$(f + g)(x) = f(x) + g(x) \quad (rf)(x) = rf(x)$$

See Figure 3.17 for illustrations. By theorems in calculus, $f + g$ and rf are also in $C[a, b]$. Thus $C[a, b]$ is closed under addition and scalar multiplication, and verification of the axioms is left to the exercises. See Exercises 14 and 15. Thus $C[a, b]$ is a vector space, and in this context a "vector" is simply a function. The zero vector $\mathbf{0}$ is the zero function $\mathbf{0}$ whose rule is $\mathbf{0}(x) = 0$, for all x in $[a, b]$. If the vector $\mathbf{u}$ is the function f with rule $f(x)$, then its negative $-\mathbf{u}$ is the function $-f$ with rule $-f(x)$. See Figure 3.17c. $C[a, b]$ is a very important vector space for many applications. ■

Figure 3.17

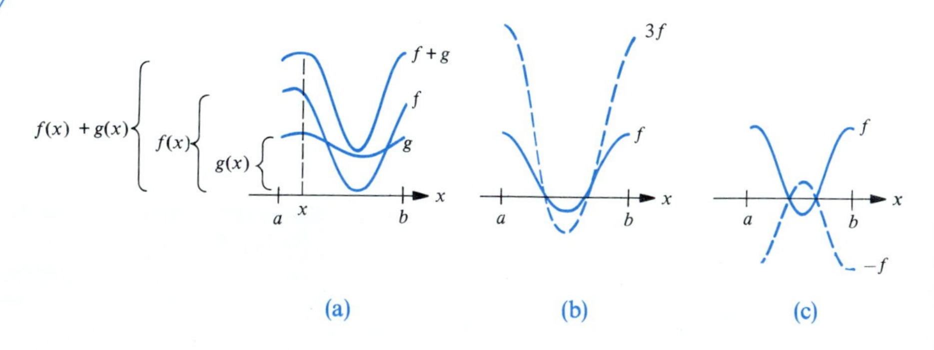

It is sometimes (but not always) possible to introduce a dot or inner product on abstract vector spaces in a meaningful way. We briefly introduce one example here because it is both fun and very important. We shall refer to it again in Chapter 4, and in Chapter 6 we shall discuss it and its application more thoroughly. For now just think of it as a mental exercise.

Example 10 In $C[a, b]$, define the inner product $f \cdot g$ by

$$f \cdot g = \int_a^b f(x)g(x)\,dx$$

In particular, in $C[0, 2\pi]$ let $f(x) = x$ and $g(x) = \sin x$. Then

$$\begin{aligned} f \cdot g = x \cdot \sin x &= \int_0^{2\pi} x \sin x\,dx \\ &= -x \cos x + \sin x\Big]_0^{2\pi} \\ &= -2\pi \cos 2\pi + \sin 2\pi - (-0 \cos 0 + \sin 0) = -2\pi \end{aligned}$$

We can define the "length" or "norm" of a function $\|f\|$ by

$$\|f\| = \sqrt{f \cdot f}$$

So for $g(x) = \sin x$,

$$\begin{aligned} g \cdot g = \int_0^{2\pi} \sin^2 x\,dx &= \tfrac{1}{2} \int_0^{2\pi} (1 - \cos 2x)\,dx \\ &= \tfrac{1}{2}(x - \tfrac{1}{2} \sin 2x)_0^{2\pi} = \pi \end{aligned}$$

Thus $\|g\| = \|\sin x\| = \sqrt{\pi}$.

Finally, what at first seems incredibly ridiculous but turns out to be very important for approximation theory, is projection. If we use Theorem (3.18), Section 3.2, we can define the projection of f on g by

$$\frac{f \cdot g}{g \cdot g}\, g$$

Thus by the previous computations the projection of $f(x) = x$ on $g(x) = \sin x$ is

$$\frac{x \cdot \sin x}{\sin x \cdot \sin x} \sin x = \frac{-2\pi}{\pi} \sin x = -2 \sin x$$

It is, of course, quite absurd at this point to try to understand this geometrically. In fact you will have to wait to see the applications to understand the example completely. ■

When dealing with a new concept, the most trivial example often plays at least a small role.

Example 11 The **zero vector space** consists of a single vector $\mathbf{0}$, that is $V = \{\mathbf{0}\}$, and addition and scalar multiplication are defined by

$$\mathbf{0} + \mathbf{0} = \mathbf{0} \qquad \text{and} \qquad r\mathbf{0} = \mathbf{0}$$

for any scalar r. All the axioms are easily verified. ■

There are several useful facts that follow from the axioms (e.g., the number zero times any vector is always the zero vector) and are always true in any vector space. The following are a few that we shall find useful later.

(3.24)

THEOREM Let V be a vector space, let $\mathbf{u}$ be a vector in V, and let r be a scalar. Then

(a) $0\mathbf{u} = \mathbf{0}$
(b) $r\mathbf{0} = \mathbf{0}$
(c) $(-1)\mathbf{u} = -\mathbf{u}$
(d) If $r\mathbf{u} = \mathbf{0}$, then either $r = 0$ or $\mathbf{u} = \mathbf{0}$

We shall prove (a) and leave the remainder to the exercises with hints. (See Exercises 31–33.)

Proof of Theorem (3.24a) Whatever vector $0\mathbf{u}$ is, it has a negative $-0\mathbf{u}$ by axiom (3.19d). Thus

$$
\begin{aligned}
\mathbf{0} &= 0\mathbf{u} + (-0\mathbf{u}) && \text{by axiom (3.19d)}\\
&= (0 + 0)\mathbf{u} + (-0\mathbf{u}) &&\\
&= [0\mathbf{u} + 0\mathbf{u}] + (-0\mathbf{u}) && \text{by axiom (3.19f)}\\
&= 0\mathbf{u} + [0\mathbf{u} + (-0\mathbf{u})] && \text{by axiom (3.19b)}\\
&= 0\mathbf{u} + \mathbf{0} && \text{by axiom (3.19d)}\\
&= 0\mathbf{u} && \text{by axiom (3.19c)}
\end{aligned}
$$
■

Admittedly this proof is a little slick. Slick proofs often have the disadvantage of being unintuitive but the advantage of being short and sweet. You are encouraged, before doing Exercises 31–33, to examine what really makes this proof work and to come up with a variation that is more intuitive and satisfactory for you.

Exercise 3.3

1. In Example 2, use the definition of scalar multiplication given there to compute
   (a) $(r + s)(x_1, x_2)$ (b) $r(x_1, x_2) + s(x_1, x_2)$
   and show these are *not* equal.

Exercises 2, 3, and 4 refer to Example 5, a line ℓ that does *not* go through the origin.

2. Explain why ℓ is the set of all $\mathbf{x} = \mathbf{x}_0 + t\mathbf{v}$, where t is any real number and $\mathbf{x}_0 \neq \mathbf{0}$ and $\mathbf{v}$ is not a multiple of $\mathbf{x}_0$.
3. Use Exercise 2 to show ℓ is not closed under addition.
4. Use Exercise 2 to show ℓ is not closed under scalar multiplication.

Exercises 5, 6, and 7 refer to Example 6, a plane P through the origin.

5. Show that P is closed under scalar multiplication
6. Show that axioms (3.19c, e) are satisfied for P.
7. Show that axioms (3.19d, f) are satisfied for P.

Exercises 8–10 refer to Example 3, a straight line ℓ through the origin.

8. Show that ℓ is closed under addition and scalar multiplication
9. Show that axioms (3.19a, c, g) are satisfied for ℓ.
10. Show that axioms (3.19b, d, h) are satisfied for ℓ.

Exercises 11 and 12 refer to Example 8, P_n, the set of all polynomials of degree less than or equal to n.

11. Show that axioms (3.19a, e, g) are satisfied for P_n.
12. Show that axioms (3.19b, f, h) are satisfied for P_n.

Exercises 13–16 refer to Example 9, $C[a, b]$.

13. For $f(x) = \dfrac{1}{x}$, explain why f is in $C[1, 2]$ but f is not in $C[-1, 1]$.
14. Show that axioms (3.19a, e, g) are satisfied for $C[a, b]$.
15. Show that axioms (3.19b, f, h) are satisfied for $C[a, b]$.
16. If f and g are in $C[a, b]$ and $g \neq 0$ (i.e., g is not the zero function so $g(x) \neq 0$ for all x in $[a, b]$) is $\dfrac{f}{g}$ in $C[a, b]$ $\left(\text{where } \left(\dfrac{f}{g}\right)(x) = \dfrac{f(x)}{g(x)}, \text{ all } x \text{ in } [a, b]\right)$? If so, explain why; if not, under what conditions on g is it?

In Exercises 17–30 a set of objects is given together with operations of addition and scalar multiplication. Decide which are vector spaces. For those that are not, find at least one property that fails to hold.

17. The set of all ordered triples of real numbers, $\{(x_1, x_2, x_3)\}$, usual addition, and

$$r(x_1, x_2, x_3) = (0, 0, 0), \qquad \text{all numbers } r$$

18. Same as Exercise 17 except

$$r(x_1, x_2, x_3) = (2rx_1, 2rx_2, 2rx_3)$$

19. Same as Exercise 17 except

$$r(x_1, x_2, x_3) = (x_1, x_2, rx_3)$$

20. Same as Exercise 17 except

$$r(x_1, x_2, x_3) = (x_1, x_2, x_3)$$

21. The set of all real numbers, usual addition and scalar multiplication.

22. The set of all n-tuples of the form $(x, x, \ldots, x)$, usual addition and scalar multiplication on $\mathbb{R}^n$.

23. The upper half-plane $= \{(x, y) \mid y \geq 0\}$, usual addition and scalar multiplication.

24. The set of all pairs of real numbers, usual scalar multiplication, but

$$(x_1, x_2) + (y_1, y_2) = (x_1 + y_1 + 1, x_2 + y_2 + 1)$$

25. The set of all upper triangular 3×3 matrices

$$\begin{bmatrix} a & b & c \\ 0 & d & e \\ 0 & 0 & f \end{bmatrix}$$

usual addition and scalar multiplication from M_{33}.

26. The set of all unit lower triangular 3×3 matrices

$$\begin{bmatrix} 1 & 0 & 0 \\ a & 1 & 0 \\ b & c & 1 \end{bmatrix}$$

usual addition and scalar multiplication.

27. The set of all 2×2 matrices of the form

$$\begin{bmatrix} a & 1 \\ 1 & b \end{bmatrix}$$

usual addition and scalar multiplication from M_{22}.

28. The set of all 2×2 matrices of the form

$$\begin{bmatrix} a & a+b \\ a+b & b \end{bmatrix}$$

usual addition and scalar multiplication from M_{22}.

29. The set of all pairs of real numbers of the form $(1, a)$

Addition: $(1, a) + (1, b) = (1, a + b)$
Scalar multiplication: $r(1, a) = (1, ra)$

30. The set of all positive real numbers, $\mathbb{R}^+$

Addition: $x + y = xy$
Scalar multiplication: $rx = x^r$

31. Prove Theorem (3.24b), $r\mathbf{0} = \mathbf{0}$. [HINT Modify the proof of Theorem (3.24a).]

32. Prove Theorem (3.24c), $(-1)\mathbf{u} = -\mathbf{u}$. [HINT Show $\mathbf{u} + (-1)\mathbf{u} = \mathbf{0}$.]

33. Prove Theorem (3.24d), if $r\mathbf{u} = \mathbf{0}$, then $r = 0$ or $\mathbf{u} = \mathbf{0}$. [HINT If $r \neq 0$, multiply by r^{-1} and use Theorem (3.24b).]

Exercises 34–40 refer to Example 10, $C[a, b]$ with the inner product

$$f \cdot g = \int_a^b f(x)g(x)\,dx$$

For Exercises 34–36, $f(x) = x$, $g(x) = x^2$ in $C[0, 1]$.

34. Show that $f \cdot g = \frac{1}{4}$.

35. Show that $\|f\| = \dfrac{1}{\sqrt{3}}$ and $\|g\| = \dfrac{1}{\sqrt{5}}$.

36. Find the projection of f on g.

For Exercises 37–40, $f(x) = \sin x$, $g(x) = \cos x$, and $h(x) = x$ in $C[0, 2\pi]$.

37. Show that f and g are perpendicular. Hence what is the projection of f on g?

38. Find $\|g\|$ and $\|h\|$.

39. Compute $g \cdot g$.

40. Find the projection of h on g.

3.4 Subspaces, Span, Null Spaces

There are certain very special subsets of $\mathbb{R}^n$ that satisfy all the axioms listed in the definition of a vector space (3.19) in Section 3.3. Such subsets are invaluable in helping us to understand linear problems, and in this section we study them in detail. Likewise such special subsets of general vector spaces are also very useful, and it is natural to consider them at the same time.

We begin with a preliminary definition.

(3.25)

DEFINITION (a) A subset S of a vector space is **closed under addition** if whenever two vectors $\mathbf{u}$ and $\mathbf{v}$ are in S, their sum $\mathbf{u} + \mathbf{v}$ is also in S.

(b) A subset S of a vector space is **closed under scalar multiplication** if whenever any vector $\mathbf{u}$ is in S, the scalar multiple $r\mathbf{u}$ is in S for any number r.

Example 1 Determine if

(a) $S =$ the first quadrant, and
(b) $T =$ the first and third quadrants.

as subsets of $\mathbb{R}^2$, are closed under either addition or scalar multiplication.

Solution The sets S and T are shaded in Figure 3.18.

Figure 3.18

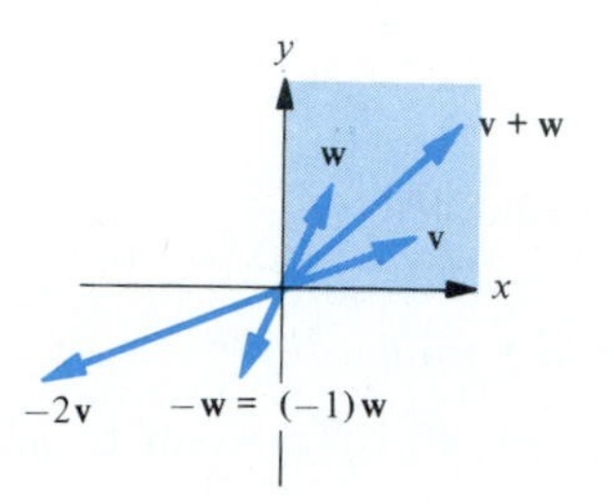

(a) *S = the first quadrant*

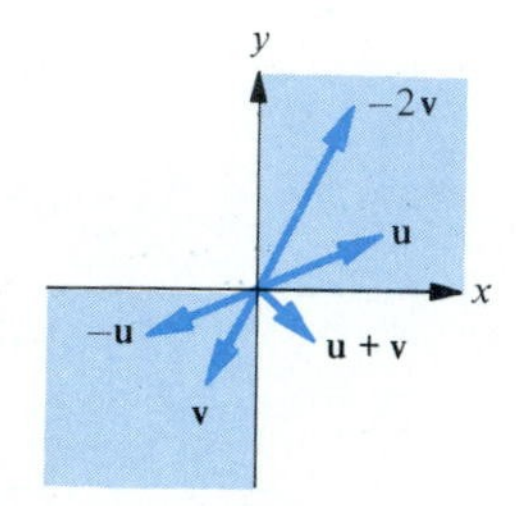

(b) *T = first and third quadrants*

As is illustrated in Figure 3.18a, the sum of two vectors in S is also in S, but multiplying a vector in S by a negative number gives a vector not in S. Thus S is closed under addition, but it is not closed under scalar multiplication.

(b) As is illustrated in Figure 3.18b, multiplying a vector in T by a scalar yields another vector in T, but the sum of two vectors in T may not be in T. Thus T is closed under scalar multiplication, but it is not closed under addition. ■

We are now ready for the main definition.

(3.26)

DEFINITION A nonempty subset S of a vector space, V, is called a **subspace** of V if S is closed under addition and scalar multiplication so that all axioms of a vector space given in Definition (3.19), Section 3.3, are satisfied.

Example 2 The set S of Example 1(a) is closed under addition but not closed under scalar multiplication. Hence S cannot be a subspace of $\mathbb{R}^2$. The set T of Example 1(b) is closed under scalar multiplication but not under addition. Hence T cannot be a subspace of $\mathbb{R}^2$ either.

Let V be any vector space, let $\mathbf{0}$ be the zero vector, and let $S = \{\mathbf{0}\}$. Then S is closed under addition (since $\mathbf{0} + \mathbf{0} = \mathbf{0}$) and scalar multiplication (since $r\mathbf{0} = \mathbf{0}$), and all the axioms of a vector space are satisfied (see Example 11, Section 3.3). Thus S is a subspace of V. ■

The definition of a subspace (3.26) requires that if a subset S is a subspace, then S must be closed under both addition and scalar multiplication. What is very surprising, and what makes these concepts so important, is that these two conditions are sufficient for a subset to be a subspace.

(3.27)

THEOREM If a nonempty subset, S, of a vector space, V, is closed under addition and scalar multiplication, then S is a subspace. Therefore, to prove S is a subspace, it is sufficient to show:

(a) If $\mathbf{u}$ and $\mathbf{v}$ are in S, then $\mathbf{u} + \mathbf{v}$ is in S.
(b) If $\mathbf{u}$ is in S and r is any number, then $r\mathbf{u}$ is in S.

Sketch of Proof The proof of this theorem is to observe that we get all eight conditions of Definition (3.19), Section 3.3, for free [once we know (a) and (b) of Theorem (3.27)]. First, conditions (a) and (b), say addition and scalar multiplication, are defined on S. Next, all the associative, commutative, and distributive properties hold because they hold in V (and we are just restricting the operations to S). Finally, the zero vector $\mathbf{0}$ is in S since $\mathbf{0} = 0\mathbf{v}$ [S is nonempty, so pick any $\mathbf{v}$ in S and apply (b) and Theorem (3.24a), Section 3.3], and if $\mathbf{v}$ is in S, then $-\mathbf{v} = (-1)\mathbf{v}$ is in S [again apply (b) and Theorem (3.24c), Section 3.3]. Thus we are done. ■

Example 3 Determine if the following subsets of the given vector spaces are subspaces.

(a) S is the x- and y-axis (only) in $\mathbb{R}^2$. Thus

$$S = \{(x, 0) \quad \text{or} \quad (0, y)\}$$

(b) S is the set of all 2×3 matrices with the second row all zero. Thus

$$S = \left\{\begin{bmatrix} a_{11} & a_{12} & a_{13} \\ 0 & 0 & 0 \end{bmatrix}\right\}$$

(c) S is the xy-plane in $\mathbb{R}^3$, $S = \{(x, y, 0) \mid x \text{ and } y \text{ are real numbers}\}$.
(d) S is the set of all functions f in $C[0, 3]$ with $f(2) = 0$.
(e) Same as (d) except that $f(2) = 3$.

Solution (a) S is not a subspace of $\mathbb{R}^2$ because it is not closed under addition (although it is closed under scalar multiplication). For example,

$$\mathbf{u} = (2, 0) \qquad \text{and} \qquad \mathbf{v} = (0, 3)$$

are in S but

$$\mathbf{u} + \mathbf{v} = (2, 0) + (0, 3) = (2, 3)$$

is not in S.

(b) S is a subspace of M_{23}. Let

$$\mathbf{u} = \begin{bmatrix} u_{11} & u_{12} & u_{13} \\ 0 & 0 & 0 \end{bmatrix} \quad \text{and} \quad \mathbf{v} = \begin{bmatrix} v_{11} & v_{12} & v_{13} \\ 0 & 0 & 0 \end{bmatrix}$$

be in S and let r be a scalar. Then

$$\mathbf{u} + \mathbf{v} = \begin{bmatrix} u_{11} & u_{12} & u_{13} \\ 0 & 0 & 0 \end{bmatrix} + \begin{bmatrix} v_{11} & v_{12} & v_{13} \\ 0 & 0 & 0 \end{bmatrix}$$
$$= \begin{bmatrix} u_{11} + v_{11} & u_{12} + v_{12} & u_{13} + v_{13} \\ 0 & 0 & 0 \end{bmatrix}$$

is obviously in S, as is

$$r\mathbf{u} = r\begin{bmatrix} u_{11} & u_{12} & u_{13} \\ 0 & 0 & 0 \end{bmatrix} = \begin{bmatrix} ru_{11} & ru_{12} & ru_{13} \\ 0 & 0 & 0 \end{bmatrix}$$

Thus S is closed under addition and scalar multiplication, and so it is a subspace of M_{23} by Theorem (3.27).

(c) S is a subspace of $\mathbb{R}^3$, since it is closed under addition and scalar multiplication, as is easy to see.

(d) S is a subspace of $C[0, 3]$. If f and g are in S and r is a scalar, then $f(2) = 0$ and $g(2) = 0$ so that

$$(f + g)(2) = f(2) + g(2) = 0 + 0 = 0$$
$$(rf)(2) = rf(2) = r0 = 0$$

Theorem (3.27) now applies.

(e) This S is not a subsubspace. For example,

$$(f + g)(2) = f(2) + g(2) = 3 + 3 = 6 \neq 3$$

Thus S is not closed under addition, and hence it is not a subspace. ■

The following useful criterion is usually very easy to use, if and when it applies. For example, it gives an even easier way to see that the S of Example 3(e) is not a subspace.

USEFUL FACT Since a subspace is a vector space, it must contain the zero vector. Thus when determining if a subset S is a subspace, first check to see if the zero vector is in S. This is usually quick and easy to do, and if $\mathbf{0}$ is not in S, you are done.

Example 4 For the S in Example 3(e), the zero vector is the zero function $\mathbf{0}$ (see Example 9, Section 3.3). Since $\mathbf{0}(x) = 0$ for all x, $\mathbf{0}(2) = 0$, so $\mathbf{0}(2) \neq 3$. Thus $\mathbf{0}$ is not in S, and hence S is not a subspace. ■

Null Spaces

One of our eventual goals in linear algebra is to understand how to solve the equation $A\mathbf{x} = \mathbf{b}$ under various circumstances. We shall see that several subspaces are associated with an $m \times n$ matrix A and that these subspaces provide insight into solving linear equations. We introduce one subspace here, both as a preview and an illustration of the concepts just presented.

In the following we consider a vector $\mathbf{x}$ in $\mathbb{R}^n$ to be an $n \times 1$ matrix or "column vector."

(3.28) DEFINITION Let A be an $m \times n$ matrix. The set $\{\mathbf{x} | A\mathbf{x} = \mathbf{0}\}$ is called the **null space** of A and is denoted by $NS(A)$.*

Example 5 Find the null space of

(a) $A = \begin{bmatrix} 1 & 2 \\ 0 & 3 \end{bmatrix}$ (b) $B = \begin{bmatrix} 1 & 2 \\ 3 & 6 \end{bmatrix}$ (c) $\mathbf{0} = \begin{bmatrix} 0 & 0 \\ 0 & 0 \end{bmatrix}$

Solution (a) If $A\mathbf{x} = \mathbf{0}$, then

$$\begin{bmatrix} 1 & 2 \\ 0 & 3 \end{bmatrix}\begin{bmatrix} x_1 \\ x_2 \end{bmatrix} = \begin{bmatrix} 0 \\ 0 \end{bmatrix} \quad \text{or} \quad \begin{aligned} x_1 + 2x_2 &= 0 \\ 3x_2 &= 0 \end{aligned}$$

Easily, $x_1 = x_2 = 0$ so $NS(A) = \{\mathbf{0}\}$. See Figure 3.19a.

(b) If $B\mathbf{x} = \mathbf{0}$, then

$$\begin{bmatrix} 1 & 2 \\ 3 & 6 \end{bmatrix}\begin{bmatrix} x_1 \\ x_2 \end{bmatrix} = \begin{bmatrix} 0 \\ 0 \end{bmatrix} \quad \text{or} \quad \begin{aligned} x_1 + 2x_2 &= 0 \\ 3x_1 + 6x_2 &= 0 \end{aligned}$$

Observe that the second row of B is a multiple of the first row. Thus, as in Chapter 1, if we let $x_2 = s$, we see $x_1 = -2s$. Thus

$$\begin{bmatrix} x_1 \\ x_2 \end{bmatrix} \quad \text{is in } NS(B) \text{ if and only if} \quad \begin{bmatrix} x_1 \\ x_2 \end{bmatrix} = s\begin{bmatrix} -2 \\ 1 \end{bmatrix}$$

That is, $NS(B)$ is the whole line through the origin and the vector $[-2 \quad 1]^T$. See Figure 3.19b.

(c) Since $0\mathbf{x} = \mathbf{0}$, for all x in $\mathbb{R}^2$, $NS(\mathbf{0})$ is all of $\mathbb{R}^2$. See Figure 3.19c. ■

* *In some contexts $NS(A)$ is also called the **kernel** of A and is denoted by $\ker(A)$.*

Figure 3.19

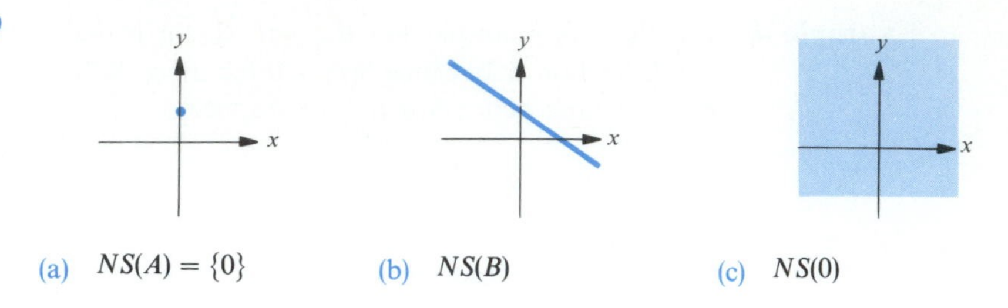

(a) $NS(A) = \{0\}$ (b) $NS(B)$ (c) $NS(0)$

In each of the three cases of Example 5, the null space is a subspace. This must always be the case.

(3.29) **THEOREM** If A is an $m \times n$ matrix, then $NS(A)$ is a subspace of $\mathbb{R}^n$.

Proof Pick any $m \times n$ matrix A. We shall work toward applying Theorem (3.27). First, $NS(A)$ is nonempty since $A\mathbf{0} = \mathbf{0}$, so the zero vector $\mathbf{0}$ is in $NS(A)$. Now suppose $\mathbf{x}$ and $\mathbf{y}$ are in $NS(A)$ and r is a scalar. Then $A\mathbf{x} = \mathbf{0}$ and $A\mathbf{y} = \mathbf{0}$. So

$$A(\mathbf{x} + \mathbf{y}) = A\mathbf{x} + A\mathbf{y} = \mathbf{0} + \mathbf{0} = \mathbf{0} \qquad \text{and} \qquad A(r\mathbf{x}) = rAx = r\mathbf{0} = \mathbf{0}$$

Thus $NS(A)$ is nonempty and closed under addition and scalar multiplication. By Theorem (3.27) $NS(A)$ is a subspace. ■

After examining Examples 5(a) and 5(b), we can readily see that finding the null space of an $m \times n$ matrix $A = (a_{ij})$ is equivalent to solving the linear system

$$\begin{aligned} a_{11}x_1 + \cdots + a_{1n}x_n &= 0 \\ &\vdots \\ a_{m1}x_1 + \cdots + a_{mn}x_n &= 0 \end{aligned}$$

Such a linear system, with all the constants on the right-hand side equal to zero, is called **homogeneous**. We shall see homogeneous systems naturally arising again is Sections 3.5, 5.2, and elsewhere.

Linear Combinations

Null spaces will be discussed further in Section 3.7. Meanwhile other subspaces will come up in the following situation. We shall be given vectors $\mathbf{v}_1, \ldots, \mathbf{v}_k$ (of some vector space V), and we shall need to find the smallest subspace (of V) that contains them. To define this subspace, we need the concept of linear combinations, which was introduced as a preview in a special case in Section 1.3 [see Theorem (1.32) and the following discussion]. That introduction should make the concept of linear combinations of vectors in an arbitrary vector space easier to understand now.

(3.30) **DEFINITION** A vector $\mathbf{w}$ is called a **linear combination** of the vectors $\mathbf{v}_1, \ldots, \mathbf{v}_k$ if there are numbers $a_1, \ldots, a_k$ such that

$$\mathbf{w} = a_1\mathbf{v}_1 + \cdots + a_k\mathbf{v}_k$$

Example 6 Let $\mathbf{v}_1 = (2, 1)$ and $\mathbf{v}_2 = (-4, -2)$. Determine if either

$$\text{(a)}\quad \mathbf{w} = (6, 3) \qquad \text{or} \qquad \text{(b)}\quad \mathbf{x} = (3, 5)$$

is a linear combination of $\mathbf{v}_1$ and $\mathbf{v}_2$.

Solution (a) We must try to determine if there are scalars a_1 and a_2 such that

$$a_1\mathbf{v}_1 + a_2\mathbf{v}_2 = \mathbf{w}$$

Thus we must solve

$$a_1\begin{bmatrix} 2 \\ 1 \end{bmatrix} + a_2\begin{bmatrix} -4 \\ -2 \end{bmatrix} = \begin{bmatrix} 6 \\ 3 \end{bmatrix} \qquad \text{or} \qquad \begin{bmatrix} 2 & -4 \\ 1 & -2 \end{bmatrix}\begin{bmatrix} a_1 \\ a_2 \end{bmatrix} = \begin{bmatrix} 6 \\ 3 \end{bmatrix}$$

for a_1 and a_2. The augmented matrix is

$$\left[\begin{array}{cc|c} 2 & -4 & 6 \\ 1 & -2 & 3 \end{array}\right] \qquad \text{which reduces to} \qquad \left[\begin{array}{cc|c} 2 & -4 & 6 \\ 0 & 0 & 0 \end{array}\right]$$

Thus there are infinitely many solutions for a_1 and a_2. For example,

$$3\begin{bmatrix} 2 \\ 1 \end{bmatrix} + 0\begin{bmatrix} -4 \\ -2 \end{bmatrix} = \begin{bmatrix} 6 \\ 3 \end{bmatrix} \qquad \text{or} \qquad 0\begin{bmatrix} 2 \\ 1 \end{bmatrix} + \frac{-3}{2}\begin{bmatrix} -4 \\ -2 \end{bmatrix} = \begin{bmatrix} 6 \\ 3 \end{bmatrix}$$

or

$$1\begin{bmatrix} 2 \\ 1 \end{bmatrix} + (-1)\begin{bmatrix} -4 \\ -2 \end{bmatrix} = \begin{bmatrix} 6 \\ 3 \end{bmatrix}, \quad \text{etc.}$$

Thus $\mathbf{w}$ is a linear combination of $\mathbf{v}_1$ and $\mathbf{v}_2$ (in many ways).

(b) As in part (a), we try to determine if there are scalars a_1 and a_2 such that

$$a_1\mathbf{v}_1 + a_2\mathbf{v}_2 = \mathbf{x}$$

We obtain the system

$$a_1\begin{bmatrix} 2 \\ 1 \end{bmatrix} + a_2\begin{bmatrix} -4 \\ -2 \end{bmatrix} = \begin{bmatrix} 3 \\ 5 \end{bmatrix} \qquad \text{or} \qquad \begin{bmatrix} 2 & -4 \\ 1 & -2 \end{bmatrix}\begin{bmatrix} a_1 \\ a_2 \end{bmatrix} = \begin{bmatrix} 3 \\ 5 \end{bmatrix}$$

The augmented matrix is

$$\left[\begin{array}{cc|c} 2 & -4 & 3 \\ 1 & -2 & 5 \end{array}\right] \qquad \text{which reduces to} \qquad \left[\begin{array}{cc|c} 2 & -4 & 3 \\ 0 & 0 & \frac{7}{2} \end{array}\right]$$

This system is inconsistent. Hence there is no solution and $\mathbf{x}$ is not a linear combination of $\mathbf{v}_1$ and $\mathbf{v}_2$. ■

There is a very important fact to be observed in Example 6(a).

(3.31) If a vector $\mathbf{w}$ is a linear combination of $\mathbf{v}_1, \ldots, \mathbf{v}_k$, then that combination may not be unique.

That is, there may be many sets of numbers $a_1, \ldots, a_k$ such that $\mathbf{w} = a_1\mathbf{v}_1 + \cdots + a_k\mathbf{v}_k$. The importance of whether the a's are unique or not will be discussed in Section 3.5.

The next two observations are straightforward and surprisingly useful. The first essentially has been covered in Section 1.3.

(3.32) If A is an $m \times n$ matrix and $\mathbf{x}$ is an $n \times 1$ vector, then the product $\mathbf{y} = A\mathbf{x}$ is a linear combination of the columns of A.

Proof See the proof of Theorem (1.32), Section 1.3. ■

The second observation is new and will be used suprisingly often.

(3.33) Suppose a vector $\mathbf{x}$ is a linear combination of the vectors $\mathbf{v}_1, \ldots, \mathbf{v}_k$ and each $\mathbf{v}_i$ is a linear combination of the vectors $\mathbf{w}_1, \ldots, \mathbf{w}_n$. Then $\mathbf{x}$ is a linear combination of $\mathbf{w}_1, \ldots, \mathbf{w}_n$.

Proof By hypothesis there are numbers $a_1, \ldots, a_k$ such that

$$\mathbf{x} = a_1\mathbf{v}_1 + \cdots + a_k\mathbf{v}_k \tag{3.34}$$

and other numbers c_{ij} such that

$$\mathbf{v}_1 = c_{11}\mathbf{w}_1 + \cdots + c_{n1}\mathbf{w}_n, \quad \ldots, \quad \mathbf{v}_k = c_{1k}\mathbf{w}_1 + \cdots + c_{nk}\mathbf{w}_n \tag{3.35}$$

Substituting all the equations of (3.35) into Equation (3.34) and rearranging, we obtain

$$\begin{aligned} \mathbf{x} &= a_1(c_{11}\mathbf{w}_1 + \cdots + c_{n1}\mathbf{w}_n) + \cdots + a_k(c_{1k}\mathbf{w}_1 + \cdots + c_{nk}\mathbf{w}_n) \\ &= (a_1c_{11} + \cdots + a_kc_{1k})\mathbf{w}_1 + \cdots + (a_1c_{n1} + \cdots + a_kc_{nk})\mathbf{w}_n \end{aligned} \tag{3.36}$$

This expresses $\mathbf{x}$ as a linear combination of the $\mathbf{w}$'s, so we are done. ■

Example 7 Let

$$\mathbf{x} = \begin{bmatrix} -1 \\ 23 \end{bmatrix}, \quad \mathbf{v}_1 = \begin{bmatrix} -7 \\ 1 \end{bmatrix}, \quad \mathbf{v}_2 = \begin{bmatrix} 10 \\ 10 \end{bmatrix}, \quad \mathbf{w}_1 = \begin{bmatrix} 1 \\ 2 \end{bmatrix}, \quad \mathbf{w}_2 = \begin{bmatrix} -3 \\ -1 \end{bmatrix}$$

Then you can easily check that

$$\mathbf{x} = 3\mathbf{v}_1 + 2\mathbf{v}_2 \tag{3.37}$$

so that $\mathbf{x}$ is a linear combination of $\mathbf{v}_1$ and $\mathbf{v}_2$. Furthermore you can see that

(3.38) $$\mathbf{v}_1 = 2\mathbf{w}_1 + 3\mathbf{w}_2 \quad \text{and} \quad \mathbf{v}_2 = 4\mathbf{w}_1 - 2\mathbf{w}_2$$

so that each $\mathbf{v}_i$ is a linear combination of $\mathbf{w}_1$ and $\mathbf{w}_2$. If we substitute the expressions in (3.38) into the equation for $\mathbf{x}$ in (3.37) and then rearrange the terms, we obtain

$$\begin{aligned}\mathbf{x} = 3\mathbf{v}_1 + 2\mathbf{v}_2 &= 3(2\mathbf{w}_1 + 3\mathbf{w}_2) + 2(4\mathbf{w}_1 - 2\mathbf{w}_2)\\ &= 14\mathbf{w}_1 + 5\mathbf{w}_2\end{aligned}$$

You can check that indeed $\mathbf{x} = 14\mathbf{w}_1 + 5\mathbf{w}_2$. This expresses $\mathbf{x}$ as a linear combination of $\mathbf{w}_1$ and $\mathbf{w}_2$. ■

The Span of Vectors

(3.39) DEFINITION Let $\mathbf{v}_1, \ldots, \mathbf{v}_n$ be vectors in a vector space V. The set S of all linear combinations of $\mathbf{v}_1, \ldots, \mathbf{v}_n$ is called the **span** of $\mathbf{v}_1, \ldots, \mathbf{v}_n$. We also say that these vectors **span** S.

Example 8 In $\mathbb{R}^3$ let

$$\mathbf{i} = \begin{bmatrix}1\\0\\0\end{bmatrix}, \quad \mathbf{j} = \begin{bmatrix}0\\1\\0\end{bmatrix}, \quad \text{and} \quad \mathbf{k} = \begin{bmatrix}0\\0\\1\end{bmatrix}$$

Then $\mathbf{i}$, $\mathbf{j}$, and $\mathbf{k}$ span $\mathbb{R}^3$ because any vector $\mathbf{x} = [x_1 \quad x_2 \quad x_3]^T$ can be expressed as

$$\mathbf{x} = x_1\mathbf{i} + x_2\mathbf{j} + x_3\mathbf{k}$$ ■

Example 9 Let $V = C[a, b]$ and consider the $k + 1$ functions $f_0, \ldots, f_k$ in V given by

$$f_0(x) = 1, \quad f_1(x) = x, \quad f_2(x) = x^2, \quad \ldots, \quad f_k(x) = x^k$$

Then the span of $f_0, \ldots, f_k$ is simply P_k, the set of all polynomials of degree less than or equal to k. This is so because if p is any such polynomial, $p(x) = a_0 + a_1x + a_2x^2 + \cdots + a_kx^k$, then

$$p = a_0f_0 + a_1f_1 + a_2f_2 + \cdots + a_kf_k$$ ■

(3.40) THEOREM Let $\mathbf{v}_1, \ldots, \mathbf{v}_n$ be vectors in a vector space V and let S be their span. Then,

(a) S is a subspace of V.
(b) S is the smallest* subspace of V that contains $\mathbf{v}_1, \ldots, \mathbf{v}_n$.

* *Here "smallest" means every subspace of V that contains $\mathbf{v}_1, \ldots, \mathbf{v}_n$ also contains all of S.*

Proof of (a) Pick any two vectors $\mathbf{u}$ and $\mathbf{w}$ of S and any number r. Then $\mathbf{u}$ and $\mathbf{w}$ are linear combinations of $\mathbf{v}_1, \ldots, \mathbf{v}_n$ so

$$\mathbf{u} = a_1\mathbf{v}_1 + \cdots + a_n\mathbf{v}_n \qquad \text{and} \qquad \mathbf{w} = b_1\mathbf{v}_1 + \cdots + b_n\mathbf{v}_n$$

where $a_1, b_1, \ldots, a_n, b_n$ are scalars. Now

$$\begin{aligned}\mathbf{u} + \mathbf{w} &= (a_1\mathbf{v}_1 + \cdots + a_n\mathbf{v}_n) + (b_1\mathbf{v}_1 + \cdots + b_n\mathbf{v}_n) \\ &= (a_1 + b_1)\mathbf{v}_1 + \cdots + (a_n + b_n)\mathbf{v}_n\end{aligned}$$

This expresses $\mathbf{u} + \mathbf{w}$ as a linear combination of $\mathbf{v}_1, \ldots, \mathbf{v}_n$, and consequently $\mathbf{u} + \mathbf{w}$ is in S. Also,

$$r\mathbf{u} = r(a_1\mathbf{v}_1 + \cdots + a_n\mathbf{v}_n) = (ra_1)\mathbf{v}_1 + \cdots + (ra_n)\mathbf{v}_n$$

Since this expresses $r\mathbf{u}$ as a linear combination of $\mathbf{v}_1, \ldots, \mathbf{v}_n$, $r\mathbf{u}$ is also in S.

Altogether, S is closed under addition and scalar multiplication, so S is a subspace.

Proof of (b) Let W be any subspace of V that contains $\mathbf{v}_1, \ldots, \mathbf{v}_n$. Let $\mathbf{u}$ be any vector in S so that $\mathbf{u}$ is a linear combination of $\mathbf{v}_1, \ldots, \mathbf{v}_n$, $\mathbf{u} = a_1\mathbf{v}_1 + \cdots + a_n\mathbf{v}_n$. Since $\mathbf{v}_1, \ldots, \mathbf{v}_n$ are in W,

$$a_1\mathbf{v}_1, \ldots, a_n\mathbf{v}_n \qquad \text{are in } W \ (W \text{ is closed under scalar multiplication})$$

Therefore,

$$a_1\mathbf{v}_1 + \cdots + a_n\mathbf{v}_n \qquad \text{is also in } W \ (W \text{ is closed under addition})$$

This says that $\mathbf{u}$ is in W. Therefore, every vector in S is also in W, that is, W contains all of S, so we are done. ■

Example 10 Let $\mathbf{j} = [0 \quad 1 \quad 0]^T$ and $\mathbf{k} = [0 \quad 0 \quad 1]^T$ be the vectors in $\mathbb{R}^3$ given in Example 8. Then their span is the yz-plane in $\mathbb{R}^3$, which is a subspace of $\mathbb{R}^3$. ■

Example 11 Let M_{22} be the space of all 2×2 matrices. Let

$$E_1 = \begin{bmatrix} 1 & 0 \\ 0 & 0 \end{bmatrix} \qquad \text{and} \qquad E_2 = \begin{bmatrix} 0 & 0 \\ 0 & 1 \end{bmatrix}$$

which are in M_{22}. Then the span of E_1 and E_2 is the set of all 2×2 matrices of the form

$$aE_1 + bE_2 = a\begin{bmatrix} 1 & 0 \\ 0 & 0 \end{bmatrix} + b\begin{bmatrix} 0 & 0 \\ 0 & 1 \end{bmatrix} = \begin{bmatrix} a & \\ & b \end{bmatrix}$$

Thus the span of E_1 and E_2 is the set of all 2×2 diagonal matrices, which is a subspace of M_{22}. ■

Exercise 3.4 In Exercises 1–24, a vector space V and a subset S of V are given.

(a) Is S closed under addition?
(b) Is S closed under scalar multiplication?
(c) Is S a subspace of V?

1. $V = \mathbb{R}^2$, $S = \{(s, 2s) \mid s \text{ is a real number}\}$
2. $V = \mathbb{R}^3$, $S = \{(0, s, t) \mid s, t \text{ are real numbers}\}$
3. $V = \mathbb{R}^2$, $S = \{(n, n) \mid n \text{ is an integer}\}$
4. $V = \mathbb{R}^3$, $S = \{(x, y, z) \mid x, y, z \geq 0\}$
5. $V = \mathbb{R}^3$, $S = \{(x, y, z) \mid z = x + y\}$
6. $V = \mathbb{R}^3$, $S = \{(x, y, z) \mid z = x + y + 1\}$
7. $V = \mathbb{R}^n$, $S = \{(x, 2x, 3x, \ldots, nx) \mid x \text{ is a real number}\}$
8. $V = \mathbb{R}^n$, $S = \{(x, x^2, x^3, \ldots, x^n) \mid x \text{ is a real number}\}$
9. $V = M_{22}$, $S = \left\{\begin{bmatrix} 0 & b \\ a & 0 \end{bmatrix} \middle| a \text{ and } b \text{ are real numbers}\right\}$
10. $V = M_{23}$, $S = \left\{\begin{bmatrix} a & b & c \\ d & e & f \end{bmatrix} \middle| a, b, c, d, e, f \text{ are integers}\right\}$
11. $V = M_{33}$, $S = \{A \mid A = A^T\}$
12. $V = M_{33}$, $S = \{A \mid A \text{ is invertible}\}$
13. $V = M_{44}$, $S = \{A \mid a_{11} + a_{22} + a_{33} + a_{44} = 0\}$
14. $V = M_{22}$, $S = \{A \mid A \text{ is singular}\}$
15. $V = C[a, b]$, $S = \{f \mid f(a) = 0\}$
16. $V = C[a, b]$, $S = \{f \mid f(a) = 1\}$
17. $V = C[a, b]$, $S = \left\{f \middle| \int_a^b f(x)\,dx = 0\right\}$
18. $V = C[a, b]$, $S = \{f \mid f(x) \leq 0, \quad \text{all } x\}$
19. $V = C[a, b]$, $S = \{f \mid f(x) = 0, \quad \text{for at least one } x \text{ in } [a, b]\}$
20. $V = C[a, b]$, $S = \{f \mid f = a + b \sin x + c \cos x, \quad a, b, c \text{ are any real numbers}\}$
21. $V = P_3$, $S = \{ax + bx^3 \mid a, b \text{ are any real numbers}\}$
22. $V = P_2$, $S = \{1 + ax + bx^2 \mid a, b \text{ are any real numbers}\}$
23. $V = P_4$, $S = \{a_0 + a_1x + a_2x^2 + a_3x^3 + a_4x^4 \mid a_0 + \cdots + a_4 = 0\}$

24. $V = P_4, \quad S = \{a_0 + a_1x + a_2x^2 + a_3x^3 + a_4x^4 | a_0a_1a_2a_3a_4 = 0\}$

In Exercises 25–32, find $NS(A)$ for the given matrix A. For which n is $NS(A)$ a subspace of $\mathbb{R}^n$? Sketch $NS(A)$ in $\mathbb{R}^2$ or $\mathbb{R}^3$.

25. $A = \begin{bmatrix} 1 & -3 \\ -2 & 6 \end{bmatrix}$

26. $\begin{bmatrix} 1 & 2 \\ 3 & 4 \end{bmatrix}$

27. $A = \begin{bmatrix} 0 & 0 & 0 \\ 0 & 0 & 0 \end{bmatrix}$

28. $\begin{bmatrix} 1 & 3 & 2 \\ 2 & 6 & 4 \end{bmatrix}$

29. $[1 \quad 2 \quad 3]$

30. $\begin{bmatrix} 1 \\ 2 \\ 3 \\ 4 \end{bmatrix}$

31. $A = \begin{bmatrix} 2 & -1 \\ -4 & 2 \end{bmatrix}$

32. $A = \begin{bmatrix} 3 & -1 & 1 \\ -6 & 2 & -2 \\ -3 & 1 & -1 \end{bmatrix}$

33. Determine if the given vector $\mathbf{w}$ is a linear combination of $\mathbf{v}_1 = (2, -1)$ and $\mathbf{v}_2 = (-4, 2)$. If it is, find a_1, a_2 such that $\mathbf{w} = a_1\mathbf{v}_1 + a_2\mathbf{v}_2$.

(a) $(-6, 3)$ (b) $(1, 1)$ (c) $(0, 0)$ (d) $(10, -5)$

34. Same question as Exercise 33 for $\mathbf{v}_1 = (2, -1, 3)$, $\mathbf{v}_2 = (1, 1, 2)$.

(a) $(1, -5, 0)$ (b) $(1, 1, 1)$ (c) $(0, 0, 0)$ (d) $(1, 7, 4)$

35. Same question as Exercise 33 for

$$\mathbf{v}_1 = \begin{bmatrix} 1 & 2 \\ -2 & 1 \end{bmatrix}, \quad \mathbf{v}_2 = \begin{bmatrix} 3 & 2 \\ -1 & 1 \end{bmatrix}$$

(a) $\begin{bmatrix} 1 & 0 \\ 0 & 1 \end{bmatrix}$ (b) $\begin{bmatrix} 0 & 0 \\ 0 & 0 \end{bmatrix}$ (c) $\begin{bmatrix} -3 & 2 \\ -4 & 1 \end{bmatrix}$ (d) $\begin{bmatrix} -13 & -6 \\ 1 & -3 \end{bmatrix}$

36. Similar question to Exercise 33 for $p_1 = x + x^2$, $p_2 = x + x^3$, and $p_3 = x + x^2 + x^3$.

(a) $2x + x^2$ (b) $2 - 3x + 4x^2 + x^3$
(c) $\mathbf{0}$ (d) x

37. Similar question to Exercise 33 for $f_1(x) = \sin^2 x$, $f_2(x) = \cos^2 x$.

(a) $\cos 2x$ (b) $1 + 2x^2$ (c) $\mathbf{0}$ (d) 1 (e) $\sin x$

38. Similar question to Exercise 33 for $\mathbf{v}_1 = (1, -1, 0, 0)$, $\mathbf{v}_2 = (0, 1, -1, 0)$, $\mathbf{v}_3 = (0, 0, 1, -1)$.

(a) $(1, 1, 1, 1)$ (b) $(2, -1, -2, 1)$
(c) $(1, 0, 0, 1)$ (d) $(2, 1, 1, 2)$

In Exercises 39–46, describe the span of the given vectors.

39. $\mathbf{v} = \begin{bmatrix} -3 \\ -3 \end{bmatrix}$

40. $\mathbf{v} = \begin{bmatrix} 4 \\ 0 \end{bmatrix}$

41. $\mathbf{v} = \begin{bmatrix} 2 \\ 1 \\ 0 \end{bmatrix}, \quad \mathbf{w} = \begin{bmatrix} 3 \\ -1 \\ 0 \end{bmatrix}$

42. $\mathbf{u} = \begin{bmatrix} 4 \\ 0 \\ 1 \end{bmatrix}, \quad \mathbf{v} = \begin{bmatrix} 2 \\ 0 \\ -3 \end{bmatrix}, \quad \mathbf{w} = \begin{bmatrix} 8 \\ 0 \\ -1 \end{bmatrix}$

43. $A = \begin{bmatrix} 2 & 0 \\ 0 & -1 \end{bmatrix}, \quad B = \begin{bmatrix} 3 & 0 \\ 0 & 2 \end{bmatrix}$

44. $A = \begin{bmatrix} 1 & 0 \\ 2 & -1 \end{bmatrix}, \quad B = \begin{bmatrix} 0 & 0 \\ 3 & 2 \end{bmatrix}, \quad C = \begin{bmatrix} 0 & 0 \\ 0 & -2 \end{bmatrix}$

45. $p_1 = 2x + 3$, $p_2 = -3x - 5$

46. $p_1 = 3x + 4x^2$, $p_2 = 2x - 5x^2$, $p_3 = x + x^2$

3.5 Linear Independence

If $\mathbf{v}_1$ and $\mathbf{v}_2$ are two vectors in $\mathbb{R}^3$, then they determine a plane unless one of them is a multiple of the other. This is illustrated in Figure 3.20.

Figure 3.20

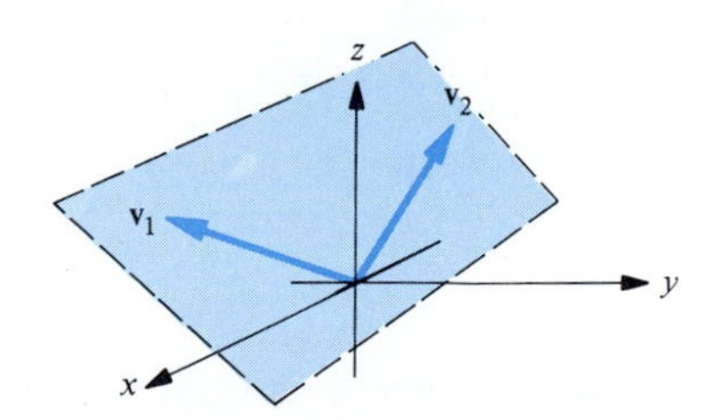

(a) *Neither of* $\mathbf{v}_1$ *or* $\mathbf{v}_2$ *is a multiple of the other. Their span is a plane.*

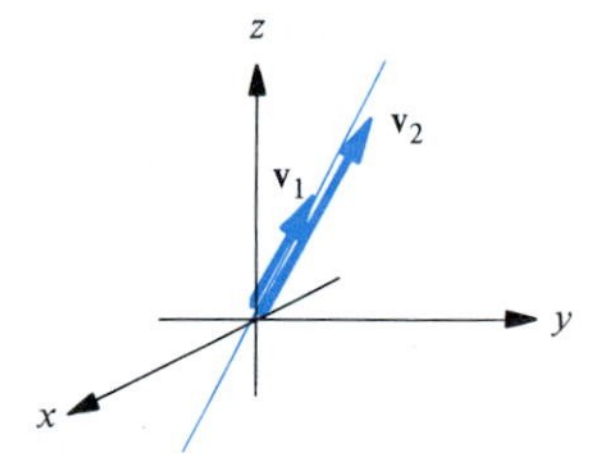

(b) $\mathbf{v}_1$ *and* $\mathbf{v}_2$ *are multiples of each other. Their span is a line.*

This illustrates a geometric consequence of a "dependency" among vectors. There are algebraic consequences, too. Suppose you reduce a matrix to echelon form and the resulting matrix has a zero row. Then there is a similar kind of dependency among the rows of the original matrix. For example,

$$\begin{bmatrix} 2 & 3 & -1 & 2 \\ -6 & 1 & 4 & 5 \\ 12 & 8 & -7 & 1 \end{bmatrix} \rightarrow \begin{bmatrix} 2 & 3 & -1 & 2 \\ 0 & 10 & 1 & 11 \\ 0 & -10 & -1 & -11 \end{bmatrix} \rightarrow \begin{bmatrix} 2 & 3 & -1 & 2 \\ 0 & 10 & 1 & 11 \\ 0 & 0 & 0 & 0 \end{bmatrix}$$

and we shall see the dependency in Example 3.

In this section we shall discuss dependency; this is an important topic, and it will grow in importance over the next several sections.

(3.41)

DEFINITION (a) A set of two or more vectors is **linearly dependent** if one vector in the set is a linear combination of the others. A set of one vector is **linearly dependent** if that vector is the zero vector.

(b) A (nonempty) set of vectors is **linearly independent** if it is not linearly dependent.

NOTE It is common to abuse the language slightly and call the vectors themselves linearly dependent or independent whereas it is really the set containing those vectors that has the property.

Example 1 In Figure 3.20a, neither $\mathbf{v}_1$ nor $\mathbf{v}_2$ is a multiple of the other, so they are linearly independent (and their span is a plane). In Figure 3.20b, $\mathbf{v}_2$ is a multiple of $\mathbf{v}_1$, so $\mathbf{v}_1$ and $\mathbf{v}_2$ are linearly dependent (and their span is a line). ■

Example 2 Suppose a collection S vectors contains the zero vector. Then it is linearly dependent, because

$$\mathbf{0} = 0\mathbf{v}_1 + 0\mathbf{v}_2 + \cdots + 0\mathbf{v}_k \qquad \text{for any } \mathbf{v}_1, \ldots, \mathbf{v}_k \in S$$

In other words, the zero vector is always a linear combination of any collection of vectors, so a set containing the zero vector must be linearly dependent. ■

Example 3 The rows of the matrix (considered as vectors in $\mathbb{R}^4$)

$$\begin{bmatrix} 2 & 3 & -1 & 2 \\ -6 & 1 & 4 & 5 \\ 12 & 8 & -7 & 1 \end{bmatrix}$$

are linearly dependent because the last row is a linear combination of the first two:

$$[12 \quad 8 \quad -7 \quad 1] = 3[2 \quad 3 \quad -1 \quad 2] - 1[-6 \quad 1 \quad 4 \quad 5]$$ ■

Example 4 The polynomials

$$3x^2 - 2x + 5, \qquad 1, \qquad x, \qquad x^2$$

are linearly dependent because the first is a linear combination of the last three

$$3x^2 - 2x + 5 = 5(1) + (-2)(x) + 3(x^2)$$

■

Example 5 The vectors $\boldsymbol{\varepsilon}_1 = (1, 0, 0)$, $\boldsymbol{\varepsilon}_2 = (0, 1, 0)$, and $\boldsymbol{\varepsilon}_3 = (0, 0, 1)$ are linearly independent. For example, suppose $\boldsymbol{\varepsilon}_1$ were a linear combination of the other two; that is, suppose

$$\boldsymbol{\varepsilon}_1 = c_1\boldsymbol{\varepsilon}_2 + c_2\boldsymbol{\varepsilon}_3$$

Then

$$(1, 0, 0) = c_1(0, 1, 0) + c_2(0, 0, 1)$$

or

$$(1, 0, 0) = (0, c_1, c_2)$$

By examining the first coordinates, we see that the equality cannot hold since $1 \neq 0$. Thus $\boldsymbol{\varepsilon}_1$ is not a linear combination of the other two. Similarly, neither $\boldsymbol{\varepsilon}_2$ nor $\boldsymbol{\varepsilon}_3$ is a linear combination of the remaining two vectors. ■

It is not difficult to imagine that the argument just given can become extremely cumbersome when the vectors are more complicated than $\boldsymbol{\varepsilon}_1$, $\boldsymbol{\varepsilon}_2$, and $\boldsymbol{\varepsilon}_3$. Therefore, we need an alternative condition for linear independence. Suppose we have a set of vectors $S = \{\mathbf{v}_1, \ldots, \mathbf{v}_k\}$ and we wish to express the zero vector $\mathbf{0}$ as a linear combination of the $\mathbf{v}_i$'s

$$\mathbf{0} = c_1\mathbf{v}_1 + \cdots + c_k\mathbf{v}_k$$

There is, of course, the trivial way of doing this, namely, taking $c_1 = 0, \ldots, c_k = 0$, and this is called the **trivial solution**. A **nontrivial solution** is a set of numbers $c_1, \ldots, c_k$, at least one of which is nonzero, such that $\mathbf{0} = c_1\mathbf{v}_1 + \cdots + c_k\mathbf{v}_k$. The surprising fact is that the existence of nontrivial solutions characterizes linear dependence.

Equivalently a set of vectors $S = \{\mathbf{v}_1, \ldots, \mathbf{v}_k\}$ is linearly independent if and only if the only solution to the equation

$$\mathbf{0} = c_1\mathbf{v}_1 + \cdots + c_k\mathbf{v}_k$$

is the trivial solution.

(3.42) THEOREM A set of vectors $S = \{\mathbf{v}_1, \ldots, \mathbf{v}_k\}$ is linearly dependent if and only if there are numbers $c_1, \ldots, c_k$, at least one of which is nonzero, such that

$$\mathbf{0} = c_1\mathbf{v}_1 + \cdots + c_k\mathbf{v}_k$$

Note that this formulation does not require that awkward separation into two cases, when S contains two or more vectors or when S contains only one vector. Of course, the proof must treat these two cases separately since the definition does.

Proof Suppose a set of vectors $S = \{\mathbf{v}_1, \ldots, \mathbf{v}_k\}$ is linearly dependent.

If S contains two or more vectors, then one of them, say $\mathbf{v}_i$, is a linear combination of the remaining vectors

$$\mathbf{v}_i = a_1\mathbf{v}_1 + \cdots + a_{i-1}\mathbf{v}_{i-1} + a_{i+1}\mathbf{v}_{i+1} + \cdots + a_k\mathbf{v}_k$$

Then

$$\mathbf{0} = a_1\mathbf{v}_1 + \cdots + a_{i-1}\mathbf{v}_{i-1} + (-1)\mathbf{v}_i + a_{i+1}\mathbf{v}_{i+1} + \cdots + a_k\mathbf{v}_k$$

If we let $c_j = a_j$ for $j \neq i$ and $c_i = -1$, then

$$\mathbf{0} = c_1\mathbf{v}_1 + \cdots + c_k\mathbf{v}_k$$

with at least one of the c's, namely, c_i, being nonzero.

If $S = \{\mathbf{v}_1\}$, then $\mathbf{v}_1 = 0$ by definition. Therefore, if we let $c_1 = 1$, we have

$$\mathbf{0} = c_1\mathbf{v}_1$$

with $c_1 \neq 0$.

Conversely, suppose $S = \{\mathbf{v}_1, \ldots, \mathbf{v}_k\}$ is a set of vectors and there are numbers $c_1, \ldots, c_k$, at least one of which is nonzero, such that

$$\mathbf{0} = c_1\mathbf{v}_1 + \cdots + c_k\mathbf{v}_k \tag{3.43}$$

If $S = \{\mathbf{v}_1\}$, then $\mathbf{0} = c_1\mathbf{v}_1$ and $c_1 \neq 0$. Thus $\mathbf{v}_1 = \mathbf{0}$ [by Theorem (3.24d), Section 3.3, if $a\mathbf{v} = \mathbf{0}$ then either $a = 0$ or $\mathbf{v} = \mathbf{0}$] and hence S is linearly dependent by definition.

If S contains more than one vector, pick any c that is nonzero, say c_i and rewrite Equation (3.43) as

$$c_i\mathbf{v}_i = -c_1\mathbf{v}_1 - \cdots - c_{i-1}\mathbf{v}_{i-1} - c_{i+1}\mathbf{v}_{i+1} - \cdots - c_k\mathbf{v}_k$$

Since $c_i \neq 0$, we can divide this equation by c_i, obtaining

$$\mathbf{v}_i = -\frac{c_1}{c_i}\mathbf{v}_1 - \cdots - \frac{c_{i-1}}{c_i}\mathbf{v}_{i-1} - \frac{c_{i+1}}{c_i}\mathbf{v}_{i+1} - \cdots - \frac{c_k}{c_i}\mathbf{v}_k$$

This expresses $\mathbf{v}_i$ as a linear combination of the remaining vectors, so we are done. ■

Example 6 Determine if the following vectors are linearly dependent or independent. If they are dependent, express one of them as a linear combination of the rest.

(a) $\mathbf{v}_1 = (2, 4, 14)$, $\mathbf{v}_2 = (7, -3, 15)$, $\mathbf{v}_3 = (-1, 4, 7)$
(b) $\mathbf{v}_1 = (2, 4, 14)$, $\mathbf{v}_2 = (7, -3, 15)$, $\mathbf{v}_3 = (-1, 4, 5)$

Solution (a) We wish to determine if there are any nontrivial solutions to the equation

$$c_1\mathbf{v}_1 + c_2\mathbf{v}_2 + c_3\mathbf{v}_3 = \mathbf{0} \tag{3.44}$$

If we write the $\mathbf{v}_i$'s and the zero vector as column vectors, Equation (3.44) is equivalent to

$$c_1\begin{bmatrix}2\\4\\14\end{bmatrix} + c_2\begin{bmatrix}7\\-3\\15\end{bmatrix} + c_3\begin{bmatrix}-1\\4\\7\end{bmatrix} = \begin{bmatrix}0\\0\\0\end{bmatrix}$$

which in turn is equivalent to the homogeneous system

$$\begin{aligned}2c_1 + 7c_2 - c_3 &= 0\\ 4c_1 - 3c_2 + 4c_3 &= 0\\ 14c_1 + 15c_2 + 7c_3 &= 0\end{aligned} \quad \text{or} \quad \begin{bmatrix}2 & 7 & -1\\4 & -3 & 4\\14 & 15 & 7\end{bmatrix}\begin{bmatrix}c_1\\c_2\\c_3\end{bmatrix} = \begin{bmatrix}0\\0\\0\end{bmatrix} \tag{3.45}$$

We now find the augmented matrix and reduce it to row echelon form.

$$\begin{bmatrix}2 & 7 & -1 & 0\\4 & -3 & 4 & 0\\14 & 15 & 7 & 0\end{bmatrix} \to \begin{bmatrix}2 & 7 & -1 & 0\\0 & -17 & 6 & 0\\0 & -34 & 14 & 0\end{bmatrix} \to \begin{bmatrix}2 & 7 & -1 & 0\\0 & -17 & 6 & 0\\0 & 0 & 2 & 0\end{bmatrix}$$

The associated system is

$$\begin{aligned}2c_1 + 7c_2 - c_3 &= 0\\ -17c_2 + 6c_3 &= 0\\ 2c_3 &= 0\end{aligned}$$

and this has only the trivial solution $c_1 = c_2 = c_3 = 0$. Hence the vectors $\mathbf{v}_1, \mathbf{v}_2, \mathbf{v}_3$ are linearly independent.

(b) We proceed as in (a) and obtain

$$\begin{bmatrix}2 & 7 & -1\\4 & -3 & 4\\14 & 15 & 5\end{bmatrix}\begin{bmatrix}c_1\\c_2\\c_3\end{bmatrix} = \begin{bmatrix}0\\0\\0\end{bmatrix}$$

Next we find the augmented matrix and reduce it to row echelon form.

$$\begin{bmatrix}2 & 7 & -1 & 0\\4 & -3 & 4 & 0\\14 & 15 & 5 & 0\end{bmatrix} \to \begin{bmatrix}2 & 7 & -1 & 0\\0 & -17 & 6 & 0\\0 & -34 & 12 & 0\end{bmatrix} \to \begin{bmatrix}2 & 7 & -1 & 0\\0 & -17 & 6 & 0\\0 & 0 & 0 & 0\end{bmatrix}$$

The associated system now is

$$\begin{aligned}2c_1 + 7c_2 - c_3 &= 0\\ -17c_2 + 6c_3 &= 0\\ 0 &= 0\end{aligned}$$

This has (infinitely many) nontrivial solutions, so the vectors $\mathbf{v}_1, \mathbf{v}_2, \mathbf{v}_3$ are linearly dependent. We can obtain one nontrivial solution by setting $c_3 = 1$

and solving this system for c_1 and c_2:

$$\begin{aligned} 2c_1 + 7c_2 &= 1 \\ -17c_2 &= -6 \end{aligned}$$

We obtain $(c_1, c_2, c_3) = (-\frac{25}{34}, \frac{6}{17}, 1)$. Thus

$$-\tfrac{25}{34}\mathbf{v}_1 + \tfrac{6}{17}\mathbf{v}_2 + \mathbf{v}_3 = \mathbf{0} \qquad \text{or} \qquad \mathbf{v}_3 = \tfrac{25}{34}\mathbf{v}_1 - \tfrac{6}{17}\mathbf{v}_2$$

This expresses $\mathbf{v}_3$ as a linear combination of the other two, so we are done. ■

An Important Theorem

Example 6 illustrates the kind of computation that is often required to determine if a set of vectors is linearly dependent or independent. However, there is an important situation in which we can tell whether or not a set of vectors is linearly dependent simply by counting the number of vectors.

(3.46) THEOREM Let S be a set of k vectors in $\mathbb{R}^n$. Then,

if $k > n$, then S is linearly dependent

Example 7 Let $\mathbf{v}_1 = (2, -3)$, $\mathbf{v}_2 = (4, -2)$, and $\mathbf{v}_3 = (-4, 3)$. These are three vectors in $\mathbb{R}^2$, so they must be linearly dependent, by Theorem (3.46). Indeed it is easy to check that $2\mathbf{v}_1 + 3\mathbf{v}_2 + 4\mathbf{v}_3 = \mathbf{0}$. ■

We shall need the next lemma for the proof of Theorem (3.46).

(3.47) LEMMA A homogeneous system of linear equations with more unknowns than equations always has a nontrivial solution.

Proof Consider the homogeneous system

$$\begin{aligned} a_{11}x_1 + a_{12}x_2 + \cdots + a_{1k}x_k &= 0 \\ a_{21}x_1 + a_{22}x_2 + \cdots + a_{2k}x_k &= 0 \\ &\vdots \\ a_{n1}x_1 + a_{n2}x_2 + \cdots + a_{nk}x_k &= 0 \end{aligned}$$

where $k > n$. When we reduce this system to row echelon form, there will be at most n pivots (perhaps fewer). Thus the resulting system will look something like

$$\begin{aligned} a_{1k_1}x_{k_1} + \cdots \qquad\qquad\qquad &= 0 \\ a_{2k_2}x_{k_2} + \cdots \qquad\quad &= 0 \\ &\vdots \\ a_{nk_n}x_{k_n} + \cdots &= 0 \end{aligned}$$

Since there are k variables, at most n pivots, and $k > n$, this system must have at least one free variable. Therefore, if we set one of the free variables equal to 1 and the remaining free variables equal to 0, we can solve the system and obtain a nontrivial solution. ■

Example 8 Find a nontrivial solution to

$$\begin{aligned} x_1 - 2x_2 + x_3 + 4x_4 - x_5 &= 0 \\ -2x_1 + 4x_2 + 3x_3 - 2x_4 + x_5 &= 0 \\ 3x_1 - 6x_2 + 13x_3 - 8x_4 - 2x_5 &= 0 \\ 2x_1 - 4x_2 + 17x_3 - 6x_4 - 2x_5 &= 0 \end{aligned}$$

Solution Reduce the system to row echelon form obtaining

$$\begin{aligned} x_1 - 2x_2 + x_3 + 4x_4 - x_5 &= 0 \\ 5x_3 + 6x_4 - x_5 &= 0 \\ -32x_4 + 3x_5 &= 0 \end{aligned}$$

as you can easily check. The leading variables are x_1, x_3, and x_4, and the free variables are x_2 and x_5. If we set $x_2 = 1$, $x_5 = 0$, and solve, we obtain $(x_1, x_2, x_3, x_4, x_5) = (2, 1, 0, 0, 0)$ as one nontrivial solution. (Of course, we could let x_2 or x_5 be any numbers we want, provided at least one is nonzero, and obtain a nontrivial solution.) ■

We are now ready to prove the main theorem.

Proof of Theorem (3.46) Let $S = \{\mathbf{v}_1, \ldots, \mathbf{v}_k\}$ be a set of vectors in $\mathbb{R}^n$, where $k > n$. We wish to show there are numbers $c_1, \ldots, c_k$, at least one of which is nonzero, such that

$$c_1\mathbf{v}_1 + \cdots + c_k\mathbf{v}_k = \mathbf{0} \tag{3.48}$$

Let

$$\begin{aligned} \mathbf{v}_1 &= (v_{11}, v_{21}, \ldots, v_{n1}) \\ \mathbf{v}_2 &= (v_{12}, v_{22}, \ldots, v_{n2}) \\ &\vdots \\ \mathbf{v}_k &= (v_{1k}, v_{2k}, \ldots, v_{nk}) \end{aligned}$$

Then Equation (3.48) is equivalent to

$$\begin{aligned} v_{11}c_1 + v_{12}c_2 + \cdots + v_{1k}c_k &= 0 \\ v_{21}c_1 + v_{22}c_2 + \cdots + v_{2k}c_k &= 0 \\ &\vdots \\ v_{n1}c_1 + v_{n2}c_2 + \cdots + v_{nk}c_k &= 0 \end{aligned} \tag{3.49}$$

This is now a homogeneous linear system and, since $k > n$, it has more unknowns than equations. By Lemma (3.47), the system (3.49) has a nontrivial solution $c_1, \ldots, c_k$, so we are done. ■

Exercise 3.5 In Exercises 1–6, show *by inspection* that the vectors are linearly dependent.

1. $\mathbf{u}_1 = (2, -1)$, $\mathbf{u}_2 = (6, -3)$, in $\mathbb{R}^2$
2. $\mathbf{v}_1 = (4, -1, 3)$, $\mathbf{v}_2 = (2, 3, -1)$, $\mathbf{v}_3 = (-1, 2, -1)$, $\mathbf{v}_4 = (5, 2, 3)$, in $\mathbb{R}^3$
3. $\mathbf{w}_1 = (2, -1, 4)$, $\mathbf{w}_2 = (5, 2, 3)$, $\mathbf{w}_3 = (0, 0, 0)$, in $\mathbb{R}^3$
4. $p_1 = -x + 3x^3$, $p_2 = 2x - 6x^3$, in P_3
5. $A_1 = \begin{bmatrix} 1 & 2 \\ 0 & 0 \end{bmatrix}$, $A_2 = \begin{bmatrix} 0 & 0 \\ 3 & -1 \end{bmatrix}$, $A_3 = \begin{bmatrix} 2 & 4 \\ -3 & 1 \end{bmatrix}$, in M_{22}
6. $f_1(x) = \sin^2 x$, $f_2(x) = \cos^2 x$, $f_3(x) = 1$, in $C[0, 1]$

In Exercises 7–12, determine if the given vectors span a line, or a plane, or something larger, and relate this to the fact that they are linearly dependent or independent.

7. $\mathbf{u}_1 = (1, 2)$, $\mathbf{u}_2 = (-2, -4)$
8. $\mathbf{v}_1 = (1, 2)$ $\mathbf{v}_2 = (2, 1)$
9. $\mathbf{w}_1 = (4, 0, 3, -2)$, $\mathbf{w}_2 = (-8, 0, -6, 4)$, $\mathbf{w}_3 = (-2, 0, -\frac{3}{2}, 1)$
10. $\mathbf{x}_1 = (4, 0, 3, -2)$, $\mathbf{x}_2 = (-8, 0, -6, 4)$, $\mathbf{x}_3 = (-2, 0, -3, 1)$
11. $\mathbf{y}_1 = (4, 0, 3, -2)$, $\mathbf{y}_2 = (-8, 1, -6, 4)$, $\mathbf{y}_3 = (-2, 0, -3, 1)$
12. $\mathbf{z}_1 = (3, -1, -2)$, $\mathbf{z}_2 = (2, 3, -1)$, $\mathbf{z}_3 = (0, -11, -1)$

In Exercises 13–26, determine if the given vectors are linearly dependent or independent. Do this an easy way, if possible.

13. $(2, -1)$, $(3, 4)$, $(2, -3)$, in $\mathbb{R}^2$
14. $(2, -1, 3)$, $(3, 4, 1)$, $(2, -3, 4)$, in $\mathbb{R}^3$
15. $(1, 1, 0, 0)$, $(0, 0, 2, 2)$, $(3, 1, 4, 1)$, in $\mathbb{R}^4$
16. $(1, 1, 0, 0)$, $(0, 1, 1, 0)$, $(0, 0, 1, 1)$, $(1, 0, 0, 1)$, in $\mathbb{R}^4$
17. $(3, -1, 2, -1)$, $(1, 2, 5, 2)$, $(3, -1, 2, -1)$, in $\mathbb{R}^4$
18. $(1, 0, 0, 0)$, $(1, 1, 0, 0)$, $(1, 1, 1, 0)$, $(1, 1, 1, 1)$, in $\mathbb{R}^4$
19. $p_1(x) = 1 + x$, $p_2(x) = x^2 + x^3$, $p_3(x) = -2 - 2x + 3x^2 + 3x^3$, in P_3
20. $p_1(x) = 1 + x + x^2$, $p_2(x) = 2 - x + 3x^2$, $p_3(x) = -1 + 5x - 3x^2$, in P_2

21. $p_1(x) = 3 + x - x^2, \quad p_2(x) = 2 - 3x + 2x^2, \quad p_3(x) = 1 + x + x^2, \qquad$ in P_2

22. $p_1(x) = 1, \quad p_2(x) = 1 + x, \quad p_3(x) = 1 + x + x^2, \quad p_4(x) = 1 + x + x^2 + x^3,$ in P_3

23. $f_1(x) = x, \quad f_2(x) = \sin x, \quad f_3(x) = \sin 2x, \qquad$ in $C[0, 1]$

24. $A_1 = \begin{bmatrix} 1 & 2 \\ 3 & 0 \end{bmatrix}, \quad A_2 = \begin{bmatrix} -2 & 4 \\ 1 & 0 \end{bmatrix}, \quad A_3 = \begin{bmatrix} 3 & -1 \\ 2 & 0 \end{bmatrix}, \qquad$ in M_{22}

25. $A_1 = \begin{bmatrix} -1 & 3 & 2 \\ 0 & 0 & 0 \end{bmatrix}, \quad A_2 = \begin{bmatrix} 1 & 0 & 0 \\ 2 & 3 & 1 \end{bmatrix}, \quad A_3 = \begin{bmatrix} 0 & 6 & 4 \\ 4 & 6 & 2 \end{bmatrix}, \qquad$ in M_{23}

26. $A_1 = \begin{bmatrix} 2 & 1 \\ 0 & -3 \\ 4 & 0 \end{bmatrix}, \quad A_2 = \begin{bmatrix} 5 & -1 \\ 0 & 4 \\ 3 & 0 \end{bmatrix}, \quad A_3 = \begin{bmatrix} 3 & -2 \\ 0 & 7 \\ -1 & 0 \end{bmatrix}, \qquad$ in M_{32}

27. If $\{\mathbf{u}_1, \mathbf{u}_2, \mathbf{u}_3\}$ is a linearly independent set, show $\{\mathbf{u}_1, \mathbf{u}_2\}$, $\{\mathbf{u}_1, \mathbf{u}_3\}$, $\{\mathbf{u}_1\}$, $\{\mathbf{u}_2\}$ are linearly independent.

28. More generally than Exercise 27, if $\{\mathbf{u}_1, \ldots, \mathbf{u}_n\}$ is linearly independent, then any nonempty subset is linearly independent.

29. Show if $\{\mathbf{u}_1, \mathbf{u}_2, \mathbf{u}_3\}$ is a linearly dependent set, then $\{\mathbf{u}_1, \mathbf{u}_2, \mathbf{u}_3, \mathbf{u}_4\}$ is also a linearly dependent set.

30. More generally than Exercise 29, if $\{\mathbf{u}_1, \ldots, \mathbf{u}_n\}$ is linearly dependent, then $\{\mathbf{u}_1, \ldots, \mathbf{u}_n, \mathbf{u}_{n+1}, \ldots, \mathbf{u}_m\}$ is also linearly dependent.

31. Suppose $\mathbf{b}_1, \ldots, \mathbf{b}_k$ is a linearly independent set of vectors in $\mathbb{R}^m$ and suppose A is an $m \times n$ matrix. If $\mathbf{x}_1, \ldots, \mathbf{x}_k$ are each solutions to the respective equation

$$A\mathbf{x} = \mathbf{b}_1, \ldots, A\mathbf{x} = \mathbf{b}_k$$

Then show that $\mathbf{x}_1, \ldots, \mathbf{x}_k$ are linearly independent vectors in $\mathbb{R}^n$.

32. Suppose the vectors $\mathbf{u}_1$, $\mathbf{u}_2$, and $\mathbf{u}_3$ are linearly dependent. Are the vectors $\mathbf{v}_1 = \mathbf{u}_1 + \mathbf{u}_2$, $\mathbf{v}_2 = \mathbf{u}_1 + \mathbf{u}_3$, and $\mathbf{v}_3 = \mathbf{u}_2 + \mathbf{u}_3$ also linearly dependent? (HINT Assume that $a_1\mathbf{v}_1 + a_2\mathbf{v}_2 + a_3\mathbf{v}_3 = \mathbf{0}$, and see what the a_i's can be.)

33. Suppose A is a 3×5 matrix with linearly independent rows. Then its columns are/might be/are definitely not linearly independent. Why?

34. Suppose that $\mathbf{v}_1, \ldots, \mathbf{v}_k$ are linearly independent and that $\mathbf{v}_{k+1}$ is not in the span of $\mathbf{v}_1, \ldots, \mathbf{v}_k$. Show that $\mathbf{v}_1, \ldots, \mathbf{v}_k, \mathbf{v}_{k+1}$ are still linearly independent. (HINT Assume that $a_1\mathbf{v}_1 + \cdots + a_k\mathbf{v}_k + a_{k+1}\mathbf{v}_{k+1} = \mathbf{0}$, and consider the separate cases $a_{k+1} = 0$ and $a_{k+1} \neq 0$.)

35. Suppose that $\mathbf{v}_k$ is a linear combination of $\mathbf{v}_1, \ldots, \mathbf{v}_{k-1}$. Show that the span of $\{\mathbf{v}_1, \ldots, \mathbf{v}_k\}$ equals the span of $\{\mathbf{v}_1, \ldots, \mathbf{v}_{k-1}\}$. (HINT Use (3.33), Section 3.4, where $\mathbf{w}_1 = \mathbf{v}_1, \ldots, \mathbf{w}_{k-1} = \mathbf{v}_{k-1}$.)

3.6 Basis and Dimension

The purpose of this section is to introduce the fundamental concepts of basis and dimension for vector spaces. Intuitively we think of lines as one dimensional, planes as two dimensional, and the space around us as three dimensional. We will now formalize these intuitive ideas; the resulting concepts of basis and dimension will be integral throughout the text.

(3.50) DEFINITION A **basis** for a vector space V is a set S of vectors of V such that

(a) S is linearly independent, and
(b) S spans V.

The plural of basis is **bases**.

Example 1 In $\mathbb{R}^n$ let $\boldsymbol{\varepsilon}_1 = (1, 0, \ldots, 0)$, $\boldsymbol{\varepsilon}_2 = (0, 1, 0, \ldots, 0), \ldots, \boldsymbol{\varepsilon}_n = (0, \ldots, 0, 1)$. The set $S = \{\boldsymbol{\varepsilon}_1, \ldots, \boldsymbol{\varepsilon}_n\}$ is linearly independent. For suppose $c_1, \ldots, c_n$ are numbers such that $c_1\boldsymbol{\varepsilon}_1 + \cdots + c_n\boldsymbol{\varepsilon}_n = \mathbf{0}$. Then

$$c_1(1, 0, \ldots, 0) + c_2(0, 1, 0, \ldots, 0) + \cdots + c_n(0, \ldots, 0, 1) = (0, \ldots, 0)$$

or

$$(c_1, \ldots, c_n) = (0, \ldots, 0)$$

Thus $c_1 = 0, \ldots, c_n = 0$, so that S is linearly independent by Theorem (3.42), Section 3.5.

To show S spans $\mathbb{R}^n$, pick any $\mathbf{x} = (x_1, \ldots, x_n)$ in $\mathbb{R}^n$. Then easily

$$\mathbf{x} = x_1\boldsymbol{\varepsilon}_1 + \cdots + x_n\boldsymbol{\varepsilon}_n$$

Thus $\mathbf{x}$ is a linear combination of the $\boldsymbol{\varepsilon}_i$'s so S spans $\mathbb{R}^n$. Thus S is both linearly independent and spans $\mathbb{R}^n$, so S is a basis for $\mathbb{R}^n$. Its simplicity makes it very special, and it has its own name; this S is called the **standard basis** for $\mathbb{R}^n$. ■

The next example illustrates the fact that $\mathbb{R}^n$ has many bases besides the standard basis.

Example 2 Let $S = \{\mathbf{v}_1, \mathbf{v}_2, \mathbf{v}_3\}$, where $\mathbf{v}_1 = (2, 1, 0)$, $\mathbf{v}_2 = (-3, -3, 1)$, and $\mathbf{v}_3 = (-2, 1, -1)$. Show S is a basis for $\mathbb{R}^3$.

Solution We need to show (a) S is linearly independent and (b) S spans $\mathbb{R}^3$. We show (b) first.

Let $\mathbf{x} = (x_1, x_2, x_3)$ be any vector in $\mathbb{R}^3$. We need to know there are numbers c_1, c_2, and c_3 such that

$$c_1\mathbf{v}_1 + c_2\mathbf{v}_2 + c_3\mathbf{v}_3 = \mathbf{x} \tag{3.51}$$

Solving Equation (3.51) is equivalent to solving

$$\begin{bmatrix} 2 & -3 & -2 \\ 1 & -3 & 1 \\ 0 & 1 & -1 \end{bmatrix}\begin{bmatrix} c_1 \\ c_2 \\ c_3 \end{bmatrix} = \begin{bmatrix} x_1 \\ x_2 \\ x_3 \end{bmatrix} \tag{3.52}$$

where we know the x_i's and are solving for the c_i's. We augment the matrix

$$\left[\begin{array}{ccc:c} 2 & -3 & -2 & x_1 \\ 1 & -3 & 1 & x_2 \\ 0 & 1 & -1 & x_3 \end{array}\right] \tag{3.53}$$

and reduce it to triangular form, obtaining

$$\left[\begin{array}{ccc:c} 2 & -3 & -2 & x_1 \\ 0 & -\frac{3}{2} & 2 & -\frac{1}{2}x_1 + x_2 \\ 0 & 0 & \frac{1}{3} & -\frac{1}{3}x_1 + \frac{2}{3}x_2 + x_3 \end{array}\right] \tag{3.54}$$

as you can check. The associated system is

$$\begin{aligned} 2c_1 - 3c_2 - 2c_3 &= x_1 \\ -\tfrac{3}{2}c_2 + 2c_3 &= -\tfrac{1}{2}x_1 + x_2 \\ \tfrac{1}{3}c_3 &= -\tfrac{1}{3}x_1 + \tfrac{2}{3}x_2 + x_3 \end{aligned} \tag{3.55}$$

Now remember, we only have to show we can solve this for the c_i's; *we do not actually have to do it*! Since the system (3.55) is in triangular form and the pivots on the diagonal of (3.54) are all nonzero, we can see that no matter what values the x_i's take the system can be solved for the c_i's. [In fact, with this reasoning we could have stopped at (3.54).]

To show (a), that S is linearly independent, we need to show the only solution to

$$c_1\mathbf{v}_1 + c_2\mathbf{v}_2 + c_3\mathbf{v}_3 = \mathbf{0} \tag{3.56}$$

is $c_1 = c_2 = c_3 = 0$. But we have done virtually all the work in the preceding part. Equation (3.56) is Equation (3.51) with $\mathbf{x} = \mathbf{0}$. So we set all $x_i = 0$ and proceed to (3.55), or again stop at (3.54). By the triangular form and the pivots being nonzero, we can see the only solution is $c_1 = c_2 = c_3 = 0$, so we are done. ■

Example 3 The set $S = \{1, x, x^2, \ldots, x^n\}$ is a basis for the vector space P_n of all polynomials of degree less than or equal to n. (See Example 8, Section 3.3.) S spans P_n because if $p(x) = a_0 + a_1x + \cdots + a_nx^n$, then

$$p = a_0(1) + a_1(x) + \cdots + a_n(x^n)$$

S is linearly independent, for suppose some linear combination of the vectors in S is the zero polynomial

$$c_0(1) + c_1(x) + \cdots + c_n(x^n) = \mathbf{0}$$

Then the following polynomials are equal

$$c_0 + c_1x + \cdots + c_nx^n = 0 + 0x + \cdots + 0x^n$$

It now follows that every $c_i = 0$ since corresponding coefficients must be equal. This S is called the **standard basis** for P_n. ■

Example 4 Let M_{23} be the vector space of all 2×3 matrices. (See Example 7, Section 3.3.) Let $S = \{M_1, \ldots, M_6\}$, where

$$M_1 = \begin{bmatrix} 1 & 0 & 0 \\ 0 & 0 & 0 \end{bmatrix}, \quad M_2 = \begin{bmatrix} 0 & 1 & 0 \\ 0 & 0 & 0 \end{bmatrix}, \quad M_3 = \begin{bmatrix} 0 & 0 & 1 \\ 0 & 0 & 0 \end{bmatrix}$$
$$M_4 = \begin{bmatrix} 0 & 0 & 0 \\ 1 & 0 & 0 \end{bmatrix}, \quad M_5 = \begin{bmatrix} 0 & 0 & 0 \\ 0 & 1 & 0 \end{bmatrix}, \quad M_6 = \begin{bmatrix} 0 & 0 & 0 \\ 0 & 0 & 1 \end{bmatrix}$$

We will show S is a basis for M_{23}.

To show S spans M_{23}, we pick any matrix M in M_{23},

$$M = \begin{bmatrix} a_{11} & a_{12} & a_{13} \\ a_{21} & a_{22} & a_{23} \end{bmatrix}$$

But then M is a linear combination of the "vectors" in S, since

$$M = a_{11}M_1 + a_{12}M_2 + a_{13}M_3 + a_{21}M_4 + a_{22}M_5 + a_{23}M_6$$

as you can verify. Thus S spans M_{23}.

To show S is linearly independent, suppose some linear combination of the M_i's is the zero 2×3 matrix

$$c_1M_1 + \cdots + c_6M_6 = \mathbf{0}$$

Then

$$\begin{bmatrix} c_1 & 0 & 0 \\ 0 & 0 & 0 \end{bmatrix} + \begin{bmatrix} 0 & c_2 & 0 \\ 0 & 0 & 0 \end{bmatrix} + \cdots + \begin{bmatrix} 0 & 0 & 0 \\ 0 & 0 & c_6 \end{bmatrix} = \begin{bmatrix} 0 & 0 & 0 \\ 0 & 0 & 0 \end{bmatrix}$$

or

$$\begin{bmatrix} c_1 & c_2 & c_3 \\ c_4 & c_5 & c_6 \end{bmatrix} = \begin{bmatrix} 0 & 0 & 0 \\ 0 & 0 & 0 \end{bmatrix}$$

Then every $c_i = 0$, by definition of matrix equality (see Section 1.3), so that S is linearly independent. This S is called the **standard basis** for M_{23}. ■

There is one trivial but very important situation that occurs often.

(3.57) THEOREM Let V be a vector space, S a set of vectors of V, and W the subspace spanned by S. If S is linearly independent, then S is a basis for W.

Proof S is linearly independent, by hypothesis, and spans W by definition. That does it! ■

Example 5 Let $S = \{\mathbf{v}_1, \mathbf{v}_2\}$, where $\mathbf{v}_1 = (1, -1, 2)$ and $\mathbf{v}_2 = (2, 0, 1)$. Then S is obviously linearly independent (neither $\mathbf{v}_i$ is a multiple of the other), so S spans a plane W in $\mathbb{R}^3$. By Theorem (3.57) S is a basis for W. ■

Example 6 Let $\mathbf{v}_1, \mathbf{v}_2$ be as in Example 5, $\mathbf{v}_3 = 2\mathbf{v}_1 - 3\mathbf{v}_2 = (-4, -2, 1)$, and $T = \{\mathbf{v}_1, \mathbf{v}_2, \mathbf{v}_3\}$. Then T spans the same subspace W of Example 5 but it is not a basis for W since T is not linearly independent. ■

Definition of Dimension

The definition of dimension is itself really quite simple: the dimension is the number of vectors in a basis. What takes a little work is to show that the definition makes sense, that is, to show that any two bases for a given vector space have the same number of vectors. Hence we set out to prove this first.

Theorem (3.46), Section 3.5, states that if we have m vectors in $\mathbb{R}^n$ and $m > n$, then those m vectors must be linearly dependent. For instance, from Example 7, Section 3.5, $\mathbf{v}_1 = (2, -3)$, $\mathbf{v}_2 = (4, -2)$, and $\mathbf{v}_3 = (-4, 3)$ are three vectors in $\mathbb{R}^2$, so they must be linearly dependent (and indeed $2\mathbf{v}_1 + 3\mathbf{v}_2 + 4\mathbf{v}_3 = \mathbf{0}$). We now generalize Theorem (3.46) by replacing $\mathbb{R}^n$ with any vector space that has a basis of n vectors.

(3.58) **THEOREM** Suppose a vector space V has a basis S with n vectors. If $m > n$, then any set with m vectors of V is linearly dependent.

Proof Let $S = \{\mathbf{v}_1, \ldots, \mathbf{v}_n\}$ be a basis of V and let $T = \{\mathbf{w}_1, \ldots, \mathbf{w}_m\}$ be any collection of m vectors of V where $m > n$. We want to show T is linearly dependent, so we need to find numbers $c_1, \ldots, c_m$, at least one of which is nonzero, such that

$$c_1\mathbf{w}_1 + c_2\mathbf{w}_2 + \cdots + c_m\mathbf{w}_m = \mathbf{0} \tag{3.59}$$

Since S is a basis, each $\mathbf{w}_j$ is a linear combination of the $\mathbf{v}_i$'s.

$$\begin{aligned} \mathbf{w}_1 &= a_{11}\mathbf{v}_1 + a_{21}\mathbf{v}_2 + \cdots + a_{n1}\mathbf{v}_n \\ \mathbf{w}_2 &= a_{12}\mathbf{v}_1 + a_{22}\mathbf{v}_2 + \cdots + a_{n2}\mathbf{v}_n \\ &\vdots \\ \mathbf{w}_m &= a_{1m}\mathbf{v}_1 + a_{2m}\mathbf{v}_2 + \cdots + a_{nm}\mathbf{v}_n \end{aligned} \tag{3.60}$$

We substitute the equations in (3.60) into Equation (3.59), obtaining

$$\begin{aligned} c_1(a_{11}\mathbf{v}_1 + \cdots + a_{n1}\mathbf{v}_n) + c_2(a_{12}\mathbf{v}_1 + \cdots + a_{n2}\mathbf{v}_n) \\ + \cdots + c_m(a_{1m}\mathbf{v}_1 + \cdots + a_{nm}\mathbf{v}_n) = \mathbf{0} \end{aligned}$$

and rewrite this as

$$(a_{11}c_1 + a_{12}c_2 + \cdots + a_{1m}c_m)\mathbf{v}_1 + \cdots + (a_{n1}c_1 + a_{n2}c_2 + \cdots + a_{nm}c_m)\mathbf{v}_n = \mathbf{0}$$

Since the $\mathbf{v}_i$'s are linearly independent, these coefficients must all be zero

(3.61)
$$\begin{aligned} a_{11}c_1 + a_{12}c_2 + \cdots + a_{1m}c_m &= 0 \\ &\vdots \\ a_{n1}c_1 + a_{n2}c_2 + \cdots + a_{nm}c_m &= 0 \end{aligned}$$

At this point we know the a_{ij}'s and we want to determine if there are c_i's, not all zero, that satisfy this system (3.61). However, $m > n$, so there are more unknowns than equations. Hence by Lemma (3.47), Section 3.5, there is a nontrivial solution $c_1, \ldots, c_m$, so we are done. ■

We can now prove the result that we need.

(3.62)

THEOREM If a vector space V has a basis S with n elements, then any other basis for V also has n elements.

Proof Let S and T be two bases of V and suppose S has n elements and T has m elements.

Suppose $m > n$. Since S is a basis, T must be linearly dependent, by Theorem (3.58), and hence not a basis. Contradiction.

Similarly, if $n > m$, T is a basis so S cannot be linearly independent [also by Theorem (3.58)], another contradiction.

Thus we must have $m = n$. ■

We now know that the following definition makes sense.

(3.63)

DEFINITION Let V be a vector space.

(a) The **dimension** of V is n (a positive integer) if V has a basis of n elements. [Hence by Theorem (3.62), every basis of V has n elements.]
(b) The **dimension** of the zero vector space is zero (even though the zero vector space has no basis).
(c) V is **finite dimensional** if the dimension of V is a non-negative integer.
(d) V is **infinite dimensional** if it is not finite dimensional.

NOTATION We write "dim V" for "the dimension of V."

Example 7 The dim $\mathbb{R}^n = n$ (it would be more surprising if this were not true!) since it has a basis, the standard basis $\{\boldsymbol{\varepsilon}_1, \ldots, \boldsymbol{\varepsilon}_n\}$, of n elements. ■

Example 8 If W is the subspace of $\mathbb{R}^3$ given in Example 5, then $\dim W = 2$ since $S = \{\mathbf{v}_1, \mathbf{v}_2\}$ is a basis. ■

Example 9 Let W be any straight line through the origin in $\mathbb{R}^n$ and let $\mathbf{v}$ be any nonzero vector of W. Then W is the span of $\{\mathbf{v}\}$ (see Example 4, Section 3.3) and $\{\mathbf{v}\}$ is linearly independent [by Definition (3.41), Section 3.5]. Thus $\{\mathbf{v}\}$ is a basis for W and $\dim W = 1$. ■

Example 10 Let P_n be the vector space of all polynomials of degree $\leq n$. Then $\dim P_n = n + 1$ by Example 3. ■

Example 11 If M_{23} is the vector space of all 2×3 matrices, then $\dim M_{23} = 6$ by Example 4. ■

The vector spaces in Examples 7–11 are all finite dimensional. The next one is not.

Example 12 Recall from Example 9, Section 3.3, that $C[a, b]$ is the set of all continuous real-valued functions defined on the interval $[a, b]$, where $a < b$. Suppose $C[a, b]$ is finite dimensional, so that $\dim C[a, b] = n$, for some n. Then by Theorem (3.58), any set of $n + 1$ functions in $C[a, b]$ is linearly dependent. However, the $n + 1$ functions $1, x, \ldots, x^n$ are all in $C[a, b]$ and they are linearly independent by Example 3. This is a contradiction, so $C[a, b]$ must be infinite dimensional. ■

Subsets in a Finite Dimensional Space

We now give several useful relationships between certain subsets and bases.

(3.64)

THEOREM Let V be a finite dimensional vector space and $S = \{\mathbf{v}_1, \ldots, \mathbf{v}_k\}$ be a set of vectors of V.

(a) If S is linearly independent, then we can extend S to a basis $\{\mathbf{v}_1, \ldots, \mathbf{v}_k, \mathbf{v}_{k+1}, \ldots, \mathbf{v}_m\}$ for V.

(b) If S spans V and V is not the zero vector space, then some subset of S is a basis for V. (There may be more than one such subset.)

Suppose, in addition, we know that $\dim V = n$:

(c) If S is linearly independent and $k = n$, then S must be a basis for V.

(d) If S spans V and $k = n$, then S must be a basis for V.

The proof follows Example 14. Example 13 is a precursor for Section 3.7.

Example 13 Let $A = \begin{bmatrix} -2 & 1 & 3 \\ 4 & -2 & 5 \end{bmatrix}$.

(a) The three columns $\mathbf{c}_1 = \begin{bmatrix} -2 \\ 4 \end{bmatrix}$, $\mathbf{c}_2 = \begin{bmatrix} 1 \\ -2 \end{bmatrix}$, $\mathbf{c}_3 = \begin{bmatrix} 3 \\ 5 \end{bmatrix}$ of A are vectors in $\mathbb{R}^2$, and we know that dim $\mathbb{R}^2 = 2$. If we also know that the three columns $\mathbf{c}_1$, $\mathbf{c}_2$, and $\mathbf{c}_3$ span $\mathbb{R}^2$, then Theorem (3.64b) states that we can find a subset that is a basis for $\mathbb{R}^2$. We easily can see that $\mathbf{c}_1 = -2\mathbf{c}_2$, so we could discard one of them. We also can see that the sets $\{\mathbf{c}_1, \mathbf{c}_3\}$ and $\{\mathbf{c}_2, \mathbf{c}_3\}$ are linearly independent. (Neither $\mathbf{c}_1$ nor $\mathbf{c}_3$ is a multiple of the other; a similar case can be made for $\{\mathbf{c}_2, \mathbf{c}_3\}$.) Thus either set gives us two linearly independent vectors in a two-dimensional space. By Theorem (3.64c), each of the sets $\{\mathbf{c}_1, \mathbf{c}_3\}$ and $\{\mathbf{c}_2, \mathbf{c}_3\}$ is a basis for $\mathbb{R}^2$. (Of course, $\{\mathbf{c}_1, \mathbf{c}_2\}$ is not a basis for $\mathbb{R}^2$. Why?)

(b) The two rows $\mathbf{v}_1 = [-2 \quad 1 \quad 3]$, $\mathbf{v}_2 = [4 \quad -2 \quad 5]$ of matrix A are vectors in $\mathbb{R}^3$, and we know that dim $\mathbb{R}^3 = 3$. It is easy to see that neither $\mathbf{v}_1$ nor $\mathbf{v}_2$ is a multiple of the other, so $\mathbf{v}_1$ and $\mathbf{v}_2$ are linearly independent. By Theorem (3.64a), we can extend $\{\mathbf{v}_1, \mathbf{v}_2\}$ to a basis for $\mathbb{R}^3$. How? Any way we can! There are an infinite number of possibilities, and there is no standard procedure. In general, we pick a vector $\mathbf{v}_3$ not in the span of $\mathbf{v}_1$ and $\mathbf{v}_2$ (see Exercise 34, Section 3.5); for simple problems, we often pick the simplest vector for $\mathbf{v}_3$ that we can. The standard basis contains simple vectors, so we try $\mathbf{v}_3 = \boldsymbol{\varepsilon}_1$, $\boldsymbol{\varepsilon}_2$, or $\boldsymbol{\varepsilon}_3$. It turns out that the first two choices work (but $2\mathbf{v}_1 + \mathbf{v}_2 = [0 \quad 0 \quad 11]$, so $\boldsymbol{\varepsilon}_3 = \frac{2}{11}\mathbf{v}_1 + \frac{1}{11}\mathbf{v}_2$). Thus $\{\mathbf{v}_1, \mathbf{v}_2, \boldsymbol{\varepsilon}_1\}$ and $\{\mathbf{v}_1, \mathbf{v}_2, \boldsymbol{\varepsilon}_2\}$ are linearly independent sets of three vectors, so each set is a basis for $\mathbb{R}^3$ [by Theorem (3.64c)], which extends $\{\mathbf{v}_1, \mathbf{v}_2\}$. ■

There is a pitfall associated with Theorem (3.64) that even some graduate students occasionally fall into.

WARNING Suppose that W is a subspace of V. If S is a basis for W, then S can be extended to a basis for V [by Theorem (3.64a)]. However, if B is a basis for V, there is most likely *not* a subset of B that forms a basis for W, since B may contain no vectors from W (or an insufficient number).

Example 14 Let $V = \mathbb{R}^2$, $W = \left\{\begin{bmatrix} x \\ x \end{bmatrix}\right\}$, $\mathbf{v} = \begin{bmatrix} 1 \\ 1 \end{bmatrix}$, and $\mathbf{w} = \begin{bmatrix} -1 \\ 1 \end{bmatrix}$. (See Figure 3.21.) Then $\{\mathbf{v}\}$ is a basis for W and can be extended to a basis for $\mathbb{R}^2$ in infinitely many ways; $\{\mathbf{v}, \boldsymbol{\varepsilon}_1\}$, $\{\mathbf{v}, \boldsymbol{\varepsilon}_2\}$, and $\{\mathbf{v}, \mathbf{w}\}$ are examples. But note that the standard basis $\{\boldsymbol{\varepsilon}_1, \boldsymbol{\varepsilon}_2\}$ is a basis for $\mathbb{R}^2$ that does not contain a basis for W. ■

Figure 3.21

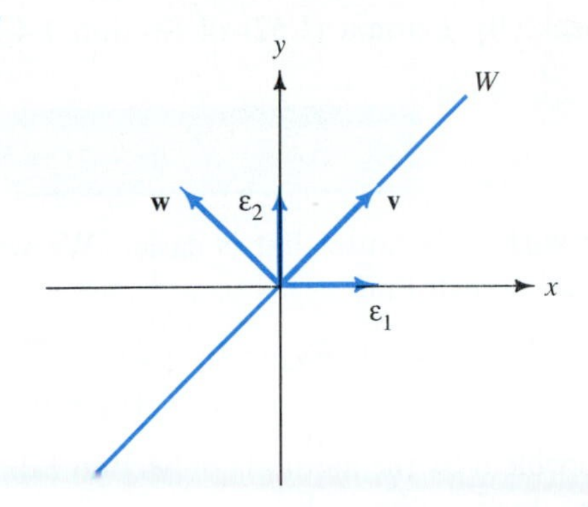

Proof of Theorem (3.64) (a) If S spans V, then S is a basis, so we are done. If S does not span V, pick any $\mathbf{v}_{k+1}$ that is not in the span of S. Then $\{\mathbf{v}_1, \ldots, \mathbf{v}_k, \mathbf{v}_{k+1}\}$ is again linearly independent (see Exercise 34, Section 3.5), and we start over (does the new S span V, and so on). This process must end, since V is a finite dimensional vector space.

(b) If S is linearly independent, then S is a basis, so we are done. If S is not linearly independent, then one of the $\mathbf{v}_i$'s is a linear combination of the others. Throw that $\mathbf{v}_i$ out, and the span remains the same (see Exercise 35, Section 3.5). We continue in this way until we reach a subset that both spans *and* is linearly independent. This process must end, since $V \neq \{\mathbf{0}\}$. (See Exercise 50.)

(c) Apply (a). If $m > n$, it contradicts Theorem (3.64) and the fact that we know there is some basis with n elements.

(d) Apply (b). If the subset is strictly smaller than S, it contradicts Theorem (3.64). ■

Basis and Dimension of $NS(A)$

In the next section, and indeed throughout the rest of the text and all of linear algebra, it is often necessary to find a basis for the null space of an $m \times n$ matrix A and hence determine dim $NS(A)$. The process is straightforward and is illustrated in the next example.

Example 15 Find a basis for $NS(A)$ and compute dim $NS(A)$ for

$$A = \begin{bmatrix} 0 & 0 & 1 & -4 \\ 2 & -4 & -1 & -2 \\ 4 & -8 & 0 & -12 \end{bmatrix}$$

Solution First, reduce A to row echelon form, obtaining

$$U = \begin{bmatrix} 2 & -4 & -1 & -2 \\ 0 & 0 & 1 & -4 \\ 0 & 0 & 0 & 0 \end{bmatrix}$$

as you can check. By Lemma (1.52) of Section 1.4, $A\mathbf{x} = \mathbf{0}$ if and only if $U\mathbf{x} = \mathbf{0}$, or

$$NS(A) = NS(U) \tag{3.65}$$

We work with U because that is easier. We wish to describe all $\mathbf{x}$ in $NS(U)$, that is, all $\mathbf{x}$ such that

$$\begin{aligned} 2x_1 - 4x_2 - x_3 - 2x_4 &= 0 \\ x_3 - 4x_4 &= 0 \end{aligned}$$

We solve this problem by the methods of Section 1.2. We move the free variables, x_2 and x_4, to the other side.

$$\begin{aligned} 2x_1 - x_3 &= 4x_2 + 2x_4 \\ x_3 &= 4x_4 \end{aligned}$$

Next we let $x_2 = s$ and $x_4 = t$

$$\begin{aligned} 2x_1 - x_3 &= 4s + 2t \\ x_3 &= 4t \end{aligned}$$

and solve by backsubstitution, obtaining

$$x_3 = 4t \qquad \text{and} \qquad x_1 = 2s + 3t$$

Altogether we have

$$\begin{aligned} x_1 &= 2s + 3t \\ x_2 &= s \\ x_3 &= 4t \\ x_4 &= t \end{aligned}$$

We rewrite this as

$$\begin{bmatrix} x_1 \\ x_2 \\ x_3 \\ x_4 \end{bmatrix} = \begin{bmatrix} 2s + 3t \\ s \\ 4t \\ t \end{bmatrix} = \begin{bmatrix} 2s \\ s \\ 0 \\ 0 \end{bmatrix} + \begin{bmatrix} 3t \\ 0 \\ 4t \\ t \end{bmatrix}$$

$$= s\begin{bmatrix} 2 \\ 1 \\ 0 \\ 0 \end{bmatrix} + t\begin{bmatrix} 3 \\ 0 \\ 4 \\ 1 \end{bmatrix}$$

Therefore, the two vectors

$$\mathbf{v}_1 = \begin{bmatrix} 2 \\ 1 \\ 0 \\ 0 \end{bmatrix}, \qquad \mathbf{v}_2 = \begin{bmatrix} 3 \\ 0 \\ 4 \\ 1 \end{bmatrix}$$

span $NS(U)$ [you can verify directly that they are in both $NS(U)$ and $NS(A)$] and $\mathbf{v}_1$ and $\mathbf{v}_2$ are linearly independent (you can check this directly, or take a peek at the argument associated with Example 3 of the next section to see an easy way of seeing these vectors are linearly independent). Therefore $\{\mathbf{v}_1, \mathbf{v}_2\}$ form a basis for $NS(U) = NS(A)$, and hence dim $NS(A) = 2$. ■

This example illustrates the following very important relationship:

(3.66) If U is a matrix in row echelon form, then dim $NS(U)$ is equal to the number of free variables in the equation $U\mathbf{x} = \mathbf{0}$.

Exercise 3.6

In Exercises 1–6, *by inspection*, explain why the given vectors do not form a basis for the given vector space.

1. $\mathbf{u}_1 = (1, -2), \quad \mathbf{u}_2 = (2, -1), \quad \mathbf{u}_3 = (3, 5), \qquad$ for $\mathbb{R}^2$
2. $\mathbf{v}_1 = (3, -1, 2), \quad \mathbf{v}_2 = (0, 2, 5), \qquad$ for $\mathbb{R}^3$
3. $\mathbf{w}_1 = (8, 7, 6, 5), \quad \mathbf{w}_2 = (-3, 4, 1, 2), \quad \mathbf{w}_3 = (-1, 3, 8, 2), \qquad$ for $\mathbb{R}^4$
4. $p_1 = 1 + 2x + 3x^2, \quad p_2 = 2 - x - x^2, \quad p_3 = 1 + x, \quad p_4 = x + x^2, \qquad$ for P_2
5. $q_1 = 1 + 2x + 3x^2, \quad q_2 = 2 - x - x^2, \qquad$ for P_2
6. $M_1 = \begin{bmatrix} 1 & 2 & 3 \\ -3 & 0 & 2 \end{bmatrix}, \quad M_2 = \begin{bmatrix} -2 & 1 & 3 \\ 0 & 4 & 2 \end{bmatrix}, \quad M_3 = \begin{bmatrix} 0 & 2 & 1 \\ -2 & 1 & 0 \end{bmatrix},$
$M_4 = \begin{bmatrix} -1 & 0 & 1 \\ 2 & 2 & 0 \end{bmatrix}, \quad M_5 = \begin{bmatrix} 3 & 1 & -1 \\ 0 & 1 & 0 \end{bmatrix}, \qquad$ for M_{23}

In Exercises 7–20, determine whether the given vectors form a basis for the given vector space.

7. $\mathbf{u}_1 = (1, 2), \quad \mathbf{u}_2 = (2, 1), \qquad$ for $\mathbb{R}^2$
8. $\mathbf{v}_1 = (2, 2), \quad \mathbf{u}_2 = (-3, -3), \qquad$ for $\mathbb{R}^2$
9. $\mathbf{w}_1 = (-3, 1), \quad \mathbf{w}_2 = (6, -2), \qquad$ for $\mathbb{R}^2$
10. $\mathbf{x}_1 = (2, 0), \quad \mathbf{x}_2 = (3, 3), \qquad$ for $\mathbb{R}^2$
11. $\mathbf{u}_1 = (1, 1, 1), \quad \mathbf{u}_2 = (0, 1, 1), \quad \mathbf{u}_3 = (0, 0, 1), \qquad$ for $\mathbb{R}^3$
12. $\mathbf{v}_1 = (3, -2, 1), \quad \mathbf{v}_2 = (2, 3, 1), \quad \mathbf{v}_3 = (2, 1, -3), \qquad$ for $\mathbb{R}^3$
13. $\mathbf{w}_1 = (2, 1, 2), \quad \mathbf{w}_2 = (1, -2, -3), \quad \mathbf{w}_3 = (5, 0, 1), \qquad$ for $\mathbb{R}^3$
14. $\mathbf{x}_1 = (1, 2, 1), \quad \mathbf{x}_2 = (3, -1, 2), \quad \mathbf{x}_3 = (0, -7, -1), \qquad$ for $\mathbb{R}^3$

15. $\mathbf{u}_1=(1, 0, 0, 0)$, $\mathbf{u}_2=(1, 1, 0, 0)$, $\mathbf{u}_3=(1, 1, 1, 0)$, $\mathbf{u}_4=(1, 1, 1, 1)$, for $\mathbb{R}^4$
16. $\mathbf{v}_1=(1, 1, 0, 0)$, $\mathbf{v}_2=(0, 1, 1, 0)$, $\mathbf{v}_3=(0, 0, 1, 1)$, $\mathbf{v}_4=(1, 0, 0, 1)$, for $\mathbb{R}^4$
17. $p_1 = 1 + x$, $p_2 = 2 - x$, for P_1
18. $q_1 = 1 + x + x^2$, $q_2 = x + 2x^2$, $q_3 = 3x^2$, for P_2
19. $r_1 = 2 - 3x + x^2$, $r_2 = 4 + 3x - 2x^2$, $r_3 = 9x - 4x^2$, for P_2
20. $s_1 = 1 + x^2$, $s_2 = x + x^3$, $s_3 = x + x^2$, $s_4 = 1 + x^3$, for P_3
21. Follow the pattern of Example 4 and find a basis for M_{32}. What is dim M_{32}?
22. Same question as Exercise 21 for M_{22}.
23. Same question as Exercise 21 for M_{mn}.
24. Determine if the matrices

$$M_1 = \begin{bmatrix} 1 & 0 \\ 0 & 0 \end{bmatrix}, \quad M_2 = \begin{bmatrix} 1 & 1 \\ 0 & 0 \end{bmatrix}, \quad M_3 = \begin{bmatrix} 1 & 1 \\ 1 & 0 \end{bmatrix}, \quad M_4 = \begin{bmatrix} 1 & 1 \\ 1 & 1 \end{bmatrix}$$

form a basis for M_{22}.

25. Same question as Exercise 24 for

$$N_1 = \begin{bmatrix} 1 & 2 \\ 0 & 3 \end{bmatrix}, \quad N_2 = \begin{bmatrix} -2 & 3 \\ 2 & 0 \end{bmatrix}, \quad N_3 = \begin{bmatrix} 1 & 0 \\ 2 & -1 \end{bmatrix}, \quad N_4 = \begin{bmatrix} 0 & 5 \\ 4 & 2 \end{bmatrix}$$

26. Let W be the subspace of $C[0, 1]$ spanned by $S = \{\sin^2 x, \cos^2 x, \cos 2x\}$.
    (a) Explain why S is not a basis for W.
    (b) Find a basis for W.
    (c) What is dim W?

In Exercises 27–32, find a basis for, and the dimension of, the null space of the given matrix.

27. $\begin{bmatrix} 1 & 2 & 1 \\ -2 & -2 & 1 \\ -1 & 0 & 2 \end{bmatrix}$

28. $\begin{bmatrix} 2 & 1 & -1 & 1 \\ 4 & -2 & -2 & 1 \end{bmatrix}$

29. $\begin{bmatrix} 2 & 3 & -2 \\ 4 & 1 & 2 \\ 2 & -2 & 4 \\ 0 & -5 & 6 \end{bmatrix}$

30. $\begin{bmatrix} 2 & -1 & 4 & -2 \end{bmatrix}$

31. $\begin{bmatrix} 2 & 3 & -1 \\ 1 & 0 & 2 \\ 0 & 2 & 3 \end{bmatrix}$

32. $\begin{bmatrix} 1 & 1 & 1 \\ 2 & 3 & -2 \\ -1 & 0 & -5 \\ 5 & 6 & 1 \end{bmatrix}$

In Exercises 33–38, find a basis for the given vector space (if possible) and determine its dimension (if possible).

33. All vectors in $\mathbb{R}^2$ whose components add to zero.

34. $NS(I_3)$, I_3 is the 3×3 identity matrix.

35. All 3×3 symmetric matrices.

36. All 3×3 skew-symmetric matrices.

37. All polynomials, p, such that $p(1) = 0$.

38. All functions f in $C[0, 2\pi]$ such that $\int_0^{2\pi} f(x)\,dx = 0$.

39. Let $W = \{(x_1, x_2, x_3, x_4) | x_1 + x_2 + x_3 + x_4 = 0\}$, $A = [1 \quad 1 \quad 1 \quad 1]$.

    (a) Explain why $W = NS(A)$.
    (b) Find a basis for W and determine its dimension. [Use (a)].

40. Let $W = \{(x_1, x_2, x_3) | x_1 + 2x_2 + 3x_3 = 0\}$.

    (a) Find a matrix A such that $W = NS(A)$.
    (b) Find a basis for W and determine its dimension.

41. Same question as Exercise 40 for $W = \{(x_1, x_2, x_3, x_4) | x_4 = x_1 + x_2, x_3 = x_1 - x_2\}$.

42. Same question as Exercise 40 for $W = \{(x_1, x_2, x_3, x_4) | x_1 = x_2 = x_3 = x_4\}$.

In Exercises 43–46, an $m \times n$ matrix is given with columns that span $\mathbb{R}^m$ and rows that are linearly independent vectors in $\mathbb{R}^n$. (WARNING Often this is not the case!) As in Example 13, contract the columns to a basis for $\mathbb{R}^m$ and expand the rows to a basis for $\mathbb{R}^n$.

43. $\begin{bmatrix} 3 & -4 & -6 \\ 0 & 2 & 3 \end{bmatrix}$

44. $\begin{bmatrix} 0 & 2 & 3 & -6 \\ 0 & 0 & -3 & 6 \end{bmatrix}$

45. $\begin{bmatrix} 3 & 4 & -2 & 2 \\ 0 & 2 & -1 & 1 \\ 0 & 0 & 0 & 2 \end{bmatrix}$

46. $\begin{bmatrix} 1 & 2 & 3 & 6 \\ 2 & 4 & 5 & 10 \\ 0 & 0 & 1 & 3 \end{bmatrix}$

47. Let W be the plane in $\mathbb{R}^3$ that contains the line $x = y$ in the xy-plane and that also contains the z-axis.

    (a) Find a basis for W.
    (b) Extend that basis to a basis for $\mathbb{R}^3$.
    (c) Let B be the standard basis for $\mathbb{R}^3$. Can you contract B to a basis for W? Explain.

48. Repeat Exercise 47 if W is the line $x = y = z$ in $\mathbb{R}^3$.

49. Suppose that V is a vector space of dimension 5 and W is a subspace of dimension 3. Are the following statements true or false.
    (a) Every basis for W can be extended to a basis for V by (judiciously) adding two more vectors.
    (b) Every basis for V can be contracted to a basis for W by (judiciously) deleting two vectors.

50. In the proof of Theorem (3.64b), why must the process end if $V \neq \{\mathbf{0}\}$? Why is the theorem false if $V = \{\mathbf{0}\}$?

51. Assume $x^2 + 2x - 3$, $x^2 - 3x + 4$, and $x^2 - x + 5$ are linearly independent polynomials in P_2. They must be/may be/are not a basis for P_2. Why?

52. Repeat Exercise 51 if you know that the following matrices span M_{22}:

$$\begin{bmatrix} 1 & 1 \\ 1 & 1 \end{bmatrix}, \quad \begin{bmatrix} 0 & 1 \\ 1 & 1 \end{bmatrix}, \quad \begin{bmatrix} 0 & 0 \\ 1 & 1 \end{bmatrix}, \quad \begin{bmatrix} 0 & 0 \\ 0 & 1 \end{bmatrix}$$

3.7 The Fundamental Subspaces of a Matrix; Rank

If we are given a vector space, it is often quite useful to have a basis for it, for example, to determine its dimension. But what do we do if we do not yet have a basis? For example, we are often given a vector space described as the span of a collection of vectors (which are not linearly independent). This section will provide us with techniques for efficiently extracting a basis in such cases. Even more important than this, we will be describing certain fundamental properties about matrices that will be central to our overall goal of understanding linear problems and how to solve them effectively and efficiently.

Row and Column Space

Let A be an $m \times n$ matrix. We have already defined the null space of A, $NS(A)$, and shown it to be a subspace of $\mathbb{R}^n$. As you may have guessed, $NS(A)$ is a subspace that is fundamental to A. In this section we introduce other subspaces that are also fundamental to A, and we prove relationships among them. As usual let us start with

$$A = \begin{bmatrix} a_{11} & a_{12} & \cdots & a_{1n} \\ a_{21} & a_{22} & \cdots & a_{2n} \\ \vdots & \vdots & & \vdots \\ a_{m1} & a_{m2} & \cdots & a_{mn} \end{bmatrix}$$

The m vectors

$$\mathbf{r}_1 = [a_{11} \quad a_{12} \quad \cdots \quad a_{1n}], \qquad \mathbf{r}_2 = [a_{21} \quad a_{22} \quad \cdots \quad a_{2n}], \quad \ldots, \quad \mathbf{r}_m = [a_{m1} \quad a_{m2} \quad \cdots \quad a_{mn}]$$

are called the **row vectors** of A, and the n vectors

$$\mathbf{c}_1 = \begin{bmatrix} a_{11} \\ a_{21} \\ \vdots \\ a_{m1} \end{bmatrix}, \quad \mathbf{c}_2 = \begin{bmatrix} a_{12} \\ a_{22} \\ \vdots \\ a_{m2} \end{bmatrix}, \quad \ldots, \quad \mathbf{c}_n = \begin{bmatrix} a_{1n} \\ a_{2n} \\ \vdots \\ a_{mn} \end{bmatrix}$$

are called the **column vectors** of A.

Example 1 Let $A = \begin{bmatrix} 4 & 2 & 7 \\ -1 & -3 & -1 \end{bmatrix}$. Then

$$\mathbf{r}_1 = [4 \quad 2 \quad 7] \quad \text{and} \quad \mathbf{r}_2 = [-1 \quad -3 \quad -1]$$

are the row vectors of A and

$$\mathbf{c}_1 = \begin{bmatrix} 4 \\ -1 \end{bmatrix}, \quad \mathbf{c}_2 = \begin{bmatrix} 2 \\ -3 \end{bmatrix}, \quad \text{and} \quad \mathbf{c}_3 = \begin{bmatrix} 7 \\ -1 \end{bmatrix}$$

are the column vectors of A. ■

(3.67) DEFINITION (a) The subspace of $\mathbb{R}^n$ spanned by the row vectors of A is called the **row space** of A and is denoted by $RS(A)$.

(b) The subspace of $\mathbb{R}^m$ spanned by the column vectors of A is called the **column space** of A and is denoted by $CS(A)$.

Example 2 Let A be the same as in Example 1. By inspection describe the $CS(A)$ and $RS(A)$, find a basis for each, and determine the dimension of each.

Solution The $CS(A)$ is contained in $\mathbb{R}^2$, so $\dim CS(A) \leq 2$. But $\mathbf{c}_1$ and $\mathbf{c}_2$ are obviously linearly independent (neither is a multiple of the other), so $\dim CS(A) \geq 2$. Altogether $\dim CS(A) = 2$, so we have $CS(A) = \mathbb{R}^2$ and $\{\mathbf{c}_1, \mathbf{c}_2\}$ is a basis for $CS(A)$.

Now $\mathbf{r}_1$ and $\mathbf{r}_2$ are linearly independent, and they span $RS(A)$. Hence $\{\mathbf{r}_1, \mathbf{r}_2\}$ is a basis for $RS(A)$. Thus $\dim RS(A) = 2$, and $RS(A)$ is the plane in $\mathbb{R}^3$ spanned by $\mathbf{r}_1$ and $\mathbf{r}_2$. ■

Usually a basis for $RS(A)$ or $CS(A)$ cannot be found by inspection. We now describe how to find such bases in general. Recall that we observed in the previous section that $NS(A) = NS(U)$ where U has been obtained by reducing A to row echelon form [see (3.65), Section 3.6]. Fortunately the same is true for row spaces (but be careful as it is *not* true for column spaces).

(3.68)

> THEOREM Let A be an $m \times n$ matrix and let U be a matrix in row echelon form obtained from A by row operations. Then
>
> $$RS(A) = RS(U)$$

The proof is straightforward but the details are a little fussy, so we put the proof off until the end of this section. The power and extreme usefulness of this theorem arise because we can easily read off from U a basis for $RS(U)$. This is illustrated in the next example.

Example 3 Find a basis for the row space of

$$A = \begin{bmatrix} 2 & -3 & 4 & 1 & 2 \\ 4 & -6 & 5 & 3 & -4 \\ -10 & 15 & -14 & 3 & 7 \\ -8 & 12 & -10 & 4 & 9 \end{bmatrix}$$

Solution We first reduce A to echelon form U

$$\begin{bmatrix} 2 & -3 & 4 & 1 & 2 \\ 4 & -6 & 5 & 3 & -4 \\ -10 & 15 & -14 & 3 & 7 \\ -8 & 12 & -10 & 4 & 9 \end{bmatrix} \rightarrow \begin{bmatrix} 2 & -3 & 4 & 1 & 2 \\ 0 & 0 & -3 & 1 & -8 \\ 0 & 0 & 6 & 8 & 17 \\ 0 & 0 & 6 & 8 & 17 \end{bmatrix}$$

$$\rightarrow \begin{bmatrix} 2 & -3 & 4 & 1 & 2 \\ 0 & 0 & -3 & 1 & -8 \\ 0 & 0 & 0 & 10 & 1 \\ 0 & 0 & 0 & 0 & 0 \end{bmatrix} = U$$

Since the last row of U is all zeros, the first three rows $\mathbf{r}_1$, $\mathbf{r}_2$, and $\mathbf{r}_3$ span $RS(U)$. We now describe how to look at the three vectors $\mathbf{r}_1$, $\mathbf{r}_2$, and $\mathbf{r}_3$ so that you can easily see they are linearly independent. Suppose $c_1\mathbf{r}_1 + c_2\mathbf{r}_2 + c_3\mathbf{r}_3 = \mathbf{0}$. Then

$$\begin{aligned} & c_1[2 \quad -3 \quad 4 \quad 1 \quad 2] \\ + & c_2[0 \quad 0 \quad -3 \quad 1 \quad -8] \\ + & c_3[0 \quad 0 \quad 0 \quad 10 \quad 1] \\ = & \ \ [0 \quad 0 \quad 0 \quad 0 \quad 0] \end{aligned} \tag{3.69}$$

Observe that the nonzero entries lie in a "staircase pattern" coming from the staircase pattern of U. The pivots lie at the "tip" of each "stair." Focus on the columns that contain the pivots, that is, on the first, third, and fourth columns.

The first column gives us

$$2c_1 + 0c_2 + 0c_3 = 0$$

or, easily, $c_1 = 0$. The third column gives us

$$4c_1 - 3c_2 + 0c_3 = 0$$

Since we now know $c_1 = 0$, we see $c_2 = 0$. Finally the fourth column gives us

$$c_1 + c_2 + 10c_3 = 0$$

Since we now know $c_1 = 0$ and $c_2 = 0$, we see $c_3 = 0$. Thus all the c_i's are 0, so the rows are linearly independent.

NOTE This argument, that vectors whose nonzero entries form a staircase pattern are linearly independent, comes up in several contexts in linear algebra, and we shall see it often.

We now know that the nonzero rows of U are linearly independent. They also span $RS(U)$, since the nonzero rows of any matrix span the row space of that matrix. Thus the nonzero rows are a basis for $RS(U)$, and they are also a basis for $RS(A)$, since $RS(U) = RS(A)$. ■

The solution to Example 3 illustrates the following fact, with the actual proof being left as an exercise.

(3.70) Let U be a matrix in row echelon form. Then the nonzero rows of U form a basis for $RS(U)$.

In particular, since the nonzero rows of such a U correspond to the pivots of U,

(3.71) If U is a matrix in row echelon form and

$$r = \text{the number of pivots of } U$$

then $r = \dim RS(U)$.

Equivalently, since each pivot corresponds to a leading variable in the equation $U\mathbf{x} = \mathbf{0}$,

(3.72) If U is a matrix in row echelon form, then $\dim RS(U)$ is equal to the number of leading variables in the equation $U\mathbf{x} = \mathbf{0}$.

The observations in (3.70)–(3.72) will turn out to be very important.

We now use the technique in Example 3 to find a basis for a subspace spanned by a finite number of vectors.

Example 4 Find a basis for the subspace W of $\mathbb{R}^5$ spanned by

$$\mathbf{v}_1 = \begin{bmatrix} 2 \\ -3 \\ 4 \\ 1 \\ 2 \end{bmatrix}, \quad \mathbf{v}_2 = \begin{bmatrix} 4 \\ -6 \\ 5 \\ 3 \\ -4 \end{bmatrix}, \quad \mathbf{v}_3 = \begin{bmatrix} -10 \\ 15 \\ -14 \\ 3 \\ 7 \end{bmatrix}, \quad \text{and} \quad \mathbf{v}_4 = \begin{bmatrix} -8 \\ 12 \\ -10 \\ 4 \\ 9 \end{bmatrix}$$

Solution First we write these vectors as row vectors of a matrix A,

$$A = \begin{bmatrix} 2 & -3 & 4 & 1 & 2 \\ 4 & -6 & 5 & 3 & -4 \\ -10 & 15 & -14 & 3 & 7 \\ -8 & 12 & -10 & 4 & 9 \end{bmatrix}$$

Then the span $\{\mathbf{v}_1, \mathbf{v}_2, \mathbf{v}_3, \mathbf{v}_4\} = RS(A)$. We proceed as in the previous example to find a basis for $RS(A)$. In fact this is the A of the previous example, so we have just computed a basis. Hence a basis for W (which we write as column vectors since we started with column vectors) is

$$\mathbf{u}_1 = \begin{bmatrix} 2 \\ -3 \\ 4 \\ 1 \\ 2 \end{bmatrix}, \quad \mathbf{u}_2 = \begin{bmatrix} 0 \\ 0 \\ -3 \\ 1 \\ -8 \end{bmatrix}, \quad \mathbf{u}_3 = \begin{bmatrix} 0 \\ 0 \\ 0 \\ 10 \\ 1 \end{bmatrix}$$ ■

Dimension of Column Space and Rank

To find a basis for the $CS(A)$, we could proceed in a similar manner as we just did with the $RS(A)$ and reduce A to "column echelon form." However, we choose a different tack, which will yield the most fundamental relationship so far discussed in this text.

For the moment we restrict our attention to matrices U in row echelon form. It is easy to find a basis for the column space of such a U, and this will be useful to us later.

(3.73) **THEOREM** Let U be an $m \times n$ matrix in row echelon form. Then the columns of U that contain the pivots form a basis for the $CS(U)$. Thus if r is the number of pivots of U, then $r = \dim CS(U)$.

The proof of this theorem will be given at the end of this section.

Example 5 Find a basis for the $CS(U)$ for the U given in Example 3.

Solution The matrix U is repeated below. The pivots are in columns 1, 3, and 4, so a basis for $CS(U)$ is $\{\mathbf{c}_1, \mathbf{c}_3, \mathbf{c}_4\}$ where

$$\mathbf{c}_1 = \begin{bmatrix} 2 \\ 0 \\ 0 \\ 0 \end{bmatrix}, \quad \mathbf{c}_3 = \begin{bmatrix} 4 \\ -3 \\ 0 \\ 0 \end{bmatrix}, \quad \mathbf{c}_4 = \begin{bmatrix} 1 \\ 1 \\ 10 \\ 0 \end{bmatrix} \qquad \blacksquare$$

Look again at the matrices A and U from Example 3.

$$A = \begin{bmatrix} 2 & -3 & 4 & 1 & 2 \\ 4 & -6 & 5 & 3 & -4 \\ -10 & 15 & -14 & 3 & 7 \\ -8 & 12 & -10 & 4 & 9 \end{bmatrix}, \quad U = \begin{bmatrix} 2 & -3 & 4 & 1 & 2 \\ 0 & 0 & -3 & 1 & -8 \\ 0 & 0 & 0 & 10 & 1 \\ 0 & 0 & 0 & 0 & 0 \end{bmatrix} \tag{3.74}$$

It is obvious by looking at A and U that

$$CS(A) \neq CS(U) \tag{3.75}$$

since all vectors in $CS(U)$ must have zero in the bottom component (and column vectors of A do not have zeros there). However, there is a very subtle relationship between $CS(A)$ and $CS(U)$.

(3.76) **THEOREM** Let A be an $m \times n$ matrix and let U be an $m \times n$ matrix that results from reducing A to row echelon form. Then whenever certain column vectors of U form a basis for $CS(U)$, the corresponding column vectors of A form a basis for the $CS(A)$.

The proof of this theorem, one of the most subtle arguments given so far in this text, will also be given at the end of this section.

Example 6 Find a basis for the $CS(A)$ where A is given in Example 3.

Solution By Example 5, the first, third, and fourth columns of U form a basis for $CS(U)$. By Theorem (3.76), the first, third, and fourth columns of A form a basis for $CS(A)$. Thus $S = \{\mathbf{c}_1, \mathbf{c}_3, \mathbf{c}_4\}$ is a basis for $CS(A)$, where

$$\mathbf{c}_1 = \begin{bmatrix} 2 \\ 4 \\ -10 \\ -8 \end{bmatrix}, \quad \mathbf{c}_3 = \begin{bmatrix} 4 \\ 5 \\ -14 \\ -10 \end{bmatrix}, \quad \mathbf{c}_4 = \begin{bmatrix} 1 \\ 3 \\ 3 \\ 4 \end{bmatrix}$$ ■

A very surprising and important result, which is an immediate corollary of Theorem (3.76), is that even though in general $CS(A) \neq CS(U)$, their dimensions are the same.

(3.77) COROLLARY Let A and U be as in Theorem (3.76). Then

$$\dim CS(A) = \dim CS(U)$$

Proof By Theorem (3.76), corresponding column vectors from A and U form bases for $CS(A)$ and $CS(U)$ respectively. Hence the number of vectors in each basis is the same, so the dimensions are the same. ■

Example 7 For the A and U of Example 3, $\dim CS(A) = 3 = \dim CS(U)$. ■

We are now ready for our fundamental relationship.

(3.78) THEOREM Let A be any $m \times n$ matrix. Then

$$\dim CS(A) = \dim RS(A)$$

DEFINITION The **rank** of A, denoted by $rk(A)$, is this common number, namely,

$$rk(A) = \dim CS(A) = \dim RS(A)$$

Proof Let A be any $m \times n$ matrix and let U be an $m \times n$ matrix that results from reducing A to echelon form. By Corollary (3.77),

$$\dim CS(A) = \dim CS(U) \tag{3.79}$$

Let r equal the number of pivots of U. By (3.73) and (3.71),

$$\dim CS(U) = r = \dim RS(U) \tag{3.80}$$

Finally, by Theorem (3.68), $RS(U) = RS(A)$ so

(3.81) $$\dim RS(U) = \dim RS(A)$$

Putting Equations (3.79)–(3.81) together, we have $\dim CS(A) = \dim RS(A)$, so we are done. ■

Example 8 For the A of Example 3, $rk(A) = 3$ by either Example 3 or Example 7. ■

Relationships with the Null Space

Let A and U be as usual. We now know the following facts.

$NS(A) = NS(U)$ and $RS(A) = RS(U)$

$\dim NS(U) =$ the number of free variables in the equation $U\mathbf{x} = \mathbf{0}$

$\dim RS(U) =$ the number of leading variables in the equation $U\mathbf{x} = \mathbf{0}$

[See (3.65), Section 3.6, (3.68), (3.66), Section 3.6, and (3.72).] However, each variable in the equation $U\mathbf{x} = \mathbf{0}$ is either free or leading, and altogether the number of these variables is n (if A and U are $m \times n$). Thus we have another fundamental relationship, which is listed first in the following theorem.

(3.82) **THEOREM** If A is an $m \times n$ matrix, then

(a) $\dim RS(A) + \dim NS(A) = n$
(b) $\dim CS(A) + \dim NS(A) = n$
(c) $rk(A) + \dim NS(A) = n$

Proof Part (a) follows from the above argument. Parts (b) and (c) follow from (a), since $\dim RS(A) = \dim CS(A) = rk(A)$. ■

We illustrate all the relationships we have discussed in this section in a final example.

Example 9 Find the dimensions of and bases for $RS(A)$, $CS(A)$, and $NS(A)$, and find $rk(A)$ for

$$A = \begin{bmatrix} 3 & 2 & -4 & 1 & 5 \\ 6 & 4 & -7 & 3 & 1 \\ -3 & -2 & 6 & 1 & 2 \\ 9 & 6 & -11 & 4 & 6 \end{bmatrix}$$

Solution Reduce A to echelon form, obtaining

$$U = \begin{bmatrix} 3 & 2 & -4 & 1 & 5 \\ 0 & 0 & 1 & 1 & -9 \\ 0 & 0 & 0 & 0 & 25 \\ 0 & 0 & 0 & 0 & 0 \end{bmatrix}$$

We can see immediately that

$$\dim RS(A) = \dim CS(A) = rk(A) = 3$$

From $\dim RS(A) + \dim NS(A) = n$ and $n = 5$, we see

$$\dim NS(A) = 2$$

The nonzero rows of U form a basis for $RS(U) = RS(A)$, so $\{\mathbf{r}_1, \mathbf{r}_2, \mathbf{r}_3\}$ is a basis for $RS(A)$, where

$$\mathbf{r}_1 = [3 \quad 2 \quad -4 \quad 1 \quad 5], \quad \mathbf{r}_2 = [0 \quad 0 \quad 1 \quad 1 \quad -9], \quad \mathbf{r}_3 = [0 \quad 0 \quad 0 \quad 0 \quad 25]$$

The first, third, and fifth columns of U contain the pivots of U, and the corresponding columns of A form a basis for $CS(A)$. Thus $\{\mathbf{c}_1, \mathbf{c}_3, \mathbf{c}_5\}$ is a basis for $CS(A)$, where

$$\mathbf{c}_1 = \begin{bmatrix} 3 \\ 6 \\ -3 \\ 9 \end{bmatrix}, \quad \mathbf{c}_3 = \begin{bmatrix} -4 \\ -7 \\ 6 \\ -11 \end{bmatrix}, \quad \mathbf{c}_5 = \begin{bmatrix} 5 \\ 1 \\ 2 \\ 6 \end{bmatrix}$$

To find a basis for $NS(A) = NS(U)$, we form the equation $U\mathbf{x} = \mathbf{0}$ and move the free variables to the other side, obtaining

$$\begin{aligned} 3x_1 - 4x_3 + 5x_5 &= 0 - 2x_2 - x_4 \\ x_3 - 9x_5 &= \qquad\quad\;\; - x_4 \\ 25x_5 &= 0 \end{aligned}$$

Let $x_2 = s$, $x_4 = t$, solve, obtaining

$$x_5 = 0, \qquad x_3 = -t, \qquad x_1 = -\tfrac{2}{3}s - \tfrac{5}{3}t$$

Altogether we have for $\mathbf{x}$ in the null space

$$\mathbf{x} = \begin{bmatrix} x_1 \\ x_2 \\ x_3 \\ x_4 \\ x_5 \end{bmatrix} = \begin{bmatrix} -\tfrac{2}{3}s - \tfrac{5}{3}t \\ s \\ -t \\ t \\ 0 \end{bmatrix} = s\begin{bmatrix} -\tfrac{2}{3} \\ 1 \\ 0 \\ 0 \\ 0 \end{bmatrix} + t\begin{bmatrix} -\tfrac{5}{3} \\ 0 \\ -1 \\ 1 \\ 0 \end{bmatrix}$$

Therefore, a basis for $NS(A) = NS(U)$ is $\{\mathbf{w}_1, \mathbf{w}_2\}$, where

$$\mathbf{w}_1 = \begin{bmatrix} -\tfrac{2}{3} \\ 1 \\ 0 \\ 0 \\ 0 \end{bmatrix}, \quad \mathbf{w}_3 = \begin{bmatrix} -\tfrac{5}{3} \\ 0 \\ -1 \\ 1 \\ 0 \end{bmatrix}$$ ■

See Exercises 33–35 for yet another fundamental subspace associated with an $m \times n$ matrix.

Square Matrices

Recall, in Theorem (1.50), Section 1.4, that we listed several properties that are equivalent to invertibility for square matrices. We are now ready to add to that list.

(3.83)

THEOREM Let A be an $n \times n$ matrix. Then the following are equivalent.

(a) A is invertible.
(b) $NS(A) = \{0\}$.
(c) $\dim NS(A) = 0$.
(d) $rk(A) = n$.
(e) $\dim CS(A) = n$.
(f) $\dim RS(A) = n$.

Proof By Theorem (1.50), Section 1.4, A is invertible $\Leftrightarrow$ A is nonsingular, that is, the only solution to $A\mathbf{x} = \mathbf{0}$ is $\mathbf{x} = \mathbf{0}$. But this is clearly equivalent to $NS(A) = \{\mathbf{0}\}$. Easily, $NS(A) = \{\mathbf{0}\} \Leftrightarrow \dim NS(A) = 0$. Since $rk(A) + \dim NS(A) = n$, by Theorem (3.82), we see $\dim NS(A) = 0 \Leftrightarrow rk(A) = n$. The rest follows since $rk(A) = \dim CS(A) = \dim RS(A)$. ■

OPTIONAL Proof of Theorem (3.68). We first prove a lemma.

(3.84)

LEMMA Let B be any $m \times n$ matrix and e an elementary operation. Then $RS(B) = RS(eB)$.

Proof of Lemma We prove this only for the case that e adds a multiple of one row to another row, and we leave the other two cases [see (1.6), Section 1.1] as exercises. (See Exercises 37 and 38.) Let $\mathbf{r}_1, \ldots, \mathbf{r}_m$ be the rows of B and $\mathbf{s}_1, \ldots, \mathbf{s}_m$ be the rows of $e(B)$. Suppose e adds m times $\mathbf{r}_j$ to $\mathbf{r}_k$. Then

$$\mathbf{s}_i = \mathbf{r}_i \quad \text{for } i \neq k \qquad \text{and} \qquad \mathbf{s}_k = \mathbf{r}_k + m\mathbf{r}_j$$

We need to show every linear combination of the **r**'s is also a linear combination of the **s**'s, and vice versa. But if $\mathbf{v} = a_1\mathbf{r}_1 + \cdots + a_n\mathbf{r}_n$, then

$$\begin{aligned} \mathbf{v} &= a_1\mathbf{r}_1 + \cdots + a_j\mathbf{r}_j - ma_k\mathbf{r}_j + \cdots + a_k\mathbf{r}_k + ma_k\mathbf{r}_j + \cdots + a_m\mathbf{r}_m \\ &= a_1\mathbf{s}_1 + \cdots + (a_j - ma_k)\mathbf{s}_j + \cdots + a_k\mathbf{s}_k + \cdots + a_m\mathbf{s}_m \end{aligned}$$

which is a linear combination of the **s**'s. Similarly, if $\mathbf{w} = b_1\mathbf{s}_1 + \cdots + b_m\mathbf{s}_m$, then

$$\begin{aligned} \mathbf{w} &= b_1\mathbf{r}_1 + \cdots + b_j\mathbf{r}_j + \cdots + b_k(\mathbf{r}_k + m\mathbf{r}_j) + \cdots + b_m\mathbf{r}_m \\ &= b_1\mathbf{r}_1 + \cdots + (b_j + b_km)\mathbf{r}_j + \cdots + b_k\mathbf{r}_k + \cdots + b_m\mathbf{r}_m \end{aligned}$$

which is a linear combination of the **r**'s, so we are done with the lemma. ■

Proof of Theorem (3.68) Since $U = e_1 \cdots e_k A$, we just apply Lemma (3.84) k times to get

$$RS(A) = RS(e_k A) = RS(e_{k-1} e_k A) = \cdots = RS(e_1 \cdots e_k A) = RS(U) \quad \blacksquare$$

OPTIONAL *Proof of Theorem (3.73)* The columns with the pivots are linearly independent because their nonzero entries form a staircase pattern (see argument given in Example 3).

To show these columns span $CS(U)$, pick any vector $\mathbf{b}$ in the $CS(U)$. Let

$$r = \text{the number of pivots of } U \tag{3.85}$$

Then the first r rows of U are nonzero and the last $m - r$ rows of U are all zero. Hence every column vector of U has its last $m - r$ entries all zero. Since $\mathbf{b}$ is a linear combination of all the columns of U, $\mathbf{b}$ has its last $m - r$ entries all zero,

$$\mathbf{b} = \begin{bmatrix} b_1 \\ \vdots \\ b_r \\ 0 \\ \vdots \\ 0 \end{bmatrix}$$

Form an $m \times n$ matrix B by taking the columns of U that contain the pivots and placing them in order in B, so

$$B = \begin{bmatrix} * & * & * & \cdots & * \\ 0 & * & * & \cdots & * \\ 0 & 0 & * & \cdots & * \\ & & \cdots & & \\ 0 & 0 & 0 & \cdots & * \\ & & \cdots & & \\ 0 & 0 & 0 & \cdots & 0 \end{bmatrix}$$

where the specially marked entries are the pivots of U and hence are nonzero. By the special forms of B and $\mathbf{b}$, the equation

$$B \begin{bmatrix} x_1 \\ \vdots \\ x_r \end{bmatrix} = \mathbf{b}$$

can be solved by backsubstitution for $x_1, \ldots, x_r$. This expresses $\mathbf{b}$ as a linear combination of the columns of B and hence of the desired columns of U, so we are done. $\blacksquare$

OPTIONAL *Proof of Theorem (3.76)* The proof follows from the fact that

$$A\mathbf{x} = \mathbf{0} \qquad \text{if and only if} \qquad U\mathbf{x} = \mathbf{0} \tag{3.86}$$

Suppose the column vectors of U, $\mathbf{c}_{i_1}, \ldots, \mathbf{c}_{i_k}$, form a basis for $CS(U)$ and $\mathbf{d}_{i_1}, \ldots, \mathbf{d}_{i_k}$ are the corresponding columns of A.

These **d**'s are linearly independent, for suppose

$$x_{i_1}\mathbf{d}_{i_1} + \cdots + x_{i_k}\mathbf{d}_{i_k} = \mathbf{0}$$

Let $x_i = 0$ for $i \neq i_1, \ldots, i_k$. Then $x_1\mathbf{d}_1 + \cdots + x_n\mathbf{d}_n = \mathbf{0}$ or

$$A\begin{bmatrix} x_1 \\ \vdots \\ x_n \end{bmatrix} = \mathbf{0} \qquad \text{which implies} \qquad U\begin{bmatrix} x_1 \\ \vdots \\ x_n \end{bmatrix} = \mathbf{0}$$

by (2.86). Thus $x_1\mathbf{c}_1 + \cdots + x_n\mathbf{c}_n = \mathbf{0}$, so

$$x_{i_1}\mathbf{c}_{i_1} + \cdots + x_{i_k}\mathbf{c}_{i_k} = \mathbf{0}$$

since $x_i = 0$ for $i \neq i_1, \ldots, i_k$. But $\mathbf{c}_{i_1}, \ldots, \mathbf{c}_{i_k}$ are linearly independent. Therefore, all of $x_{i_1}, \ldots, x_{i_k}$ must be zero, and we are done.

To show $\mathbf{d}_{i_1}, \ldots, \mathbf{d}_{i_k}$ span $CS(A)$, pick any $\mathbf{y}$ in $CS(A)$. Then $\mathbf{y}$ is a linear combination of all the **d**'s, and we wish to show $\mathbf{y}$ is a linear combination of only $\mathbf{d}_{i_1}, \ldots, \mathbf{d}_{i_k}$, also.

Pick any column $\mathbf{d}_j, j \neq i_1, \ldots, i_k$, of A and consider the corresponding column $\mathbf{c}_j$ of U. Since $\mathbf{c}_{i_1}, \ldots, \mathbf{c}_{i_k}$ span $CS(U)$, $\mathbf{c}_j$ is a linear combination of them,

$$\mathbf{c}_j = g_{i_1}\mathbf{c}_{i_1} + \cdots + g_{i_k}\mathbf{c}_{i_k} \qquad \text{or} \qquad \mathbf{0} = -\mathbf{c}_j + g_{i_1}\mathbf{c}_{i_1} + \cdots + g_{i_k}\mathbf{c}_{i_k}$$

Let $g_j = -1$ and $g_i = 0$ for $i \neq j, i_1, \ldots, i_k$. Then

$$\mathbf{0} = g_1\mathbf{c}_1 + \cdots + g_n\mathbf{c}_n \qquad \text{or} \qquad U\mathbf{g} = \mathbf{0} \text{ for } \mathbf{g} = \begin{bmatrix} g_1 \\ \vdots \\ g_n \end{bmatrix}$$

Thus $A\mathbf{g} = \mathbf{0}$ by (3.86), so

$$\mathbf{0} = g_1\mathbf{d}_1 + \cdots + g_n\mathbf{d}_n \qquad \text{or} \qquad \mathbf{d}_j = g_{i_1}\mathbf{d}_{i_1} + \cdots + g_{i_k}\mathbf{d}_{i_k}$$

In this way every $\mathbf{d}_j$ is a linear combination of $\mathbf{d}_{i_1}, \ldots, \mathbf{d}_{i_k}$. Hence $\mathbf{y}$ is a linear combination of only these vectors by Theorem (3.33), Section 3.4, so we are done. ■

Exercise 3.7 In Exercises 1 and 2, list the row vectors and column vectors of the given matrix

1. $\begin{bmatrix} 2 & 4 & 3 \\ 0 & -2 & 4 \\ 3 & -5 & -4 \\ -1 & 0 & 2 \end{bmatrix}$

2. $\begin{bmatrix} 1 & 2 & 0 & -3 & 4 \\ 5 & 1 & -3 & 2 & -2 \end{bmatrix}$

In Exercises 3–6, a matrix in row echelon form is given. Find a basis for its row space, find a basis for its column space, and determine its rank.

3. $\begin{bmatrix} 1 & -2 & 5 & 2 & 1 & 3 \\ 0 & 0 & 2 & 3 & 0 & -1 \\ 0 & 0 & 0 & 0 & 0 & 4 \end{bmatrix}$

4. $\begin{bmatrix} 2 & -3 & 5 \\ 0 & 0 & 2 \\ 0 & 0 & 0 \\ 0 & 0 & 0 \end{bmatrix}$

5. $\begin{bmatrix} 2 & -4 & 3 & 1 \\ 0 & -3 & -2 & 7 \\ 0 & 0 & -4 & 1 \\ 0 & 0 & 0 & 5 \end{bmatrix}$

6. $\begin{bmatrix} 0 & 2 & -3 & 4 & 1 & 2 & 1 & 7 \\ 0 & 0 & 3 & -2 & 0 & 4 & -5 & 3 \\ 0 & 0 & 0 & 0 & 0 & 6 & -2 & 2 \\ 0 & 0 & 0 & 0 & 0 & 0 & 2 & 1 \\ 0 & 0 & 0 & 0 & 0 & 0 & 0 & 0 \end{bmatrix}$

In Exercises 7–10, for the given matrix find a basis for its row space, find a basis for its column space, and determine its rank.

7. $\begin{bmatrix} 1 & 2 & -3 \\ 2 & 4 & 5 \\ -3 & -6 & 0 \end{bmatrix}$

8. $\begin{bmatrix} 3 & 2 & -1 \\ 6 & 3 & 5 \\ -3 & -1 & -6 \\ 0 & -1 & 7 \end{bmatrix}$

9. $\begin{bmatrix} 0 & 2 & -3 & 1 & 2 \\ 0 & -2 & 3 & 3 & 1 \\ 0 & 4 & -6 & 6 & 7 \end{bmatrix}$

10. $\begin{bmatrix} 1 & 1 & 0 & 0 \\ 1 & 0 & 1 & 0 \\ 1 & 0 & 0 & 1 \end{bmatrix}$

In Exercises 11–16, use the method of Example 4 to find a basis for the subspace spanned by the given vectors.

11. $(2, -1, 6), \quad (4, 3, -2), \quad (8, 1, 10)$

12. $(-2, 4, 1, 2), \quad (4, 2, 3, -1), \quad (2, 6, 4, 1)$

13. $\begin{bmatrix} 4 \\ 1 \\ -4 \\ 2 \end{bmatrix}, \begin{bmatrix} 2 \\ 3 \\ 2 \\ -4 \end{bmatrix}, \begin{bmatrix} 0 \\ -5 \\ -8 \\ 10 \end{bmatrix}, \begin{bmatrix} 6 \\ -1 \\ -10 \\ 8 \end{bmatrix}$

14. $\begin{bmatrix} -1 \\ 2 \\ -6 \\ 3 \end{bmatrix}, \begin{bmatrix} 5 \\ 1 \\ 2 \\ 1 \end{bmatrix}, \begin{bmatrix} 2 \\ 2 \\ 0 \\ 4 \end{bmatrix}, \begin{bmatrix} 2 \\ 1 \\ -4 \\ 0 \end{bmatrix}$

In Exercises 15 and 16, first write the coefficients as n-tuples. Then use the method of Example 4 to find a basis for the subspace spanned by those n-tuples. Finally, find a basis for the subspace spanned by the given polynomials.

15. $2 - 3x + 4x^2 - 5x^3$, $4 - x + 15x^2 - 2x^3$, $2 + 2x + 11x^2 + 3x^3$

16. $1 + x^2 - x^3$, $2 + x - 3x^3$, $8 + 3x + 2x^2 - 11x^3$, $-1 + x - 3x^2$

In Exercises 17–20, for the given matrix, find a basis for the row space, find a basis for the null space, find dim RS, find dim NS, and verify dim RS + dim $NS = n$.

17. $\begin{bmatrix} 2 & -1 & 3 \\ 4 & 2 & 1 \\ 2 & 3 & -2 \\ 8 & 0 & 7 \end{bmatrix}$

18. $\begin{bmatrix} 1 & -2 & 4 & 1 \\ 3 & 1 & -3 & -1 \\ 5 & -3 & 5 & 1 \end{bmatrix}$

19. $\begin{bmatrix} 1 & 2 & 3 \\ 2 & -1 & 2 \\ 3 & 1 & 4 \end{bmatrix}$

20. $\begin{bmatrix} 2 & 4 & 1 & 2 & 1 \\ 4 & 2 & 3 & 2 & 2 \\ 2 & -2 & 2 & 0 & 1 \\ 6 & 6 & 4 & 4 & 3 \end{bmatrix}$

In Exercises 21–24, determine if **b** lies in the column space of the given matrix. If it does, express **b** as a linear combination of the columns.

21. $\begin{bmatrix} 1 & -3 \\ 2 & 5 \end{bmatrix}$, $\mathbf{b} = \begin{bmatrix} 2 \\ 15 \end{bmatrix}$

22. $\begin{bmatrix} 2 & -3 \\ -4 & 6 \end{bmatrix}$, $\mathbf{b} = \begin{bmatrix} 4 \\ -6 \end{bmatrix}$

23. $\begin{bmatrix} 1 & 1 & -1 \\ 1 & 2 & 2 \\ 2 & 3 & 1 \end{bmatrix}$, $\mathbf{b} = \begin{bmatrix} 2 \\ -3 \\ 1 \end{bmatrix}$

24. $\begin{bmatrix} 1 & 1 & -1 \\ 1 & 2 & 2 \\ 2 & 3 & 1 \end{bmatrix}$, $\mathbf{b} = \begin{bmatrix} 2 \\ -3 \\ -1 \end{bmatrix}$

In Exercises 25–28, describe geometrically the RS, CS, NS of the given matrix.

25. I_3

26. $\mathbf{0}$, the 3×3 zero matrix.

27. $\begin{bmatrix} 0 & 1 & 1 \\ 0 & 0 & 1 \\ 0 & 0 & 0 \end{bmatrix}$

28. $\begin{bmatrix} 1 & 1 & 1 \\ 1 & 1 & 1 \\ 1 & 1 & 1 \end{bmatrix}$

29. Show that $A\mathbf{x} = \mathbf{b}$ has a solution if and only if $\mathbf{b}$ is in $CS(A)$.

30. Show that if the product of two matrices is the zero matrix, $AB = \mathbf{0}$, then the column space of B is contained in the null space of A. [HINT Pick an arbitrary $\mathbf{y} \in CS(B)$. Then use Exercise 29 and the hypothesis to show that $\mathbf{y} \in NS(A)$.]

31. Show that $A\mathbf{x} = \mathbf{b}$ has a solution if and only if $rk(A) = rk[A \vdots \mathbf{b}]$. [HINT $rk(A) = \dim CS(A)$ and use (3.32), Section 3.4.]

32. Show that $RS(A) = CS(A^T)$ and $CS(A) = RS(A^T)$.

DEFINE The **left null space** of $A = \{\mathbf{x} | \mathbf{x}A = \mathbf{0}\}$. Denote this by $LNS(A)$.

33. Show that $LNS(A) = NS(A^T)$ and $LNS(A^T) = NS(A)$.

34. Show that if A is $m \times n$, then

$$\dim LNS(A) + \dim CS(A) = m$$

[HINT Use Exercises 32 and 33 and Theorem (3.76).]

In Exercises 35 and 36, find bases for RS, NS, CS, LNS, and find the rank and verify $\dim RS + \dim NS = n$, $\dim CS + \dim LNS = m$, for the given matrix.

35. $\begin{bmatrix} 1 & 3 & -2 & 4 \\ 0 & 0 & 5 & 1 \\ 0 & 0 & 0 & 0 \end{bmatrix}$

36. $\begin{bmatrix} 2 & 1 & -3 \\ -2 & -1 & 3 \\ 4 & 2 & -6 \\ -6 & -3 & 9 \end{bmatrix}$

37. Let B be any $m \times n$ matrix and let e be the elementary operation that interchanges two rows. Show $RS(B) = RS(eB)$. (HINT The rows of B are the same as the rows of eB.)

38. Let B be any $m \times n$ matrix and let e be the elementary operation that multiplies a row by a nonzero m. Show that $RS(B) = RS(eB)$.

39. Are the following statements true or false? Give a brief reason or counterexample for each.

    (a) If A is an $n \times n$ matrix, then the row space of A is equal to the column space of A.
    (b) Even if A is square, the column space of A can never equal the null space of A.
    (c) If A is an $m \times n$ matrix and the columns of A are linearly independent, then $A\mathbf{x} = \mathbf{b}$ has exactly one solution for every $\mathbf{b}$.
    (d) A 3×4 matrix never has linearly independent columns.
    (e) A 4×3 matrix must have linearly independent columns.

40. Suppose that a matrix A is formed by taking n vectors from $\mathbb{R}^m$ as its columns.

    (a) If these vectors are linearly independent, what is the rank of A and what is the relationship between m and n?
    (b) If these vectors span $\mathbb{R}^m$ instead, what is the rank of A and what is the relationship between m and n?
    (c) If these vectors form a basis for $\mathbb{R}^m$, what is the relationship between m and n then?

41. Add more entries to the list in Theorem (3.83) by showing that for an $n \times n$ matrix A,

$$\begin{aligned} A \text{ is invertible} &\Leftrightarrow \text{the rows of } A \text{ are linearly independent} \\ &\Leftrightarrow \text{the columns of } A \text{ are linearly independent} \\ &\Leftrightarrow \text{the rows of } A \text{ span } \mathbb{R}^n \\ &\Leftrightarrow \text{the columns of } A \text{ span } \mathbb{R}^n \end{aligned}$$

[HINT See Theorem (3.64), Section 3.6.]

DEFINITION If X and Y are matrices such that $XY = I$, then Y is called a **right inverse** of X and X is called a **left inverse** of Y.

42. (a) Find a 1×2 matrix X and a 2×1 matrix Y such that $XY = I_1$.
(b) Is it possible to find matrices X and Y that satisfy (a) and that also satisfy $YX = I_2$? Why or why not?

43. Generalize Exercise 42 to the $m \times n$ case if $m \neq n$.

44. Suppose that A is an $n \times n$ matrix and has a right inverse B that is also $n \times n$.
(a) Show that $CS(A) = CS(I) = \mathbb{R}^n$.
(b) Show that A is invertible.
(c) Show that $A^{-1} = B$. Hence, if a square matrix has a square right inverse, then that right inverse is also a left inverse and is therefore the inverse of the original matrix.

45. Suppose that A is an $n \times n$ matrix and has a left inverse B that is also $n \times n$.
(a) Show that $NS(A) = NS(I) = \{\mathbf{0}\}$.
(b) Show that A is invertible.
(c) Show $A^{-1} = B$. Hence, if a square matrix has a square left inverse, then that left inverse is also a right inverse and is therefore the inverse of the original matrix.

3.8 Coordinates and Change of Basis*

In Section 3.6, we introduced the standard basis for $\mathbb{R}^n$ (and P_n and M_{mn}). All other things being equal, the standard basis is the easiest base to use. In some situations, unfortunately, we must use other bases to make our computations manageable or understandable. The use of a basis of eigenvectors for a given matrix (see Chapter 5) or the use of Fourier series (see almost any engineer or physicist or Section 6.1) are two such cases. In this section, we develop the tools for using different bases easily.

* *The subsection on pages 214–218 is optional. It is a prerequisite only for Section 4.7.*

Figure 3.22

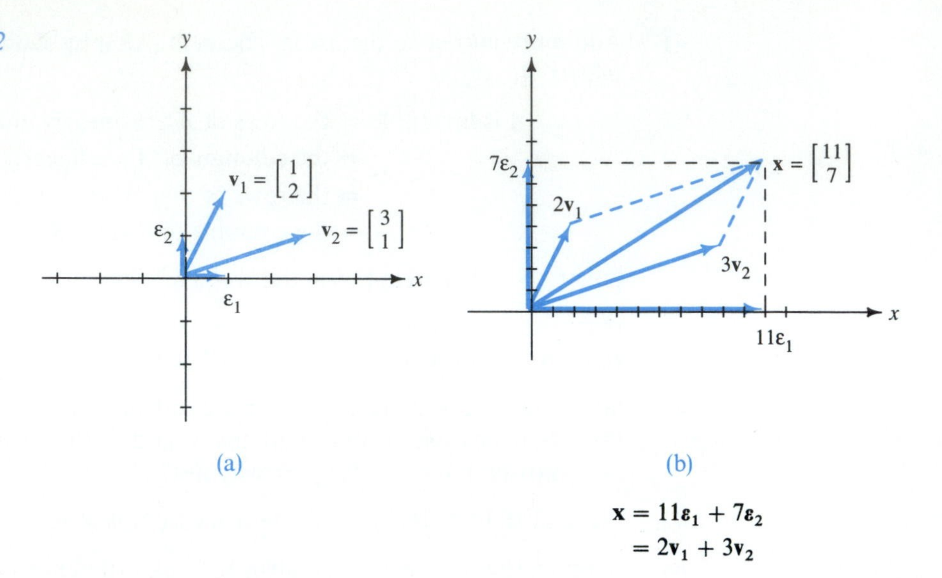

$$\mathbf{x} = 11\boldsymbol{\varepsilon}_1 + 7\boldsymbol{\varepsilon}_2$$
$$= 2\mathbf{v}_1 + 3\mathbf{v}_2$$

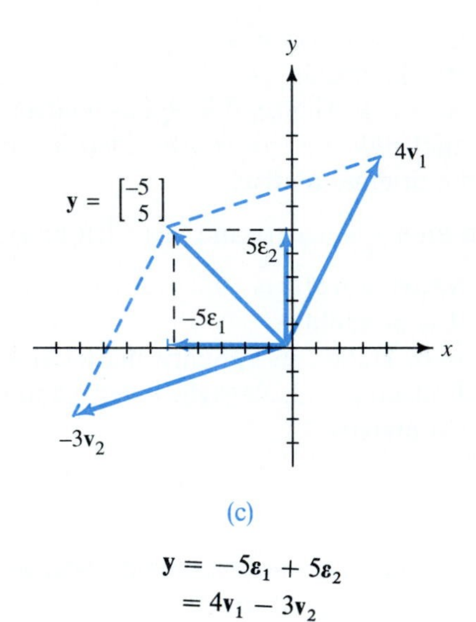

$$\mathbf{y} = -5\boldsymbol{\varepsilon}_1 + 5\boldsymbol{\varepsilon}_2$$
$$= 4\mathbf{v}_1 - 3\mathbf{v}_2$$

Coordinates

Consider two different bases for $\mathbb{R}^2$: the standard basis $S = \{\boldsymbol{\varepsilon}_1, \boldsymbol{\varepsilon}_2\}$, and the basis $B = \{\mathbf{v}_1, \mathbf{v}_2\}$, where $\mathbf{v}_1 = [1 \quad 2]^T$ and $\mathbf{v}_2 = [3 \quad 1]^T$. (See Figure 3.22a). We can quickly verify that B is a basis for $\mathbb{R}^2$. Since each basis spans $\mathbb{R}^2$, any vector in $\mathbb{R}^2$ can be expressed as a linear combination of the basis vectors from either basis.

Example 1 Let $\mathbf{x} = \begin{bmatrix} 11 \\ 7 \end{bmatrix}$ and $\mathbf{y} = \begin{bmatrix} -5 \\ 5 \end{bmatrix}$. Then

$$\begin{aligned} \mathbf{x} &= 11\varepsilon_1 + 7\varepsilon_2 = 2\mathbf{v}_1 + 3\mathbf{v}_2 \\ \mathbf{y} &= -5\varepsilon_1 + 5\varepsilon_2 = 4\mathbf{v}_1 - 3\mathbf{v}_2 \end{aligned}$$

as can be checked easily. (See Figures 3.22b, c.) ■

Two things are important here: the numbers used in the linear combinations, and the fact that those numbers are unique.

(3.87)

Let $S = \{\mathbf{v}_1, \ldots, \mathbf{v}_n\}$ be a basis for a vector space V and let $\mathbf{x}$ be a vector in V.

THEOREM The vector $\mathbf{x}$ can be expressed in one *and only one* way as a linear combination of the $\mathbf{v}_i$'s. Thus if

$$\mathbf{x} = a_1\mathbf{v}_1 + \cdots + a_n\mathbf{v}_n \quad \text{and}$$
$$\mathbf{x} = b_1\mathbf{v}_1 + \cdots + b_n\mathbf{v}_n$$

then $a_i = b_i$ for all i.

DEFINITION If $\mathbf{x} = a_1\mathbf{v}_1 + \cdots + a_n\mathbf{v}_n$, then the numbers $a_1, \ldots, a_n$ are called the **coordinates** of $\mathbf{x}$ relative to the basis S. The $n \times 1$ matrix made up of these coordinates is called the **coordinate matrix** of $\mathbf{x}$ relative to S and is denoted by $[\mathbf{x}]_S$:

$$[\mathbf{x}]_S = \begin{bmatrix} a_1 \\ \vdots \\ a_n \end{bmatrix}$$

The proof follows Example 3.

Example 2 From Example 1, the coordinates of $\mathbf{x}$ in S are 11 and 7, the coordinates of $\mathbf{x}$ in B are 2 and 3, and

$$[\mathbf{x}]_S = \begin{bmatrix} 11 \\ 7 \end{bmatrix}, \quad [\mathbf{x}]_B = \begin{bmatrix} 2 \\ 3 \end{bmatrix}$$

Similarly,

$$[\mathbf{y}]_S = \begin{bmatrix} -5 \\ 5 \end{bmatrix}, \quad [\mathbf{y}]_B = \begin{bmatrix} 4 \\ -3 \end{bmatrix}$$ ■

WARNING *Technical Detail* Order is important, since order is important in matrices. Thus bases are really **ordered bases** (and we shall use this phrase when we need to emphasize order). If we merely rearrange the vectors in a basis, we must consider it a different (ordered) basis.

Example 3 Let $\mathbf{x}$ be the $\mathbf{x}$ in Examples 1 and 2, and consider the ordered bases $T = \{\boldsymbol{\varepsilon}_2, \boldsymbol{\varepsilon}_1\}$ and $C = \{\mathbf{v}_2, \mathbf{v}_1\}$. Then

$$\mathbf{x} = 7\boldsymbol{\varepsilon}_2 + 11\boldsymbol{\varepsilon}_1 = 3\mathbf{v}_2 + 2\mathbf{v}_1$$

and here the importance of order is not apparent. But

$$[\mathbf{x}]_T = \begin{bmatrix} 7 \\ 11 \end{bmatrix} \neq \begin{bmatrix} 11 \\ 7 \end{bmatrix} = [\mathbf{x}]_S$$

$$[\mathbf{x}]_C = \begin{bmatrix} 3 \\ 2 \end{bmatrix} \neq \begin{bmatrix} 2 \\ 3 \end{bmatrix} = [\mathbf{x}]_B$$

and now the importance of order is quite apparent. ■

Proof of Theorem (3.87) Suppose

$$\mathbf{x} = a_1\mathbf{v}_1 + \cdots + a_n\mathbf{v}_n \qquad \text{and} \qquad \mathbf{x} = b_1\mathbf{v}_1 + \cdots + b_n\mathbf{v}_n$$

Then

$$\begin{aligned} \mathbf{0} = \mathbf{x} - \mathbf{x} &= (a_1\mathbf{v}_1 + \cdots + a_n\mathbf{v}_n) - (b_1\mathbf{v}_1 + \cdots + b_n\mathbf{v}_n) \\ &= (a_1 - b_1)\mathbf{v}_1 + \cdots + (a_n - b_n)\mathbf{v}_n \end{aligned}$$

Since the $\mathbf{v}_i$'s are linear independent,

$$a_i - b_i = 0 \text{ for all } i \qquad \text{or equivalently} \qquad a_i = b_i \text{ for all } i$$ ■

The next example illustrates the role of uniqueness.

Example 4 Let W be the subspace of $\mathbb{R}^3$ described in Example 5, Section 3.6, and $S = \{\mathbf{v}_1, \mathbf{v}_2\}$ be the basis for W also described there. Thus $\mathbf{v}_1 = (1, -1, 2)$ and $\mathbf{v}_2 = (2, 0, 1)$. If $\mathbf{x} = (-1, -3, 4)$ then $\mathbf{x}$ is in W and indeed

$$\mathbf{x} = 3\mathbf{v}_1 + (-2)\mathbf{v}_2$$

as you can check. Moreover, if we somehow obtained two numbers a_1 and a_2 with the property $\mathbf{x} = a_1\mathbf{v}_1 + a_2\mathbf{v}_2$, then we would also know $a_1 = 3$ and $a_2 = -2$ [by Theorem (3.87)]. On the other hand, let $T = \{\mathbf{v}_1, \mathbf{v}_2, \mathbf{v}_3\}$ (as described in Example 6, Section 3.6). Then T spans W but is not a basis for W. Thus we can express $\mathbf{x}$ as a linear combination of the vectors in T but *not* in a *unique* way. For example.

$$\begin{aligned} \mathbf{x} &= 3\mathbf{v}_1 + (-2)\mathbf{v}_2 + 0\mathbf{v}_3 = \mathbf{v}_1 + \mathbf{v}_2 + \mathbf{v}_3 \\ &= 5\mathbf{v}_1 - 5\mathbf{v}_2 - \mathbf{v}_3 = 2\mathbf{v}_1 - \tfrac{1}{2}\mathbf{v}_2 + \tfrac{1}{2}\mathbf{v}_3 = \cdots \end{aligned}$$

as you can check. There are, of course, infinitely many ways of expressing $\mathbf{x}$ as a linear combination of $\mathbf{v}_1, \mathbf{v}_2, \mathbf{v}_3$. ■

We give one example involving a general vector space.

Example 5 Let $S = \{1, x, x^2\}$ be the standard (ordered) basis for P_2 (see Example 3, Section 3.6), and let $B = \{p_1, p_2, p_3\}$, where $p_1 = 1 + x + x^2$, $p_2 = x + 2x^2$, and $p_3 = 3x^2$. Let $f(x) = \frac{3}{2} + \frac{7}{2}x - 2x^2$. Then we easily can check that $f(x) = \frac{3}{2}p_1 + 2p_2 - \frac{5}{2}p_3$, so that

$$[f]_S = \begin{bmatrix} \frac{3}{2} \\ \frac{7}{2} \\ -2 \end{bmatrix} \quad \text{and} \quad [f]_B = \begin{bmatrix} \frac{3}{2} \\ 2 \\ -\frac{5}{2} \end{bmatrix}$$ ■

Finding the coordinates of a particular vector usually involves solving a system of equations, unless there is something special about the basis. It may have nice form, like the standard basis, or it may be orthogonal (see Section 4.4) or exhibit some other property.

Example 6 Let $S = \{\mathbf{v}_1, \mathbf{v}_2, \mathbf{v}_3\}$ be the basis for $\mathbb{R}^3$ discussed in Example 2, Section 3.6. Then $\mathbf{v}_1 = (2, 1, 0)$, $\mathbf{v}_2 = (-3, -3, 1)$, and $\mathbf{v}_3 = (-2, 1, -1)$. If we need to find the coordinate of $\mathbf{x} = (4, 13, -6)$ relative to S, then we need to find numbers c_1, c_2, and c_3 such that

$$c_1\mathbf{v}_1 + c_2\mathbf{v}_2 + c_3\mathbf{v}_3 = \mathbf{x}$$

Thus we must solve the system

$$\begin{aligned} 2c_1 - 3c_2 - 2c_3 &= 4 \\ c_1 - 3c_2 + c_3 &= 13 \\ c_2 - c_3 &= -6 \end{aligned} \quad \text{or} \quad \begin{bmatrix} 2 & -3 & -2 \\ 1 & -3 & 1 \\ 0 & 1 & -1 \end{bmatrix} \begin{bmatrix} c_1 \\ c_2 \\ c_3 \end{bmatrix} = \begin{bmatrix} 4 \\ 13 \\ -6 \end{bmatrix}$$

and obtain $(c_1, c_2, c_3) = (3, -2, 4)$. It is straightforward to check that

$$3\mathbf{v}_1 - 2\mathbf{v}_2 + 4\mathbf{v}_3 = \mathbf{x} \quad \text{so that} \quad [\mathbf{x}]_S = \begin{bmatrix} 3 \\ -2 \\ 4 \end{bmatrix}$$ ■

One last theorem about coordinate matrices is needed.

(3.88) THEOREM Let S be a basis for a vector space V, let $\mathbf{x}$ and $\mathbf{y}$ be vectors in V, and let r be any number. Then

$$[\mathbf{x}]_B + [\mathbf{y}]_B = [\mathbf{x} + \mathbf{y}]_B \quad \text{and} \quad r[\mathbf{x}]_B = [r\mathbf{x}]_B$$

The proof is left to Exercises 15, 16.

Change of Basis; Transition Matrices*

Suppose that we change bases in a vector space from some old basis B to some new basis C. How is the old coordinate matrix $[\mathbf{v}]_B$ of a vector $\mathbf{v}$ related to the new coordinate matrix $[\mathbf{v}]_C$? It turns out that the answer is easy to state, so we give it first.

(3.89)

THEOREM Let V be an n-dimensional vector space with bases B and C. Then there is a unique, invertible $n \times n$ matrix P with the property

$$P[\mathbf{x}]_B = [\mathbf{x}]_C \qquad \text{for all } \mathbf{x} \text{ in } V$$

DEFINITION This matrix P is called the **transition matrix** from B to C.

The proof follows Example 7. First, we give an easy way of finding P in the case $V = \mathbb{R}^n$. Recall the method we used to find A^{-1} (presented in Example 9, Section 1.4). We formed an $n \times 2n$ matrix $[A \,\vdots\, I]$, performed elementary row operations to reduce the left side to I, and obtained A^{-1} on the right side:

$$[A \,\vdots\, I] \xrightarrow[\text{operations}]{\text{row}} [I \,\vdots\, A^{-1}]$$

Our method for finding P essentially extends this to

$$[C \,\vdots\, B] \xrightarrow[\text{operations}]{\text{row}} [I \,\vdots\, P]$$

WARNING Be careful of the order. C and B are reversed!

(3.91)

THEOREM Let $B = \{\mathbf{u}_1, \ldots, \mathbf{u}_n\}$ and $C = \{\mathbf{v}_1, \ldots, \mathbf{v}_n\}$ be two (ordered) bases of $\mathbb{R}^n$. Form the $n \times 2n$ matrix by using the vectors first from C and then from B (in that order) as its columns. If row operations are used on this $n \times 2n$ matrix to reduce the left side to I_n, then the transition matrix P from B to C is found on the right side:

$$[\mathbf{v}_1 \cdots \mathbf{v}_n \,\vdots\, \mathbf{u}_1 \cdots \mathbf{u}_n] \xrightarrow[\text{operations}]{\text{row}} [I \,\vdots\, P]$$

The proof follows Example 8. It is now time for an example.

* *This subsection is optional. It is a prerequisite only for Section 4.7.*

Example 7 In $\mathbb{R}^3$, let $\mathbf{u}_1 = (2, 3, -1)$, $\mathbf{u}_2 = (2, 1, 2)$, $\mathbf{u}_3 = (4, 3, 1)$, $\mathbf{v}_1 = (1, 1, 1)$, $\mathbf{v}_2 = (0, 1, 1)$, $\mathbf{v}_3 = (0, 0, 1)$, and $\mathbf{x} = (10, 13, -5)$. It is straightforward to check that $B = \{\mathbf{u}_1, \mathbf{u}_2, \mathbf{u}_3\}$ and $C = \{\mathbf{v}_1, \mathbf{v}_2, \mathbf{v}_3\}$ are bases for $\mathbb{R}^3$ and that $\mathbf{x} = 3\mathbf{u}_1 - 2\mathbf{u}_2 + 2\mathbf{u}_3$. Thus $[\mathbf{x}]_B = [3 \quad -2 \quad 2]^T$. We find P and then compute $[\mathbf{x}]_C = P[\mathbf{x}]_B$. We form

$$[C \vdots B] = \left[\begin{array}{ccc:ccc} 1 & 0 & 0 & 2 & 2 & 4 \\ 1 & 1 & 0 & 3 & 1 & 3 \\ 1 & 1 & 1 & -1 & 2 & 1 \end{array}\right]$$

$$\rightarrow \left[\begin{array}{ccc:ccc} 1 & 0 & 0 & 2 & 2 & 4 \\ 0 & 1 & 0 & 1 & -1 & -1 \\ 0 & 1 & 1 & -3 & 0 & -3 \end{array}\right]$$

$$\rightarrow \left[\begin{array}{ccc:ccc} 1 & 0 & 0 & 2 & 2 & 4 \\ 0 & 1 & 0 & 1 & -1 & -1 \\ 0 & 0 & 1 & -4 & 1 & -2 \end{array}\right] = [I \vdots P]$$

Thus

$$P = \begin{bmatrix} 2 & 2 & 4 \\ 1 & -1 & -1 \\ -4 & 1 & -2 \end{bmatrix}$$

Next

$$[\mathbf{x}]_C = P[\mathbf{x}]_B = \begin{bmatrix} 2 & 2 & 4 \\ 1 & -1 & -1 \\ -4 & 1 & -2 \end{bmatrix} \begin{bmatrix} 3 \\ -2 \\ 2 \end{bmatrix} = \begin{bmatrix} 10 \\ 3 \\ -18 \end{bmatrix}$$

To check that this is correct, we compute

$$\begin{aligned} 10\mathbf{v}_1 + 3\mathbf{v}_2 - 18\mathbf{v}_3 &= 10(1, 1, 1) + 3(0, 1, 1) - 18(0, 0, 1) \\ &= (10, 13, -5) = \mathbf{x} \end{aligned}$$

■

Proof of Theorem (3.89) We must prove that P exists, that P is unique, and that P is invertible. Although the order may seem strange, we must prove uniqueness first in order to determine how to construct P for existence.

UNIQUENESS Suppose that P is an $n \times n$ matrix such that $P[\mathbf{x}]_B = [\mathbf{x}]_C$ for all $\mathbf{x}$ in V. Let $B = \{\mathbf{u}_1, \ldots, \mathbf{u}_n\}$. Now $\mathbf{u}_i = 0\mathbf{u}_1 + \cdots + 0\mathbf{u}_{i-1} + 1\mathbf{u}_i + 0\mathbf{u}_{i+1} + \cdots + 0\mathbf{u}_n$, so

$$[\mathbf{u}_i]_B = \begin{bmatrix} 0 \\ \vdots \\ 1 \\ \vdots \\ 0 \end{bmatrix} = \varepsilon_i$$

Since each $\mathbf{u}_i$ is an $\mathbf{x}$ in V, we have

$$P[\mathbf{u}_i]_B = [\mathbf{u}_i]_C \qquad \text{or} \qquad P\varepsilon_i = [\mathbf{u}_i]_C$$

But $P\varepsilon_i$ is the ith column of P. Thus P must be the matrix in which $[\mathbf{u}_i]_C$ is the ith column. Since there is only one such matrix, P is unique.

EXISTENCE We form the P we just found and see if it always works. Let P be the matrix in which $[\mathbf{u}_i]_C$ is the ith column, and then see if $P[\mathbf{x}]_B = [\mathbf{x}]_C$ for all $\mathbf{x}$ in V. Let $\mathbf{x} = a_1\mathbf{u}_1 + \cdots + a_n\mathbf{u}_n$. Then $[\mathbf{x}]_B = [a_1 \cdots a_n]^T$ and

$$\begin{aligned} P[\mathbf{x}]_B &= \big[[\mathbf{u}_1]_C \cdots [\mathbf{u}_n]_C\big]\begin{bmatrix} a_1 \\ \vdots \\ a_n \end{bmatrix} \\ &= a_1[\mathbf{u}_1]_C + \cdots + a_n[\mathbf{u}_n]_C \\ &= [a_1\mathbf{u}_1 + \cdots + a_n\mathbf{u}_n]_C \qquad \text{[by Theorem (3.88)]} \\ &= [\mathbf{x}]_C \end{aligned}$$

It works, so we have proved existence.

INVERTIBILITY Since B is a basis, the vectors $\mathbf{u}_1, \ldots, \mathbf{u}_n$ are linearly independent. This implies that $[\mathbf{u}]_C, \ldots, [\mathbf{u}]_C$ are linearly independent vectors in $\mathbb{R}^n$. (See Exercise 15.) Thus P is an $n \times n$ matrix with linearly independent columns, so that P is invertible (by Exercise 41, Section 3.7). ■

The existence part of the proof of Theorem (3.89) contains a description of P that is useful and should be emphasized.

(3.92) THEOREM Let V, B, C, and P be as in Theorem (3.89) and let $B = \{\mathbf{u}_1, \ldots, \mathbf{u}_n\}$. Then

$$P = \big[[\mathbf{u}_1]_C \,\vdots\, [\mathbf{u}_2]_C \,\vdots\, \cdots \,\vdots\, [\mathbf{u}_n]_C\big]$$

The following example is a little bizarre, but working your way through it will help you understand what is going on.

Example 8 Let $V = M_{22}$ and let

$$B = \left\{\begin{bmatrix} 1 & 1 \\ 1 & 1 \end{bmatrix}, \begin{bmatrix} 0 & 1 \\ 1 & 1 \end{bmatrix}, \begin{bmatrix} 0 & 0 \\ 1 & 1 \end{bmatrix}, \begin{bmatrix} 0 & 0 \\ 0 & 1 \end{bmatrix}\right\} = \{\mathbf{u}_1, \mathbf{u}_2, \mathbf{u}_3, \mathbf{u}_4\}$$

$$C = \left\{\begin{bmatrix} 0 & 1 \\ 1 & 0 \end{bmatrix}, \begin{bmatrix} 1 & 0 \\ 0 & 1 \end{bmatrix}, \begin{bmatrix} 1 & 0 \\ 1 & 0 \end{bmatrix}, \begin{bmatrix} 1 & 0 \\ 0 & 0 \end{bmatrix}\right\} = \{\mathbf{v}_1, \mathbf{v}_2, \mathbf{v}_3, \mathbf{v}_4\}$$

We can check that B and C are bases for V. Further,

$$\begin{aligned} &\mathbf{u}_1 = \mathbf{v}_1 + \mathbf{v}_2 && \text{so that} \quad [\mathbf{u}_1]_C = [1 \quad 1 \quad 0 \quad 0]^T \\ &\mathbf{u}_2 = \mathbf{v}_1 + \mathbf{v}_2 - \mathbf{v}_4 && \text{so that} \quad [\mathbf{u}_2]_C = [1 \quad 1 \quad 0 \quad -1]^T \\ &\mathbf{u}_3 = \mathbf{v}_2 + \mathbf{v}_3 - 2\mathbf{v}_4 && \text{so that} \quad [\mathbf{u}_3]_C = [0 \quad 1 \quad 1 \quad -2]^T \\ &\mathbf{u}_4 = \mathbf{v}_2 - \mathbf{v}_4 && \text{so that} \quad [\mathbf{u}_4]_C = [0 \quad 1 \quad 0 \quad -1]^T \end{aligned}$$

Therefore, by Theorem (3.92),

$$P = \begin{bmatrix} 1 & 1 & 0 & 0 \\ 1 & 1 & 1 & 1 \\ 0 & 0 & 1 & 0 \\ 0 & -1 & -2 & -1 \end{bmatrix}$$

Let $\mathbf{x} = \begin{bmatrix} 2 & -1 \\ 0 & -2 \end{bmatrix}$. Then we can check that

$$[\mathbf{x}]_B = \begin{bmatrix} 2 \\ -3 \\ 1 \\ -2 \end{bmatrix}, \qquad [\mathbf{x}]_C = \begin{bmatrix} -1 \\ -2 \\ 1 \\ 3 \end{bmatrix}, \qquad \text{and} \qquad [\mathbf{x}]_C = P[\mathbf{x}]_B$$

Exercise 24 asks you to verify the details. ■

Proof of Theorem (3.91) We are given $B = \{\mathbf{u}_1, \ldots, \mathbf{u}_n\}$, $C = \{\mathbf{v}_1, \ldots, \mathbf{v}_n\}$ bases for $\mathbb{R}^n$. Let $U = [\mathbf{u}_1 \cdots \mathbf{u}_n]$ and $V = [\mathbf{v}_1 \cdots \mathbf{v}_n]$ be the $n \times n$ matrices with columns consisting of the basis vectors $\mathbf{u}_i$ and $\mathbf{v}_i$, respectively. The proof that P can be obtained as described in the statement of the theorem is broken up into two parts. Part 1 shows that $P = V^{-1}U$; part 2 shows that $[V \,\vdots\, U] \to [I \,\vdots\, ?]$ yields $? = V^{-1}U$.

PART 1 Pick an arbitrary $\mathbf{x}$ in $\mathbb{R}^n$, and let $[\mathbf{x}]_B = [a_1 \cdots a_n]^T$. Then

$$U[\mathbf{x}]_B = [\mathbf{u}_1 \cdots \mathbf{u}_n] \begin{bmatrix} a_1 \\ \vdots \\ a_n \end{bmatrix} = a_1\mathbf{u}_1 + \cdots + a_n\mathbf{u}_n = \mathbf{x}$$

Similarly, $V[\mathbf{x}]_C = \mathbf{x}$. Since V is invertible (by Exercise 41, Section 3.7) and since $\mathbf{x} = \mathbf{x}$, we have

$$U[\mathbf{x}]_B = V[\mathbf{x}]_C \qquad \text{or} \qquad V^{-1}U[\mathbf{x}]_B = [\mathbf{x}]_C$$

Since $\mathbf{x}$ is an arbitrary vector in $\mathbb{R}^n$, we have $V^{-1}U[\mathbf{x}]_B = [\mathbf{x}]_C$ for all $\mathbf{x}$ in $\mathbb{R}^n$. Thus $V^{-1}U = P$ by the uniqueness part of Theorem (3.89).

PART 2 In Section 1.4, we showed that

(3.93) $$\text{If } [V \,\vdots\, I] \xrightarrow[\text{operations}]{\text{row}} [I \,\vdots\, ?] \qquad \text{then} \qquad ? = V^{-1}$$

Now let e be an elementary row operation and E be its corresponding elementary matrix. Then applying e to a product,

$$e(AB) = E(AB) = (EA)B = e(A)B$$

is equivalent to applying e to the first factor only and then multiplying. Thus

$$[V \,\vdots\, U] = [V \,\vdots\, IU] \xrightarrow[\text{operations}]{\text{row}} [I \,\vdots\, ?U] = [I \,\vdots\, V^{-1}U]$$

by (3.93). This completes part 2.

Putting the two parts together proves the theorem. ■

We have one last important fact to state and prove.

(3.94) **THEOREM** If P is the transition matrix from B to C, then P^{-1} is the transition matrix from C to B.

Proof By Theorem (3.89), P has the property

$$P[\mathbf{x}]_B = [\mathbf{x}]_C \qquad \text{for every } \mathbf{x} \text{ in } V$$

Since P is invertible,

$$[\mathbf{x}]_B = P^{-1}[\mathbf{x}]_C \qquad \text{for every } \mathbf{x} \text{ in } V$$

Thus P^{-1} is the transition matrix from C to B by the uniqueness part of Theorem (3.89). ■

Examples are given in the exercises.

Exercise 3.8

In Exercises 1–12, you are given a vector space V, a basis B for V, a second basis C for V, which is a reordering of B, and a vector $\mathbf{x}$ in V. Find $[\mathbf{x}]_B$ and $[\mathbf{x}]_C$. In Exercises 1–4, sketch B and $\mathbf{x}$.

1. $\mathbb{R}^2$, $B = \{(3, -1), (2, 3)\}$, $C = \{(2, 3), (3, -1)\}$, $\mathbf{x} = (5, -9)$.
2. $\mathbb{R}^2$, $B = \{(\frac{1}{2}, 1), (3, -1)\}$, $C = \{(3, -1), (\frac{1}{2}, 1)\}$, $\mathbf{x} = (8, 9)$.
3. $\mathbb{R}^2$, $B = \{(-2, -3), (2, -1)\}$, $C = \{(2, -1), (-2, -3)\}$, $\mathbf{x} = (-2, 10)$.
4. $\mathbb{R}^2$, $B = \{(-3, 2), (2, -3)\}$, $C = \{(2, -3), (-3, 2)\}$, $\mathbf{x} = (-8, 9)$.
5. $\mathbb{R}^3$, $B = \{(-1, 2, 1), (2, 1, 0), (3, 4, -2)\}$, $C = \{(2, 1, 0), (3, 4, -2), (-1, 2, 1)\}$, $\mathbf{x} = (3, 14, -4)$.
6. $\mathbb{R}^3$, $B = \{(1, 1, 2), (1, 2, 1), (1, 3, 1)\}$, $C = \{(1, 3, 1), (1, 2, 1), (1, 1, 2)\}$, $\mathbf{x} = (2, 2, 5)$.
7. $\mathbb{R}^3$, $B = \{(-2, -1, 0), (0, 1, 3), (0, 1, 0)\}$, $C = \{(0, 1, 0), (-2, 1, 0), (0, 1, 3)\}$, $\mathbf{x} = (-8, -3, 9)$.
8. $\mathbb{R}^3$, $B = \{(1, 4, -2), (3, -1, -2), (2, -5, 1)\}$, $C = \{(3, -1, -2), (1, 4, -2), (2, 5, -1)\}$, $\mathbf{x} = (-1, -13, 9)$.

9. P_2, $B = \{1 + 2x - x^2, 1 - 3x, 2\}$, $C = \{(2, 1 - 3x, 1 + 2x - x^2\}$, $\mathbf{x} = 3 - 2x^2$.

10. P_2, $B = \{2 + x, 3 - 2x, 2 + 3x + x^2\}$, $C = \{2 + 3x + x^2, 2 + x, 3 - 2x\}$, $\mathbf{x} = -5 - 5x - 3x^2$.

11. M_{22}, $B = \left\{\begin{bmatrix} 1 & 1 \\ 1 & 1 \end{bmatrix}, \begin{bmatrix} 1 & 1 \\ 0 & 1 \end{bmatrix}, \begin{bmatrix} 0 & 1 \\ 0 & 1 \end{bmatrix}, \begin{bmatrix} 0 & 0 \\ 0 & 1 \end{bmatrix}\right\}$
$C = \left\{\begin{bmatrix} 0 & 1 \\ 0 & 1 \end{bmatrix}, \begin{bmatrix} 1 & 1 \\ 0 & 1 \end{bmatrix}, \begin{bmatrix} 1 & 1 \\ 1 & 1 \end{bmatrix}, \begin{bmatrix} 0 & 0 \\ 0 & 1 \end{bmatrix}\right\}$, $\mathbf{x} = \begin{bmatrix} 1 & 4 \\ 2 & -2 \end{bmatrix}$

12. M_{22}, $B = \left\{\begin{bmatrix} 1 & 2 \\ 0 & 1 \end{bmatrix}, \begin{bmatrix} 0 & 1 \\ 1 & 0 \end{bmatrix}, \begin{bmatrix} 1 & 0 \\ 2 & 0 \end{bmatrix}, \begin{bmatrix} 0 & 1 \\ 0 & 0 \end{bmatrix}\right\}$
$C = \left\{\begin{bmatrix} 0 & 1 \\ 0 & 0 \end{bmatrix}, \begin{bmatrix} 1 & 2 \\ 0 & 1 \end{bmatrix}, \begin{bmatrix} 0 & 1 \\ 1 & 0 \end{bmatrix}, \begin{bmatrix} 1 & 0 \\ 2 & 0 \end{bmatrix}\right\}$, $\mathbf{x} = \begin{bmatrix} 2 & -2 \\ 4 & -1 \end{bmatrix}$

In Exercises 13 and 14, three vectors $\mathbf{v}_1$, $\mathbf{v}_2$, $\mathbf{v}_3$ that span $\mathbb{R}^2$ are given (but are not a basis—why?). Express $\mathbf{x}$ and $\mathbf{y}$ as a linear combination of the $\mathbf{v}_i$'s in three different ways. Sketch two of the ways you found.

13. $\mathbf{v}_1 = (2, 3)$, $\mathbf{v}_2 = (-1, 1)$, $\mathbf{v}_3 = (2, 1)$, $\mathbf{x} = (1, 1)$, $\mathbf{y} = (1, 0)$.

14. $\mathbf{v}_1 = (-1, -2)$, $\mathbf{v}_2 = (3, 1)$, $\mathbf{v}_3 = (-1, 1)$, $\mathbf{x} = (1, 2)$, $\mathbf{y} = (0, 0)$.

In Exercises 15 and 16, let $B = \{\mathbf{v}_1, \ldots, \mathbf{v}_n\}$ be a basis for a vector space V, and let $\mathbf{x} = a_1\mathbf{v}_1 + \cdots + a_n\mathbf{v}_n$ and $\mathbf{y} = b_1\mathbf{v}_1 + \cdots + b_n\mathbf{v}_n$ be arbitrary vectors in V.

15. Find $[\mathbf{x}]_B$, $[\mathbf{y}]_B$, and $[\mathbf{x} + \mathbf{y}]_B$. Then show that the first equality in Theorem (3.88) holds.

16. Let r be an arbitrary real number, and find $[\mathbf{x}]_B$ and $[r\mathbf{x}]_B$. Then show that the second equality in Theorem (3.88) holds.

In Exercises 17–23, find the transition matrix from B to C and find $[\mathbf{x}]_C$.

17. $B = \{(1, 1), (1, 0)\}$, $C = \{(2, 3), (4, 2)\}$, $[\mathbf{x}]_B = [-1 \quad 2]^T$.

18. $B = \{(3, 1), (-1, -2)\}$, $C = \{(1, -3), (5, 0)\}$, $[\mathbf{x}]_B = [-1 \quad -2]^T$.

19. $B =$ the standard basis, $C = \{(2, 3), (-1, -2)\}$, $[\mathbf{x}]_B = [-1 \quad -3]^T$.

20. $B = \{(2, -1), (2, 2)\}$, $C =$ the standard basis, $[\mathbf{x}]_B = [1 \quad 1]^T$.

21. $B = \{(-1, 2, 1), (-1, 1, -1), (2, -3, -1)\}$, $C = \{(1, 0, -1), (1, 1, 1), (0, 1, 3)\}$, $[\mathbf{x}]_B = [-1 \quad 2 \quad 1]^T$.

22. $B = \{(1, 1, 1), (-2, -1, 0), (2, 1, 2)\}$, $C = \{(-6, -2, 1), (-1, 1, 5), (-1, -1, 1)\}$, $[\mathbf{x}]_B = [-3 \quad 2 \quad 4]^T$.

23. $B =$ the standard basis, $C = \{(1, 1, 1), (0, 1, 1), (0, 0, 1)\}$, $[\mathbf{x}]_B = [2 \quad -1 \quad 1]^T$.

24. Verify all of the details of Example 8.

In Exercises 25 and 26, find the transition matrix from B to C and find $[\mathbf{x}]_C$.

25. $B = \{1 + x + x^2, 1 + x, x + x^2\}$, C = the standard basis (for P_2), $[\mathbf{x}]_B = [2 \quad 3 \quad -2]^T$.

26. $B = \left\{\begin{bmatrix} 1 & 2 \\ -1 & 0 \end{bmatrix}, \begin{bmatrix} 2 & -1 \\ 0 & 1 \end{bmatrix}, \begin{bmatrix} -1 & 1 \\ 0 & 0 \end{bmatrix}, \begin{bmatrix} 1 & 0 \\ 0 & 0 \end{bmatrix}\right\}$,
C = the standard basis, $[\mathbf{x}]_B = [2 \quad -1 \quad 0 \quad 1]^T$.

In Exercises 27–30, compute the transition matrix from B to C and the transition matrix from C to B. Verify that the product is I.

27. B and C in Exercise 17.

28. B and C in Exercise 18.

29. B and C in Exercise 21.

30. B and C in Exercise 22.

31. Suppose that B, C, and D are bases for a vector space V. Also suppose that P is the transition matrix from B to C and that Q is the transition matrix from C to D. Is PQ or QP the transition matrix from B to D? Prove your answer, using the uniqueness part of Theorem (3.89).

32. Given the hypotheses in Exercise 31, find the transition matrix from D to B. Prove your answer, using Theorem (3.94).

3.9 An Application: Error-Correcting Codes*

The purpose of this section is to present an extremely useful application of the linear algebra presented in this chapter. It is an application that requires only a minimal amount of additional background and has the further advantage of hinting at higher abstract algebra and its concrete applications.

Computers and digital communications are permeating the whole fabric of our technological society. Examples include satellite transmission of data, intercontinental communications, and, already in the experimental stage, direct computer-aided home-to-store shopping. By **digital communication** we mean that information is transmitted in strings of 0's and 1's. Such strings are called **binary messages** and are coded in such a way as to convey information. For example, when a black and white photograph is transmitted, say from a satellite, what might get transmitted is a sequence of loca-

* *This section is, of course, optional, but all students are encouraged at least to skim this section to get an idea of this fascinating application.*

tions on the photograph and a gradation level from white to black at each location. If the photograph scanner measured "23," the satellite would transmit 10111, the binary representation for 23, for that location.

It is not hard to imagine that errors are sometimes introduced into such messages, by static or other types of interference. A first attempt to determine an error in the message was made by augmenting the message with an extra numeral, 0 or 1, to make the number of 1's even, as indicated in Figure 3.23.

Figure 3.23

(a) *The number 23 has an even number of 1's in its binary expansion. The add-on numeral is "0."*

(b) *The number 22 has an odd number of 1's in its binary expansion. The add-on numeral is "1."*

This is called a **parity check**. If you looked at an augmented message and found an odd number of 1's, you knew there was an error. However, you could not tell where the error was, nor could you tell if in fact there were three or five errors instead of just one. Worst of all, an even number of errors went undetected.

Error-correcting codes generalize the idea of parity checks in such a way that you can tell where the errors are (and hence correct them). The theory was pioneered by Richard W. Hamming in the early 1950s when he was working at Bell Laboratories. We will study one of the simplest of all such codes, one that will detect the existence and location of a single error in a message of four 0's or 1's.

A New Number System and Its Vector Spaces

(3.95) DEFINITION A **word** (of length n) is an n-tuple of 0's and 1's. Such a word is also called a **string of length n**.

Since we shall be using only 0's and 1's, we shall need a number system that reflects this. One such number system is the **integers mod 2**, Z_2 (also called "zee-two"). $Z_2 = \{0, 1\}$ and addition and multiplication are as usual, except $1 + 1 = 0$. Thus the addition and multiplication tables are

+	0	1
0	0	1
1	1	0

×	0	1
0	0	0
1	0	1

All the arithmetic properties (associativity, commutativity, distributivity, etc.) hold, except (be careful!) the number 1 is its own negative (since $1 + 1 = 0$).

Just as $\mathbb{R}^n$ is the set of all n-tuples of real numbers, we denote by Z_2^n the set of all n-tuples of numbers from Z_2, that is, n-tuples of 0's or 1's. Thus

(3.96) Z_2^n is the set of all words of length n.

Addition and scalar multiplication are defined in Z_2^n exactly as in $\mathbb{R}^n$, except that scalars must come from Z_2 (i.e., scalars are either 0 or 1).

Example 1 For $\mathbf{v} = (1, 1, 0, 0, 1, 0)$ and $\mathbf{w} = (0, 1, 1, 0, 1, 1)$ in Z_2^6, find (a) $\mathbf{v} + \mathbf{w}$ (b) $-\mathbf{v}$ (c) $r\mathbf{v}$ for all r in Z_2

Solution (a) $\mathbf{v} + \mathbf{w} = (1, 1, 0, 0, 1, 0) + (0, 1, 1, 0, 1, 1) = (1, 0, 1, 0, 0, 1)$

(b) Since $\mathbf{v} + \mathbf{v} = (1, 1, 0, 0, 1, 0) + (1, 1, 0, 0, 1, 0) = (0, 0, 0, 0, 0, 0) = \mathbf{0}$, $\mathbf{v}$ is its own negative. Thus $-\mathbf{v} = \mathbf{v}$, and in fact this is true for any vector in Z_2^n.

(c) There are only two r's in Z_2, $r = 0$ and $r = 1$. Of course,

$$0\mathbf{v} = \mathbf{0} \qquad \text{and} \qquad 1\mathbf{v} = \mathbf{v} \qquad (\text{for any } \mathbf{v} \text{ in } Z_2^n)$$ ■

It is left as an exercise to verify the following.

(3.97) Under the above definitions of addition and scalar multiplication, if we restrict scalars in Z_2, then Z_2^n satisfies all of the axioms of a vector space given in Definition (3.19), Section 3.3. We abbreviate this by saying

"Z_2^n is a vector space over Z_2"

Just as in $\mathbb{R}^n$, the standard basis

$$\boldsymbol{\varepsilon}_1 = (1, 0, \ldots, 0), \quad \boldsymbol{\varepsilon}_2 = (0, 1, 0, \ldots, 0), \quad \ldots, \quad \boldsymbol{\varepsilon}_n = (0, \ldots, 0, 1)$$

is a basis for Z_2^n. Thus

(3.98) THEOREM The vector space Z_2^n has dimension n.

At this point we could launch into a study of the various properties of Z_2^n. A little of this will be done in the exercises. But in the text of this section, we shall restrict ourselves to accomplishing our stated purpose: How do we encode a message so that if a single error occurs in transmission, then that error can be detected and corrected at the receiving end? In some sense what is done is to take a word of length four (a longer word could be broken up into words of length four) and add on three parity checks in such a way as to accomplish our purpose. The algebra tells us how to add on those checks.

A message of length four plus three parity checks yields a word of length seven. For this reason we work in Z_2^7. There are four particular vectors that are important to us:

(3.99)
$$\mathbf{u}_1 = (1, 0, 0, 0, 0, 1, 1), \quad \mathbf{u}_2 = (0, 1, 0, 0, 1, 0, 1)$$
$$\mathbf{u}_3 = (0, 0, 1, 0, 1, 1, 0), \quad \mathbf{u}_4 = (0, 0, 0, 1, 1, 1, 1)$$

(3.100) THEOREM The vectors $\mathbf{u}_1, \mathbf{u}_2, \mathbf{u}_3, \mathbf{u}_4$ are linearly independent.

Proof Their leading nonzero entries form a staircase pattern, so they are linearly independent (by the argument given in Example 3, Section 3.7). ■

Thus $\mathbf{u}_1, \ldots, \mathbf{u}_4$ span (and form a basis for) a four-dimensional subspace of Z_2^7, which is denoted by $C_{7,4}$.

(3.101) DEFINITION A **code** is a k-dimensional subspace of Z_2^n. The code $C_{7,4}$ is called a **(7, 4) Hamming code.**

Encoding Messages

To **encode** a message means to convert that message to a vector in a particular code. In our case we wish to convert a message of length four to a 7-tuple in $C_{7,4}$. However, if you examine the vectors $\mathbf{u}_1, \ldots, \mathbf{u}_4$, you will see each $\mathbf{u}_i = \boldsymbol{\varepsilon}_i$ in Z_4 with "parity checks" attached at the end. Thus we send an arbitrary message

$$(x_1, x_2, x_3, x_4) = x_1\boldsymbol{\varepsilon}_1 + x_2\boldsymbol{\varepsilon}_2 + x_3\boldsymbol{\varepsilon}_3 + x_4\boldsymbol{\varepsilon}_4$$

to

$$x_1\mathbf{u}_1 + x_2\mathbf{u}_2 + x_3\mathbf{u}_3 + x_4\mathbf{u}_4$$

Example 2 Encode $\mathbf{w} = (1, 0, 1, 1)$.

Solution We send $\mathbf{w} = \boldsymbol{\varepsilon}_1 + \boldsymbol{\varepsilon}_3 + \boldsymbol{\varepsilon}_4$ to $\mathbf{u}_1 + \mathbf{u}_3 + \mathbf{u}_4$, or

$$(1, 0, 0, 0, 0, 1, 1) + (0, 0, 1, 0, 1, 1, 0) + (0, 0, 0, 1, 1, 1, 1) = (1, 0, 1, 1, 0, 1, 0)$$

Note that what we get is just $\mathbf{w}$ with "parity checks" attached. ■

Decoding Messages

To **decode** a message means to check the message to determine if there has been an error, to correct any error, and finally to extract the original message. We first discuss the check. Consider the 3×7 matrix

$$H = \begin{bmatrix} 0 & 0 & 0 & 1 & 1 & 1 & 1 \\ 0 & 1 & 1 & 0 & 0 & 1 & 1 \\ 1 & 0 & 1 & 0 & 1 & 0 & 1 \end{bmatrix}$$

with entries from Z_2. A quick glance at columns 1, 2, and 4 reveals $rk(H) = 3$. From

$$rk(H) + \dim NS(H) = 7$$

we conclude $\dim NS(H) = 4$. Amazingly, a quick check reveals

$$H\mathbf{u}_1 = \mathbf{0}, \quad H\mathbf{u}_2 = \mathbf{0}, \quad H\mathbf{u}_3 = \mathbf{0}, \quad H\mathbf{u}_4 = \mathbf{0}$$

(when the **u**'s are written as column vectors). Thus all of the four-dimensional code $C_{7,4}$ is in the four-dimensional subspace $NS(H)$, so we conclude

$$NS(H) = C_{7,4} \tag{3.102}$$

It is now time to list all of the 16 vectors in $C_{7,4}$

$$\begin{aligned} \mathbf{0} &= (0, 0, 0, 0, 0, 0, 0) \\ \mathbf{u}_1 &= (1, 0, 0, 0, 0, 1, 1) \\ \mathbf{u}_2 &= (0, 1, 0, 0, 1, 0, 1) \\ \mathbf{u}_3 &= (0, 0, 1, 0, 1, 1, 0) \\ \mathbf{u}_4 &= (0, 0, 0, 1, 1, 1, 1) \\ \mathbf{u}_1 + \mathbf{u}_2 &= (1, 1, 0, 0, 1, 1, 0) \\ \mathbf{u}_1 + \mathbf{u}_3 &= (1, 0, 1, 0, 1, 0, 1) \\ \mathbf{u}_1 + \mathbf{u}_4 &= (1, 0, 0, 1, 1, 0, 0) \\ \mathbf{u}_2 + \mathbf{u}_3 &= (0, 1, 1, 0, 0, 1, 1) \\ \mathbf{u}_2 + \mathbf{u}_4 &= (0, 1, 0, 1, 0, 1, 0) \\ \mathbf{u}_3 + \mathbf{u}_4 &= (0, 0, 1, 1, 0, 0, 1) \\ \mathbf{u}_1 + \mathbf{u}_2 + \mathbf{u}_3 &= (1, 1, 1, 0, 0, 0, 0) \\ \mathbf{u}_1 + \mathbf{u}_2 + \mathbf{u}_4 &= (1, 1, 0, 1, 0, 0, 1) \\ \mathbf{u}_1 + \mathbf{u}_3 + \mathbf{u}_4 &= (1, 0, 1, 1, 0, 1, 0) \\ \mathbf{u}_2 + \mathbf{u}_3 + \mathbf{u}_4 &= (0, 1, 1, 1, 1, 0, 0) \\ \mathbf{u}_1 + \mathbf{u}_2 + \mathbf{u}_3 + \mathbf{u}_4 &= (1, 1, 1, 1, 1, 1, 1) \end{aligned} \tag{3.103}$$

The next theorem helps us detect a word with a single error.

(3.104)

THEOREM If any vector in $C_{7,4}$ is altered in exactly one coordinate, then the resulting vector is not in $C_{7,4}$.

Proof If a $\mathbf{c}$ in $C_{7,4}$ is altered in only the ith coordinate, the result is $\mathbf{c} + \boldsymbol{\varepsilon}_i$, where $\boldsymbol{\varepsilon}_i$ is in Z_2^7. Since $C_{7,4} = NS(H)$, by (3.102), we have

$$\begin{aligned} H(\text{altered } \mathbf{c}) &= H(\mathbf{c} + \boldsymbol{\varepsilon}_i) = H(\mathbf{c}) + H(\boldsymbol{\varepsilon}_i) \\ &= \mathbf{0} + H(\boldsymbol{\varepsilon}_i) \neq \mathbf{0} \end{aligned}$$

so that the altered $\mathbf{c}$ is not in $C_{7,4}$. ■

An alternative proof, which is more advanced but develops a powerful tool, is presented in Exercises 20–24.

Since $C_{7,4} = NS(H)$, Theorem (3.104) tells us that a quick way to determine if an error has been introduced into a word that was originally in the code $C_{7,4}$ is to determine if that word is still in $NS(H)$ (assuming, again, any error is a change in exactly one entry).

Example 3 Determine if the following words are in $C_{7,4}$.

$$\mathbf{x} = \begin{bmatrix} 0 \\ 1 \\ 1 \\ 0 \\ 0 \\ 1 \\ 1 \end{bmatrix}, \quad \mathbf{y} = \begin{bmatrix} 0 \\ 1 \\ 1 \\ 1 \\ 0 \\ 0 \\ 1 \end{bmatrix}, \quad \mathbf{z} = \begin{bmatrix} 1 \\ 0 \\ 1 \\ 1 \\ 0 \\ 0 \\ 0 \end{bmatrix}$$

Solution Compute (using the operations of Z_2!)

$$H\mathbf{x} = \begin{bmatrix} 0 \\ 0 \\ 0 \end{bmatrix}, \quad H\mathbf{y} = \begin{bmatrix} 0 \\ 1 \\ 0 \end{bmatrix}, \quad H\mathbf{z} = \begin{bmatrix} 1 \\ 1 \\ 0 \end{bmatrix}$$

Thus $\mathbf{x}$ is in $C_{7,4}$, but $\mathbf{y}$ and $\mathbf{z}$ are not. ■

Of course, we might have compared $\mathbf{x}$, $\mathbf{y}$, and $\mathbf{z}$ with the list (3.103) and seen that $\mathbf{x}$ is on the list ($\mathbf{x} = \mathbf{u}_2 + \mathbf{u}_3$) but $\mathbf{y}$ and $\mathbf{z}$ are not, but the computations given in Example 3 will tell us how to correct any word with a sign error!

Again examine the matrix:

$$H = \begin{bmatrix} 0 & 0 & 0 & 1 & 1 & 1 & 1 \\ 0 & 1 & 1 & 0 & 0 & 1 & 1 \\ 1 & 0 & 1 & 0 & 1 & 0 & 1 \end{bmatrix}$$

Notice that the columns represent, in order, the binary representations of the numbers 1 through 7. In addition, since $rk(H) = 3$, $CS(H) = Z_2^3$ and *every single nonzero vector of* Z_2^3 *is a column of* H. Thus if $\mathbf{w}$ is a word in Z_2^7 and $H\mathbf{w} \neq \mathbf{0}$, then $H\mathbf{w}$, as a vector in Z_2^3, is a column of H. The incredible, almost magical, fact is

(3.105)

> THEOREM Suppose $\mathbf{w}$ is a vector in Z_2^7 and $H\mathbf{w}$ is the kth column of H (thus $H\mathbf{w} \neq \mathbf{0}$). If the kth entry of $\mathbf{w}$ is changed, then the corrected vector $\mathbf{w}_c$ will be in $NS(H) = C_{7,4}$.

We shall prove this theorem after an example.

Example 4 Change one entry in each of the vectors $\mathbf{y}$ and $\mathbf{z}$ of Example 3 so that the resulting vectors are in $C_{7,4}$.

Solution By Example 3,

$$\mathbf{y} = \begin{bmatrix} 0 \\ 1 \\ 1 \\ 1 \\ 0 \\ 0 \\ 1 \end{bmatrix}, \qquad H\mathbf{y} = \begin{bmatrix} 0 \\ 1 \\ 0 \end{bmatrix}, \qquad \mathbf{z} = \begin{bmatrix} 1 \\ 0 \\ 1 \\ 1 \\ 0 \\ 0 \\ 0 \end{bmatrix}, \qquad H\mathbf{z} = \begin{bmatrix} 1 \\ 1 \\ 0 \end{bmatrix}$$

Now $H\mathbf{y}$ is the second column of H so we change the second entry of $\mathbf{y}$, a 1, to a 0; $H\mathbf{z}$ is the sixth column of H so we change the sixth entry of $\mathbf{z}$, a 0, to a 1. The corrected vectors are

$$\mathbf{y}_c = \begin{bmatrix} 0 \\ 0 \\ 1 \\ 1 \\ 0 \\ 0 \\ 1 \end{bmatrix} \quad \text{and} \quad \mathbf{z}_c = \begin{bmatrix} 1 \\ 0 \\ 1 \\ 1 \\ 0 \\ 1 \\ 0 \end{bmatrix}$$

which are $\mathbf{u}_3 + \mathbf{u}_4$ and $\mathbf{u}_1 + \mathbf{u}_3 + \mathbf{u}_4$ respectively. ■

Proof of Theorem (3.105) By hypothesis, $H\mathbf{w} = \mathbf{c}_i$, the ith column of H. If $\mathbf{w}$ is altered by changing the ith component, the *corrected* vector, $\mathbf{w}_c$, is $\mathbf{w}_c = \mathbf{w} + \boldsymbol{\varepsilon}_i$, where $\boldsymbol{\varepsilon}_i$ is in Z_2^7. Since for any matrix A, $A\boldsymbol{\varepsilon}_i$ is the ith column of A, we have

$$H(\mathbf{w}_c) = H(\mathbf{w} + \boldsymbol{\varepsilon}_i) = H\mathbf{w} + H\boldsymbol{\varepsilon}_i = \mathbf{c}_i + \mathbf{c}_i = \mathbf{0}$$

so that $\mathbf{w}_c$ is in $NS(H) = C_{7,4}$. ■

We can now write down the complete decoding process. Suppose a word $\mathbf{w}$ of length seven is received (which was encoded from the word $\mathbf{c}$ of length four, but $\mathbf{w}$ might have a single error).

1. Compute $H\mathbf{w}$.
2. If $H\mathbf{w} = \mathbf{0}$, $\mathbf{c}$ is the first four entries of $\mathbf{w}$.
3. If $H\mathbf{w} \neq \mathbf{0}$, then $H\mathbf{w}$ is the kth column of H for some k. Change the kth entry of $\mathbf{w}$, obtaining $\mathbf{w}_c$.
4. Then $\mathbf{c}$ is the first four entries of $\mathbf{w}_c$.

It must be emphasized that this code cannot handle all possible combinations of errors (and indeed no code can). If there are two errors, the decoding process for this code gives the wrong message, and certain combinations of three or more errors will go undetected. This section is only a brief introduction to a fascinating branch of mathematics. If you are intrigued and wish to find out more information on coding theory, here are some suggested references.

1. Rice, B. F., and Wilde, C. O., "Error Correcting Codes I," UMAP—Modules and Monographs in Undergraduate Mathematics and its Applications Project, Unit 346.
2. Pless, Vera, *Introduction to the Theory of Error-Correcting Codes*, Wiley, 1982.
3. Lin, Shu, *An Introduction to Error-Correcting Codes*, Prentice-Hall, 1970.

The author would like to express his extreme gratitude to Professor M. James Stewart, Lansing Community College, for suggesting this topic as an excellent application of the theory in this chapter.

Exercise 3.9

In Exercises 1–4, encode the given message into a code word in $C_{7,4}$.

1. (0, 1, 1, 0)
2. (1, 1, 0, 1)
3. (0, 1, 1, 1)
4. (1, 0, 0, 1)

In Exercises 5–12, consider each 7-tuple, $\mathbf{v}$, to be a message with at most one error in Z_2^7. Determine if $\mathbf{v}$ is in $C_{7,4}$. If it is, decode it. If it is not, correct it and decode the corrected message.

5. (0, 1, 1, 0, 0, 1, 1)
6. (0, 1, 1, 1, 1, 0, 0)
7. (0, 1, 1, 1, 0, 1, 1)
8. (1, 0, 1, 0, 0, 1, 1)
9. (1, 0, 0, 1, 1, 1, 1)
10. (1, 1, 1, 0, 0, 1, 0)
11. (1, 1, 0, 0, 1, 1, 1)
12. (1, 1, 1, 1, 0, 1, 1)

Properties of Z_2^n.

13. List all the vectors in Z_2^3. Count them.
14. How many vectors are in Z_2^4? In Z_2^n?

DEFINITION A **line** in a vector space is all scalar multiples of a given nonzero vector; that is, a one-dimensional subspace.

15. How many vectors are there on a line in Z_2^3? In Z_2^n?
16. How many lines are there in Z_2^3? In Z_2^n?

DEFINITION A **plane** in a vector space is a two-dimensional subspace.

17. How many vectors are there in a plane in Z_2^3? In Z_2^n?
18. List all the planes in Z_2^3 that contain $\boldsymbol{\varepsilon}_1 = (1, 0, 0)$.
19. How many planes are there in Z_2^3? In Z_2^n?

Outline of an alternate proof of Theorem (3.104).

DEFINITION If $\mathbf{u}$, $\mathbf{v}$ are in Z_2^n, then the distance between $\mathbf{u}$ and $\mathbf{v}$, $d(\mathbf{u}, \mathbf{v})$, is given by

$$d(\mathbf{u}, \mathbf{v}) \qquad \text{is the number of entries at which } \mathbf{u} \text{ and } \mathbf{v} \text{ differ}$$

EXAMPLE If $\mathbf{u} = (1, 0, 1, 0)$, $\mathbf{v} = (0, 0, 1, 1)$ in Z_2^4, then $d(\mathbf{u}, \mathbf{v}) = 2$.

NOTE This is called the **Hamming distance** function.

20. Show that $d(\mathbf{u}, \mathbf{v})$ satisfies the following properties.
   (a) $d(\mathbf{u}, \mathbf{v}) \geq 0$ and $= 0$ if and only if $\mathbf{u} = \mathbf{v}$.
   (b) $d(\mathbf{u}, \mathbf{v}) = d(\mathbf{v}, \mathbf{u})$.
   (c) $d(\mathbf{u}, \mathbf{v}) \leq d(\mathbf{u}, \mathbf{w}) + d(\mathbf{w}, \mathbf{v})$ (the triangle inequality).

NOTE A function $d(x, y)$ defined for all x, y in a set X is called a **distance function** on X if d satisfies $a - c$.

21. Show that $d(\mathbf{u}, \mathbf{v}) = d(\mathbf{0}, \mathbf{u} - \mathbf{v}) = d(\mathbf{0}, \mathbf{u} + \mathbf{v})$.
22. Show $d(\mathbf{0}, \mathbf{u}) \geq 3$, for all $\mathbf{u}$ in $C_{7,4}$, $\mathbf{u} \neq 0$. (Technically this says the minimum *weight* of the subspace $C_{7,4}$ is 3.) [HINT Just look at Equations (3.97) and compute.]
23. Show that $d(\mathbf{u}, \mathbf{v}) \geq 3$, for all $\mathbf{u}$, $\mathbf{v}$ in $C_{7,4}$. (HINT Use Exercises 21 and 22.)

24. Explain why Exercise 23 proves Theorem (3.104).

DEFINITION The **sphere of radius r about a vector v** is the set of all vectors **w** such that $d(\mathbf{v}, \mathbf{w}) \leq r$.

25. Suppose that $0 \leq r \leq 1$. Show the sphere of radius r about **v** is $\{\mathbf{v}\}$.

26. Let $\boldsymbol{\varepsilon}_1 = (1, 0, 0)$.

    (a) List all the elements in the sphere of radius 1.5 about $\boldsymbol{\varepsilon}_1$.
    (b) List all the elements in the sphere of radius 1.5 about **0**.
    (c) Count the number of elements in the sets obtained in (a) and (b) and see the numbers are the same.

27. Use Exercise 21 to show that the number of elements in a sphere of radius r about **v** is equal to the number of elements in a sphere of radius r about **0**.

28. Take two spheres of radius 1 about two distinct vectors in $C_{7,4}$. Show these spheres are disjoint. (HINT Suppose not. Pick a **w** in the intersection. Use Exercises 20(c) and 23.)

29. (a) Show that the sphere of radius 1 about **0** in Z_2^7 contains eight vectors.
    (b) Show that any sphere of radius 1 about any vector in Z_2^7 contains eight vectors.

30. Let $W = \{\mathbf{x} \text{ in } Z_2^7 \mid \mathbf{x} \text{ is in a sphere of radius 1 about some vector } \mathbf{v} \text{ in } C_{7,4}\}$.

    (a) By counting elements (and using Exercises 28 and 29), show that W has 128 elements.
    (b) Show that $W = Z_2^7$.

31. Prove the following theorem, which is stronger than Theorem (3.104). Any message in Z_2^7 is either in $C_{7,4}$ or differs from a vector in $C_{7,4}$ in exactly one coordinate, and not both.

Review Exercises

In Exercises 1–3, let

$$\mathbf{u} = \begin{bmatrix} 2 \\ 1 \\ -3 \\ 4 \end{bmatrix} \quad \text{and} \quad \mathbf{v} = \begin{bmatrix} 3 \\ -1 \\ 3 \\ -2 \end{bmatrix}$$

be vectors in $\mathbb{R}^4$, with the Euclidean inner product.

1. Find $\mathbf{u} + \mathbf{v}$ and $2\mathbf{u} - 3\mathbf{v}$.

2. Find $\|\mathbf{u}\|$, $\|\mathbf{v}\|$, $\mathbf{u} \cdot \mathbf{v}$, and $\cos\theta$, where θ is the angle between **u** and **v**.

3. Find any unit vector perpendicular to both **u** and **v**.

In Exercises 4–6, let S be the set of all vectors $\mathbf{x} = (x_1, x_2, x_3, x_4)$ in $\mathbb{R}^4$ for which $x_2 = x_3 = x_4$.

4. Show S is a subspace of $\mathbb{R}^4$.

5. Find a basis for S.

6. Find a matrix A such that $\mathbf{x}$ is in the $NS(A) \Leftrightarrow \mathbf{x}$ is in S.

7. Let

$$A = \begin{bmatrix} 2 & 1 & -1 & 0 & 3 \\ -2 & -1 & 2 & 1 & -1 \\ 2 & 1 & 1 & 2 & 7 \end{bmatrix}.$$

Reduce A to echelon form and determine the dimension of $RS(A)$, $CS(A)$, and $NS(A)$.

8. Do the vectors (1, 1, 0), (1, 2, 1), and (1, −1, −2) form a basis for $\mathbb{R}^3$?

9. Is (1, −1, 1) in the span of (1, 1, 0), (1, 2, 1), and (1, −1, −2)?

10. Find the rank and a basis for the null space of

$$\begin{bmatrix} 0 & 0 & 1 \\ 1 & 2 & 3 \\ 0 & 0 & 2 \end{bmatrix}$$

11. Are the matrices in $M_{2,2}$

$$A_1 = \begin{bmatrix} 1 & 1 \\ 0 & 1 \end{bmatrix}, \quad A_2 = \begin{bmatrix} 2 & 1 \\ 1 & 0 \end{bmatrix}, \quad A_3 = \begin{bmatrix} 0 & 1 \\ -1 & 2 \end{bmatrix}, \quad A_4 = \begin{bmatrix} 1 & 0 \\ 2 & 1 \end{bmatrix}$$

linearly dependent? If so, write one of them as a linear combination of the rest.

12. Let $p_1(x) = 1 - x + x^2$, $p_2(x) = 2 + x$, and $p_3 = 2 - 2x + x^2$. Do p_1, p_2, and p_3 form a basis for P_2?

13. Find all solutions to

$$\begin{bmatrix} 1 & 2 & 3 \\ 1 & 2 & 4 \\ 1 & 2 & 5 \end{bmatrix}\begin{bmatrix} x_1 \\ x_2 \\ x_3 \end{bmatrix} = \begin{bmatrix} 5 \\ 6 \\ 7 \end{bmatrix} \quad \text{and} \quad \begin{bmatrix} 1 & 2 & 3 \\ 1 & 2 & 4 \\ 1 & 2 & 5 \end{bmatrix}\begin{bmatrix} x_1 \\ x_2 \\ x_3 \end{bmatrix} = \begin{bmatrix} 4 \\ 6 \\ 7 \end{bmatrix}$$

if possible.

14. Let $S = \{\boldsymbol{\alpha}_1, \boldsymbol{\alpha}_2, \boldsymbol{\alpha}_3, \boldsymbol{\alpha}_4, \boldsymbol{\alpha}_5\}$ where $\boldsymbol{\alpha}_1 = (1, 0, 0)$, $\boldsymbol{\alpha}_2 = (2, 0, 0)$, $\boldsymbol{\alpha}_3 = (0, 1, 1)$, $\boldsymbol{\alpha}_4 = (2, 2, 2)$, and $\boldsymbol{\alpha}_5 = (0, 3, 3)$. List all possible bases for span(S) using just the $\boldsymbol{\alpha}$'s.

15. Prove: If A is $n \times n$, $rk(A) = n$, $A^2 = A$, then $A = I$.

16. Prove: If A is $m \times n$, $CS(A) = NS(A)$, then $n = m$ and n is even.

17. Are the following true or false? The solutions $\mathbf{x}$ of

$$A\mathbf{x} = \begin{bmatrix} 1 & 2 & 3 \\ 1 & 1 & 1 \end{bmatrix} \begin{bmatrix} x_1 \\ x_2 \\ x_3 \end{bmatrix} = \begin{bmatrix} 0 \\ 0 \end{bmatrix}$$

(a) are a point; (b) are a line; (c) are a plane; (d) $= NS(A)$; (e) $= CS(A)$.

In Exercises 18–20, let $\mathbb{R}^\infty$ be the set of all "∞-tuples" like $(x_1, x_2, \ldots)$ with component-wise addition and scalar multiplication.

18. With this addition and scalar multiplication, $\mathbb{R}^\infty$ is a vector space. Verify three of the axioms.

19. Which of the following are subspaces of $\mathbb{R}^\infty$?

    (a) All sequences like $(1, 0, 2, 0, \ldots)$ that contain infinitely many zeros.
    (b) All sequences $(x_1, x_2, \ldots)$ that have $x_i = 0$ from some point onward.
    (c) All increasing sequences; that is, $x_i \leq x_{i+1}$ for all i.
    (d) All convergent sequences; that is, the x_i's have a limit as $i \to \infty$.
    (e) All arithmetic progressions $(a, a + k, a + 2k, \ldots)$.
    (f) All geometric progressions $(a, ak, ak^2, \ldots)$.

20. Is $\mathbb{R}^\infty$ finite or infinite dimensional? Why?

21. Which of the following are subspaces of M_{22}?

    (a) $\{A \mid A \text{ is nonsingular}\}$ (b) $\{A \mid A \text{ is singular}\}$
    (c) $\{A \mid A \text{ is upper triangular}\}$ (d) $\{A \mid a_{11} = a_{22}\}$

22. If A is an $n \times n$ matrix, what is the relationship between $NS(A)$ and $NS(A^2)$? Give a 2×2 example in which they are not equal.

23. Let $U = \begin{bmatrix} 0 & 1 & 2 & 4 \\ 0 & 0 & 3 & 3 \\ 0 & 0 & 0 & 0 \end{bmatrix}$.

    (a) Find a basis for the column space of U.
    (b) Express each column of U as a linear combination of the basis vectors found in (a).
    (c) Find a matrix A that has this row echelon form but a different column space.

24. Find a counterexample to the following statement: If $B = \{\mathbf{v}_1, \mathbf{v}_2, \mathbf{v}_3\}$ is a basis for $\mathbb{R}^3$ and W is a subspace of $\mathbb{R}^3$, then some subset of B is a basis for W.

25. In M_{22}, find a basis for the subspace of all matrices in which the row sums and column sums are all equal. (For extra credit, find five linearly independent 3×3 matrices with this property.)

26. Let A be an $m \times n$ matrix, $\mathbf{b}$ be a vector in $\mathbb{R}^m$, and $A' = [A \,\vdots\, \mathbf{b}]$, which is an $m \times (n+1)$ matrix. Show that $A\mathbf{x} = \mathbf{b}$ has a solution $\Leftrightarrow rk(A) = rk(A')$. (HINT Rank is equal to the dimension of the column space. When does adding an extra column leave the dimension unchanged?)

27. Let A be an $m \times n$ matrix of rank r. What is the relationship between m, n, and r in each of the following?

    (a) A has an inverse.
    (b) The equation $A\mathbf{x} = \mathbf{b}$ has a unique solution for every $\mathbf{b}$ in $\mathbb{R}^m$.
    (c) The equation $A\mathbf{x} = \mathbf{b}$ has a unique solution for some, but *not* for all, $\mathbf{b}$'s in $\mathbb{R}^m$.
    (d) The equation $A\mathbf{x} = \mathbf{b}$ has infinitely many solutions for every $\mathbf{b}$ in $\mathbb{R}^m$.

28.* Let A be an $m \times n$ matrix and $\mathbf{b}$ be a vector in $\mathbb{R}^m$. Suppose the *augmented* matrix $[A \,\vdots\, \mathbf{b}]$ is reduced to row echelon form U.

    (a) How can you tell by looking at U if the system $A\mathbf{x} = \mathbf{b}$ is consistent?
    (b) If the system $A\mathbf{x} = \mathbf{b}$ is consistent, how can you tell by looking at U if the solution is unique?

Cumulative Review Exercises

NOTE Answers to *all* Cumulative Review Exercises appear in the *Instructor's Manual*. These problems are not answered in the back of the textbook.

1. If A is a square matrix, then A is invertible $\Leftrightarrow$. Make eight other equivalent statements.

2. Let $A = \begin{bmatrix} 1 & 2 \\ 4 & 8 \end{bmatrix}$. In the xy-plane, sketch $RS(A)$, $CS(A)$, and $NS(A)$. Then give a basis for each of these subspaces.

3. Reduce

$$A = \begin{bmatrix} 0 & 1 & 5 & 3 \\ 0 & 3 & 13 & 7 \\ 0 & 5 & 23 & 13 \end{bmatrix}$$

to a matrix U in row echelon form. Then find bases for $RS(U)$, $CS(U)$, and $NS(U)$ and for $RS(A)$, $CS(A)$, and $NS(A)$. Once you have a basis for $CS(A)$, express each of the columns of A as a linear combination of the basis vectors.

4. Find the dimensions of the following vector spaces:

    (a) The vector space of all vectors in $\mathbb{R}^3$ with components that add up to zero.

* *Exercise 28 and the last four exercises of the review exercises to Chapter 4 were suggested by Prof. Thomas W. Cairns, The University of Tulsa.*

(b) $NS(I_3)$.
(c) The vector space of all 3×3 matrices.

5. (a) Solve the following system by Gaussian elimination and back-substitution:

$$\begin{aligned} x \phantom{{}+y} + z &= -1 \\ x + y \phantom{{}+z} &= 3 \\ x + y + z &= 0 \end{aligned}$$

(b) Write the system in the form $AX = B$, and factor $A = LU$.

6. If

$$(1)\ E = \begin{bmatrix} 1 & & \\ 0 & 3 & \\ 2 & 0 & 1 \end{bmatrix} \qquad (2)\ E = \begin{bmatrix} 0 & 0 & 0 \\ 1 & 1 & 1 \end{bmatrix} \qquad (3)\ E = \begin{bmatrix} & & 1 \\ & 1 & \\ 1 & & \end{bmatrix}$$

(a) How are the *rows* of EA related to the rows of A?
(b) How are the *columns* of AE related to the columns of A?

7. Answer true or false. Give a reason if true or a counterexample if false.

(a) If all the diagonal elements of the square matrix A are zero, then A is singular.
(b) If A is invertible and the rows of B are the rows A in *reverse* order, then B is invertible.
(c) If A and B are symmetric $n \times n$ matrices, then AB is a symmetric $n \times n$ matrix.
(d) If A and B are invertible $n \times n$ matrices, then AB is an invertible $n \times n$ matrix.
(e) If A^T is invertible, then A is invertible.
(f) If $AB = \mathbf{0}$, then either $A = \mathbf{0}$ or $B = \mathbf{0}$.

8. (a) Do the vectors $\mathbf{v}_1 = (1, 1, 4)$, $\mathbf{v}_2 = (2, 5, 1)$, and $\mathbf{v}_3 = (1, -7, 10)$ form a basis for $\mathbb{R}^3$? If so, express $\boldsymbol{\varepsilon}_1 = (1, 0, 0)$ as a linear combination of these vectors. If not, show that they are linearly dependent.
(b) Repeat (a) for $\mathbf{w}_1 = (2, -1, 1)$, $\mathbf{w}_2 = (3, 1, 2)$, and $\mathbf{w}_3 = (4, 1, 1)$.

9. In P_2 (the vector space of all polynomials of degree ≤ 2), let $S = \{1, x, x^2\}$ be the standard basis and let $T = \{2 - x, 2 + x^2, 1 + x + x^2\}$ be another basis. Find the transition matrix from S to T; then use this matrix to find $[1 + 2x + 3x^2]_T$.

10. Evaluate $\det\begin{bmatrix} 2 & -3 \\ 1 & 2 \end{bmatrix}$ and $\det\begin{bmatrix} 4 & -1 & 1 \\ 2 & 1 & 0 \\ 3 & 0 & -1 \end{bmatrix}$.

Chapter 4

LINEAR TRANSFORMATIONS, ORTHOGONAL PROJECTIONS, AND LEAST SQUARES

In this chapter we develop some fascinating mathematics one purpose of which is to study a very important application. The application is least-squares fitting of data, which goes by several names. One such name is statistical *regression analysis.*

A typical application might arise as follows: A biologist observes that as the temperature falls (say from 60°F to 40°F), animals eat more in order to stay warm. The scientist runs an experiment to verify the observation and charts the results from one animal, as in Figure 4.1a. Some theories of metabolism imply that the data should lie on a straight line, but because of experimental error beyond the scientist's control, they clearly do not. Therefore, the scientist tries to "fit" a straight line to the data, but there are many lines

Figure 4.1

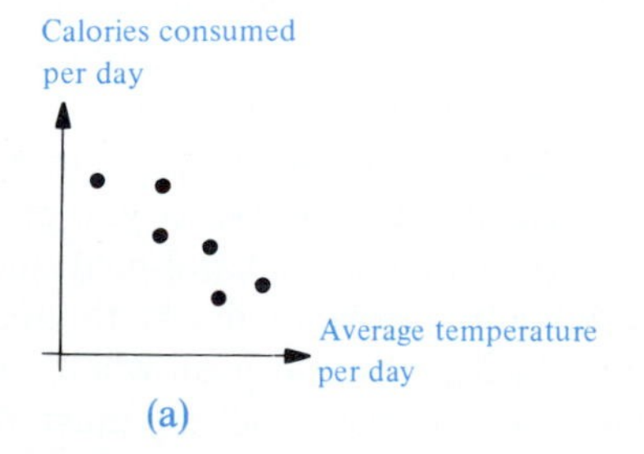

(a)

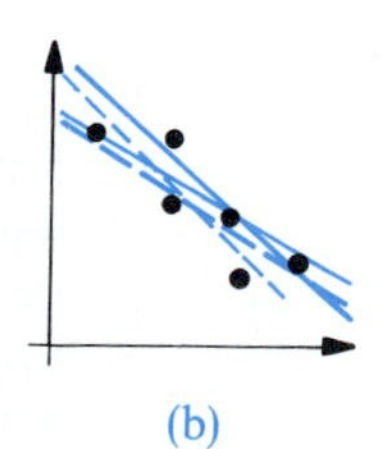

(b)

that look reasonable, as indicated in Figure 4.1b. Fortunately for the scientist, there are statistical reasons for picking one particular straight line (the one that minimizes the squares of certain errors), and there are very nice mathematical procedures for finding that line.

These mathematical procedures, and the mathematical tools developed for handling them, are the topics of this chapter. The tools themselves are widely useful in other areas of linear algebra; in particular, they will be used in the study of eigenvalues in Chapter 5.

4.1 Matrices as Linear Transformations

This section begins the development of the tools described in the introduction. We begin by reviewing the definition of a function.

(4.1)

DEFINITION A **function** f consists of two sets, D and R, and a rule, $f(x)$, that assigns to each element of D a unique element of R.

The set D is called the **domain** of f and the set R is called the **range** of f.

NOTATION We shall write "$f: D \to R$" for "f is a function from the set D to the set R," that is, f is a function with D as its domain and R as its range.

Our main examples will be functions from $\mathbb{R}^n$ to $\mathbb{R}^m$ that are associated with matrices. First we examine an application from computer-aided design (CAD) that both motivates the mathematics you are about to encounter and illustrates its diversity. Figure 4.2a shows the frame of an experimental race car. Figures 4.2b–e illustrate various changes in the perspective of the frame; the matrix that governs each change from the original position is also given. All of these changes were calculated by a computer and the pictures were taken off the computer screen. The computer stores the coordinates of points and then performs transformations of these points. It is this type of transformation that we are about to study. As you progress through this section, remember that a person who uses a hand-held calculator to add up numbers or solve a trigonometry problem needs to understand the mathematics involved to know which buttons to push when. Similarly, a person who uses a CAD computer to visualize a design must understand the underlying mathematical principles to write programs or give proper commands to the computer. We now embark into that mathematics.

*Figure 4.2**

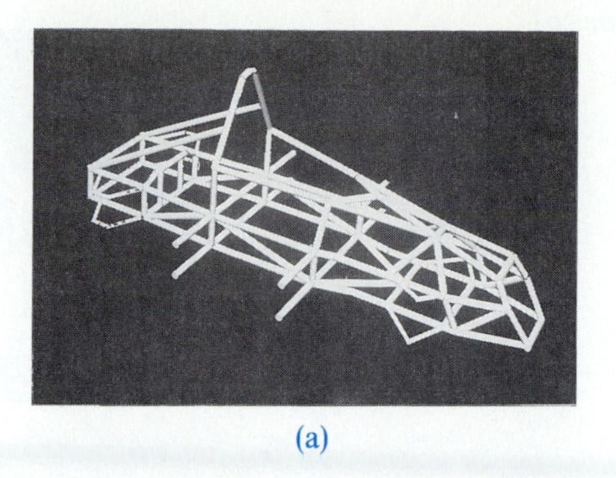

(a)

Figure (a) *is the starting point for each of the following operations.*

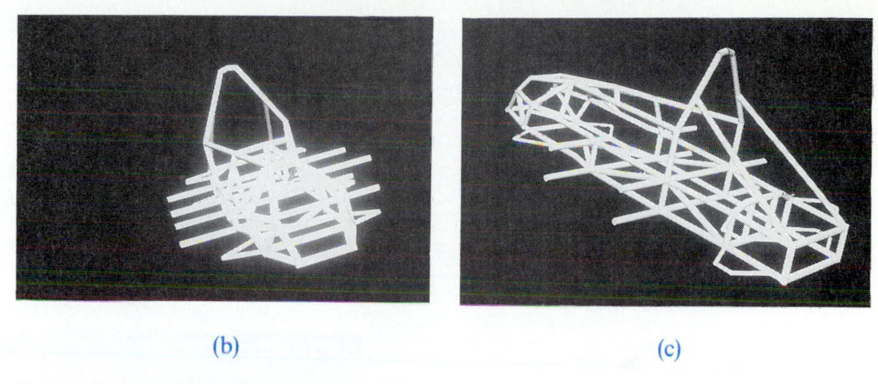

(b) (c)

Figure (b) *was obtained by rotating Figure* (a) *45° in the xy-plane, produced by the matrix*

$$\begin{bmatrix} \cos 45^\circ & -\sin 45^\circ & \\ \sin 45^\circ & \cos 45^\circ & \\ & & 1 \end{bmatrix} = \begin{bmatrix} \frac{1}{2}\sqrt{2} & -\frac{1}{2}\sqrt{2} & \\ \frac{1}{2}\sqrt{2} & \frac{1}{2}\sqrt{2} & \\ & & 1 \end{bmatrix}$$

Figure (c) *was obtained by rotating Figure* (a) *180° in the xy-plane, produced by the matrix*

$$\begin{bmatrix} \cos 180^\circ & -\sin 180^\circ & \\ \sin 180^\circ & \cos 180^\circ & \\ & & 1 \end{bmatrix} = \begin{bmatrix} -1 & 0 & \\ 0 & -1 & \\ & & 1 \end{bmatrix}$$

(continued)

* *These graphics were produced by a project of the MSU Student Section of the Society of Automotive Engineering in the Case Center for Computer-Aided Engineering and Manufacturing, Michigan State University.*

Figure 4.2 (Continued)

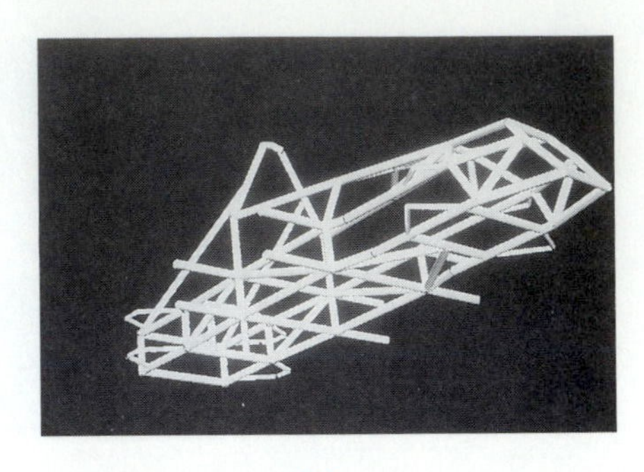

(d)

Figure (d) *was obtained by rotating Figure* (a) *first 60° in the xz-plane and then 60° in the yz-plane, produced by the respective matrices*

$$\begin{bmatrix} 1 & & \\ & \cos 60^\circ & -\sin 60^\circ \\ & \sin 60^\circ & \cos 60^\circ \end{bmatrix} \begin{bmatrix} \cos 60^\circ & 0 & -\sin 60^\circ \\ 0 & 1 & 0 \\ \sin 60^\circ & 0 & \cos 60^\circ \end{bmatrix}$$

$$= \begin{bmatrix} 1 & & \\ & \frac{1}{2} & -\frac{1}{2}\sqrt{3} \\ & \frac{1}{2}\sqrt{3} & \frac{1}{2} \end{bmatrix} \begin{bmatrix} \frac{1}{2} & 0 & -\frac{1}{2}\sqrt{3} \\ 0 & 1 & 0 \\ \frac{1}{2}\sqrt{3} & 0 & \frac{1}{2} \end{bmatrix}$$

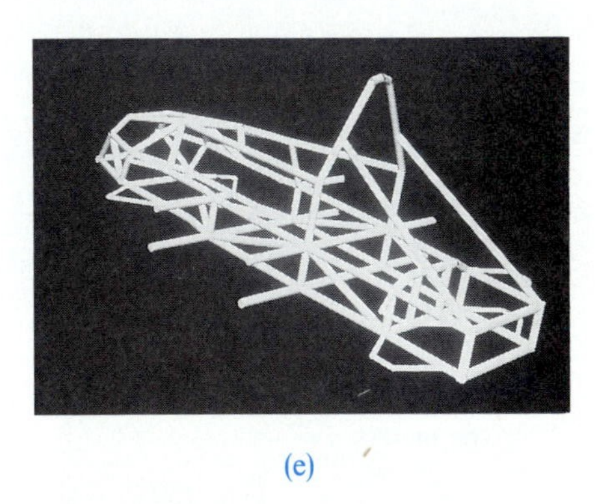

(e)

Figure (e) *was reflected through the xz-plane; that is, $x \to x$, $y \to -y$, $z \to z$, produced by the matrix*

$$\begin{bmatrix} 1 & & \\ & -1 & \\ & & 1 \end{bmatrix}$$

Note that this result looks the same as the result in (c), *since figure* (e) *is symmetric to the yz-plane.*

We begin by seeing how a matrix induces a function. Let A be an $m \times n$ matrix and $\mathbf{x}$ be a column vector in $\mathbb{R}^n$. Then the product $\mathbf{y} = A\mathbf{x}$ is an $m \times 1$ column vector in $\mathbb{R}^m$. In this way an $\mathbf{x}$ in $\mathbb{R}^n$ going to $\mathbf{y} = A\mathbf{x}$ in $\mathbb{R}^m$ forms the rule of the induced function f_A.

(4.2)

DEFINITION If A is an $m \times n$ matrix, its **induced function**

$$f_A : \mathbb{R}^n \to \mathbb{R}^m \qquad \text{is defined by} \qquad f_A(\mathbf{x}) = A\mathbf{x}$$

Example 1

$$A = \begin{bmatrix} 1 & -2 & 3 \\ 2 & 1 & -2 \end{bmatrix}$$

Then $f_A : \mathbb{R}^3 \to \mathbb{R}^2$ by

$$f_A(x_1, x_2, x_3) = \begin{bmatrix} 1 & -2 & 3 \\ 2 & 1 & -2 \end{bmatrix} \begin{bmatrix} x_1 \\ x_2 \\ x_3 \end{bmatrix} = \begin{bmatrix} x_1 - 2x_2 + 3x_3 \\ 2x_1 + x_2 - 2x_3 \end{bmatrix}$$

■

Example 2 Let A be the $n \times n$ identity matrix, I. Then $f_I(\mathbf{x}) = I\mathbf{x} = \mathbf{x}$, so $f_I : \mathbb{R}^n \to \mathbb{R}^n$ is just the identity function. ■

Example 3 Let A be the $m \times n$ zero matrix, $\mathbf{0}$. Then $f_{\mathbf{0}}(\mathbf{x}) = \mathbf{0x} = \mathbf{0}$, so $f_{\mathbf{0}} : \mathbb{R}^n \to \mathbb{R}^m$ is the zero function. ■

Example 4 Let $A = \begin{bmatrix} -1 & 0 \\ 0 & 2 \end{bmatrix}$. Then

$$f_A \begin{bmatrix} x \\ y \end{bmatrix} = \begin{bmatrix} -1 & 0 \\ 0 & 2 \end{bmatrix} \begin{bmatrix} x \\ y \end{bmatrix} = \begin{bmatrix} -x \\ 2y \end{bmatrix}$$

As illustrated in Figure 4.3, the function f_A reflects the x-axis, expands or stretches the y-axis by multiplying by 2, and an arbitrary vector is transformed by a combination of the two effects. ■

Figure 4.3

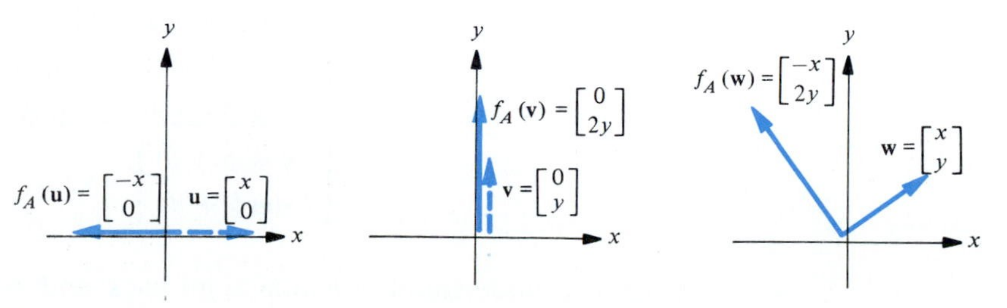

(a) *f_A reflects the x-axis* (b) *f_A expands the y-axis* (c) *f_A combines (a) and (b) for a general vector*

Once you understand Example 4, go back and examine Figure 4.2c to see why the matrix given there has the effect pictured.

We shall see in Chapter 5 that if A is any symmetric $n \times n$ matrix, then f_A acts on $\mathbb{R}^n$ very much as the f_A in Example 4 acts on $\mathbb{R}^2$. The next example introduces a fundamental building block, which was used in Figure 4.2.

Example 5 Let

$$A = \begin{bmatrix} \cos\theta & -\sin\theta \\ \sin\theta & \cos\theta \end{bmatrix}$$

Then f_A rotates the plane $\mathbb{R}^2$ about the origin through the angle θ. To see this, let $\mathbf{u} = [x \quad y]^T$, let ϕ be the angle between $\mathbf{u}$ and the x-axis, and let $\mathbf{v}$ be the vector that results when $\mathbf{u}$ is rotated through θ. See Figure 4.4.

Figure 4.4

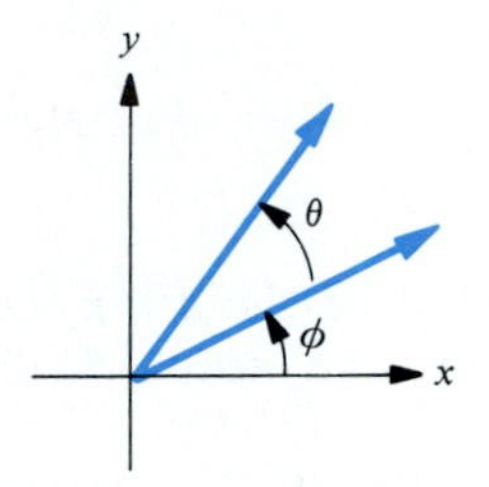

Thus if $r = \|\mathbf{u}\|$, then

$$\mathbf{u} = \begin{bmatrix} x \\ y \end{bmatrix} = \begin{bmatrix} r\cos\phi \\ r\sin\phi \end{bmatrix} \quad \text{and} \quad \mathbf{v} = \begin{bmatrix} r\cos(\phi+\theta) \\ r\sin(\phi+\theta) \end{bmatrix}$$

Computing f_A, we find

$$\begin{aligned} f_A(\mathbf{u}) = A\mathbf{u} &= \begin{bmatrix} \cos\theta & -\sin\theta \\ \sin\theta & \cos\theta \end{bmatrix}\begin{bmatrix} x \\ y \end{bmatrix} \\ &= \begin{bmatrix} x\cos\theta - y\sin\theta \\ x\sin\theta + y\cos\theta \end{bmatrix} \\ &= \begin{bmatrix} r\cos\phi\cos\theta - r\sin\phi\sin\theta \\ r\cos\phi\sin\theta + r\sin\phi\cos\theta \end{bmatrix} \\ &= \begin{bmatrix} r\cos(\phi+\theta) \\ r\sin(\phi+\theta) \end{bmatrix} = \mathbf{v} \end{aligned}$$

■

Once you understand Example 5, go back and examine Figures 4.2b and 4.2d to see why the matrices given there have the effects pictured.

Linear Transformations

Functions induced by matrices form a very special class of functions from $\mathbb{R}^n$ to $\mathbb{R}^m$.

(4.3)

DEFINITION Let V and W be vector spaces. A function $T: V \to W$ is called a **linear transformation*** if for all **u**, **v** in V and for all scalars r,

$$T(\mathbf{u} + \mathbf{v}) = T(\mathbf{u}) + T(\mathbf{v})$$
$$T(r\mathbf{u}) = rT(u)$$

THEOREM If A is an $m \times n$ matrix, then its induced function $f_A: \mathbb{R}^n \to \mathbb{R}^m$ is a linear transformation.

Proof Let A be any $m \times n$ matrix, let **u** and **v** be any vectors in $\mathbb{R}^n$ considered as column vectors, and let r be any scalar in $\mathbb{R}$. Then

$$f_A(\mathbf{u} + \mathbf{v}) = A(\mathbf{u} + \mathbf{v}) = A\mathbf{u} + A\mathbf{v} = f_A(\mathbf{u}) + f_A(\mathbf{v})$$

and

$$f_A(r\mathbf{u}) = A(r\mathbf{u}) = rA\mathbf{u} = rf_A(\mathbf{u})$$

by the elementary properties of matrix addition and multiplication. ■

Example 6

$$\text{Let } A = \begin{bmatrix} 1 & -2 & 3 \\ 2 & 1 & -2 \end{bmatrix}$$

of Example 1, let

$$\mathbf{u} = \begin{bmatrix} -3 \\ 2 \\ 4 \end{bmatrix}, \qquad \mathbf{v} = \begin{bmatrix} -1 \\ 1 \\ -3 \end{bmatrix},$$

and let $r = -2$. Then

$$f_A(\mathbf{u} + \mathbf{v}) = f_A \begin{bmatrix} -4 \\ 3 \\ 1 \end{bmatrix} = \begin{bmatrix} 1 & -2 & 3 \\ 2 & 1 & -2 \end{bmatrix} \begin{bmatrix} -4 \\ 3 \\ 1 \end{bmatrix} = \begin{bmatrix} -7 \\ -7 \end{bmatrix}$$

* *Vector spaces have the properties that you can add and scalar multiply in them. Linear transformations are exactly those functions that preserve those important properties.*

$$f_A(\mathbf{u}) + f_A(\mathbf{v}) = \begin{bmatrix} 1 & -2 & 3 \\ 2 & 1 & -2 \end{bmatrix}\begin{bmatrix} -3 \\ 2 \\ 4 \end{bmatrix} + \begin{bmatrix} 1 & -2 & 3 \\ 2 & 1 & -2 \end{bmatrix}\begin{bmatrix} -1 \\ 1 \\ -3 \end{bmatrix}$$

$$= \begin{bmatrix} 5 \\ -12 \end{bmatrix} + \begin{bmatrix} -12 \\ 5 \end{bmatrix} = \begin{bmatrix} -7 \\ -7 \end{bmatrix}$$

$$f_A(r\mathbf{u}) = f_A\begin{bmatrix} 6 \\ -4 \\ -8 \end{bmatrix} = \begin{bmatrix} 1 & -2 & 3 \\ 2 & 1 & -2 \end{bmatrix}\begin{bmatrix} 6 \\ -4 \\ -8 \end{bmatrix} = \begin{bmatrix} -10 \\ 24 \end{bmatrix}$$

$$rf_A(\mathbf{u}) = -2\begin{bmatrix} 1 & -2 & 3 \\ 2 & 1 & -2 \end{bmatrix}\begin{bmatrix} -3 \\ 2 \\ 4 \end{bmatrix} = -2\begin{bmatrix} 5 \\ -12 \end{bmatrix} = \begin{bmatrix} -10 \\ 24 \end{bmatrix}$$

We can see that

$$f_A(\mathbf{u} + \mathbf{v}) = f_A(\mathbf{u}) + f_A(\mathbf{v}) \qquad \text{and} \qquad f_A(r\mathbf{u}) = rf_A(\mathbf{u})$$ ■

An important property of a linear transformation from $\mathbb{R}^n$ to $\mathbb{R}^m$ is that it is induced by an $m \times n$ matrix.

(4.4)

> THEOREM Let $T: \mathbb{R}^n \to \mathbb{R}^m$ be a linear transformation. Then there is an $m \times n$ matrix A such that $T = f_A$.
>
> DEFINITION We shall call this A the matrix **induced** by T and denote it by $A = [T]$.

In other words, if you are given a linear transformation T, the following proof shows how to construct a matrix A such that $T = f_A$. In fact the importance of this theorem cannot be overemphasized. It forms one of the cornerstones of the interaction between the algebra of matrices and the geometry of $\mathbb{R}^n$.

Proof of Theorem (4.4) Let $\boldsymbol{\varepsilon}_1, \ldots, \boldsymbol{\varepsilon}_n$ be the standard basis for $\mathbb{R}^n$. Let $T: \mathbb{R}^n \to \mathbb{R}^m$ be a linear transformation, and let A be the $m \times n$ matrix whose ith column is $T(\boldsymbol{\varepsilon}_i)$, that is,

$$A = [T(\boldsymbol{\varepsilon}_1) \vdots T(\boldsymbol{\varepsilon}_2) \vdots \cdots \vdots T(\boldsymbol{\varepsilon}_n)]$$

Let $\mathbf{x}$ be any vector in $\mathbb{R}^n$ so that

$$\mathbf{x} = \begin{bmatrix} x_1 \\ x_2 \\ \vdots \\ x_n \end{bmatrix} = x_1\boldsymbol{\varepsilon}_1 + x_2\boldsymbol{\varepsilon}_2 + \cdots + x_n\boldsymbol{\varepsilon}_n$$

Then by the definition of a linear transformation (4.3),

$$\begin{aligned} T(\mathbf{x}) &= T(x_1\boldsymbol{\varepsilon}_1 + x_2\boldsymbol{\varepsilon}_2 + \cdots + x_n\boldsymbol{\varepsilon}_n) \\ &= T(x_1\boldsymbol{\varepsilon}_1) + T(x_2\boldsymbol{\varepsilon}_2) + \cdots + T(x_n\boldsymbol{\varepsilon}_n) \\ &= x_1 T(\boldsymbol{\varepsilon}_1) + x_2 T(\boldsymbol{\varepsilon}_2) + \cdots + x_n T(\boldsymbol{\varepsilon}_n) \\ &= [T(\boldsymbol{\varepsilon}_1) \vdots T(\boldsymbol{\varepsilon}_2) \vdots \cdots \vdots T(\boldsymbol{\varepsilon}_n)] \begin{bmatrix} x_1 \\ x_2 \\ \vdots \\ x_n \end{bmatrix} \\ &= A\mathbf{x} = f_A(\mathbf{x}) \end{aligned}$$

■

Example 7 Determine if each of the following functions is a linear transformation. If it is, find the matrix that induces the function.

(a) $T(x_1, x_2, x_3) = (x_1^2 + x_3x_2, x_1x_3)$
(b) $T(x_1, x_2, x_3) = (3x_1 - 2x_2 + x_3, -x_1 + 5x_2 + 4x_3)$

Solution Let $\mathbf{u} = (x_1, x_2, x_3)$ and $\mathbf{v} = (y_1, y_2, y_3)$.

(a) Then,

$$\begin{aligned} T(r\mathbf{u}) &= T(rx_1, rx_2, rx_3) = ((rx_1)^2 + rx_3rx_2, rx_1rx_3) \\ &= r^2(x_1^2 + x_3x_2, x_1x_3) = r^2T(\mathbf{u}) \end{aligned}$$

Thus $T(r\mathbf{u}) \neq rT(\mathbf{u})$ (for $r \neq 0, 1$) so T cannot be a linear transformation. It is also true that $T(\mathbf{u} + \mathbf{v}) \neq T(\mathbf{u}) + T(\mathbf{v})$, but we only need to show that one part of Definition (4.3) does not hold for a function to fail to be a linear transformation.

(b) This time,

$$\begin{aligned} T(\mathbf{u} + \mathbf{v}) &= T(x_1 + y_1, x_2 + y_2, x_3 + y_3) \\ &= (3[x_1 + y_1] - 2[x_2 + y_2] + [x_3 + y_3], \\ &\quad -[x_1 + y_1] + 5[x_2 + y_2] + 4[x_3 + y_3]) \\ &= (3x_1 - 2x_2 + x_3, -x_1 + 5x_2 + 4x_3) \\ &\quad + (3y_1 - 2y_2 + y_3, -y_1 + 5y_2 + 4y_3) \\ &= T(\mathbf{u}) + T(\mathbf{v}) \\ T(r\mathbf{u}) &= T(rx_1, rx_2, rx_3) \\ &= (3rx_1 - 2rx_2 + rx_3, -rx_1 + 5rx_2 + 4rx_3) \\ &= r(3x_1 - 2x_2 + x_3, -x_1 + 5x_2 + 4x_3) = rT(\mathbf{u}) \end{aligned}$$

Since T is a linear transformation, we move on to the second part of the problem and find A. To find A, we compute $T(\boldsymbol{\varepsilon}_1)$, $T(\boldsymbol{\varepsilon}_2)$, and $T(\boldsymbol{\varepsilon}_3)$

$$\begin{aligned} T(\boldsymbol{\varepsilon}_1) &= T(1, 0, 0) = (3, -1) \\ T(\boldsymbol{\varepsilon}_2) &= T(0, 1, 0) = (-2, 5) \\ T(\boldsymbol{\varepsilon}_3) &= T(0, 0, 1) = (1, 4) \end{aligned}$$

and then arrange them as column vectors in A:

$$A = \begin{bmatrix} 3 & -2 & 1 \\ -1 & 5 & 4 \end{bmatrix}$$

To check that $f_A = T$, we compute

$$f_A(\mathbf{u}) = A\mathbf{u} = \begin{bmatrix} 3 & -2 & 1 \\ -1 & 5 & 4 \end{bmatrix} \begin{bmatrix} x_1 \\ x_2 \\ x_3 \end{bmatrix} = \begin{bmatrix} 3x_1 - 2x_2 + x_3 \\ -x_1 + 5x_2 + 4x_3 \end{bmatrix} = T(\mathbf{u})$$ ■

This section just begins the study of the relationship between matrices and linear transformations. We shall examine this relationship more closely as we proceed through this book.

For now, however, we conclude this section with a very important linear transformation defined on an infinite-dimensional vector space.

Example 8 Let $C^\infty(\mathbb{R})$ be the vector space of all functions $f:\mathbb{R} \to \mathbb{R}$ such that the nth derivative of f, $f^{(n)}$, exists for all positive integers n. Show that

$$D: C^\infty(\mathbb{R}) \to C^\infty(\mathbb{R}) \qquad \text{defined by} \qquad D(f) = f'$$

is a linear transformation.

Solution If a function f has all of its derivatives, then its derivative f' has all of its derivatives. Hence D itself is a function (in a sense a "super function") with domain and range both $C^\infty(\mathbb{R})$. By elementary theorems about derivatives,

$$(f+g)' = f' + g' \qquad \text{which says that} \qquad D(f+g) = D(f) + D(g)$$
$$(rf)' = rf' \qquad \text{which says that} \qquad D(rf) = rD(f)$$

This is exactly what we need for D to be a linear transformation, so we are done. ■

Exercise 4.1

In Exercises 1–6 a matrix A is given. Find the rule of the induced function, f_A, and give the domain and range of f_A.

1. $\begin{bmatrix} 1 & 2 \\ -2 & 0 \\ 3 & -1 \end{bmatrix}$

2. $\begin{bmatrix} 2 & -1 & 1 \\ 3 & 0 & -3 \end{bmatrix}$

3. $\begin{bmatrix} 4 & -1 \\ 1 & 2 \\ 3 & 0 \\ -1 & 1 \end{bmatrix}$

4. $\begin{bmatrix} 8 & -1 & 1 & 2 \\ 1 & 0 & -1 & 3 \end{bmatrix}$

5. $\begin{bmatrix} 1 & 2 & -1 & 3 \end{bmatrix}$

6. $\begin{bmatrix} 5 \\ -1 \\ 2 \end{bmatrix}$

In Exercises 7–12, for the given matrix A, describe geometrically what the induced function, f_A, does to the x-axis and the y-axis.

7. $\begin{bmatrix} 2 & 0 \\ 0 & \frac{1}{2} \end{bmatrix}$

8. $\begin{bmatrix} -2 & 0 \\ 0 & -\frac{1}{2} \end{bmatrix}$

9. $\begin{bmatrix} 0 & 1 \\ 1 & 0 \end{bmatrix}$

10. $\begin{bmatrix} 1 & 0 \\ 0 & 0 \end{bmatrix}$

11. $\begin{bmatrix} \frac{\sqrt{2}}{2} & -\frac{\sqrt{2}}{2} \\ \frac{\sqrt{2}}{2} & \frac{\sqrt{2}}{2} \end{bmatrix}$

12. $\begin{bmatrix} 0 & 1 \\ -1 & 0 \end{bmatrix}$

In Exercises 13–16, find the 3×3 matrix that induces the described action on $\mathbb{R}^3$. (See Figure 4.2 and Examples 4 and 5.)

13. Rotates the xy-plane by 30°; leaves the z-axis fixed.
14. Rotates the xz-plane by 60°; leaves the y-axis fixed.
15. Rotates the yz-plane by 45°; reflects the x-axis.
16. Rotates the xy-plane by 90°; reflects the z-axis.

In Exercises 17–40, determine if the given function T is a linear transformation. In Exercises 17–26, also give the domain and range of T; if T is linear, find the A such that $T = f_A$.

17. $T(x, y) = (2x, y)$
18. $T(x, y) = (y, x)$
19. $T(x, y) = (x^2, y)$
20. $T(x, y) = (x + y^2, \sqrt[3]{xy})$
21. $T(x, y, z) = (2x + y, x - y + z)$
22. $T(x, y, z) = (z - x, z - y)$
23. $T(x, y, z) = (0, 0, 0)$
24. $T(x, y, z) = (1, 1, 1)$
25. $T(x, y) = (x - y, x + y, 2x - 3y)$
26. $T(x, y, z, w) = x + y - z - w$
27. $T: M_{22} \to M_{22}$ by $T\begin{bmatrix} a & b \\ c & d \end{bmatrix} = \begin{bmatrix} a + d & 0 \\ 0 & b + c \end{bmatrix}$
28. $T: M_{22} \to M_{22}$ by $T\begin{bmatrix} a & b \\ c & d \end{bmatrix} = \begin{bmatrix} 2ab & 3cd \\ 0 & 0 \end{bmatrix}$
29. $T: M_{22} \to \mathbb{R}$ by $T\begin{bmatrix} a & b \\ c & d \end{bmatrix} = ad - bc$
30. $T: M_{22} \to \mathbb{R}$ by $T\begin{bmatrix} a & b \\ c & d \end{bmatrix} = a + b + c + d$
31. $T: P_2 \to P_2$ by $T(a_0 + a_1x + a_2x^2) = (a_0 + a_2) + 2a_1x + (a_1 + a_2)x^2$
32. $T: P_2 \to P_2$ by $T(a_0 + a_1x + a_2x^2) = \mathbf{0}$

33. $T: P_2 \to P_2$ by $T(a_0 + a_1 x + a_2 x^2) = a_0 + a_1(x - 1) + a_2(x - 1)^2$

34. $T: P_2 \to P_2$ by $T(a_0 + a_1 x + a_2 x^2) = (a_0 - 1) + a_1 x + a_2 x^2$

In Exercises 35 and 36, A is a fixed 2×3 matrix.

35. $T: M_{22} \to M_{23}$ by $T(B) = BA$

36. $T: M_{33} \to M_{23}$ by $T(C) = AC$

37. $T: M_{mn} \to M_{nm}$ by $T(A) = A^T$

38. $T: M_{nn} \to \mathbb{R}$ by $T(A) = a_{11} + a_{22} + \cdots + a_{nn}$

39. $T: M_{nn} \to \mathbb{R}$ by $T(A) = a_{11}a_{22} \cdots a_{nn}$

40. $T: M_{nn} \to M_{nn}$ by $T(A) = \begin{cases} A^{-1}, & \text{if } A \text{ is invertible} \\ \mathbf{0}, & \text{if } A \text{ is not invertible} \end{cases}$

41. For $f \in C[a, b]$, define

$$J(f) = \int_a^b f(x)\,dx$$

(a) Is J linear?
(b) What are the domain and range of J?

42. For $f \in C^\infty[a, b]$, define $D^2 f = D(Df)$. See Example 8.

(a) Is D^2 linear?
(b) Describe its "null space."
(c) Explain why the fundamental theorem of calculus states that D (and hence D^2) is **onto**; that is, every f is a Dg, for some g.

4.2 Relationships Involving Inner Products

In this section we prove several theorems about inner products that we shall need in the next several sections. These relationships are quite fundamental to a great deal of theoretical and applied mathematics.

In Section 3.2, we defined an inner product, or dot product, on $\mathbb{R}^n$ by $\mathbf{u} \cdot \mathbf{v} = u_1 v_1 + \cdots + u_n v_n$. This product has the following properties.

(4.5) THEOREM If $\mathbf{u}$, $\mathbf{v}$, and $\mathbf{w}$ are vectors in $\mathbb{R}^n$ and r is any scalar, then

(a) $(\mathbf{u} + \mathbf{v}) \cdot \mathbf{w} = \mathbf{u} \cdot \mathbf{w} + \mathbf{v} \cdot \mathbf{w}$
(b) $\mathbf{u} \cdot \mathbf{v} = \mathbf{v} \cdot \mathbf{u}$
(c) $(r\mathbf{u}) \cdot \mathbf{v} = r(\mathbf{u} \cdot \mathbf{v})$
(d) $\mathbf{u} \cdot \mathbf{u} \geq 0$, and $\mathbf{u} \cdot \mathbf{u} = 0$ if and only if $\mathbf{u} = \mathbf{0}$

Proof (a) Let $\mathbf{u} = (u_1, \ldots, u_n)$, $\mathbf{v} = (v_1, \ldots, v_n)$, and $\mathbf{w} = (w_1, \ldots, w_n)$. Then

$$\begin{aligned}(\mathbf{u} + \mathbf{v}) \cdot \mathbf{w} &= (u_1 + v_1, \ldots, u_n + v_n) \cdot (w_1, \ldots, w_n) \\ &= (u_1 + v_1)w_1 + \cdots + (u_n + v_n)w_n \\ &= (u_1 w_1 + \cdots + u_n w_n) + (v_1 w_1 + \cdots + v_n w_n) \\ &= \mathbf{u} \cdot \mathbf{w} + \mathbf{v} \cdot \mathbf{w}\end{aligned}$$

The proofs of (b) and (c) are similar and are left to the exercises. See Exercises 29 and 30.

(d) If r is any real number, then $r^2 \geq 0$, and $r^2 = 0$ if and only if $r = 0$. Using this,

$$\mathbf{u} \cdot \mathbf{u} = u_1^2 + \cdots + u_n^2 \geq 0$$

and

$$\begin{aligned}u_1^2 + \cdots + u_n^2 = 0 \quad &\Leftrightarrow \quad \text{each } u_i = 0,\ 1 \leq i \leq n \\ &\Leftrightarrow \quad \mathbf{u} = (0, \ldots, 0) = \mathbf{0}\end{aligned}$$

■

Recall, in Example 10 in Section 3.3, we defined an inner product on $C[a, b]$, $a < b$, by

$$f \cdot g = \int_a^b f(x)g(x)\,dx$$

(4.7) **THEOREM** The inner product on $C[a, b]$ given by Equation (4.6) satisfies properties (a)–(d) of Theorem (4.5).

Proof These all follow from basic theorems about integrals. For example,

$$\begin{aligned}(f + g) \cdot h &= \int_a^b [f(x) + g(x)]h(x)\,dx \\ &= \int_a^b [f(x)h(x) + g(x)h(x)]\,dx \\ &= \int_a^b f(x)h(x)\,dx + \int_a^b g(x)h(x)\,dx \\ &= f \cdot h + g \cdot h\end{aligned}$$

The remaining properties are left as exercises. See Exercises 31–33.

■

Theorems (4.5) and (4.7) motivate the following definition.

(4.8) **DEFINITION** An **inner product space** is a vector space V equipped with an inner product. This inner product associates with every pair of vectors $\mathbf{v}$ and $\mathbf{w}$ in V a real number denoted by $\mathbf{v} \cdot \mathbf{w}$ so that all properties (a)–(d) of Theorem (4.5) are satisfied.

The principal purpose of making the abstract Definition (4.8) is to help us understand the properties of inner products better. In particular we are able to see which properties follow from axioms (a)–(d) of Theorem (4.5) and which properties require more specific information. For example, there are many simple properties, such as $\mathbf{u} \cdot (\mathbf{v} + \mathbf{w}) = \mathbf{u} \cdot \mathbf{v} + \mathbf{u} \cdot \mathbf{w}$, that follow directly from the axioms. Two such properties are given in Exercises 34 and 35. The following inequalities also follow from the axioms, so they hold in any inner product space. The relationships with matrices given at the end of this section, however, hold only in $\mathbb{R}^n$, and defining their properties requires more specific information.

Cauchy–Schwarz and Triangle Inequalities

A norm is defined in an inner product space by

(4.9) **DEFINITION** $\|\mathbf{u}\| = \sqrt{\mathbf{u} \cdot \mathbf{u}}$

Compare Definition (4.9) with Equation (3.16), Section 3.2, and Example 10 in Section 3.3, and see Exercises 36 and 37 for various properties of norms defined this way. The next two theorems are more easily stated using norms rather than the inner product $\mathbf{u} \cdot \mathbf{u}$.

The first theorem was proved independently by Augustin Louis (Baron de) Cauchy (1789–1857) and Hermann Amandus Schwarz (1843–1921), both of whom did fundamental work in analysis.

(4.10) **THEOREM (CAUCHY–SCHWARZ INEQUALITY)** Suppose $\mathbf{u}$ and $\mathbf{v}$ are vectors in an inner product space V. Then

$$|\mathbf{u} \cdot \mathbf{v}| \leq \|\mathbf{u}\|\,\|\mathbf{v}\|$$

Proof If either $\mathbf{u}$ or $\mathbf{v}$ is $\mathbf{0}$, then both sides $\mathbf{u} \cdot \mathbf{v}$ and $\|\mathbf{u}\|\,\|\mathbf{v}\|$ are zero, so the inequality holds. So suppose $\mathbf{u}$ and $\mathbf{v}$ are nonzero. By Theorem (4.5d), if $\mathbf{w}$ is any vector, then $\|\mathbf{w}\|^2 = \mathbf{w} \cdot \mathbf{w} \geq 0$. Let t be any scalar and let $\mathbf{w} = \mathbf{u} - t\mathbf{v}$. Then

$$0 \leq (\mathbf{u} - t\mathbf{v}) \cdot (\mathbf{u} - t\mathbf{v})$$

$$0 \leq \mathbf{u} \cdot \mathbf{u} - t\mathbf{v} \cdot \mathbf{u} - t\mathbf{u} \cdot \mathbf{v} + t^2\mathbf{v} \cdot \mathbf{v}$$

$$0 \leq \mathbf{u} \cdot \mathbf{u} - 2t\mathbf{u} \cdot \mathbf{v} + t^2\mathbf{v} \cdot \mathbf{v} \tag{4.11}$$

by Theorem (4.5b). Now let $t = \dfrac{\mathbf{u} \cdot \mathbf{v}}{\mathbf{v} \cdot \mathbf{v}}$. Then from (4.11),

$$0 \leq \mathbf{u} \cdot \mathbf{u} - 2\frac{\mathbf{u} \cdot \mathbf{v}}{\mathbf{v} \cdot \mathbf{v}}\mathbf{u} \cdot \mathbf{v} + \frac{(\mathbf{u} \cdot \mathbf{v})^2}{(\mathbf{v} \cdot \mathbf{v})^2}\mathbf{v} \cdot \mathbf{v}$$

$$0 \leq \frac{(\mathbf{u} \cdot \mathbf{u})(\mathbf{v} \cdot \mathbf{v}) - (\mathbf{u} \cdot \mathbf{v})^2}{\mathbf{v} \cdot \mathbf{v}}$$

$$0 \leq (\mathbf{u} \cdot \mathbf{u})(\mathbf{v} \cdot \mathbf{v}) - (\mathbf{u} \cdot \mathbf{v})^2 \tag{4.12}$$

since $\mathbf{v} \cdot \mathbf{v} > 0$. Since $\|\mathbf{w}\|^2 = \mathbf{w} \cdot \mathbf{w}$, (4.12) can be rewritten as

$$(\mathbf{u} \cdot \mathbf{v})^2 \leq \|\mathbf{u}\|^2 \|\mathbf{v}\|^2$$

Taking square roots gives the result

$$|\mathbf{u} \cdot \mathbf{v}| \leq \|\mathbf{u}\| \, \|\mathbf{v}\|$$ ■

Example 1 Let $\mathbf{u} = (-2, 4, -1, 3)$ and $\mathbf{v} = (5, -1, 2, -1)$. Then

$$\|\mathbf{u}\| = \sqrt{(-2)^2 + 4^2 + (-1)^2 + 3^2} = \sqrt{30}$$
$$\|\mathbf{v}\| = \sqrt{5^2 + (-1)^2 + 2^2 + (-1)^2} = \sqrt{31}$$
$$\mathbf{u} \cdot \mathbf{v} = -2(5) + 4(-1) + (-1)2 + 3(-1) = -19$$

We can see that $|\mathbf{u} \cdot \mathbf{v}| = 19$ is less than $\|\mathbf{u}\| \, \|\mathbf{v}\| = \sqrt{30}\sqrt{31}$ (which is slightly greater than 30). ■

The statements of the Cauchy–Schwarz inequality in $\mathbb{R}^n$ and in $C[a, b]$ are rather startling. Exercises 38 and 39 suggest a proof of the following.

(4.13) Equality holds in the Cauchy–Schwarz inequality if and only if one vector is a multiple of the other.

One of the many uses of the Cauchy–Schwarz inequality is to prove the triangle inequality, which says the sum of the lengths of two sides of a triangle is greater than or equal to the length of the third side, as shown in Figure 4.5.

Figure 4.5

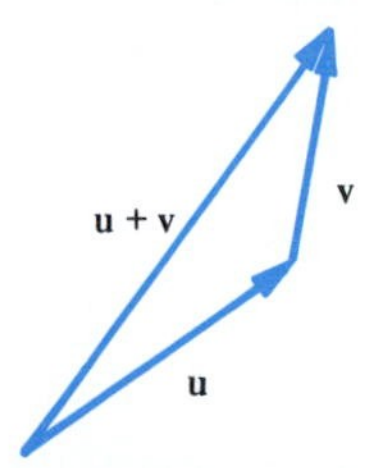

(4.14) THEOREM (TRIANGLE INEQUALITY)
If $\mathbf{u}$ and $\mathbf{v}$ are vectors in an inner product space V, then

$$\|\mathbf{u} + \mathbf{v}\| \leq \|\mathbf{u}\| + \|\mathbf{v}\|$$

Proof By the Cauchy–Schwarz inequality, $\mathbf{u} \cdot \mathbf{v} \leq |\mathbf{u} \cdot \mathbf{v}| \leq \|\mathbf{u}\| \|\mathbf{v}\|$. Thus

$$\begin{aligned} \|\mathbf{u} + \mathbf{v}\|^2 &= (\mathbf{u} + \mathbf{v}) \cdot (\mathbf{u} + \mathbf{v}) \\ &= \mathbf{u} \cdot \mathbf{u} + 2\mathbf{u} \cdot \mathbf{v} + \mathbf{v} \cdot \mathbf{v} \\ &\leq \|\mathbf{u}\|^2 + 2\|\mathbf{u}\| \|\mathbf{v}\| + \|\mathbf{v}\|^2 \\ &= (\|\mathbf{u}\| + \|\mathbf{v}\|)^2 \end{aligned}$$

Thus $\|\mathbf{u} + \mathbf{v}\|^2 \leq (\|\mathbf{u}\| + \|\mathbf{v}\|)^2$, and taking square roots gives the result. ■

Example 2 If $\mathbf{u} = (-2, 4, -1, 3)$ and $\mathbf{v} = (5, -1, 2, -1)$ are from Example 1, then

$$\|\mathbf{u} + \mathbf{v}\| = \|(3, 3, 1, 2)\| = \sqrt{9 + 9 + 1 + 4} = \sqrt{23}$$

By Example 1, $\|\mathbf{u}\| = \sqrt{30}$ and $\|\mathbf{v}\| = \sqrt{31}$, so we can see

$$\|\mathbf{u} + \mathbf{v}\| = \sqrt{23} \leq \|\mathbf{u}\| + \|\mathbf{v}\| = \sqrt{30} + \sqrt{31}$$ ■

For the remainder of this section we shall work only in $\mathbb{R}^n$.

Matrices and Dot Products

There is a fundamental relationship between a matrix, its transpose, and dot products.

(4.15)

> THEOREM Let A be an $m \times n$ matrix, let B be an $n \times m$ matrix, let $\mathbf{x}$ be a (column) vector in $\mathbb{R}^n$, and let $\mathbf{y}$ be a (column) vector in $\mathbb{R}^m$. Then
>
> (a) $A\mathbf{x} \cdot \mathbf{y} = \mathbf{x} \cdot A^T\mathbf{y}$ (b) $\mathbf{x} \cdot B\mathbf{y} = B^T\mathbf{x} \cdot \mathbf{y}$

Of course, (b) follows from (a) by letting $A = B^T$ (and observing $A^T = (B^T)^T = B$). We shall prove part (a) after an example and proving a preliminary result.

Example 3 Let

$$A = \begin{bmatrix} 2 & -1 & 3 \\ 4 & 5 & -2 \end{bmatrix}, \quad \mathbf{x} = \begin{bmatrix} 6 \\ -1 \\ -3 \end{bmatrix}, \quad \mathbf{y} = \begin{bmatrix} 5 \\ -4 \end{bmatrix}$$

Then,

$$A\mathbf{x} \cdot \mathbf{y} = \begin{bmatrix} 2 & -1 & 3 \\ 4 & 5 & -2 \end{bmatrix} \begin{bmatrix} 6 \\ -1 \\ -3 \end{bmatrix} \cdot \begin{bmatrix} 5 \\ -4 \end{bmatrix} = \begin{bmatrix} 4 \\ 25 \end{bmatrix} \cdot \begin{bmatrix} 5 \\ -4 \end{bmatrix} = -80$$

$$\mathbf{x} \cdot A^T\mathbf{y} = \begin{bmatrix} 6 \\ -1 \\ -3 \end{bmatrix} \cdot \begin{bmatrix} 2 & 4 \\ -1 & 5 \\ 3 & -2 \end{bmatrix} \begin{bmatrix} 5 \\ -4 \end{bmatrix} = \begin{bmatrix} 6 \\ -1 \\ -3 \end{bmatrix} \cdot \begin{bmatrix} -6 \\ -25 \\ 23 \end{bmatrix} = -80$$ ■

The proof of Theorem (4.15) is very short if inner products are represented as described in the following theorem.

(4.16)

THEOREM Suppose we identify a 1×1 matrix $[r]$ with its entry r. If $\mathbf{u}$ and $\mathbf{v}$ are vectors in $\mathbb{R}^m$ represented as column vectors, then

$$\mathbf{u} \cdot \mathbf{v} = \mathbf{u}^T\mathbf{v}$$

This formula is useful in solving problems in several contexts, not the least of which is the least-squares problem.

Example 4 Let

$$\mathbf{u} = \begin{bmatrix} 4 \\ 7 \\ -1 \\ 2 \end{bmatrix}, \qquad \mathbf{v} = \begin{bmatrix} -3 \\ 2 \\ 1 \\ 3 \end{bmatrix}$$

in $\mathbb{R}^4$. Then

$$\mathbf{u} \cdot \mathbf{v} = \begin{bmatrix} 4 \\ 7 \\ -1 \\ 2 \end{bmatrix} \cdot \begin{bmatrix} -3 \\ 2 \\ 1 \\ 3 \end{bmatrix} = 7 \text{ ``=''} [7] = [4 \quad 7 \quad -1 \quad 2] \begin{bmatrix} -3 \\ 2 \\ 1 \\ 3 \end{bmatrix} = \mathbf{u}^T\mathbf{v} \quad \blacksquare$$

Example 4 illustrates the proof of Theorem (4.16).

Proof of Theorem (4.16) Pick

$$\mathbf{u} = \begin{bmatrix} u_1 \\ \vdots \\ u_m \end{bmatrix} \quad \text{and} \quad \mathbf{v} = \begin{bmatrix} v_1 \\ \vdots \\ v_m \end{bmatrix}$$

in $\mathbb{R}^m$. Then

$$\mathbf{u} \cdot \mathbf{v} = u_1v_1 + \cdots + u_mv_m \text{ ``=''} [u_1v_1 + \cdots + u_mv_m] = \mathbf{u}^T\mathbf{v} \quad \blacksquare$$

Proof of Theorem (4.15) Let A be an $m \times n$ matrix. Represent $\mathbf{x}$ in $\mathbb{R}^n$ and $\mathbf{y}$ in $\mathbb{R}^m$ as column vectors so that $A\mathbf{x}$ and $A^T\mathbf{y}$ are also column vectors. Then

$$A\mathbf{x} \cdot \mathbf{y} = (A\mathbf{x})^T\mathbf{y} = (\mathbf{x}^T A^T)\mathbf{y} = \mathbf{x}^T(A^T\mathbf{y}) = \mathbf{x} \cdot A^T\mathbf{y}$$

where the first and last equalities follow by Theorem (4.16). ■

WARNING The notation of Theorem (4.16) works very nicely for column vectors, but this notation can be extremely misleading in other cases. For example, if $\mathbf{u} = [2 \quad 1 \quad -3 \quad -1]$ and $\mathbf{v} = [-1 \quad 0 \quad 2 \quad -2]$ are *row* vectors (in $\mathbb{R}^4$), then

$$\mathbf{u}^T\mathbf{v} = \begin{bmatrix} 2 \\ 1 \\ -3 \\ -1 \end{bmatrix} [-1 \quad 0 \quad 2 \quad -2] = \begin{bmatrix} -2 & 0 & 4 & -4 \\ -1 & 0 & 2 & -2 \\ 3 & 0 & -6 & 6 \\ 1 & 0 & -2 & 2 \end{bmatrix}$$

which is a 4×4 matrix and *not* $\mathbf{u} \cdot \mathbf{v}$. For row vectors $\mathbf{u} \cdot \mathbf{v} = \mathbf{u}\mathbf{v}^T$. For example, for $\mathbf{u}$ and $\mathbf{v}$ as above

$$\mathbf{u} \cdot \mathbf{v} = -2 + 0 + (-6) + 2 = -6$$

and

$$\mathbf{u}\mathbf{v}^T = [2 \quad 1 \quad -3 \quad -1] \begin{bmatrix} -1 \\ 0 \\ 2 \\ -2 \end{bmatrix} = [-6]$$ ■

Following is a theorem involving dot products that we shall need in the next section. However, as is often the case, we first prove a lemma.

(4.17) **LEMMA** A vector $\mathbf{x}$ in $\mathbb{R}^n$ is the zero vector if and only if $\mathbf{x} \cdot \mathbf{y} = 0$ for all vectors $\mathbf{y}$ in $\mathbb{R}^n$.

Proof If $\mathbf{x} = \mathbf{0} = (0, \ldots, 0)$ then easily $\mathbf{x} \cdot \mathbf{y} = 0y_1 + \cdots + 0y_n = 0$.

Suppose $\mathbf{x} \cdot \mathbf{y} = 0$ for all vectors $\mathbf{y}$ in $\mathbb{R}^n$. Then letting $\mathbf{y} = \mathbf{x}$, we have $\mathbf{x} \cdot \mathbf{x} = 0$, which implies $\mathbf{x} = \mathbf{0}$ by Theorem (4.5d). ■

Actually, Lemma (4.17) is true in any inner product space, but the proof for the general case is slightly more involved.

(4.18) **THEOREM** Let A be an $m \times n$ matrix. Then
(a) $NS(A^TA) = NS(A)$ (b) $rk(A^TA) = rkA$

Proof (a) Pick $\mathbf{u}$ in $NS(A)$. Then $A\mathbf{u} = \mathbf{0}$, so $A^TA\mathbf{u} = A^T\mathbf{0} = \mathbf{0}$. Hence $\mathbf{u}$ is in $NS(A^TA)$.

Pick $\mathbf{v}$ in $NS(A^TA)$. Then

$$\begin{aligned} A^TA\mathbf{v} &= \mathbf{0} \\ A^TA\mathbf{v} \cdot \mathbf{v} &= \mathbf{0} \cdot \mathbf{v} = 0 && \text{by Lemma (4.17)} \\ A\mathbf{v} \cdot A\mathbf{v} &= 0 && \text{by Theorem (4.15b)} \\ A\mathbf{v} &= \mathbf{0} && \text{by Theorem (4.5d)} \end{aligned}$$

$\mathbf{v}$ is in the $NS(A)$, by definition

(b) By Theorem (3.82), Section 3.7, we know

$$rkA + \dim NS(A) = n$$
$$rk(A^TA) + \dim NS(A^TA) = n$$

Since $\dim NS(A) = \dim \mathrm{NS}(A^T A)$ by (a), we easily get $rkA = rk(A^TA)$. ■

What will be most useful to us is the following corollary.

(4.19) **COROLLARY** If A is an $m \times n$ matrix, $m \geq n$, and $rkA = n$, then the $n \times n$ matrix A^TA is invertible.

Proof Now A^TA is $n \times n$. Since $rkA = n$, $rkA^TA = n$ by Theorem (4.18), so A^TA is invertible by Theorem (3.83), Section 3.7. ■

Exercise 4.2

In Exercises 1–8, let

$$\mathbf{u}_1 = \begin{bmatrix} 1 \\ -1 \\ 2 \end{bmatrix}, \qquad \mathbf{v}_1 = \begin{bmatrix} 2 \\ 1 \\ -1 \end{bmatrix}, \qquad \mathbf{w}_1 = \begin{bmatrix} 1 \\ 1 \\ 1 \end{bmatrix},$$

$$\mathbf{u}_2 = [2 \quad -1 \quad 2 \quad 1], \quad \mathbf{v}_2 = [1 \quad -1 \quad -1 \quad 1], \quad \mathbf{w}_2 = [2 \quad 2 \quad 2 \quad -1],$$

and let $r = -5$.

1. Compute for $i = 1, 2$, $\mathbf{u}_i \cdot (\mathbf{v}_i + \mathbf{w}_i)$ and $\mathbf{u}_i \cdot \mathbf{v}_i + \mathbf{u}_i \cdot \mathbf{w}_i$.
2. Compute for $i = 1, 2$, $\mathbf{u}_i \cdot \mathbf{v}_i$, $\mathbf{u}_i \cdot \mathbf{w}_i$ and $\mathbf{v}_i \cdot \mathbf{u}_i$, $\mathbf{v}_i \cdot \mathbf{w}_i$.
3. Compute for $i = 1, 2$, $(r\mathbf{u}_i) \cdot \mathbf{v}_i$, $r(\mathbf{u}_i \cdot \mathbf{v}_i)$, $\mathbf{u}_i \cdot (r\mathbf{v}_i)$.
4. Compute for $i = 1, 2$, $\|r\mathbf{u}_i\|$, $r\|\mathbf{u}_i\|$, $|r|\,\|\mathbf{u}_i\|$.
5. Compute for $i = 1, 2$, $\|\mathbf{u}_i + \mathbf{w}_i\|$, $\|\mathbf{u}_i\| + \|\mathbf{w}_i\|$ (to two significant figures).
6. Compute for $i = 1, 2$, $\big|\|\mathbf{u}_i\| - \|\mathbf{w}_i\|\big|$, $\|\mathbf{u}_i - \mathbf{w}_i\|$.
7. Compute for $i = 1, 2$, (a) $\mathbf{u}_i \cdot \mathbf{v}_i$, (b) $\mathbf{u}_i^T\mathbf{v}_i$, (c) $\mathbf{u}_i\mathbf{v}_i^T$. Which two are equal?
8. Compute for $i = 1, 2$, (a) $\mathbf{u}_i \cdot \mathbf{w}_i$, (b) $\mathbf{u}_i^T\mathbf{w}_i$, (c) $\mathbf{u}_i\mathbf{w}_i^T$. Which two are equal?

In Exercises 9–11, let

$$U = \begin{bmatrix} u_1 & u_2 \\ u_3 & u_4 \end{bmatrix}, \qquad V = \begin{bmatrix} v_1 & v_2 \\ v_3 & v_4 \end{bmatrix}$$

be 2×2 matrices in M_{22}. Decide if the given formula defines an inner product on M_{22} by satisfying (a)–(d) of Theorem (4.5). If not, what fails?

9. $U \cdot V = u_1v_1 + u_2v_2 + u_3v_3 + u_4v_4$

10. $U \cdot V = u_1v_1 + u_2v_3 + u_3v_2 + u_4v_4$

11. $U \cdot V = |u_1v_1| + |u_2v_2| + |u_3v_3| + |u_4v_4|$

For Exercises 12–16, same instructions as for Exercises 9–11 for $p(x) = a_0 + a_1x + a_2x^2$ and $q(x) = b_0 + b_1x + b_2x^2$ and P_2.

12. $p \cdot q = a_0b_0 + a_1b_1 + a_2b_2$

13. $p \cdot q = p(0)q(0) + p(1)q(1) + p(2)q(2)$

14. $p \cdot q = p(0)q(0)$

15. $p \cdot q = a_0b_2 + a_1b_1 + a_2b_0$

16. $p \cdot q = \int_{-1}^{1} p(x)q(x)\,dx$

In Exercises 17–20, compute the number $\mathbf{u} \cdot \mathbf{v}$ and the 1×1 matrix $\mathbf{u}^T\mathbf{v}$. Hereafter, we shall take them to be the same (even though technically they are different).

17. $\mathbf{u} = \begin{bmatrix} -3 \\ 4 \end{bmatrix}, \quad \mathbf{v} = \begin{bmatrix} 2 \\ -2 \end{bmatrix}$

18. $\mathbf{u} = \begin{bmatrix} -4 \\ -1 \end{bmatrix}, \quad \mathbf{v} = \begin{bmatrix} 2 \\ -3 \end{bmatrix}$

19. $\mathbf{u} = \begin{bmatrix} 2 \\ -3 \\ 4 \end{bmatrix}, \quad \mathbf{v} = \begin{bmatrix} -2 \\ -1 \\ 1 \end{bmatrix}$

20. $\mathbf{u} = \begin{bmatrix} 2 \\ -3 \\ 1 \\ -2 \end{bmatrix}, \quad \mathbf{v} = \begin{bmatrix} -1 \\ 0 \\ 5 \\ -2 \end{bmatrix}$

In Exercises 21–24, compute (a) $A\mathbf{x} \cdot \mathbf{y}$ and $\mathbf{x} \cdot A^T\mathbf{y}$ and (b) $\mathbf{x} \cdot B\mathbf{y}$ and $B^T\mathbf{x} \cdot \mathbf{y}$.

21. $A = \begin{bmatrix} -2 & 3 & 5 \\ 1 & -2 & 2 \end{bmatrix}, \quad B = \begin{bmatrix} 2 & -2 \\ 3 & 2 \\ 0 & -3 \end{bmatrix}, \quad \mathbf{x} = \begin{bmatrix} -1 \\ 3 \\ -2 \end{bmatrix}, \quad \mathbf{y} = \begin{bmatrix} -3 \\ -2 \end{bmatrix}$

22. $A = \begin{bmatrix} 3 & -4 \\ 5 & 5 \\ 2 & -1 \end{bmatrix}, \quad B = \begin{bmatrix} -4 & -2 & 1 \\ 3 & 3 & 0 \end{bmatrix}, \quad \mathbf{x} = \begin{bmatrix} -4 \\ 1 \end{bmatrix}, \quad \mathbf{y} = \begin{bmatrix} -3 \\ 2 \\ -2 \end{bmatrix}$

23. $A = \begin{bmatrix} 4 & 1 \\ -3 & -6 \\ 5 & -2 \\ -1 & 2 \end{bmatrix}, \quad B = \begin{bmatrix} 3 & -1 & 2 & 0 \\ -2 & -5 & 2 & 3 \end{bmatrix}, \quad \mathbf{x} = \begin{bmatrix} -2 \\ 3 \end{bmatrix}, \quad \mathbf{y} = \begin{bmatrix} -2 \\ -1 \\ 5 \\ -3 \end{bmatrix}$

24. $A = \begin{bmatrix} -1 & 1 & 3 & 5 \\ -1 & -4 & 1 & 2 \end{bmatrix}, \quad B = \begin{bmatrix} 5 & 1 \\ -2 & 2 \\ -3 & -3 \\ 1 & -2 \end{bmatrix}, \quad \mathbf{x} = \begin{bmatrix} -3 \\ 5 \\ -1 \\ -2 \end{bmatrix}, \quad \mathbf{y} = \begin{bmatrix} -1 \\ -2 \end{bmatrix}$

In Exercises 25–28, for the given matrix A, find (a) A^TA; then find (b) $NS(A)$, (c) $NS(A^TA)$, (d) $rk(A)$, and (e) $rk(A^TA)$.

25. $\begin{bmatrix} 3 & -6 \\ 5 & -10 \\ -2 & 4 \end{bmatrix}$

26. $\begin{bmatrix} -4 & 8 \\ -2 & 4 \\ 3 & -6 \\ 1 & -2 \end{bmatrix}$

27. $\begin{bmatrix} -3 & 6 & -3 \\ -1 & 2 & 8 \\ 2 & -4 & 0 \\ 1 & -2 & -1 \end{bmatrix}$

28. $\begin{bmatrix} -2 & 4 & 8 \\ 3 & -6 & -12 \end{bmatrix}$

29. Prove Theorem (4.5b): If $\mathbf{u}$, $\mathbf{v}$ are in $\mathbb{R}^n$, then $\mathbf{u} \cdot \mathbf{v} = \mathbf{v} \cdot \mathbf{u}$.

30. Prove Theorem (4.5c): If $\mathbf{u}$, $\mathbf{v}$ are in $\mathbb{R}^n$, r is a scalar, then $(r\mathbf{u}) \cdot \mathbf{v} = r(\mathbf{u} \cdot \mathbf{v})$.

In Exercises 31–33, f, g are in $C[a, b]$ and $f \cdot g = \int_a^b f(x)g(x)\,dx$.

31. Show that $f \cdot g$ satisfies Theorem (4.5b): $f \cdot g = g \cdot f$.

32. Show that $f \cdot g$ satisfies Theorem (4.5c): $(rf) \cdot g = r(f \cdot g)$.

33. Show that $f \cdot g$ satisfies Theorem (4.5d): $f \cdot f \geq 0$ and $f \cdot f = 0$ if and only if $f = 0$. [HINT If $f \neq 0$, there is a point c in $[a, b]$ where $f(c) = d \neq 0$. By continuity, there is an interval $[c - \delta, c + \delta]$ where $|f(x)| \geq \frac{|d|}{2}$, all x in the interval. Go from here, using the fact $\int_a^b f(x)^2\,dx \geq \int_{c-\delta}^{c+\delta} f(x)^2\,dx$.]

In Exercises 34–39, assume V is an inner product space [Definition (4.8)] and $\mathbf{u}$, $\mathbf{v}$, $\mathbf{w}$ are vectors in V, r is a scalar. Use properties (a)–(d) of Theorem (4.5).

34. Prove: $\mathbf{u} \cdot (\mathbf{v} + \mathbf{w}) = \mathbf{u} \cdot \mathbf{v} + \mathbf{u} \cdot \mathbf{w}$.

35. Prove: $\mathbf{u} \cdot (r\mathbf{v}) = r(\mathbf{u} \cdot \mathbf{v})$.

In Exercises 36 and 37, define $\|\mathbf{v}\| = \sqrt{\mathbf{v} \cdot \mathbf{v}}$.

36. Prove: $\|\mathbf{v}\| \geq 0$ and $\|\mathbf{v}\| = 0$ if and only if $\mathbf{v} = 0$.

37. Prove: $\|r\mathbf{u}\| = |r|\,\|\mathbf{u}\|$.

38. Suppose $\mathbf{u} = r\mathbf{v}$. Substitute this into both sides of Theorem (4.10) and verify that equality holds.

39. Suppose equality holds in Theorem (4.10). Work backward through the proof to show $(\mathbf{u} - t\mathbf{v}) \cdot (\mathbf{u} - t\mathbf{v}) = 0$. Explain why it now follows that $\mathbf{u}$ is a multiple of $\mathbf{v}$.

40. By choosing the correct vector $\mathbf{v}$ in the Cauchy–Schwarz inequality, prove $(a_1 + \cdots + a_n)^2 \leq n(a_1^2 + \cdots + a_n^2)$.

41. When does equality hold in Exercise 40?

Exercises 42 and 43 give an alternate proof of the Cauchy–Schwarz inequality in $\mathbb{R}^n$.

42. Assume $\|\mathbf{v}\| = 1$, $\|\mathbf{w}\| = 1$. Prove that

$$|\mathbf{v} \cdot \mathbf{w}| \leq \sum_{i=1}^{n} |\mathbf{v}_i|\,|\mathbf{w}_i| \leq \sum_{i=1}^{n} \frac{|\mathbf{v}_i|^2 + |\mathbf{w}_i|^2}{2} = \frac{1}{2} + \frac{1}{2} = \|\mathbf{v}\|\,\|\mathbf{w}\|$$

NOTE The middle step is the difficult one. First prove $|a|\,|b| \leq \dfrac{a^2 + b^2}{2}$, where a, b are any numbers.

43. Use Exercise 42 to prove the Cauchy–Schwarz inequality in $\mathbb{R}^n$.

44. Use various results, including Theorem (4.18), to prove: If $rk(A) = r$, then $rk(A^T) = r$, $rk(A^TA) = r$, $rk(AA^T) = r$.

45. Find a matrix A such that A^TA is invertible but AA^T is not.

46. Suppose A is square. Show that A is invertible $\Leftrightarrow$ A^TA is invertible.

4.3 Least Squares and Orthogonal Projections

In this section we discuss finding functions that our scientist in the introduction would use to "best" fit data. We first give the mathematical background and a simple example, and then we move on to two typical applications of the theory.

Mathematical Background and a Simple Example

Recall that a system of equations $A\mathbf{x} = \mathbf{y}$ is called **inconsistent** if it does not have a solution. Inconsistent systems arise quite often, especially when trying to fit equations to data taken experimentally and when there are more equations than unknowns.

Example 1 Determine the existence and the equation of a straight line in the plane that goes through the origin and also through the points $(2, y_1)$, $(4, y_2)$, and $(6, y_3)$ if

(a) $\mathbf{y} = \begin{bmatrix} y_1 \\ y_2 \\ y_3 \end{bmatrix} = \begin{bmatrix} 4 \\ 8 \\ 12 \end{bmatrix}$ (b) $\mathbf{y} = \begin{bmatrix} y_1 \\ y_2 \\ y_3 \end{bmatrix} = \begin{bmatrix} 4 \\ 5 \\ 21 \end{bmatrix}$

Solution The equation for a (nonvertical) line through the origin is $y = ax$. In each of the cases, we are given three points and we need to determine if all three points can satisfy a single equation of the form $y = ax$.

(a) We need to solve the equations

$$\begin{aligned} 2a &= 4 \\ 4a &= 8 \\ 6a &= 12 \end{aligned} \qquad \text{or} \qquad \begin{bmatrix} 2 \\ 4 \\ 6 \end{bmatrix} a = \begin{bmatrix} 4 \\ 8 \\ 12 \end{bmatrix}$$

We see $a = 2$ works, so there is an equation that fits all three points, namely, $y = 2x$.

(b) We need to solve the equations

$$\begin{aligned} 2a &= 4 \\ 4a &= 5 \\ 6a &= 21 \end{aligned} \qquad \text{or} \qquad \begin{bmatrix} 2 \\ 4 \\ 6 \end{bmatrix} a = \begin{bmatrix} 4 \\ 5 \\ 21 \end{bmatrix} \tag{4.20}$$

We see there is no single number a that works for all the equations. The system is inconsistent, so there is no single equation $y = ax$ that fits all three points. ■

By examining Example 1, we see that there is an equation $y = ax$ if and only if the vector

$$\mathbf{y} = \begin{bmatrix} y_1 \\ y_2 \\ y_3 \end{bmatrix}$$

is a multiple of

$$\begin{bmatrix} 2 \\ 4 \\ 6 \end{bmatrix}$$

In general $\mathbf{y}$ is a multiple of A if and only if $\mathbf{y}$ is in the column space of A. More precisely, if A is an $m \times n$ matrix with columns $\mathbf{c}_1, \ldots, \mathbf{c}_n$, then

$$\begin{aligned} & A\mathbf{x} = \mathbf{y} \quad \text{has a solution} \\ \Leftrightarrow \quad & [\mathbf{c}_1 \vdots \cdots \vdots \mathbf{c}_n] \begin{bmatrix} x_1 \\ \vdots \\ x_n \end{bmatrix} = \mathbf{y} \quad \text{has a solution} \\ \Leftrightarrow \quad & x_1\mathbf{c}_1 + \cdots + x_n\mathbf{c}_n = \mathbf{y} \quad \text{has a solution} \\ \Leftrightarrow \quad & \mathbf{y} \text{ is a linear combination of the columns of } A \\ \Leftrightarrow \quad & \mathbf{y} \text{ is in the column space of } A \end{aligned}$$

What we have shown is

(4.21)

THEOREM If A is an $m \times n$ matrix, the equation $A\mathbf{x} = \mathbf{y}$ has a solution if and only if $\mathbf{y}$ is in the column space of A.

Thus the system $A\mathbf{x} = \mathbf{y}$ is inconsistent exactly when $\mathbf{y}$ is not in $CS(A)$. When this happens, that is, when $A\mathbf{x} = \mathbf{y}$ has no exact solution $\mathbf{x}$ and hence for any $\mathbf{x}$, $A\mathbf{x}$ differs from $\mathbf{y}$ (we say $\mathbf{x}$ produces errors), we change the rules, stop looking for an exact answer, and change the problem!

(4.22)

The problem is to find an $\mathbf{x}$ that minimizes the errors in some sense.

There are different ways of minimizing errors for different purposes. One of the most common ways, and the way we shall be using, is to minimize the sum of the squares of the errors in each individual coordinate. This is illustrated in the next example.

Example 2 Find the sums of the squares of the errors in Example 1b.

Solution We use the lower-case Greek letter epsilon ε to represent errors $y - ax = \varepsilon$:

$$\begin{aligned} 4 - 2a &= \varepsilon_1 \\ 5 - 4a &= \varepsilon_2 \\ 21 - 6a &= \varepsilon_3 \end{aligned} \tag{4.23}$$

See Figure 4.6. The sum of the squares is then

$$\varepsilon_1^2 + \varepsilon_2^2 + \varepsilon_3^2 = (4 - 2a)^2 + (5 - 4a)^2 + (21 - 6a)^2$$

Figure 4.6

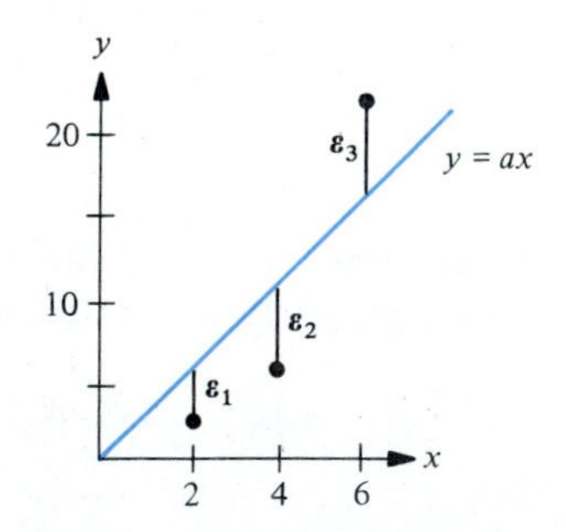

For this example, then:

(4.24)

The **least-squares problem** for Example 2 is to find the number a for which $\varepsilon_1^2 + \varepsilon_2^2 + \varepsilon_3^2$ is a minimum.

Example 3 Find the sum of squares of the errors in Example 2 for $a =$

(a) 2 (b) $\frac{5}{4}$ (c) $\frac{7}{2}$ (d) $\frac{11}{4}$

Solution (a) When $a = 2$, $\varepsilon_1 = 0$, and

$$\varepsilon_1^2 + \varepsilon_2^2 + \varepsilon_3^2 = 0 + [5 - 4(2)]^2 + [21 - 6(2)]^2 = 3^2 + 9^2 = 90$$

(b) When $a = \frac{5}{4}$, $\varepsilon_2 = 0$, and

$$\varepsilon_1^2 + \varepsilon_2^2 + \varepsilon_3^2 = [4 - 2(\tfrac{5}{4})]^2 + 0^2 + [21 - 6(\tfrac{5}{4})]^2 = \tfrac{9}{4} + \tfrac{729}{4} = 184.5$$

(c) When $a = \frac{7}{2}$, $\varepsilon_3 = 0$, and

$$\varepsilon_1^2 + \varepsilon_3^2 + \varepsilon_3^2 = [4 - 2(\tfrac{7}{2})]^2 + [5 - 4(\tfrac{7}{2})]^2 + 0^2 = 9 + 81 = 90$$

(d) When $a = \frac{11}{4}$,

$$\begin{aligned}\varepsilon_1^2 + \varepsilon_2^2 + \varepsilon_3^2 &= [4 - 2(\tfrac{11}{4})]^2 + [5 - 4(\tfrac{11}{4})]^2 + [21 - 6(\tfrac{11}{4})]^2 \\ &= \tfrac{9}{4} + 36 + \tfrac{81}{4} = \tfrac{117}{2} = 58.5\end{aligned}$$

■

We shall see in Example 4 that for all possible values of a the sum of the squares $\varepsilon_1^2 + \varepsilon_2^2 + \varepsilon_3^2$ is smallest when $a = \frac{11}{4}$. To set the stage for this, we reinterpret the problem geometrically. Let

$$A = \begin{bmatrix} 2 \\ 4 \\ 6 \end{bmatrix}, \qquad \mathbf{y} = \begin{bmatrix} 4 \\ 5 \\ 21 \end{bmatrix}, \qquad \boldsymbol{\varepsilon} = \begin{bmatrix} \varepsilon_1 \\ \varepsilon_2 \\ \varepsilon_3 \end{bmatrix} \tag{4.25}$$

We rewrite Equation (4.23) as

$$\mathbf{y} - Aa = \boldsymbol{\varepsilon} \tag{4.26}$$

We interpret this as follows:

A is both a 3×1 column vector in $\mathbb{R}^3$ as well as a 3×1 matrix.

Aa is in the subspace of $\mathbb{R}^3$ spanned by (the columns of) A (in this case, a line).

$\mathbf{y}$ is a vector, probably not in the subspace spanned by (the columns of) A (otherwise $Aa = \mathbf{y}$ has an exact solution).

$\boldsymbol{\varepsilon}$ is the error vector, the difference between $\mathbf{y}$ and Aa.

This solution is shown in Figure 4.7. Observe that $\varepsilon_1^2 + \varepsilon_2^2 + \varepsilon_3^2 = \|\boldsymbol{\varepsilon}\|^2$. Thus what we are trying to minimize is the length of the error vector. This gives us the geometric interpretation (4.24).

Figure 4.7

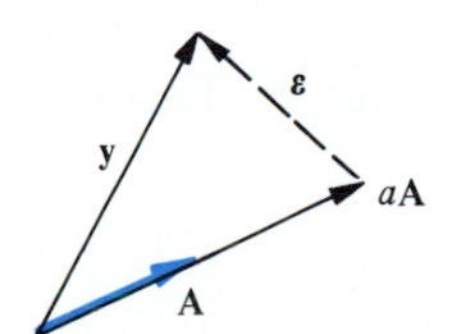

(4.27)

> The least-squares problem for Example 2 is to find the number a for which $\|\boldsymbol{\varepsilon}\|$ is a minimum.

This has a nice geometric solution.

(4.28)

> THEOREM The length $\|\boldsymbol{\varepsilon}\|$ is a minimum exactly when $\boldsymbol{\varepsilon} \perp A$, if we interpret A as being an $m \times 1$ column vector.

Proof Let $\bar{a}$ be the number such that $\bar{\boldsymbol{\varepsilon}} = \mathbf{y} - A\bar{a}$ is perpendicular to A. Let a be any other number with corresponding $\boldsymbol{\varepsilon} = \mathbf{y} - Aa$. See Figure 4.8. Then we need to show $\|\boldsymbol{\varepsilon}\| \geq \|\bar{\boldsymbol{\varepsilon}}\|$.

$$\boldsymbol{\varepsilon} = \mathbf{y} - Aa = \mathbf{y} - A\bar{a} + A\bar{a} - Aa = \bar{\boldsymbol{\varepsilon}} + A(\bar{a} - a)$$

Figure 4.8

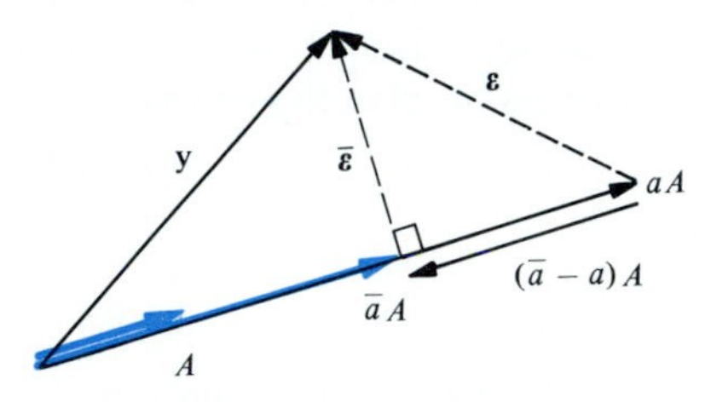

By the Pythagorean theorem, since $\bar{\boldsymbol{\varepsilon}} \perp A(\bar{a} - a)$,

$$\|\boldsymbol{\varepsilon}\|^2 = \|\bar{\boldsymbol{\varepsilon}}\|^2 + \|A(\bar{a} - a)\|^2 \geq \|\bar{\boldsymbol{\varepsilon}}\|^2 \quad \blacksquare$$

Therefore, to find the $\bar{a}$ that minimizes $\|\boldsymbol{\varepsilon}\|$, we need to solve

$$A \cdot (\mathbf{y} - A\bar{a}) = 0$$

This yields

$$A \cdot \mathbf{y} - (A \cdot A)\bar{a} = 0$$

or

(4.29)
$$\bar{a} = \frac{A \cdot \mathbf{y}}{A \cdot A}$$

However, by Theorem (3.18), Section 3.2, the vector

$$A\bar{a} = A\,\frac{\mathbf{y} \cdot A}{A \cdot A}$$

is exactly the projection of $\mathbf{y}$ onto A, that is, onto the column space of A. Thus

(4.30)

THEOREM For an $m \times 1$ matrix A, the solution $\bar{a}$ to the least-squares problem $Ax = \mathbf{y}$ is

$$\bar{a} = \frac{A \cdot \mathbf{y}}{A \cdot A}$$

so that $A\bar{a}$ is the projection of $\mathbf{y}$ onto the column space of A.

Example 4 For the A and $\mathbf{y}$ of Equations (4.25),

$$A = \begin{bmatrix} 2 \\ 4 \\ 6 \end{bmatrix}, \qquad \mathbf{y} = \begin{bmatrix} 4 \\ 5 \\ 21 \end{bmatrix}$$

and

$$\bar{a} = \frac{A \cdot \mathbf{y}}{A \cdot A} = \frac{8 + 20 + 126}{4 + 16 + 36} = \frac{154}{56} = \frac{11}{4}$$

This is why $a = \frac{11}{4}$ gives the smallest answer in Example 3. ■

We have now laid the framework for more general least-squares problems.

The General Problem and Its Solution

In general, when we have an inconsistent system $A\mathbf{x} = \mathbf{y}$, where A is an $m \times n$ matrix, we let the error vector

$$\boldsymbol{\varepsilon} = \mathbf{y} - A\mathbf{x} \tag{4.31}$$

which in matrix form is

$$\begin{bmatrix} \varepsilon_1 \\ \vdots \\ \varepsilon_m \end{bmatrix} = \begin{bmatrix} y_1 \\ \vdots \\ y_m \end{bmatrix} - \begin{bmatrix} a_{11} & \cdots & a_{1n} \\ \vdots & & \\ a_{m1} & \cdots & a_{mn} \end{bmatrix} \begin{bmatrix} x_1 \\ \vdots \\ x_n \end{bmatrix} \tag{4.32}$$

Now we can generalize (4.27).

(4.33)

The **least-squares problem** for the inconsistent system $A\mathbf{x} = \mathbf{y}$ is to find a vector $\bar{\mathbf{x}}$ for which $\|\boldsymbol{\varepsilon}\| = \|y - A\bar{x}\|$ is a minimum.

This has a very nice geometric solution, which generalizes Theorem (4.28).

(4.34)

THEOREM The length $\|\boldsymbol{\varepsilon}\|$ is a minimum exactly when $\boldsymbol{\varepsilon} \perp CS(A)$.*

* *Here, by a vector being perpendicular to a subspace, we mean that the given vector is perpendicular to every vector in the subspace.*

Proof Not unlike the proof of Theorem (4.28), we let $\bar{\mathbf{x}}$ be an $n \times 1$ vector such that $\bar{\boldsymbol{\varepsilon}} = \mathbf{y} - A\bar{\mathbf{x}}$ is perpendicular to $CS(A)$. Let $\mathbf{x}$ be any other $n \times 1$ vector with corresponding $\boldsymbol{\varepsilon} = \mathbf{y} - A\mathbf{x}$. See Figure 4.9. To show $\|\boldsymbol{\varepsilon}\| \geq \|\bar{\boldsymbol{\varepsilon}}\|$, we write

$$\boldsymbol{\varepsilon} = \mathbf{y} - A\mathbf{x} = \mathbf{y} - A\bar{\mathbf{x}} + A\bar{\mathbf{x}} - A\mathbf{x} = \bar{\boldsymbol{\varepsilon}} - A(\bar{\mathbf{x}} - \mathbf{x})$$

Figure 4.9

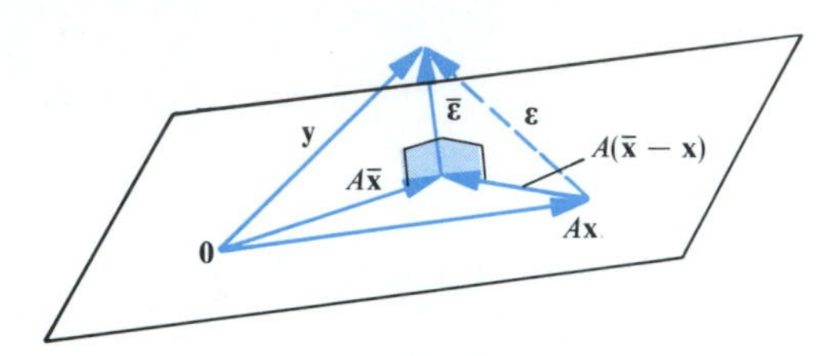

By the Pythagorean theorem, since $\bar{\boldsymbol{\varepsilon}} \perp A(\bar{\mathbf{x}} - \mathbf{x})$,

$$\|\boldsymbol{\varepsilon}\|^2 = \|\bar{\boldsymbol{\varepsilon}}\|^2 + \|A(\bar{\mathbf{x}} - \mathbf{x})\|^2 \geq \|\bar{\boldsymbol{\varepsilon}}\|$$ ■

Theorem (4.34) tells us the solution to the least-squares problem for an inconsistent system $A\mathbf{x} = \mathbf{y}$ is to find a vector $\bar{\mathbf{x}}$ for which $(\mathbf{y} - A\bar{\mathbf{x}}) \perp CS(A)$. By Theorem (4.30) we know how to find $\bar{\mathbf{x}}$ if A is an $m \times 1$ matrix, namely, $\bar{\mathbf{x}} = \dfrac{A \cdot \mathbf{y}}{A \cdot A}$. To generalize this, we recall from Theorem (4.16), Section 4.2, that for an $m \times 1$ matrix A, we can rewrite the answer in the form

$$\bar{\mathbf{x}} = \frac{A \cdot \mathbf{y}}{A \cdot A} = \frac{A^T\mathbf{y}}{A^TA} = (A^TA)^{-1}A^T\mathbf{y}$$

This is the form that gives us the general answers.

(4.35) **THEOREM** Suppose A is an $m \times n$ matrix and $\mathbf{y}$ is in $\mathbb{R}^m$. At least-squares solution to the system $A\mathbf{x} = \mathbf{y}$ is an exact solution to the equation

(4.36)
$$A^TA\mathbf{x} = A^T\mathbf{y}$$

This one matrix equation represents n equations in n unknowns, often called the **normal equations**.

If the columns of A are linearly independent, then A^TA is invertible, so (4.36) has a unique solution $\bar{\mathbf{x}}$. This solution is often expressed *theoretically* as

(4.37)
$$\bar{\mathbf{x}} = (A^TA)^{-1}A^T\mathbf{y}$$

NOTE If the columns of A are not linearly independent, then Equation (4.36) has infinitely many solutions. One approach to this situation involves the *singular value decomposition* discussed in Section 6.2.

Proof Note that an arbitrary vector $\mathbf{z}$ in the $CS(A)$ is represented by $\mathbf{z} = A\mathbf{w}$, for some $n \times 1$ column vector $\mathbf{w}$. Thus, by Theorem (4.34),

The vector $\bar{\mathbf{x}}$ is a least-squares solution to $A\mathbf{x} = \mathbf{y}$

$$
\begin{array}{lll}
\Leftrightarrow & (\mathbf{y} - A\bar{\mathbf{x}}) \perp CS(A) & \\
\Leftrightarrow & (\mathbf{y} - A\bar{\mathbf{x}}) \cdot \mathbf{z} = 0, & \text{all } \mathbf{z} \text{ in } CS(A) \\
\Leftrightarrow & (\mathbf{y} - A\bar{\mathbf{x}}) \cdot A\mathbf{w} = 0, & \text{all } \mathbf{w} \text{ in } \mathbb{R}^n \\
\Leftrightarrow & A^T(\mathbf{y} - A\bar{\mathbf{x}}) \cdot \mathbf{w} = 0, & \text{all } \mathbf{w} \text{ in } \mathbb{R}^n\text{, by Theorem (4.15), Section 4.2} \\
\Leftrightarrow & A^T(\mathbf{y} - A\bar{\mathbf{x}}) = \mathbf{0}, & \text{by Lemma (4.17), Section 4.2} \\
\Leftrightarrow & A^T\mathbf{y} - A^TA\bar{\mathbf{x}} = \mathbf{0} & \\
\Leftrightarrow & A^TA\bar{\mathbf{x}} = A^T\mathbf{y} &
\end{array}
$$

We now have the first part of the theorem. For the second part, suppose the columns of A are linearly independent. Then $rkA = n$ and $m \geq n$ so A^TA is invertible by Corollary (4.19), Section 4.2. Multiplying both sides of Equation (4.36) by $(A^TA)^{-1}$ gives us Equation (4.37). ■

Calculating $\bar{\mathbf{x}}$

We should observe that calculating $\bar{\mathbf{x}}$ requires solving the equation $(A^TA)\bar{\mathbf{x}} = A^T\mathbf{y}$. However, for the reasons given in Chapter 1, we almost never compute $(A^TA)^{-1}$, so it seems that the preferred method here would be to use Gaussian elimination. Unfortunately a real computational tragedy occurs here:

For least-squares problems, A^TA is almost always badly ill conditioned.

Thus, as discussed in Section 1.7, if m is at all large ($m \geq 4$), Gaussian elimination tends to give very inaccurate answers when performed on a computer. We shall develop in the next two sections one of the methods for dealing with this.

Perpendicular Projections

(4.38)

DEFINITION Suppose W is a subspace of $\mathbb{R}^m$ and $\mathbf{y}$ is an arbitrary vector in $\mathbb{R}^m$. Then the **perpendicular projection** of $\mathbf{y}$ into W is the unique vector $\mathbf{p}$ in W such that $\mathbf{y} - \mathbf{p}$ is perpendicular to W.

The function P, which assigns to every vector $\mathbf{y}$ in $\mathbb{R}^m$ its perpendicular projection $\mathbf{p}$, is called the (**perpendicular**) **projection** (**function**)*

* *It is common practice also to denote by P the matrix corresponding to the function P[see Theorem (4.4), Section 4.1] and identify the two. We do this in Equation (4.41).*

When looked at properly, it is very easy to obtain perpendicular projections from the way we solved least-squares problems. Let $c_1, \ldots, c_n$ be a basis for W and let them form the columns of a matrix A:

$$A = [c_1 \vdots \cdots \vdots c_n]$$

Then $W = CS(A)$, and we can think of $\mathbf{p}$ as $A\mathbf{x}$ for some $\mathbf{x}$. Furthermore

$$(\mathbf{y} - \mathbf{p}) \perp W \Leftrightarrow (\mathbf{y} - A\mathbf{x}) \perp CS(A)$$

so that finding $\mathbf{p}$ is equivalent to solving the least-squares problem, by (4.33) and Theorem (4.34). Theorem (4.35) solves this problem, so we have shown the following.

(4.39) THEOREM Let W be a subspace of $\mathbb{R}^m$ and A be an $m \times n$ matrix whose columns form a basis for W. (Thus the columns of A are linearly independent and $W = CS(A)$.) If $\mathbf{y}$ is any vector in $\mathbb{R}^m$, then its perpendicular projection into W is given by

(4.40)
$$\mathbf{p} = A\bar{\mathbf{x}} = A[(A^TA)^{-1}A^T\mathbf{y}]$$

Thus the perpendicular projection function, P, which projects $\mathbb{R}^m$ onto W, is given by

(4.41)
$$P = A(A^TA)^{-1}A^T$$

NOTE In general, the projection P is a theoretical tool, not a computational tool. For example, when a projection vector $\mathbf{p}$ is needed, it is found most often by first computing $\bar{\mathbf{x}}$ and then computing $A\bar{\mathbf{x}}$.

Example 5 Let W be the subspace of $\mathbb{R}^4$ spanned by $\mathbf{c}_1 = (1, 2, 0, -2)$ and $\mathbf{c}_2 = (3, 1, 1, 1)$. Find the perpendicular projection of $(1, 1, 1, 1)$ into W and the perpendicular projection function P.

Solution First notice that since $\mathbf{c}_1$ and $\mathbf{c}_2$ are linearly independent, they form a basis for W. Next form the 4×2 matrix $A = [\mathbf{c}_1 \vdots \mathbf{c}_2]$ and compute

$$A^TA = \begin{bmatrix} 1 & 2 & 0 & -2 \\ 3 & 1 & 1 & 1 \end{bmatrix} \begin{bmatrix} 1 & 3 \\ 2 & 1 \\ 0 & 1 \\ -2 & 1 \end{bmatrix} = \begin{bmatrix} 9 & 3 \\ 3 & 12 \end{bmatrix}$$

$$(A^TA)^{-1} = \frac{1}{99}\begin{bmatrix} 12 & -3 \\ -3 & 9 \end{bmatrix} = \frac{1}{33}\begin{bmatrix} 4 & -1 \\ -1 & 3 \end{bmatrix}$$

By Theorem (4.39),

$$\mathbf{p} = A(A^TA)^{-1}A^T\mathbf{y} = A(A^TA)^{-1}\begin{bmatrix} 1 & 2 & 0 & -2 \\ 3 & 1 & 1 & 1 \end{bmatrix}\begin{bmatrix} 1 \\ 1 \\ 1 \\ 1 \end{bmatrix}$$

$$= A\frac{1}{33}\begin{bmatrix} 4 & -1 \\ -1 & 3 \end{bmatrix}\begin{bmatrix} 1 \\ 6 \end{bmatrix}$$

$$= \frac{1}{33}\begin{bmatrix} 1 & 3 \\ 2 & 1 \\ 0 & 1 \\ -2 & 1 \end{bmatrix}\begin{bmatrix} -2 \\ 17 \end{bmatrix} = \frac{1}{33}\begin{bmatrix} 49 \\ 13 \\ 17 \\ 21 \end{bmatrix}$$

Also,

$$P = A(A^TA)^{-1}A^T$$

$$= \frac{1}{33}A\begin{bmatrix} 4 & -1 \\ -1 & 3 \end{bmatrix}\begin{bmatrix} 1 & 2 & 0 & -2 \\ 3 & 1 & 1 & 1 \end{bmatrix}$$

$$= \frac{1}{33}\begin{bmatrix} 1 & 3 \\ 2 & 1 \\ 0 & 1 \\ -2 & 1 \end{bmatrix}\begin{bmatrix} 1 & 7 & -1 & -9 \\ 8 & 1 & 3 & 5 \end{bmatrix}$$

$$= \frac{1}{33}\begin{bmatrix} 25 & 10 & 8 & 6 \\ 10 & 15 & 1 & -13 \\ 8 & 1 & 3 & 5 \\ 6 & -13 & 5 & 23 \end{bmatrix}$$

You can check that $\mathbf{p} = P\mathbf{y}$. ■

Two Typical Applications

The first application comes from the problem discussed in the chapter introduction.

Example 6 Find the equation for the line in the plane that best fits, in the least-squares sense, the data given in Table 4.1 and graphed in Figure 4.10.

Table 4.1

i	x_i	y_i
1	40	482
2	45	467
3	50	452
4	55	433
5	60	421

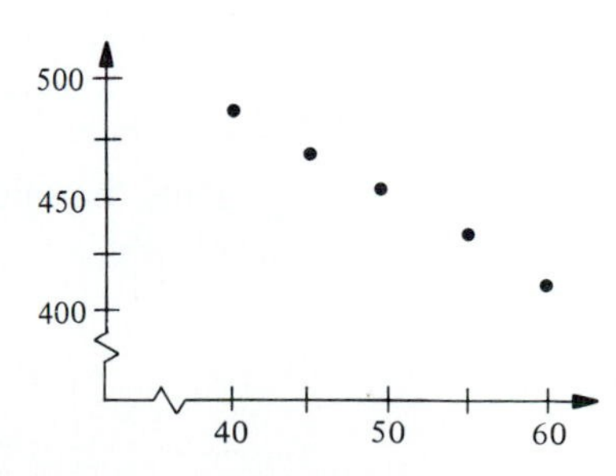

Figure 4.10

Solution The equation for a (nonvertical) straight line is $y = mx + b$. Thus the problem is to find the m and b that best fit the data. We substitute the values from Table 4.1 into the equation $mx + b = y$ and get

$$\begin{aligned} 40m + b &= 482 \\ 45m + b &= 467 \\ 50m + b &= 452 \\ 55m + b &= 433 \\ 60m + b &= 421 \end{aligned} \tag{4.42}$$

These five equations are inconsistent, so to solve the problem, we apply our least-squares methods and we rewrite (4.42) in matrix form:

$$\begin{bmatrix} 40 & 1 \\ 45 & 1 \\ 50 & 1 \\ 55 & 1 \\ 60 & 1 \end{bmatrix} \begin{bmatrix} m \\ b \end{bmatrix} = \begin{bmatrix} 482 \\ 467 \\ 452 \\ 433 \\ 421 \end{bmatrix} \quad \text{or} \quad A\mathbf{x} = \mathbf{y} \tag{4.43}$$

We can recognize this as a least-squares problem to find

$$\bar{\mathbf{x}} = \begin{bmatrix} \bar{m} \\ \bar{b} \end{bmatrix}$$

Using Theorem (4.35), we first form the equation $A^TA\mathbf{x} = A^T\mathbf{y}$

$$\begin{bmatrix} 40 & \cdots & 60 \\ 1 & \cdots & 1 \end{bmatrix} \begin{bmatrix} 40 & 1 \\ \vdots & \vdots \\ 60 & 1 \end{bmatrix} \begin{bmatrix} m \\ b \end{bmatrix} = \begin{bmatrix} 40 & \cdots & 60 \\ 1 & \cdots & 1 \end{bmatrix} \begin{bmatrix} 482 \\ \vdots \\ 421 \end{bmatrix}$$

or

$$\begin{bmatrix} 40^2 + \cdots + 60^2 & 40 + \cdots + 60 \\ 40 + \cdots + 60 & 1 + \cdots + 1 \end{bmatrix} \begin{bmatrix} m \\ b \end{bmatrix} = \begin{bmatrix} 40(482) + \cdots + 60(421) \\ 482 + \cdots + 421 \end{bmatrix}$$

or

$$\begin{bmatrix} 12{,}750 & 250 \\ 250 & 5 \end{bmatrix} \begin{bmatrix} m \\ b \end{bmatrix} = \begin{bmatrix} 111{,}970 \\ 2255 \end{bmatrix}$$

Since the columns of A are linearly independent, A^TA is invertible and

$$(A^TA)^{-1} = \begin{bmatrix} 12{,}750 & 250 \\ 250 & 5 \end{bmatrix}^{-1} = \frac{1}{1250} \begin{bmatrix} 5 & -250 \\ -250 & 12{,}750 \end{bmatrix}$$

Thus the solution is $\bar{\mathbf{x}} = (A^TA)^{-1}A^T\mathbf{y}$ or

$$\begin{aligned} \bar{\mathbf{x}} = \begin{bmatrix} \bar{m} \\ \bar{b} \end{bmatrix} &= \frac{1}{1250} \begin{bmatrix} 5 & -250 \\ -250 & 12750 \end{bmatrix} \begin{bmatrix} 111{,}970 \\ 2255 \end{bmatrix} \\ &= \frac{1}{1250} \begin{bmatrix} -3900 \\ 758750 \end{bmatrix} = \begin{bmatrix} -3.12 \\ 607 \end{bmatrix} \end{aligned}$$

Therefore, the equation of the line that best fits the given data in the least-squares sense is

$$y = -3.12x + 607$$

■

Example 7 Airplanes fly over the land taking infrared pictures for various reasons, among which are early disease detection in crops, mineral searches, and military reconnaissance. As the plane flies, an infrared detector swings back and forth and an image is recorded electronically.

Figure 4.11

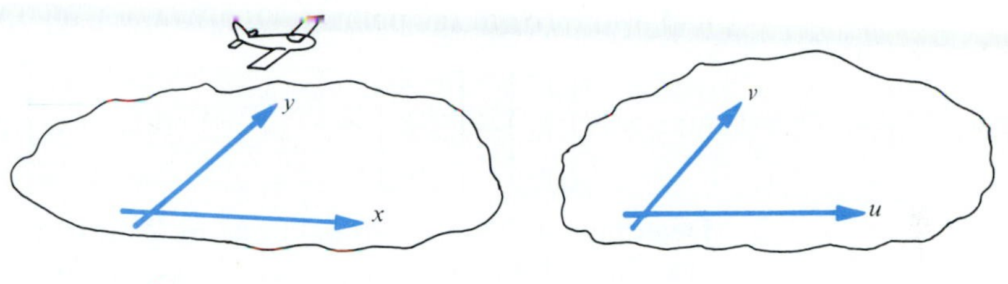

(a) *Actual coordinates* (b) *Electronic image coordinates*

However, various influences distort the electronic image, so to produce an accurate map, the electronic image coordinates need to be corrected to obtain the actual coordinates (see Figure 4.11). Under certain assumptions the relationship is

$$y = a + bu + cv + duv$$
$$x = e + fu + gv + huv$$

Suppose the coordinates of several prominent features are known on both grids, that is, suppose we know $(u_1, v_1), \ldots, (u_m, v_m)$ correspond to $(x_1, y_1), \ldots, (x_m, y_m)$ respectively. Find the equation whose solution gives us the coefficients $a, \ldots, h$ that best fit, in the sense of "least squares," the data.

Solution The two equations are similar, so we deal with only the first. We substitute the data into that equation and then translate to matrix form:

$$\begin{aligned} a + bu_1 + cv_1 + du_1v_1 &= y_1 \\ a + bu_2 + cv_2 + du_2v_2 &= y_2 \\ &\vdots \\ a + bu_m + cv_m + du_mv_m &= y_m \end{aligned}$$

or

$$\begin{bmatrix} 1 & u_1 & v_1 & u_1v_1 \\ 1 & u_2 & v_2 & u_2v_2 \\ \vdots & \vdots & \vdots & \vdots \\ 1 & u_m & v_m & u_mv_m \end{bmatrix} \begin{bmatrix} a \\ b \\ c \\ d \end{bmatrix} = \begin{bmatrix} y_1 \\ y_2 \\ \vdots \\ y_m \end{bmatrix} \tag{4.44}$$

This is an equation of the form $A\mathbf{x} = \mathbf{y}$. We can recognize this as a least-squares problem to find $\bar{\mathbf{x}} = [\bar{a} \;\; \bar{b} \;\; \bar{c} \;\; \bar{d}]^T$. By Theorem (4.35), $\bar{\mathbf{x}}$ is the solution to $A^TA\bar{\mathbf{x}} = A^T\mathbf{y}$, where A and $\mathbf{y}$ are given in Equation (4.44).

Exercise 27 gives some typical data for such a problem.

Exercise 4.3

1. Determine if there is a straight line in the plane that goes through the origin and also through the points $(-1, y_1)$, $(4, y_2)$, $(6, y_3)$ if

   (a) $\mathbf{y} = \begin{bmatrix} y_1 \\ y_2 \\ y_3 \end{bmatrix} = \begin{bmatrix} -2 \\ 8 \\ 10 \end{bmatrix}$ (b) $\mathbf{y} = \begin{bmatrix} y_1 \\ y_2 \\ y_3 \end{bmatrix} = \begin{bmatrix} 3 \\ -12 \\ -18 \end{bmatrix}$

2. Determine if there is a straight line in $\mathbb{R}^3$ that goes through the origin and also through the points $(1, y_1, z_1)$, $(2, y_2, z_2)$, $(4, y_3, z_3)$ if

   (a) $\mathbf{y} = \begin{bmatrix} 2 \\ 4 \\ 8 \end{bmatrix}$, $\mathbf{z} = \begin{bmatrix} 3 \\ 4 \\ 5 \end{bmatrix}$ (b) $\mathbf{y} = \begin{bmatrix} 2 \\ 4 \\ 6 \end{bmatrix}$, $\mathbf{z} = \begin{bmatrix} -1 \\ -2 \\ -3 \end{bmatrix}$

   (c) $\mathbf{y} = \begin{bmatrix} -2 \\ -4 \\ -8 \end{bmatrix}$, $\mathbf{z} = \begin{bmatrix} \frac{1}{2} \\ 1 \\ 2 \end{bmatrix}$

In Exercises 3–6, proceed as in Examples 2 and 3.

3. Find the sum of the squares of the errors for

$$\begin{aligned} a - 1 &= \varepsilon_1 \\ 2a - 1 &= \varepsilon_2 \\ 4a - 1 &= \varepsilon_3 \end{aligned}$$

   for $a = 1, \frac{1}{2}, \frac{1}{3}, \frac{1}{4}$. Which a gives the smallest sum?

4. Find the sum of the squares of the errors for

$$\begin{aligned} -a + 1 &= \varepsilon_1 \\ a + 0 &= \varepsilon_2 \\ 2a - 4 &= \varepsilon_3 \end{aligned}$$

   for $a = 0, 1, \frac{3}{2}, 2$. Which a gives the smallest sum?

5. Find the sum of the squares of the errors for

$$\begin{aligned} -m + b - 1 &= \varepsilon_1 \\ 0m + b - 2 &= \varepsilon_2 \\ m + b - 5 &= \varepsilon_3 \end{aligned}$$

   for $(m, b) = (2, 3), (3, 2), (1, 2), (2, \frac{8}{3})$.

6. Find the sum of the squares of the errors for

$$\begin{aligned} 0m + b - 2 &= \varepsilon_1 \\ m + b - 2 &= \varepsilon_2 \\ 2m + b - 0 &= \varepsilon_3 \end{aligned}$$

for $(m, b) = (-1, 2), (0, 2), (-2, 4), (-1, \frac{10}{3})$.

In Exercises 7–16, find the equation of the line that best fits the given points in the least-squares sense. In Exercises 7–10, the line must go through the origin.

7. $(1, 1), (2, 1), (4, 1)$. (Compare with Exercise 3.)
8. $(-1, -1), (1, 0), (2, 4)$. (Compare with Exercise 4.)
9. $(-2, -2), (-1, 0), (0, -2), (1, 0)$
10. $(-1, 1), (0, 1), (1, 0), (2, -1)$
11. Same points as Exercise 7.
12. Same points as Exercise 10.
13. $(-1, 1), (0, 2), (1, 5)$. (Compare with Exercise 5.)
14. $(0, 2), (1, 2), (2, 0)$. (Compare with Exercise 6.)
15. $(-2, 0), (-1, 0), (0, 1), (1, 3), (2, 5)$
16. $(1, 5), (2, 4), (3, 1), (4, 1), (5, -1)$

In Exercises 17–22, find the perpendicular projection $\mathbf{p}$ of the given vector $\mathbf{v}$ into the subspace spanned by the given vectors $\mathbf{c}_1, \mathbf{c}_2, \ldots, \mathbf{c}_k$. Also find the perpendicular projection function P.

17. $\mathbf{v} = (1, -2, -3)$, $\mathbf{c}_1 = (1, 1, 1)$
18. $\mathbf{v} = (-1, 2, 1, 1, 3)$, $\mathbf{c}_1 = (1, 1, 2, 0, 2)$
19. $\mathbf{v} = (-2, -1, 0)$, $\mathbf{c}_1 = (1, 0, 1)$, $\mathbf{c}_2 = (0, 1, 1)$
20. $\mathbf{v} = (1, 1, 2, -1)$, $\mathbf{c}_1 = (1, 1, -1, -1)$, $\mathbf{c}_2 = (2, 1, 2, 0)$
21. $\mathbf{v} = (1, 2, 3, 4, 5, 6)$, $\mathbf{c}_1 = \boldsymbol{\varepsilon}_1$, $\mathbf{c}_2 = \boldsymbol{\varepsilon}_2$, $\mathbf{c}_3 = \boldsymbol{\varepsilon}_3$, $\mathbf{c}_4 = \boldsymbol{\varepsilon}_4$
22. $\mathbf{v} = (1, 2, 3, 4, 5, 6)$, $\mathbf{c}_1 = \boldsymbol{\varepsilon}_4$, $\mathbf{c}_2 = \boldsymbol{\varepsilon}_5$, $\mathbf{c}_3 = \boldsymbol{\varepsilon}_6$

In Exercises 23–30 a function and several points are given. Set up the normal equations that would find the constants of the function that best fits the given points, in the least-squares sense. Do not solve the normal equations.

23. $y = a + bx + cx^2$, $(x_i, y_i) = (-1, 1), (0, 1), (1, 2), (2, 2)$
24. $y = a + bx + cx^2$, $(x_i, y_i) = (1, -1), (2, 0), (3, -1), (4, -1)$
25. $z = a + bx + cy$, $(x_i, y_i, z_i) = (1, 1, 1), (0, 1, 2), (1, 0, 2), (0, 0, 4)$

26. $z = a + bx + cy$, $(x_i, y_i, z_i) = (-2, 0, 4), (0, 2, 4), (0, -2, -2), (2, 0, -1)$

27. $y = a + bu + cv + duv$, $(u_i, v_i, y_i) = (0, 0, 1), (1, 0, 3), (2, 0, 5), (0, 1, 2), (1, 1, 4), (2, 1, 5)$

28. $x = e + fu + gv + huv$, $(u_i, v_i, x_i) = (0, 0, 0), (1, 0, 1), (2, 0, 3), (0, 1, 0), (1, 1, 2), (2, 1, 4)$

29. $y = ab^x$, $(x_i, y_i) = (0, e^2), (1, e^5), (2, e^9), (3, e^{17})$ (HINT First, take ln (= $\log_e$) of both sides, obtaining $\ln y = \ln a + x \ln b$).

30. $y = ae^{bx}$, $(x, y) = (3, e^4), (4, e^5), (5, e^6), (6, e^8)$. (See Hint for Exercise 29.)

4.4 Orthogonal Bases and the Gram–Schmidt Process

In the previous section, we observed that for least-square problems the matrix $A^T A$ tends to be ill conditioned. In these next two sections we develop tools to deal with these computational difficulties. However, the tools we develop are of fundamental importance throughout linear algebra. For example, they will be very important to us when we study eigenvalue problems in Chapter 5.

Orthogonal Vectors and Orthogonal Bases

(4.45) DEFINITION A collection of vectors in $\mathbb{R}^n$ (or in any inner product space) is called **orthogonal** if any two are perpendicular.

Thus the vectors $\mathbf{v}_1, \ldots, \mathbf{v}_m$ in $\mathbb{R}^n$ are orthogonal if $\mathbf{v}_i \cdot \mathbf{v}_j = 0$ for $i \neq j$.

Example 1 In $\mathbb{R}^4$ let $\mathbf{v}_1 = (1, 1, 1, 1)$, $\mathbf{v}_2 = (-1, -1, 1, 1)$, $\mathbf{v}_3 = (-1, 1, -1, 1)$, and $\mathbf{v}_4 = (-1, 1, 1, -1)$. Then it is straightforward to show $\mathbf{v}_i \cdot \mathbf{v}_j = 0$ for $i \neq j$, so $\mathbf{v}_1, \ldots, \mathbf{v}_4$ are orthogonal. ■

See Exercise 27 for an infinite-dimensional example.

Orthogonal vectors have many important properties; the foremost is the following.

(4.46) THEOREM If $\mathbf{v}_1, \ldots, \mathbf{v}_m$ are nonzero orthogonal vectors, then they are linearly independent.

Proof Suppose $a_1, \ldots, a_m$ are numbers such that

$$a_1\mathbf{v}_1 + \cdots + a_m\mathbf{v}_m = \mathbf{0} \tag{4.47}$$

We need to show each a_i must be zero. Pick any i, $1 \leq i \leq m$. From Equation (4.47), it follows that

$$(a_1\mathbf{v}_1 + \cdots + a_i\mathbf{v}_i + \cdots + a_m\mathbf{v}_m) \cdot \mathbf{v}_i = \mathbf{0} \cdot \mathbf{v}_i$$

Then

$$a_1\mathbf{v}_1 \cdot \mathbf{v}_i + \cdots + a_i\mathbf{v}_i \cdot \mathbf{v}_i + \cdots + a_m\mathbf{v}_m \cdot \mathbf{v}_i = 0$$
$$a_1 0 + \cdots + a_i\mathbf{v}_i \cdot \mathbf{v}_i + \cdots + a_m 0 = 0$$
$$a_i\mathbf{v}_i \cdot \mathbf{v}_i = 0$$

Since $\mathbf{v}_i \neq \mathbf{0}$, $\mathbf{v}_i \cdot \mathbf{v}_i$ is a positive number. Hence we must have $a_i = 0$, so we are done. ■

Example 2 The vectors $\mathbf{v}_1, \ldots, \mathbf{v}_4$ from Example 1 are orthogonal and nonzero. Therefore, they are linearly independent, by Theorem (4.46). Furthermore, since they are four linearly independent vectors in $\mathbb{R}^4$ and $\dim \mathbb{R}^4 = 4$, they form a basis for $\mathbb{R}^4$. ■

(4.48)

DEFINITION A vector $\mathbf{v}$ is called **normal** if $\|\mathbf{v}\| = 1$. A collection of vectors $\mathbf{v}_1, \ldots, \mathbf{v}_m$ in $\mathbb{R}^n$ is called **orthonormal** if they are orthogonal *and* each $\|\mathbf{v}_i\| = 1$. An **orthonormal basis** is a basis made up of orthonormal vectors.

Example 3 The standard basis $\{\boldsymbol{\varepsilon}_1, \ldots, \boldsymbol{\varepsilon}_n\}$ for $\mathbb{R}^n$ is the most famous orthonormal basis. ■

Example 4 The vectors $\mathbf{v}_1, \ldots, \mathbf{v}_4$ of Examples 1 and 2 do *not* form an orthonormal basis for $\mathbb{R}^n$ since $\|\mathbf{v}_i\| = 2$ and not 1. However, let $\mathbf{w}_i = (1/\|\mathbf{v}_i\|)\mathbf{v}_i$, $1 \leq i \leq 4$, so that

$$\mathbf{w}_1 = (\tfrac{1}{2}, \tfrac{1}{2}, \tfrac{1}{2}, \tfrac{1}{2}), \qquad \mathbf{w}_2 = (-\tfrac{1}{2}, -\tfrac{1}{2}, \tfrac{1}{2}, \tfrac{1}{2}),$$
$$\mathbf{w}_3 = (-\tfrac{1}{2}, \tfrac{1}{2}, -\tfrac{1}{2}, \tfrac{1}{2}), \qquad \mathbf{w}_4 = (-\tfrac{1}{2}, \tfrac{1}{2}, \tfrac{1}{2}, -\tfrac{1}{2})$$

Then $\mathbf{w}_1, \ldots, \mathbf{w}_n$ *are* orthonormal, so they form an orthonormal basis for $\mathbb{R}^4$. ■

There are many situations that are greatly simplified if we are dealing with orthogonal or orthonormal vectors. Here are two such situations.

(4.49)

THEOREM Let $\mathbf{v}_1, \ldots, \mathbf{v}_n$ be a basis for (an inner product space) V. Let $\mathbf{y}$ be a vector in V so that

(4.50)

$$\mathbf{y} = c_1\mathbf{v}_1 + \cdots + c_n\mathbf{v}_n$$

Then for $1 \leq i \leq n$,

(a) $c_i = \dfrac{\mathbf{y} \cdot \mathbf{v}_i}{\mathbf{v}_i \cdot \mathbf{v}_i}$ if the basis is orthogonal,

(b) $c_i = \mathbf{y} \cdot \mathbf{v}_i$ if the basis is orthonormal.

Thus, while computing the coordinates of a vector relative to a basis can be difficult in general, it is quite easy if the basis is orthogonal or orthonormal.

Example 5 Find the coordinates for $\mathbf{y} = (1, 2, 3, 4)$ in $\mathbb{R}^4$ relative to the orthogonal basis for $\mathbb{R}^4$ discussed in Examples 1 and 2.

Solution We need to compute $c_1, \ldots, c_4$, where $\mathbf{y} = c_1\mathbf{v}_1 + \cdots + c_4\mathbf{v}_4$ and the $\mathbf{v}_i$'s are given in Example 1. First we compute

$$\begin{aligned}\mathbf{y}\cdot\mathbf{v}_1 &= (1, 2, 3, 4)\cdot(1, 1, 1, 1) = 10\\ \mathbf{y}\cdot\mathbf{v}_2 &= (1, 2, 3, 4)\cdot(-1, -1, 1, 1) = 4\\ \mathbf{y}\cdot\mathbf{v}_3 &= (1, 2, 3, 4)\cdot(-1, 1, -1, 1) = 2\\ \mathbf{y}\cdot\mathbf{v}_4 &= (1, 2, 3, 4)\cdot(-1, 1, 1, -1) = 0\end{aligned}$$

Since $c_i = \dfrac{\mathbf{y}\cdot\mathbf{v}_i}{\mathbf{v}_i\cdot\mathbf{v}_i}$ and $\mathbf{v}_i\cdot\mathbf{v}_i = 4$, we conclude, by Theorem (4.49), that $c_1 = \frac{5}{2}$, $c_2 = 1$, $c_3 = \frac{1}{2}$ and $c_4 = 0$. Thus

$$\mathbf{y} = \tfrac{5}{2}\mathbf{v}_1 + \mathbf{v}_2 + \tfrac{1}{2}\mathbf{v}_3$$

We can verify this directly. ■

Proof of Theorem (4.49) Assume the basis is orthogonal. Pick any i, $1 \le i \le n$. From Equation (4.50), it follows that

$$\begin{aligned}(c_1\mathbf{v}_1 + \cdots + c_i\mathbf{v}_i + \cdots + c_n\mathbf{v}_n)\cdot\mathbf{v}_i &= \mathbf{y}\cdot\mathbf{v}_i\\ c_1(\mathbf{v}_1\cdot\mathbf{v}_i) + \cdots + c_i(\mathbf{v}_i\cdot\mathbf{v}_i) + \cdots + c_n(\mathbf{v}_n\cdot\mathbf{v}_i) &= \mathbf{y}\cdot\mathbf{v}_i\\ 0 + \cdots + c_i(\mathbf{v}_i\cdot\mathbf{v}_i) + \cdots + 0 &= \mathbf{y}\cdot\mathbf{v}_i\end{aligned}$$

Thus

$$c_i(\mathbf{v}_i\cdot\mathbf{v}_i) = \mathbf{y}\cdot\mathbf{v}_i \qquad \text{or} \qquad c_i = \frac{\mathbf{y}\cdot\mathbf{v}_i}{\mathbf{v}_i\cdot\mathbf{v}_i}$$

since $\mathbf{v}_i \neq 0$. This proves part (a) of Theorem (4.49). Part (b) follows directly from part (a), since now each $\mathbf{v}_i\cdot\mathbf{v}_i = 1$. ■

The second situation simplified by orthogonal vectors is the following.

(4.51)

THEOREM Let A be an $m \times n$ matrix.

(a) If the columns of A are orthogonal vectors, then A^TA is an $n \times n$ diagonal matrix.

(b) If the columns of A are orthonormal vectors, then A^TA is the $n \times n$ identity matrix I_n.

NOTE Part (b) says $A^T = A^{-1}$ if A is square with orthonormal columns. We shall study this situation further in the next section.

Proof Let $B = [b_{ij}] = A^T A$. Then, by the definition of matrix multiplication,

$$\begin{aligned} b_{ij} &= (i\text{th row of } A^T) \cdot (j\text{th column of } A) \\ &= (i\text{th column of } A) \cdot (j\text{th column of } A) \\ &= 0 \text{ if } i \neq j \end{aligned}$$

This proves part (a). Part (b) follows from part (a) and from the fact that

$$b_{ii} = (i\text{th column of } A) \cdot (i\text{th column of } A) = 1$$

if the columns of A are orthonormal. ■

The next example illustrates how Theorem (4.51) sometimes can be used to simplify least-squares problems greatly.

Example 6 Return to Example 6 in Section 4.3, and translate the origin on the x-axis to the *center* point of the x data, $x = 50$. Show that the resulting A has orthogonal columns, and show how easy it is to solve the resulting problem.

Solution

Table 4.2

i	x_i	y_i
1	−10	482
2	−5	467
3	0	452
4	5	433
5	10	421

Translating the origin to $x = 50$ changes Table 4.1 to Table 4.2. We substitute the values from Table 4.2 into the equation $y = mx + b$ and obtain the system

$$\begin{aligned} -10m + b &= 482 \\ -5m + b &= 467 \\ 0m + b &= 452 \\ 5m + b &= 433 \\ 10m + b &= 421 \end{aligned} \quad \text{or} \quad \begin{bmatrix} -10 & 1 \\ -5 & 1 \\ 0 & 1 \\ 5 & 1 \\ 10 & 1 \end{bmatrix} \begin{bmatrix} m \\ b \end{bmatrix} = \begin{bmatrix} 482 \\ 467 \\ 452 \\ 433 \\ 421 \end{bmatrix} \tag{4.52}$$

Thus we obtain a new inconsistent system $A\mathbf{x} = \mathbf{y}$, but the columns of the new A are easily seen to be orthogonal. The new normal equations,

$A^TAx = A^Ty$, are

$$\begin{bmatrix} -10 & -5 & 0 & 5 & 10 \\ 1 & 1 & 1 & 1 & 1 \end{bmatrix} \begin{bmatrix} -10 & 1 \\ -5 & 1 \\ 0 & 1 \\ 5 & 1 \\ 10 & 1 \end{bmatrix} \begin{bmatrix} m \\ b \end{bmatrix} = \begin{bmatrix} -10 & -5 & 0 & 5 & 10 \\ 1 & 1 & 1 & 1 & 1 \end{bmatrix} \begin{bmatrix} 482 \\ 467 \\ 452 \\ 433 \\ 421 \end{bmatrix}$$

or

$$\begin{bmatrix} 250 & 0 \\ 0 & 5 \end{bmatrix} \begin{bmatrix} m \\ b \end{bmatrix} = \begin{bmatrix} -780 \\ 2255 \end{bmatrix}$$

Now $(A^TA)^{-1}$ is very easy to compute, so we obtain

$$\begin{bmatrix} m \\ b \end{bmatrix} = \begin{bmatrix} \frac{1}{250} & 0 \\ 0 & \frac{1}{5} \end{bmatrix} \begin{bmatrix} -780 \\ 2255 \end{bmatrix} = \begin{bmatrix} -3.12 \\ 451 \end{bmatrix}$$

Thus the equation for the straight line, in the *translated* coordinates, is

$$y = -3.12x + 451$$

In the original coordinates the equation is

$$y = -3.12(x - 50) + 451 \qquad \text{or} \qquad y = -3.12x + 607$$

This is what we obtained in Example 6, Section 4.3, but with considerably less effort this time. ■

This last example indicates one place (of many) where it would be nice to be able to change vectors into orthogonal or orthonormal vectors. The next topic shows how to do this.

The Gram–Schmidt Process

Suppose we start with n *linearly independent* vectors $\mathbf{v}_1, \ldots, \mathbf{v}_n$ in $\mathbb{R}^m$ (or actually in any inner product space), and we wish to construct n orthonormal vectors $\mathbf{u}_1, \ldots, \mathbf{u}_n$ that span the same subspace. (Here "$\mathbf{u}$" is for "unit" vector.) One method used to obtain such $\mathbf{u}_i$'s is the Gram–Schmidt process, and we shall now go through several steps of this important procedure. In addition to being orthonormal, the $\mathbf{u}_i$'s constructed using the Gram–Schmidt process also have the following property, which can be very important.

$$\operatorname{span}\{\mathbf{u}_1, \ldots, \mathbf{u}_k\} = \operatorname{span}\{\mathbf{v}_1, \ldots, \mathbf{v}_k\}, \qquad 1 \le k \le n$$

and not just for $k = n$.

Step 1 Since $\mathbf{v}_1, \ldots, \mathbf{v}_n$ are linearly independent, $\mathbf{v}_1 \ne \mathbf{0}$ so $\|\mathbf{v}_1\| \ne 0$. To create a normal vector, let $\mathbf{u}_1 = \mathbf{v}_1/\|\mathbf{v}_1\|$. Then $\|\mathbf{u}_1\| = 1$ and $\operatorname{span}\{\mathbf{u}_1\} = \operatorname{span}\{\mathbf{v}_1\}$. To keep the notation uniform with what follows, let $\mathbf{w}_1 = \mathbf{v}_1$. Then

$$\mathbf{v}_1 = \|\mathbf{w}_1\|\mathbf{u}_1 \tag{4.53}$$

Step 2 Recall the projection of $\mathbf{v}_2$ onto $\mathbf{u}_1$ is $\dfrac{\mathbf{v}_2 \cdot \mathbf{u}_1}{\mathbf{u}_1 \cdot \mathbf{u}_1}\mathbf{u}_1$. This projection simplifies to $(\mathbf{v}_2 \cdot \mathbf{u}_1)\mathbf{u}_1$ since $\mathbf{u}_1 \cdot \mathbf{u}_1 = \|\mathbf{u}_1\|^2 = 1$. If we subtract this projection from $\mathbf{v}_2$, the resulting vector will be perpendicular to $\mathbf{u}_1$. To see this we let

$$\mathbf{w}_2 = \mathbf{v}_2 - (\mathbf{v}_2 \cdot \mathbf{u}_1)\mathbf{u}_1 \tag{4.54}$$

We check that $\mathbf{w}_2 \perp \mathbf{u}_1$:

$$\mathbf{w}_2 \cdot \mathbf{u}_1 = [\mathbf{v}_2 - (\mathbf{v}_2 \cdot \mathbf{u}_1)\mathbf{u}_1] \cdot \mathbf{u}_1 = \mathbf{v}_2 \cdot \mathbf{u}_1 - (\mathbf{v}_2 \cdot \mathbf{u}_1)(\mathbf{u}_1 \cdot \mathbf{u}_1) = 0$$

since $\mathbf{u}_1 \cdot \mathbf{u}_1 = 1$. We also know that $\mathbf{w}_2 \neq \mathbf{0}$ because by Equation (4.54),

$$\mathbf{w}_2 = \mathbf{v}_2 - \frac{\mathbf{v}_2 \cdot \mathbf{u}_1}{\|\mathbf{v}_1\|^2}\mathbf{v}_1$$

which is a linear combination of $\mathbf{v}_1$ and $\mathbf{v}_2$ (which are linearly independent) and not all of the coefficients of the $\mathbf{v}$'s are zero (the coefficient of $\mathbf{v}_2$ is 1). Thus $\|\mathbf{w}_2\| \neq 0$, so we can let $\mathbf{u}_2 = \mathbf{w}_2/\|\mathbf{w}_2\|$. Therefore, $\mathbf{w}_2 = \|\mathbf{w}_2\|\mathbf{u}_2$, and substituting this into Equation (4.54) gives us

$$\mathbf{v}_2 = \|\mathbf{w}_2\|\mathbf{u}_2 + (\mathbf{v}_2 \cdot \mathbf{u}_1)\mathbf{u}_1 \tag{4.55}$$

We next show the $\mathbf{v}$'s and $\mathbf{u}$'s have the same span. Now $\mathbf{u}_1$ and $\mathbf{u}_2$ are linearly independent since they are orthonormal, so they span a two-dimensional subspace. The $\mathbf{v}$'s are a linear combination of the $\mathbf{u}$'s [from Equations (4.53) and (4.55)], and the $\mathbf{v}$'s are linearly independent. Thus span$\{\mathbf{v}_1, \mathbf{v}_2\}$ is a two-dimensional subspace of the two-dimensional subspace span$\{\mathbf{u}_1, \mathbf{u}_2\}$, so they must be the same.

Step 3 The projections of $\mathbf{v}_3$ on $\mathbf{u}_1$ and on $\mathbf{u}_2$ are $(\mathbf{v}_3 \cdot \mathbf{u}_1)\mathbf{u}_1$ and $(\mathbf{v}_3 \cdot \mathbf{u}_2)\mathbf{u}_2$ respectively. If we subtract both of these projections from $\mathbf{v}_3$, the resulting vector, $\mathbf{w}_3$, will be perpendicular to both $\mathbf{u}_1$ and $\mathbf{u}_2$ *since* $\mathbf{u}_1$ *and* $\mathbf{u}_2$ *are orthonormal.* To see this, let

$$\mathbf{w}_3 = \mathbf{v}_3 - (\mathbf{v}_3 \cdot \mathbf{u}_1)\mathbf{u}_1 - (\mathbf{v}_3 \cdot \mathbf{u}_2)\mathbf{u}_2 \tag{4.56}$$

Then

$$\begin{aligned}
\mathbf{w}_3 \cdot \mathbf{u}_1 &= \mathbf{v}_3 \cdot \mathbf{u}_1 - (\mathbf{v}_3 \cdot \mathbf{u}_1)(\mathbf{u}_1 \cdot \mathbf{u}_1) - (\mathbf{v}_3 \cdot \mathbf{u}_2)(\mathbf{u}_2 \cdot \mathbf{u}_1) \\
&= \mathbf{v}_3 \cdot \mathbf{u}_1 - \mathbf{v}_3 \cdot \mathbf{u}_1 - 0 = 0 \\
\mathbf{w}_3 \cdot \mathbf{u}_2 &= \mathbf{v}_3 \cdot \mathbf{u}_2 - (\mathbf{v}_3 \cdot \mathbf{u}_1)(\mathbf{u}_1 \cdot \mathbf{u}_2) - (\mathbf{v}_3 \cdot \mathbf{u}_2)(\mathbf{u}_2 \cdot \mathbf{u}_2) \\
&= \mathbf{v}_3 \cdot \mathbf{u}_2 - 0 - \mathbf{v}_3 \cdot \mathbf{u}_2 = 0
\end{aligned}$$

Notice that it is crucial to know that $\mathbf{u}_1$ and $\mathbf{u}_2$ are orthonormal before constructing $\mathbf{w}_3$ this way. Following the argument in step 2, we can show $\mathbf{w}_3 \neq \mathbf{0}$ because the $\mathbf{v}$'s are linearly independent. Thus $\|\mathbf{w}_3\| \neq 0$, and we let $\mathbf{u}_3 = \mathbf{w}_3/\|\mathbf{w}_3\|$. Next we obtain $\mathbf{v}_3 = \|\mathbf{w}_3\|\mathbf{u}_3 + (\mathbf{v}_3 \cdot \mathbf{u}_2)\mathbf{u}_2 + (\mathbf{v}_3 \cdot \mathbf{u}_1)\mathbf{u}_1$, and we can show span$\{\mathbf{v}_1, \mathbf{v}_2, \mathbf{v}_3\}$ is a three-dimensional subspace of the three-dimensional space span$\{\mathbf{u}_1, \mathbf{u}_2, \mathbf{u}_3\}$. Hence the two spans are the same.

Step 4 ... Step n These proceed in exactly the pattern indicated. You are strongly encouraged to go through the details of step 4. See Exercise 32. An important fact we shall need is

$$\mathbf{v}_n = \|\mathbf{w}_n\|\mathbf{u}_n + (\mathbf{v}_n \cdot \mathbf{u}_{n-1})\mathbf{u}_{n-1} + \cdots + (\mathbf{v}_n \cdot \mathbf{u}_1)\mathbf{u}_1 \tag{4.57}$$

Example 7 Apply the Gram–Schmidt process to the vectors,

$$\mathbf{v}_1 = (1, 1, 1, 1), \qquad \mathbf{v}_2 = (0, 1, 1, 1), \qquad \mathbf{v}_3 = (0, 0, 1, 1)$$

Solution Observe first that the $\mathbf{v}$'s are linearly independent since their nonzero entries form a staircase pattern.

Step 1 $\|\mathbf{v}_1\| = \sqrt{1^2 + 1^2 + 1^2 + 1^2} = 2$, $\mathbf{w}_1 = \mathbf{v}_1$ and

$$\mathbf{u}_1 = \frac{\mathbf{v}_1}{\|\mathbf{v}_1\|} = (\tfrac{1}{2}, \tfrac{1}{2}, \tfrac{1}{2}, \tfrac{1}{2})$$

Step 2 First compute

$$\mathbf{v}_2 \cdot \mathbf{u}_1 = 0 + \tfrac{1}{2} + \tfrac{1}{2} + \tfrac{1}{2} = \tfrac{3}{2}$$

Then

$$\begin{aligned}\mathbf{w}_2 &= \mathbf{v}_2 - [\mathbf{v}_2 \cdot \mathbf{u}_1]\mathbf{u}_1 \\ &= (0, 1, 1, 1) - \tfrac{3}{2}(\tfrac{1}{2}, \tfrac{1}{2}, \tfrac{1}{2}, \tfrac{1}{2}) \\ &= (0, 1, 1, 1) - (\tfrac{3}{4}, \tfrac{3}{4}, \tfrac{3}{4}, \tfrac{3}{4}) \\ &= (-\tfrac{3}{4}, \tfrac{1}{4}, \tfrac{1}{4}, \tfrac{1}{4})\end{aligned}$$

Thus $\|\mathbf{w}_2\| = \sqrt{(-\tfrac{3}{4})^2 + (\tfrac{1}{4})^2 + (\tfrac{1}{4})^2 + (\tfrac{1}{4})^2} = \sqrt{\tfrac{12}{16}} = \tfrac{1}{2}\sqrt{3}$, so

$$\mathbf{u}_2 = \frac{\mathbf{w}_2}{\|\mathbf{w}_2\|} = (-\tfrac{1}{2}\sqrt{3}, \tfrac{1}{6}\sqrt{3}, \tfrac{1}{6}\sqrt{3}, \tfrac{1}{6}\sqrt{3}).$$

Step 3 First compute

$$\mathbf{v}_3 \cdot \mathbf{u}_1 = 0 + 0 + \tfrac{1}{2} + \tfrac{1}{2} = 1 \quad \text{and} \quad \mathbf{v}_3 \cdot \mathbf{u}_2 = 0 + 0 + \tfrac{1}{6}\sqrt{3} + \tfrac{1}{6}\sqrt{3} = \tfrac{1}{3}\sqrt{3}$$

Then

$$\begin{aligned}\mathbf{w}_3 &= \mathbf{v}_3 - [\mathbf{v}_3 \cdot \mathbf{u}_1]\mathbf{u}_1 - [\mathbf{v}_3 \cdot \mathbf{u}_2]\mathbf{u}_2 \\ &= (0, 0, 1, 1) - [1](\tfrac{1}{2}, \tfrac{1}{2}, \tfrac{1}{2}, \tfrac{1}{2}) - [\tfrac{1}{3}\sqrt{3}](-\tfrac{1}{2}\sqrt{3}, \tfrac{1}{6}\sqrt{3}, \tfrac{1}{6}\sqrt{3}, \tfrac{1}{6}\sqrt{3}) \\ &= (0, 0, 1, 1) - (\tfrac{1}{2}, \tfrac{1}{2}, \tfrac{1}{2}, \tfrac{1}{2}) - (-\tfrac{1}{2}, \tfrac{1}{6}, \tfrac{1}{6}, \tfrac{1}{6}) \\ &= (0, -\tfrac{2}{3}, \tfrac{1}{3}, \tfrac{1}{3})\end{aligned}$$

Thus $\|\mathbf{w}_3\| = \sqrt{0^2 + (-\tfrac{2}{3})^2 + (\tfrac{1}{3})^2 + (\tfrac{1}{3})^2} = \sqrt{\tfrac{6}{9}} = \tfrac{1}{3}\sqrt{6}$, so

$$\mathbf{u}_3 = \mathbf{w}_3/\|\mathbf{w}_3\| = (0, -\tfrac{1}{3}\sqrt{6}, \tfrac{1}{6}\sqrt{6}, \tfrac{1}{6}\sqrt{6})$$

You can check directly that $\mathbf{u}_1$, $\mathbf{u}_2$, and $\mathbf{u}_3$ are orthonormal. ■

One of the many uses of the Gram–Schmidt process is to extend an orthonormal set to an orthonormal basis. This process is illustrated in the next example.

Example 8 The vectors $\mathbf{u}_1 = (\frac{1}{3}, \frac{2}{3}, \frac{2}{3})$ and $\mathbf{u}_2 = \left(0, \frac{1}{\sqrt{2}}, -\frac{1}{\sqrt{2}}\right)$ are orthonormal. Find a vector $\mathbf{u}_3$ so that altogether the vectors $\mathbf{u}_1$, $\mathbf{u}_2$, and $\mathbf{u}_3$ form an orthonormal basis for $\mathbb{R}^3$. (In doing this we say we **extend** $\mathbf{u}_1$ and $\mathbf{u}_2$ to a basis.)

Solution First we find any vector $\mathbf{v}_3$ so that $\mathbf{u}_1$, $\mathbf{u}_2$ and $\mathbf{v}_3$ are linearly independent. One obvious choice is $\mathbf{v}_3 = \boldsymbol{\varepsilon}_3 = (0, 0, 1)$, since the nonzero entries of $\mathbf{u}_1$, $\mathbf{u}_2$, and $\boldsymbol{\varepsilon}_3$ form a staircase pattern.

Next we apply the Gram–Schmidt process to $\mathbf{u}_1$, $\mathbf{u}_2$, and $\mathbf{v}_3$. Since $\mathbf{u}_1$ and $\mathbf{u}_2$ are already orthonormal, we proceed immediately to step 3 and compute $\mathbf{w}_3$,

$$\begin{aligned}
\mathbf{w}_3 &= \mathbf{v}_3 - (\mathbf{v}_3 \cdot \mathbf{u}_1)\mathbf{u}_1 - (\mathbf{v}_3 \cdot \mathbf{u}_2)\mathbf{u}_2 \\
&= (0, 0, 1) - \tfrac{2}{3}(\tfrac{1}{3}, \tfrac{2}{3}, \tfrac{2}{3}) - \left(-\frac{1}{\sqrt{2}}\right)\left(0, \frac{1}{\sqrt{2}}, -\frac{1}{\sqrt{2}}\right) \\
&= (0, 0, 1) - (\tfrac{2}{9}, \tfrac{4}{9}, \tfrac{4}{9}) + (0, \tfrac{1}{2}, -\tfrac{1}{2}) \\
&= (-\tfrac{2}{9}, \tfrac{1}{18}, \tfrac{1}{18})
\end{aligned}$$

Then $\mathbf{u}_3 = \dfrac{\mathbf{w}_3}{\|\mathbf{w}_3\|} = \left(-\dfrac{4}{\sqrt{18}}, \dfrac{1}{\sqrt{18}}, \dfrac{1}{\sqrt{18}}\right)$. ■

A final remark: When working by hand we usually get numbers that are easier to work if we first find only *orthogonal* vectors and then normalize them all at once. Starting with $\mathbf{v}_1, \mathbf{v}_2, \ldots, \mathbf{v}_n$ that are linearly independent, the formulas that are alternate to Equation (4.57) are

$$\begin{aligned}
\mathbf{w}_1 &= \mathbf{v}_1 \\
\mathbf{w}_2 &= \mathbf{v}_2 - \frac{\mathbf{v}_2 \cdot \mathbf{w}_1}{\mathbf{w}_1 \cdot \mathbf{w}_1}\mathbf{w}_1 \\
\mathbf{w}_3 &= \mathbf{v}_3 - \frac{\mathbf{v}_3 \cdot \mathbf{w}_2}{\mathbf{w}_2 \cdot \mathbf{w}_2}\mathbf{w}_2 - \frac{\mathbf{v}_3 \cdot \mathbf{w}_1}{\mathbf{w}_1 \cdot \mathbf{w}_1}\mathbf{w}_1 \qquad (4.57a) \\
&\ \ \vdots \\
\mathbf{w}_n &= \mathbf{v}_n - \frac{\mathbf{v}_n \cdot \mathbf{w}_{n-1}}{\mathbf{w}_{n-1} \cdot \mathbf{w}_{n-1}}\mathbf{w}_{n-1} - \cdots - \frac{\mathbf{v}_n \cdot \mathbf{w}_1}{\mathbf{w}_1 \cdot \mathbf{w}_1}\mathbf{w}_1
\end{aligned}$$

You can check that $\mathbf{w}_1, \mathbf{w}_2, \ldots, \mathbf{w}_n$ are nonzero and orthogonal. We obtain orthonormal vectors by $\mathbf{u}_1 = \dfrac{\mathbf{w}_1}{\|\mathbf{w}_1\|}, \ldots, \mathbf{u}_n = \dfrac{\mathbf{w}_n}{\|\mathbf{w}_n\|}$.

Exercise 4.4 In Exercises 1–4, determine if the given vectors are orthogonal.

1. $\mathbf{u}_1 = (1, 1, 1, 1)$, $\mathbf{u}_2 = (1, -1, 0, 0)$, $\mathbf{u}_3 = (1, 1, -2, 0)$, $\mathbf{u}_4 = (1, 1, 1, -3)$
2. $\mathbf{v}_1 = (1, 0, 0, 1)$, $\mathbf{v}_2 = (0, 1, 1, 0)$, $\mathbf{v}_3 = (1, 1, -1, -1)$, $\mathbf{v}_4 = (1, -1, 1, -1)$
3. $\mathbf{w}_1 = (1, 2, 3, 4)$, $\mathbf{w}_2 = (-9, 1, 1, 1)$, $\mathbf{w}_3 = (1, -8, 1, 1)$, $\mathbf{w}_4 = (0, 1, -2, 1)$
4. $\mathbf{x}_1 = (1, 0, 1, 0)$, $\mathbf{x}_2 = (0, 1, 0, 1)$, $\mathbf{x}_3 = (1, 0, -1, 0)$, $\mathbf{x}_4 = (1, 1, -1, -1)$
5. Show that $S = \{\mathbf{u}_1, \mathbf{u}_2, \mathbf{u}_3\}$ is an orthonormal basis for $\mathbb{R}^3$, where
$$\mathbf{u}_1 = (\tfrac{2}{3}, \tfrac{2}{3}, \tfrac{1}{3}), \qquad \mathbf{u}_2 = (\tfrac{1}{2}\sqrt{2}, -\tfrac{1}{2}\sqrt{2}, 0), \qquad \mathbf{u}_3 = (\tfrac{1}{6}\sqrt{2}, \tfrac{1}{6}\sqrt{2}, -\tfrac{2}{3}\sqrt{2})$$
6. Show that $T = \{\mathbf{v}_1, \mathbf{v}_2, \mathbf{v}_3, \mathbf{v}_4, \mathbf{v}_5, \mathbf{v}_6\}$ is an orthonormal basis for $\mathbb{R}^6$, where
$$\mathbf{v}_1 = (-\tfrac{2}{3}, \tfrac{1}{3}, \ldots, \tfrac{1}{3}), \qquad \mathbf{v}_2 = (\tfrac{1}{3}, -\tfrac{2}{3}, \tfrac{1}{3}, \ldots, \tfrac{1}{3}), \qquad \mathbf{v}_3 = (\tfrac{1}{3}, \tfrac{1}{3}, -\tfrac{2}{3}, \tfrac{1}{3}, \tfrac{1}{3}, \tfrac{1}{3}),$$
$$\mathbf{v}_4 = (\tfrac{1}{3}, \tfrac{1}{3}, \tfrac{1}{3}, -\tfrac{2}{3}, \tfrac{1}{3}, \tfrac{1}{3}), \qquad \mathbf{v}_5 = (\tfrac{1}{3}, \ldots, \tfrac{1}{3}, -\tfrac{2}{3}, \tfrac{1}{3}), \qquad \mathbf{v}_6 = (\tfrac{1}{3}, \ldots, \tfrac{1}{3}, -\tfrac{2}{3})$$
7. Let $\mathbf{x} = (6, -4, 1)$. Find the coordinates for $\mathbf{x}$ in (a) the standard basis and and (b) the basis S of Exercise 5.
8. Same question as Exercise 7 if $\mathbf{x} = (-2, -3, 1)$.
9. Let $\mathbf{y} = (1, 2, 3, 4, 5, 6)$. Find the coordinates for $\mathbf{y}$ in (a) the standard basis and (b) the basis T of Exercise 6.
10. Same question as Exercise 9 if $\mathbf{y} = (1, 1, 1, 1, 1, 1)$.

Exercises 11–16, find the equation that best fits the given points in the least squares sense. However, first translate the x-coordinates so that in the system $A\mathbf{x} = \mathbf{b}$, A has orthogonal columns. Then solve the problem, *but* be sure to translate the answer back to the original coordinates. What is A?

11. (1, 2), (2, 4), (3, 5)
12. (−1, 4), (1, 3), (3, 0)
13. (−4, 2), (−3, 1), (−2, −1), (−1, −2), (0, −4)
14. (−3, −5), (−2, −3), (−1, −4), (0, −1), (1, −2)
15. (1, 1), (2, 1), (3, 2), (4, 2)
16. (−1, 4), (0, 2), (1, 1), (2, 1)

In Exercises 17–26, apply the Gram–Schmidt process to the given vectors.

17. (1, 1), (1, 2)
18. $\begin{bmatrix} 1 \\ -1 \end{bmatrix}, \begin{bmatrix} 2 \\ 1 \end{bmatrix}$
19. $\begin{bmatrix} 1 \\ 1 \\ 1 \end{bmatrix}, \begin{bmatrix} 1 \\ 2 \\ 2 \end{bmatrix}, \begin{bmatrix} 1 \\ 0 \\ 1 \end{bmatrix}$
20. (1, 1, 0, 0), (0, 1, 1, 0), (1, 0, 1, 1)

21. (2, 0, 2, −1), (1, 0, −2, −2), (2, 1, 1, −1)
22. (1, 1, 1, 1), (1, 2, 1, 0), (1, 3, 0, 0)

In Exercises 23–24, use $f \cdot g = \int_{-1}^{1} f(x)g(x)\,dx$.

23. $f_1(x) = 1, f_2(x) = x, f_3(x) = x^2$
24. $g_3(x) = 1 + x + x^2, g_2(x) = 1 + x, g_1(x) = 1 - x$

In Exercises 25 and 26, use $A \cdot B = a_{11}b_{11} + a_{12}b_{12} + a_{21}b_{21} + a_{22}b_{22}$.

25. $\begin{bmatrix} 1 & 1 \\ 1 & 1 \end{bmatrix}, \begin{bmatrix} -2 & 1 \\ 1 & 0 \end{bmatrix}, \begin{bmatrix} 1 & 1 \\ 0 & 0 \end{bmatrix}$
26. $\begin{bmatrix} 1 & 0 \\ 0 & 1 \end{bmatrix}, \begin{bmatrix} 1 & 1 \\ 1 & 0 \end{bmatrix}, \begin{bmatrix} 1 & 1 \\ 0 & 0 \end{bmatrix}$

In Exercises 27 and 28, use $f \cdot g = \int_0^{2\pi} f(x)g(x)\,dx$ and let $f_n(x) = \sin(nx)$, $n = 1, 2, \ldots$.

27. Show $f_n \cdot f_m = 0$ if $n \neq m$. (HINT Use the identity $\sin u \sin v = \frac{1}{2}[\cos(u - v) - \cos(u + v)]$.)
28. Compute $\|f_n\|$ and produce an infinite collection of orthonormal "vectors."
29. Show $\|\mathbf{x} + \mathbf{y}\| = \|\mathbf{x} - \mathbf{y}\|$ if and only if $\mathbf{x}$ and $\mathbf{y}$ are orthogonal.
30. Show $\mathbf{x} + \mathbf{y}$ and $\mathbf{x} - \mathbf{y}$ are orthogonal if and only if $\mathbf{x}$ and $\mathbf{y}$ have the same norms.
31. Show that the three steps

$$\mathbf{x} = \mathbf{v}_3 - [\mathbf{v}_3 \cdot \mathbf{u}_1]\mathbf{u}_1, \qquad \mathbf{w}_3 = \mathbf{x} - [\mathbf{x} \cdot \mathbf{u}_2]\mathbf{u}_2, \qquad \mathbf{u}_3 = \frac{\mathbf{w}_3}{\|\mathbf{w}_3\|}$$

produce the same $\mathbf{w}_3$ and $\mathbf{u}_3$ as described in Equation (4.55) in the text. This is a modified Gram–Schmidt orthogonalization, which is more numerically stable.

32. Carefully go through all the details of step 4 in the Gram–Schmidt process.

4.5 Orthogonal Matrices, *QR* Decompositions, and Least Squares (Revisited)

In Section 4.4, we saw how easy the least-squares problem, $A\mathbf{x} = \mathbf{b}$, is to solve if A happens to have orthogonal columns. In this section we use the Gram–Schmidt process to make the columns of A orthonormal. As often happens, the mathematics we develop for this purpose has a wide range of applications, some of which we shall see in the next chapter.

Orthogonal Matrices

We shall see there are many equivalent criteria for a matrix's being orthogonal. For the definition, we pick the one that fits the introductory discussion.

(4.58) DEFINITION A matrix is **orthogonal** if it is square and its columns are orthonormal.

WARNING A matrix with orthonormal columns that is *not* square is *not* called orthogonal.

NOTE In some sense orthogonal matrices should really be called "orthonormal matrices," but we are stuck with tradition.

Example 1 Find a third column so that the matrix

$$Q = \begin{bmatrix} \frac{1}{\sqrt{2}} & \frac{1}{\sqrt{3}} & \\ 0 & \frac{1}{\sqrt{3}} & \\ -\frac{1}{\sqrt{2}} & \frac{1}{\sqrt{3}} & \end{bmatrix}$$

is orthogonal. How many choices are there?

Solution Let the three columns of Q be $\mathbf{q}_1, \mathbf{q}_2, \mathbf{q}_3$. We can see $\|\mathbf{q}_1\| = \|\mathbf{q}_2\| = 1$ and $\mathbf{q}_1 \cdot \mathbf{q}_2 = 0$. Let $\mathbf{q}_3 = [x_1 \quad x_2 \quad x_3]^T$. Then

(a) Since $\mathbf{q}_1 \cdot \mathbf{q}_3 = 0$, $x_1 - x_3 = 0$, so $x_3 = x_1$.
(b) Since $\mathbf{q}_2 \cdot \mathbf{q}_3 = 0$, $x_1 + x_2 + x_3 = 0$, so $x_2 = -x_1 - x_3 = -2x_1$. We now know $\mathbf{q}_3 = [x_1 \quad -2x_1 \quad x_1]^T$.
(c) Since $\|\mathbf{q}_3\| = 1$, $1 = x_1^2 + (-2x_1)^2 + x_1^2 = 6x_1^2$, so $6x_1^2 = 1$ or $x_1 = \pm\frac{1}{\sqrt{6}} = \pm\frac{\sqrt{6}}{6}$.

Therefore, there are two possible answers, $\mathbf{q}_3 = \pm(\frac{1}{6}\sqrt{6}, -\frac{1}{3}\sqrt{6}, \frac{1}{6}\sqrt{6})$. ■

We shall often use the letter Q to denote an orthogonal matrix, and $\mathbf{q}_1, \ldots, \mathbf{q}_n$ to denote its column vectors. Suppose Q is orthogonal and $\mathbf{q}_1, \ldots, \mathbf{q}_n$ are its columns. Since $\mathbf{q}_i \cdot \mathbf{q}_j = \mathbf{q}_i^T\mathbf{q}_j$ and $\mathbf{q}_1, \ldots, \mathbf{q}_n$ are orthonormal, we have

$$\mathbf{q}_i^T\mathbf{q}_j = \mathbf{q}_i \cdot \mathbf{q}_j = 0 \quad \text{if } i \neq j \qquad \text{and} \qquad \mathbf{q}_i^T\mathbf{q}_i = \mathbf{q}_i \cdot \mathbf{q}_i = 1$$

Since the entries of the product Q^TQ are $\mathbf{q}_i^T\mathbf{q}_j$, we see

$$Q^TQ = \begin{bmatrix} \cdots & \mathbf{q}_1^T & \cdots \\ & \vdots & \\ \cdots & \mathbf{q}_n^T & \cdots \end{bmatrix} \begin{bmatrix} \vdots & & \vdots \\ \mathbf{q}_1 & \cdots & \mathbf{q}_n \\ \vdots & & \vdots \end{bmatrix} = \begin{bmatrix} 1 & & \\ & \ddots & \\ & & 1 \end{bmatrix} = I$$

Therefore, if Q is orthogonal, $Q^TQ = I$. Conversely, suppose A is a square matrix with columns $\mathbf{c}_1, \ldots, \mathbf{c}_n$ and $A^TA = I$. Then

$$\mathbf{c}_i \cdot \mathbf{c}_j = \mathbf{c}_i^T\mathbf{c}_j = 0 \quad \text{for } i \neq j \qquad \text{and} \qquad \mathbf{c}_i \cdot \mathbf{c}_i = \mathbf{c}_i^T\mathbf{c}_i = 1$$

and therefore A has orthonormal columns. We have just proved

(4.59)

> THEOREM If Q is a square matrix, then
>
> $$Q \text{ is orthogonal} \quad \Leftrightarrow \quad Q^TQ = I$$

Suppose A is not square but still has orthonormal columns. The above argument also applies in this case, showing A^TA is the identity matrix.

(4.60)

> THEOREM If A is an $m \times n$ matrix with orthonormal columns (so that necessarily $m \geq n$), then
>
> $$A^TA = I_n$$
>
> where I_n is the $n \times n$ identity matrix.

We shall need this fact later.

If Q is square and $Q^TQ = I$, then Q^T plays the role of Q^{-1}. By the uniqueness of matrix inverses [Theorem (1.38), Section 1.4], we know Q is invertible and $Q^T = Q^{-1}$. Since for any invertible matrix A, $A^{-1}A = I = AA^{-1}$, we must have $QQ^T = I$, also. For row vectors, $\mathbf{v} \cdot \mathbf{w} = \mathbf{v}\mathbf{w}^T$. [See the warning after the proof of Theorem (4.15), Section 4.2.] If $\mathbf{r}_1, \ldots, \mathbf{r}_n$ are the row vectors of Q, then we can argue as we did with columns to see

$$QQ^T = 1 \qquad \text{if and only if the rows of } Q \text{ are orthonormal}$$

See Exercise 21.

Therefore, we see

(4.61)

> THEOREM If Q is square, then
>
> Q is orthogonal
>
> $\Leftrightarrow \quad Q^T = Q^{-1}$
>
> $\Leftrightarrow \quad QQ^T = I$
>
> $\Leftrightarrow \quad$ the rows of Q are orthonormal.

Example 2 Let

$$Q = \begin{bmatrix} \cos\theta & -\sin\theta \\ \sin\theta & \cos\theta \end{bmatrix}$$

Then $Q^TQ = I$, so Q is orthogonal and

$$Q^T = Q^{-1} = \begin{bmatrix} \cos\theta & \sin\theta \\ -\sin\theta & \cos\theta \end{bmatrix}$$

By Example 5 in Section 4.1, Q rotates the plane $\mathbb{R}^2$ through the angle θ. Using the facts $\cos(-\theta) = \cos\theta$ and $\sin(-\theta) = -\sin\theta$, we can see $Q^T = Q^{-1}$ rotates the plane back through $-\theta$. ■

Example 3 Any permutation matrix, P, has the property that all entries are zero except for a single one in each column and in each row. Thus the columns (or rows) are orthonormal, so P is an orthogonal matrix. For example,

$$P_1 = \begin{bmatrix} 0 & 1 \\ 1 & 0 \end{bmatrix} \quad \text{so} \quad P_1^T = P_1^{-1} = \begin{bmatrix} 0 & 1 \\ 1 & 0 \end{bmatrix}$$

$$P_2 = \begin{bmatrix} 0 & 0 & 1 \\ 1 & 0 & 0 \\ 0 & 1 & 0 \end{bmatrix} \quad \text{so} \quad P_2^T = P_2^{-1} = \begin{bmatrix} 0 & 1 & 0 \\ 0 & 0 & 1 \\ 1 & 0 & 0 \end{bmatrix}$$ ■

There are two geometric characterizations of orthogonality, namely,

(4.62)

THEOREM Let Q be a square matrix. Then Q is orthogonal

(a) $\Leftrightarrow$ Q preserves inner products; that is, $Q\mathbf{u} \cdot Q\mathbf{v} = \mathbf{u} \cdot \mathbf{v}$ for all $\mathbf{u}$, $\mathbf{v}$.

(b) $\Leftrightarrow$ Q preserves norms; that is, $\|Q\mathbf{u}\| = \|\mathbf{u}\|$, for all $\mathbf{u}$.

Proof Orthogonality $\Rightarrow$ (a). If $Q^TQ = I$, then

$$Q\mathbf{u} \cdot Q\mathbf{v} = (Q^TQ\mathbf{u}) \cdot \mathbf{v} = (I\mathbf{u}) \cdot \mathbf{v} = \mathbf{u} \cdot \mathbf{v}$$

(a) $\Rightarrow$ Orthogonality. Observe first if $\boldsymbol{\varepsilon}_1, \ldots, \boldsymbol{\varepsilon}_n$ is the standard basis for $\mathbb{R}^n$ (written as column vectors) and A is any $m \times n$ matrix, then $A\boldsymbol{\varepsilon}_i$ is the ith column of A. Suppose Q is square and $Q\mathbf{u} \cdot Q\mathbf{v} = \mathbf{u} \cdot \mathbf{v}$, for all $\mathbf{u}$ and $\mathbf{v}$. Let $\mathbf{q}_1, \ldots, \mathbf{q}_n$ be the columns of Q and let $\mathbf{u} = \boldsymbol{\varepsilon}_i$, $\mathbf{v} = \boldsymbol{\varepsilon}_j$ in (a). Then

$$\mathbf{q}_i \cdot \mathbf{q}_j = Q\boldsymbol{\varepsilon}_i \cdot Q\boldsymbol{\varepsilon}_j = \boldsymbol{\varepsilon}_i \cdot \boldsymbol{\varepsilon}_j = \begin{cases} 0 & \text{if } i \neq j \\ 1 & \text{if } i = j \end{cases}$$

so the columns of Q are orthonormal.

(a) $\Rightarrow$ (b). Let $\mathbf{v} = \mathbf{u}$, so $Q\mathbf{u} \cdot Q\mathbf{u} = \mathbf{u} \cdot \mathbf{u} \Rightarrow \|Q\mathbf{u}\| = \|\mathbf{u}\|$.
(b) $\Rightarrow$ (a). Use the identity

$$\mathbf{x} \cdot \mathbf{y} = \tfrac{1}{4}(\|\mathbf{x} + \mathbf{y}\|^2 - \|\mathbf{x} - \mathbf{y}\|^2)$$

See Exercises 22 and 23. ■

Sometimes Theorem (4.62) can be used to give nice geometric arguments, as in the proof of the following.

(4.63) **THEOREM** If Q_1 and Q_2 are orthogonal $n \times n$ matrices, then so is their product Q_1Q_2.

Proof Certainly Q_1Q_2 is $n \times n$. Orthogonality follows by Theorem (4.62b) since if Q_1 and Q_2 preserve norms, their product must, that is, for all $\mathbf{u}$,

$$\begin{aligned} \|(Q_1Q_2)\mathbf{u}\| &= \|Q_1(Q_2\mathbf{u})\| \\ &= \|Q_2\mathbf{u}\|, \quad \text{since } Q_1 \text{ preserves norms} \\ &= \|\mathbf{u}\|, \quad \text{since } Q_2 \text{ preserves norms} \end{aligned}$$

■

Note that Theorem (4.63) also can be easily proved using Theorem (4.61); it is a lot harder to prove it using the definition.

The *QR* Decomposition

Suppose A is an $m \times n$ matrix with linearly independent columns. (Thus $m \geq n$.) We have seen, in Section 4.4, that it would be nice if we could replace A with a matrix with orthonormal columns. By applying the Gram–Schmidt process to the columns of A, we can almost do this.

(4.64) **THEOREM (THE *QR* DECOMPOSITION)** If A is an $m \times n$ matrix with linearly independent columns,* then A can be decomposed into the product

$$A = QR$$

where Q is an $m \times n$ matrix with orthonormal columns and R is a nonsingular upper triangular matrix.

Hence if A is square with linearly independent columns,† then $A = QR$, where Q is orthogonal and R is nonsingular and upper triangular.

---

* *Thus, necessarily, $m \geq n$.*

† *Thus, if A is invertible. See Exercise 24.*

Proof Apply the Gram–Schmidt process to the columns of A, $\mathbf{c}_1, \ldots, \mathbf{c}_n$ to obtain $\mathbf{q}_1, \ldots, \mathbf{q}_n$. By Equation (4.57), Section 4.4, we have

$$\begin{aligned}
\mathbf{c}_1 &= \|\mathbf{w}_1\|\mathbf{q}_1 \\
\mathbf{c}_2 &= (\mathbf{c}_2 \cdot \mathbf{q}_1)\mathbf{q}_1 + \|\mathbf{w}_2\|\mathbf{q}_2 \\
\mathbf{c}_3 &= (\mathbf{c}_3 \cdot \mathbf{q}_1)\mathbf{q}_1 + (\mathbf{c}_3 \cdot \mathbf{q}_2)\mathbf{q}_2 + \|\mathbf{w}_3\|\mathbf{q}_3 \\
&\vdots \\
\mathbf{c}_n &= (\mathbf{c}_n \cdot \mathbf{q}_1)\mathbf{q}_1 + (\mathbf{c}_n \cdot \mathbf{q}_2)\mathbf{q}_2 + (\mathbf{c}_n \cdot \mathbf{q}_3)\mathbf{q}_3 + \cdots + \|\mathbf{w}_n\|\mathbf{q}_n
\end{aligned} \tag{4.65}$$

In matrix form this yields

$$[\mathbf{c}_1 \vdots \mathbf{c}_2 \vdots \mathbf{c}_3 \vdots \cdots \vdots \mathbf{c}_n] =$$

$$[\mathbf{q}_1 \vdots \mathbf{q}_2 \vdots \mathbf{q}_3 \vdots \cdots \vdots \mathbf{q}_n]\begin{bmatrix} \|\mathbf{w}_1\| & \mathbf{c}_2 \cdot \mathbf{q}_1 & \mathbf{c}_3 \cdot \mathbf{q}_1 & \cdots & \mathbf{c}_n \cdot \mathbf{q}_1 \\ & \|\mathbf{w}_2\| & \mathbf{c}_3 \cdot \mathbf{q}_2 & \cdots & \mathbf{c}_n \cdot \mathbf{q}_2 \\ & & \|\mathbf{w}_3\| & \cdots & \mathbf{c}_n \cdot \mathbf{q}_3 \\ & & & & \vdots \\ & & & & \|\mathbf{w}_n\| \end{bmatrix} \tag{4.66}$$

or $A = QR$ as described. Finally, R is nonsingular since each $\|\mathbf{w}_i\| \neq 0$. ■

Example 4

Find the QR decomposition for

$$A = \begin{bmatrix} 1 & 0 & 0 \\ 1 & 1 & 0 \\ 1 & 1 & 1 \\ 1 & 1 & 1 \end{bmatrix}$$

Solution Note that the columns of A, $\mathbf{c}_1, \mathbf{c}_2, \mathbf{c}_3$, are the vectors $\mathbf{v}_1, \mathbf{v}_2, \mathbf{v}_3$, respectively, of Example 7 in Section 4.4. Then by that example,

$$\|\mathbf{w}_1\| = 2, \qquad \|\mathbf{w}_2\| = \tfrac{1}{2}\sqrt{3}, \qquad \|\mathbf{w}_3\| = \tfrac{1}{3}\sqrt{6}$$

$$\mathbf{c}_2 \cdot \mathbf{q}_1 = \mathbf{v}_2 \cdot \mathbf{u}_1 = \tfrac{3}{2}, \qquad \mathbf{c}_3 \cdot \mathbf{q}_1 = \mathbf{v}_3 \cdot \mathbf{u}_1 = 1, \qquad \mathbf{c}_3 \cdot \mathbf{q}_2 = \mathbf{v}_3 \cdot \mathbf{u}_2 = \tfrac{1}{3}\sqrt{3}$$

Substituting these values into Equation (4.65) gives us

$$\begin{aligned}
\mathbf{c}_1 &= 2\mathbf{q}_1 \\
\mathbf{c}_2 &= \tfrac{3}{2}\mathbf{q}_1 + \tfrac{1}{2}\sqrt{3}\mathbf{q}_2 \\
\mathbf{c}_3 &= \mathbf{q}_1 + \tfrac{1}{3}\sqrt{3}\mathbf{q}_2 + \tfrac{1}{3}\sqrt{6}\mathbf{q}_3
\end{aligned}$$

where

$$\mathbf{q}_1 = \begin{bmatrix} \frac{1}{2} \\ \frac{1}{2} \\ \frac{1}{2} \\ \frac{1}{2} \end{bmatrix}, \qquad \mathbf{q}_2 = \begin{bmatrix} -\frac{1}{2}\sqrt{3} \\ \frac{1}{6}\sqrt{3} \\ \frac{1}{6}\sqrt{3} \\ \frac{1}{6}\sqrt{3} \end{bmatrix}, \qquad \mathbf{q}_3 = \begin{bmatrix} 0 \\ -\frac{1}{3}\sqrt{6} \\ \frac{1}{6}\sqrt{6} \\ \frac{1}{6}\sqrt{6} \end{bmatrix}$$

Therefore the QR decomposition for A is

$$\begin{bmatrix} 1 & 0 & 0 \\ 1 & 1 & 0 \\ 1 & 1 & 1 \\ 1 & 1 & 1 \end{bmatrix} = \begin{bmatrix} \frac{1}{2} & -\frac{1}{2}\sqrt{3} & 0 \\ \frac{1}{2} & \frac{1}{6}\sqrt{3} & -\frac{1}{3}\sqrt{6} \\ \frac{1}{2} & \frac{1}{6}\sqrt{3} & \frac{1}{6}\sqrt{6} \\ \frac{1}{2} & \frac{1}{6}\sqrt{3} & \frac{1}{6}\sqrt{6} \end{bmatrix} \begin{bmatrix} 2 & \frac{3}{2} & 1 \\ & \frac{1}{2}\sqrt{3} & \frac{1}{3}\sqrt{3} \\ & & \frac{1}{3}\sqrt{6} \end{bmatrix}$$

This can be checked by multiplying out. ■

Least Squares (Revisited)

We commented in Section 4.3 that the normal equations $A^TA\mathbf{x} = A^T\mathbf{y}$ are usually ill conditioned. We shall now see how the QR decomposition helps the situation.

Suppose we have a least-squares problem $A\mathbf{x} = \mathbf{y}$, where A has linearly independent columns, and we also have the QR decomposition for A, $A = QR$. We substitute the decomposition into normal equations and simplify,

(4.67) $$A^TA\mathbf{x} = A^T\mathbf{y} \quad \text{becomes} \quad (QR)^TQR\mathbf{x} = (QR)^T\mathbf{y} \quad \text{or} \quad R^TQ^TQR\mathbf{x} = R^TQ^T\mathbf{y}$$

Now $Q^TQ = I_n$, by Theorem (4.59), and R is nonsingular. We multiply both sides of Equation (4.67) by $(R^T)^{-1}$ and simplify to obtain

$$(R^T)^{-1}R^TIR\mathbf{x} = (R^T)^{-1}R^TQ^T\mathbf{y} \qquad \text{or} \qquad R\mathbf{x} = Q^T\mathbf{y}$$

Thus we have proved the following.

(4.68) THEOREM If A has linearly independent columns, then by using the QR decomposition the normal equations can be rewritten as

$$R\mathbf{x} = Q^T\mathbf{y}$$

Since R is upper triangular, solving $R\mathbf{x} = Q^T\mathbf{y}$ is very straightforward.

Example 5 Use the QR decomposition to solve the least-squares problem

$$\begin{aligned} 2m + b &= 3 \\ 3m + b &= 4 \\ 4m + b &= 4 \\ 7m + b &= 7 \end{aligned}$$

Solution Observe that we are trying to find the equation of the straight line closest to (in the least-squares sense) the points (2, 3), (3, 4), (4, 4), (7, 7). We first write the problem in matrix form, keeping the b's first as this simplifies the computation

$$\begin{bmatrix} 1 & 2 \\ 1 & 3 \\ 1 & 4 \\ 1 & 7 \end{bmatrix}\begin{bmatrix} b \\ m \end{bmatrix} = \begin{bmatrix} 3 \\ 4 \\ 4 \\ 7 \end{bmatrix}$$

This is of the form $A\mathbf{x} = \mathbf{y}$, where A has linearly independent columns, so we apply the Gram–Schmidt process to the columns of A, $\mathbf{c}_1 = [1 \quad 1 \quad 1 \quad 1]^T$ and $\mathbf{c}_2 = [2 \quad 3 \quad 4 \quad 7]^T$.

Step 1 Here $\mathbf{w}_1 = \mathbf{c}_1$ and $\|\mathbf{c}_1\| = 2$. Thus

$$\mathbf{q}_1 = \frac{\mathbf{c}_1}{\|\mathbf{c}_1\|} = [\tfrac{1}{2} \quad \tfrac{1}{2} \quad \tfrac{1}{2} \quad \tfrac{1}{2}]^T \qquad \text{and} \qquad \mathbf{c}_1 = 2\mathbf{q}_1$$

Step 2 First compute $\mathbf{c}_2 \cdot \mathbf{q}_1 = \mathbf{c}_2 \cdot \dfrac{\mathbf{c}_1}{\|\mathbf{c}_1\|} = \dfrac{16}{2} = 8$. Thus

$$\begin{aligned} \mathbf{w}_2 &= \mathbf{c}_2 - (\mathbf{c}_2 \cdot \mathbf{q}_1)\mathbf{q}_1 \\ &= \mathbf{c}_2 - 8\mathbf{q}_1 \\ &= [2 \quad 3 \quad 4 \quad 7]^T - 8[\tfrac{1}{2} \quad \tfrac{1}{2} \quad \tfrac{1}{2} \quad \tfrac{1}{2}]^T \\ &= [-2 \quad -1 \quad 0 \quad 3]^T \end{aligned}$$

From this we see

$$\|\mathbf{w}_2\| = \sqrt{14} \qquad \text{and} \qquad \mathbf{q}_2 = \frac{\mathbf{w}_2}{\|\mathbf{w}_2\|} = \frac{1}{\sqrt{14}}[-2 \quad -1 \quad 0 \quad 3]^T$$

$$\text{and} \qquad \mathbf{c}_2 = 8\mathbf{q}_1 + \sqrt{14}\mathbf{q}_2$$

Therefore,

$$A = [\mathbf{c}_1 \vdots \mathbf{c}_2] = [2\mathbf{q}_1 \vdots 8\mathbf{q}_1 + \sqrt{14}\mathbf{q}_2] = [\mathbf{q}_1 \vdots \mathbf{q}_2]\begin{bmatrix} 2 & 8 \\ 0 & \sqrt{14} \end{bmatrix}$$

Thus the QR decomposition for A is

$$\begin{bmatrix} 1 & 2 \\ 1 & 3 \\ 1 & 4 \\ 1 & 7 \end{bmatrix} = \begin{bmatrix} \frac{1}{2} & -\dfrac{2}{\sqrt{14}} \\ \frac{1}{2} & -\dfrac{1}{\sqrt{14}} \\ \frac{1}{2} & 0 \\ \frac{1}{2} & \dfrac{3}{\sqrt{14}} \end{bmatrix}\begin{bmatrix} 2 & 8 \\ 0 & \sqrt{14} \end{bmatrix}$$

From this $Q^T\mathbf{y}$ is

$$\begin{bmatrix} \frac{1}{2} & \frac{1}{2} & \frac{1}{2} & \frac{1}{2} \\ -\frac{2}{\sqrt{14}} & -\frac{1}{\sqrt{14}} & 0 & \frac{3}{\sqrt{14}} \end{bmatrix} \begin{bmatrix} 3 \\ 4 \\ 4 \\ 7 \end{bmatrix} = \begin{bmatrix} 9 \\ \frac{11}{\sqrt{14}} \end{bmatrix}$$

so the equation $R\mathbf{x} = Q^T\mathbf{y}$ is

$$\begin{bmatrix} 2 & 8 \\ 0 & \sqrt{14} \end{bmatrix} \begin{bmatrix} b \\ m \end{bmatrix} = \begin{bmatrix} 9 \\ \frac{11}{\sqrt{14}} \end{bmatrix} \quad \text{or} \quad \begin{aligned} 2b + 8m &= 9 \\ \sqrt{14}m &= \frac{11}{\sqrt{14}} \end{aligned}$$

Therefore, $m = \frac{11}{14}$ and $b = \dfrac{9 - 8m}{2} = \dfrac{9 - \frac{44}{7}}{2} = \frac{19}{14}$, and the equation for the desired line is $y = \frac{11}{14}x + \frac{19}{14}$. ■

Exercise 4.5 In Exercises 1–6, find a last column or row so that the resulting matrix is orthogonal.

1. $\begin{bmatrix} \frac{1}{\sqrt{5}} & \\ \frac{2}{\sqrt{5}} & \end{bmatrix}$

2. $\begin{bmatrix} \frac{1}{\sqrt{2}} & -\frac{1}{\sqrt{2}} \\ & \end{bmatrix}$

3. $\begin{bmatrix} \frac{3}{5} & 0 & \\ \frac{4}{5} & 0 & \\ 0 & 1 & \end{bmatrix}$

4. $\begin{bmatrix} \frac{1}{\sqrt{3}} & \frac{1}{\sqrt{6}} & \\ \frac{1}{\sqrt{3}} & \frac{1}{\sqrt{6}} & \\ -\frac{1}{\sqrt{3}} & \frac{2}{\sqrt{6}} & \end{bmatrix}$

5. $\begin{bmatrix} -\frac{3}{7} & \frac{2}{7} & \frac{6}{7} \\ \frac{6}{7} & \frac{3}{7} & \frac{2}{7} \\ & & \end{bmatrix}$

6. $\begin{bmatrix} \frac{1}{2} & \frac{1}{2} & \frac{1}{2} & \frac{1}{2} \\ -\frac{1}{2} & -\frac{1}{2} & \frac{1}{2} & \frac{1}{2} \\ \frac{1}{2} & -\frac{1}{2} & -\frac{1}{2} & \frac{1}{2} \\ & & & \end{bmatrix}$

In Exercises 7–12, a matrix is given. Determine if it is orthogonal. If it is, find its inverse.

7. $\begin{bmatrix} 0 & -1 \\ 1 & 0 \end{bmatrix}$

8. $\begin{bmatrix} \frac{3}{5} & -\frac{4}{5} \\ \frac{4}{5} & \frac{3}{5} \end{bmatrix}$

9. $\begin{bmatrix} 0 & 1 & 0 \\ 1 & 0 & 0 \\ \dfrac{1}{\sqrt{2}} & 0 & \dfrac{1}{\sqrt{2}} \end{bmatrix}$

10. $\begin{bmatrix} \frac{1}{9} & \frac{4}{5} & \frac{3}{7} \\ \frac{4}{9} & \frac{3}{5} & -\frac{2}{7} \\ \frac{8}{9} & -\frac{2}{5} & \frac{3}{7} \end{bmatrix}$

11. $\begin{bmatrix} 1 & 0 & 0 & 0 \\ 0 & \dfrac{1}{\sqrt{2}} & \dfrac{1}{\sqrt{3}} & \dfrac{1}{\sqrt{6}} \\ 0 & 0 & \dfrac{1}{\sqrt{3}} & -\dfrac{2}{\sqrt{6}} \\ 0 & -\dfrac{1}{\sqrt{2}} & \dfrac{1}{\sqrt{3}} & \dfrac{1}{\sqrt{6}} \end{bmatrix}$

12. $\begin{bmatrix} \frac{1}{6} & -\frac{5}{6} & \frac{1}{6} & \frac{1}{2} \\ -\frac{5}{6} & \frac{1}{6} & \frac{1}{6} & \frac{1}{2} \\ \frac{1}{6} & \frac{1}{6} & -\frac{5}{6} & \frac{1}{2} \\ \frac{1}{2} & \frac{1}{2} & \frac{1}{2} & \frac{1}{2} \end{bmatrix}$

In Exercises 12–16, find the QR decomposition of the given matrix.

13. $\begin{bmatrix} 4 & 1 \\ 3 & -1 \end{bmatrix}$

14. $\begin{bmatrix} 3 & 5 \\ -4 & 0 \end{bmatrix}$

15. $\begin{bmatrix} 1 & 1 \\ 2 & 1 \\ -2 & 1 \end{bmatrix}$

16. $\begin{bmatrix} 1 & -1 & -1 \\ 1 & 1 & 0 \\ 1 & 0 & 1 \end{bmatrix}$

In Exercises 17–21, graph the points and draw the line that you estimate seems to fit them best. Then set up the least-squares problem to find that line, and solve it using the QR decomposition. Graph the solution, and compare it with your estimate.

17. $(0, -2), (1, -1), (2, 2), (3, 2)$

18. $(-1, 0), (0, 0), (1, 2), (2, 2)$

19. $(-3, 2), (-1, 1), (0, 0), (2, -1), (3, -1)$

20. $(-5, 1), (-4, 2), (-3, 1), (2, -2), (3, -2)$

21. Argue as in the proof of Theorem (4.59) to show that

$$QQ^T = I \quad \text{if and only} \quad \text{if the rows of } Q \text{ are orthonormal.}$$

22. Use $\|\mathbf{u}\|^2 = \mathbf{u} \cdot \mathbf{u}$ to show

$$\mathbf{x} \cdot \mathbf{y} = \tfrac{1}{4}(\|\mathbf{x} + \mathbf{y}\|^2 - \|\mathbf{x} - \mathbf{y}\|^2)$$

23. Use Exercise 22 to show: If $\|Q\mathbf{u}\| = \|\mathbf{u}\|$, for all $\mathbf{u}$, then $Q\mathbf{u} \cdot Q\mathbf{v} = \mathbf{u} \cdot \mathbf{v}$, for all $\mathbf{u}$, $\mathbf{v}$.

24. Show that A is invertible $\Leftrightarrow A = QR$, where Q is orthogonal and R is upper triangular with nonzero entries on its diagonal.

4.6 Encoding the QR Decomposition—A Geometric Approach*

In the previous section we used the Gram–Schmidt process to prove that matrices have QR decompositions. However, actually using the Gram–Schmidt process to encode this decomposition is fraught with danger. All the subtractions can lead to a mammoth buildup of error. Hence other approaches have been developed.

There are two facts at the core of the approach we shall cover. The first fact is that the product of orthogonal matrices is orthogonal [see Theorem (4.63), Section 4.5]. Hence if we can find orthogonal matrices $Q_1, \ldots, Q_k$ such that

(4.69)
$$Q_k \cdots Q_1 A = R, \qquad R \text{ is upper triangular}$$

then

(4.70)
$$A = Q_1^T \cdots Q_k^T R = QR$$

where $Q = Q_1^T \cdots Q_k^T$ is orthogonal (since each $Q_i^T = Q_i^{-1}$).

Note that this is a slightly different decomposition from the one obtained in Theorem (4.64), Section 4.5. Here if A is $m \times n$, Q will actually be square $m \times m$ and R will be $m \times n$ upper triangular,

$$R = \begin{bmatrix} * & * & \cdots & * \\ 0 & * & \cdots & * \\ & & \cdots & \\ 0 & 0 & \cdots & * \\ & & \cdots & \\ 0 & 0 & \cdots & 0 \end{bmatrix} = \begin{bmatrix} R' \\ \cdots \\ 0 \end{bmatrix}$$

Here, R' is $n \times n$ upper triangular and is the upper triangular R obtained in Theorem (4.64).

The second fact is that for most applications we do not need to know Q explicitly, so it is sufficient to leave Q in factored form. In addition, we shall obtain a very simple and easily storable form of the Q_i's.

The reason why it is advantageous to approach the QR decomposition this way is that orthogonal matrices preserve length, and hence multiplying by orthogonal matrices does not increase the size of errors. Technically we say this method is **stable**.

* *This section is optional. The material is nontraditional for this course, but it is in the spirit of mentioning numerical approaches to the mathematics discussed in the text. The geometry is simple and elegant in dealing with the problem of computing the QR decomposition, and mathematically the material is well within the level of this text. This section is used only in Section 5.8 and there only if you choose this way to encode the QR algorithm for finding eigenvalues.*

Geometric Background

Suppose we have two different nonzero vectors, $\mathbf{v}$ and $\mathbf{w}$, in $\mathbb{R}^n$, and we wish to find a linear transformation that interchanges $\mathbf{v}$ and $\mathbf{w}$. If we also happen to know that $\|\mathbf{v}\| = \|\mathbf{w}\|$, then there is an easy way to perform this interchange. Consider $\mathbf{v} - \mathbf{w}$ and let P be the hyperplane perpendicular to $\mathbf{v} - \mathbf{w}$, that is, $P = \{\mathbf{x} \text{ in } \mathbb{R}^n | \mathbf{x} \cdot (\mathbf{v} - \mathbf{w}) = 0\}$. See Figure 4.12.

Figure 4.12

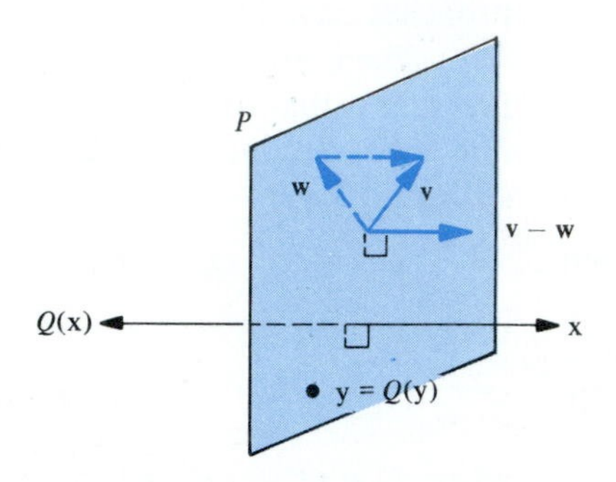

Let Q be the function that **reflects** $\mathbb{R}^n$ through P. Intuitively what this means is to find $Q(\mathbf{x})$, we go perpendicularly to P, going through P to the other side and continuing until we are the same distance from P as $\mathbf{x}$ is; this is $Q(\mathbf{x})$. If $\mathbf{y}$ is in P, $Q(\mathbf{y}) = \mathbf{y}$.

Algebraically,

$$Q(\mathbf{x}) = \mathbf{x} - 2\mathbf{p} \tag{4.71}$$

where $\mathbf{p}$ is the projection of $\mathbf{x}$ on $\mathbf{v} - \mathbf{w}$. It is not hard to see geometrically that Q is linear and Q interchanges $\mathbf{v}$ and $\mathbf{w}$. We shall call such a Q an **elementary reflection function**.

NOTE We could have projected onto $\mathbf{w} - \mathbf{v}$ instead. This freedom will be helpful to us later.

Example 1 Let $\mathbf{v} = (1, 0, 0)$ and $\mathbf{w} = (0, 1, 0)$. Let Q be the function that reflects $\mathbf{v}$ to $\mathbf{w}$ and $\mathbf{w}$ to $\mathbf{v}$. Find

(a) Find the hyperplane, P, left fixed by Q.
(b) Find $Q(0, 0, 1)$ and $Q(1, 2, 3)$.
(c) Sketch.

Solution (a) Since P is perpendicular to $\mathbf{v} - \mathbf{w} = (1, -1, 0)$, the equation for P is

$$(\mathbf{v} - \mathbf{w}) \cdot (x, y, z) = 0 \qquad \text{or} \qquad x - y = 0$$

(b) Geometrically Q interchanges the x- and y-axes and leaves the z-axis fixed. Thus $Q(x, y, z) = (y, x, z)$, so $Q(0, 0, 1) = (0, 0, 1)$ and $Q(1, 2, 3) = (2, 1, 3)$.

(c) Seee Figure 4.13. ■

Figure 4.13

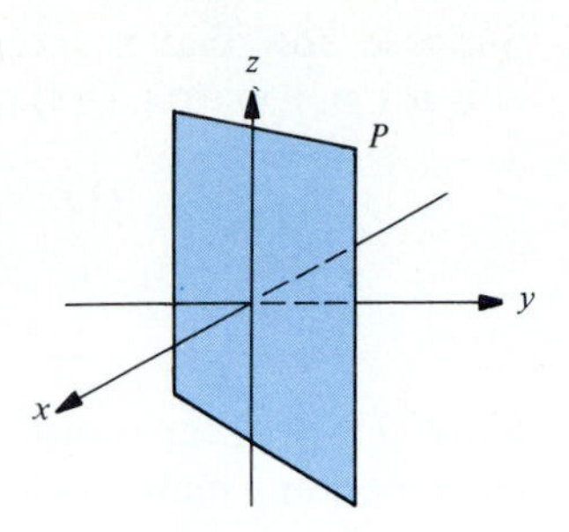

Algebraic Description

It is not hard to express the formula for Q given in Equation (4.71) in terms of $\mathbf{v}$ and $\mathbf{w}$. Since $\mathbf{v} \neq \mathbf{w}$, $\mathbf{v} - \mathbf{w} \neq \mathbf{0}$, so by Theorem (3.18), Section 3.2, the projection, $\mathbf{p}$, of $\mathbf{x}$ on $\mathbf{v} - \mathbf{w}$ is

$$\mathbf{p} = \frac{\mathbf{x} \cdot (\mathbf{v} - \mathbf{w})}{(\mathbf{v} - \mathbf{w}) \cdot (\mathbf{v} - \mathbf{w})} (\mathbf{v} - \mathbf{w})$$

Substituting this into Equation (4.71) yields

$$Q(\mathbf{x}) = \mathbf{x} - 2\frac{\mathbf{x} \cdot (\mathbf{v} - \mathbf{w})}{(\mathbf{v} - \mathbf{w}) \cdot (\mathbf{v} - \mathbf{w})} (\mathbf{v} - \mathbf{w}) \tag{4.72}$$

We could now verify directly that $Q(\mathbf{v}) = \mathbf{w}$ and $Q(\mathbf{w}) = \mathbf{v}$, using Equation (4.72) (and the assumption that $\|\mathbf{v}\| = \|\mathbf{w}\|$ and hence $\mathbf{v} \cdot \mathbf{v} = \mathbf{w} \cdot \mathbf{w}$). See Exercises 33 and 34.

For the use we shall want, $\mathbf{v}$ will be a particular nonzero vector and $\mathbf{w} = a\|\mathbf{v}\|\boldsymbol{\varepsilon}_1$, where $\boldsymbol{\varepsilon}_1 = (1, 0, \ldots, 0)$ and $a = \pm 1$. For this special situation, the denominator of the fraction in Equation (4.72) becomes $(\mathbf{v} - a\|\mathbf{v}\|\boldsymbol{\varepsilon}_1) \cdot (\mathbf{v} - a\|\mathbf{v}\|\boldsymbol{\varepsilon}_1)$. If $\mathbf{v} = (v_1, \ldots, v_n)$ and $v_1 \approx \|\mathbf{v}\|$, that is, the other components are small, then we might have disastrous cancellation in the first component of $\mathbf{v} - a\|\mathbf{v}\|\boldsymbol{\varepsilon}_1$, resulting in a denominator's being close to zero. However, if we take

$$a = -1 \quad \text{if} \quad v_1 \geq 0 \qquad \text{and} \qquad a = +1 \quad \text{if} \quad v_1 < 0 \tag{4.73}$$

then we can guarantee there is no disastrous cancellation. Now set

$$\mathbf{u} = \frac{\mathbf{v} - a\|\mathbf{v}\|\boldsymbol{\varepsilon}_1}{\|\mathbf{v} - a\|\mathbf{v}\|\boldsymbol{\varepsilon}_1\|} \tag{4.74}$$

where a is given in (4.73). Then

$$\begin{aligned} Q(\mathbf{x}) &= \mathbf{x} - 2\frac{\mathbf{x} \cdot (\mathbf{v} - a\|\mathbf{v}\|\boldsymbol{\varepsilon}_1)}{\|\mathbf{v} - a\|\mathbf{v}\|\boldsymbol{\varepsilon}_1\|^2} (\mathbf{v} - a\|\mathbf{v}\|\boldsymbol{\varepsilon}_1) \\ &= \mathbf{x} - 2\left[\mathbf{x} \cdot \frac{\mathbf{v} - a\|\mathbf{v}\|\boldsymbol{\varepsilon}_1}{\|\mathbf{v} - a\|\mathbf{v}\|\boldsymbol{\varepsilon}_1\|}\right] \frac{\mathbf{v} - a\|\mathbf{v}\|\boldsymbol{\varepsilon}_1}{\|\mathbf{v} - a\|\mathbf{v}\|\boldsymbol{\varepsilon}_1\|} \\ &= \mathbf{x} - 2[\mathbf{x} \cdot \mathbf{u}]\mathbf{u} \end{aligned}$$

Suppose now that $\mathbf{x}$ and $\mathbf{u}$ are column vectors. Using the fact $A = A^T$ if A is a 1×1 matrix, $Q(\mathbf{x})$ simplifies to

$$\begin{aligned} Q(\mathbf{x}) &= \mathbf{x} - 2(\mathbf{x}^T\mathbf{u})\mathbf{u} = \mathbf{x} - 2(\mathbf{x}^T\mathbf{u})^T\mathbf{u} \\ &= \mathbf{x} - 2(\mathbf{u}^T\mathbf{x})\mathbf{u} = \mathbf{x} - 2\mathbf{u}(\mathbf{u}^T\mathbf{x}) \\ &= \mathbf{x} - 2(\mathbf{u}\mathbf{u}^T)\mathbf{x} = (\mathbf{I} - 2\mathbf{u}\mathbf{u}^T)\mathbf{x} \end{aligned}$$

Since Q is a linear transformation, it has [by Theorem (4.4), Section 4.1] a corresponding matrix, which we also denote by Q. Thus we have just shown the matrix $Q = I - 2\mathbf{u}\mathbf{u}^T$ and have proved the first part of the following theorem.

(4.75)

THEOREM Let $\mathbf{v}$ be a nonzero column vector in $\mathbb{R}^n$, let a be the same as in (4.73), and let

$$\mathbf{u} = \frac{\mathbf{v} - a\|\mathbf{v}\|\boldsymbol{\varepsilon}_1}{\|\mathbf{v} - a\|\mathbf{v}\|\boldsymbol{\varepsilon}_1\|}$$

so that $\|\mathbf{u}\| = 1$, and let

$$Q = I - 2\mathbf{u}\mathbf{u}^T$$

Then

(a) Q is the matrix corresponding to the elementary reflection function that interchanges $\mathbf{v}$ and $a\|\mathbf{v}\|\boldsymbol{\varepsilon}_1$.

(b) Q is symmetric and orthogonal.

DEFINITION We shall call matrices Q obtained in this way **Householder** matrices.

Proof We have already proved part (a).

(b) We first show Q is symmetric by showing $Q^T = Q$.

$$Q^T = (I - 2\mathbf{u}\mathbf{u}^T)^T = I^T - 2(\mathbf{u}\mathbf{u}^T)^T = I - 2(\mathbf{u}^T)^T\mathbf{u}^T = Q$$

We next show Q is orthogonal by showing $QQ^T = I$. Since $Q^T = Q$,

$$\begin{aligned} QQ^T &= (I - 2\mathbf{u}\mathbf{u}^T)(I - 2\mathbf{u}\mathbf{u}^T) = I - 4\mathbf{u}\mathbf{u}^T + 2\mathbf{u}\mathbf{u}^T 2\mathbf{u}\mathbf{u}^T \\ &= I - 4\mathbf{u}\mathbf{u}^T + 4\mathbf{u}\mathbf{u}^T\mathbf{u}\mathbf{u}^T = I - 4\mathbf{u}\mathbf{u}^T + 4\mathbf{u}\|\mathbf{u}\|^2\mathbf{u}^T \\ &= I - 4\mathbf{u}\mathbf{u}^T + 4\mathbf{u}\mathbf{u}^T = I \end{aligned}$$

since $\|\mathbf{u}\| = 1$. ■

Example 2 Find the function Q described in Theorem (4.75) for the vectors

(a) $\mathbf{v} = (2, 1, -2)$ in $\mathbb{R}^3$ (b) $\mathbf{v} = (-5, 0, 0, 0)$ in $\mathbb{R}^4$

Solution (a) First, by (4.73), $a = -1$ since $\mathbf{v}_1 \geq 0$. Next, $\|\mathbf{v}\| = 3$, as you can check. Thus

$$\mathbf{v} - a\|\mathbf{v}\|\boldsymbol{\varepsilon}_1 = \mathbf{v} + 3\boldsymbol{\varepsilon}_1 = (5, 1, -2) \qquad \text{so} \qquad \mathbf{u} = \frac{\mathbf{v} + 3\boldsymbol{\varepsilon}_1}{\|\mathbf{v} + 3\boldsymbol{\varepsilon}_1\|} = \frac{1}{\sqrt{30}}(5, 1, -2)$$

We write $\mathbf{u}$ as a column vector and compute

$$Q = I - 2\mathbf{u}\mathbf{u}^T = I - 2\frac{1}{\sqrt{30}}\begin{bmatrix} 5 \\ 1 \\ -2 \end{bmatrix}\frac{1}{\sqrt{30}}[5 \quad 1 \quad -2]$$

$$= \begin{bmatrix} 1 & & \\ & 1 & \\ & & 1 \end{bmatrix} - \tfrac{2}{30}\begin{bmatrix} 25 & 5 & -10 \\ 5 & 1 & -2 \\ -10 & -2 & 4 \end{bmatrix} = \begin{bmatrix} -\frac{2}{3} & -\frac{1}{3} & \frac{2}{3} \\ -\frac{1}{3} & \frac{14}{15} & \frac{2}{15} \\ \frac{2}{3} & \frac{2}{15} & \frac{11}{15} \end{bmatrix}$$

(b) First, by (4.73), $a = 1$ for this $\mathbf{v}$ since $v_1 < 0$. Next, easily $\|\mathbf{v}\| = 5$. Thus

$$\mathbf{v} - a\|\mathbf{v}\|\boldsymbol{\varepsilon}_1 = \mathbf{v} - 5\boldsymbol{\varepsilon}_1 = (-5, 0, 0, 0) - (5, 0, 0, 0) = (-10, 0, 0, 0)$$

so

$$\mathbf{u} = \frac{\mathbf{v} - 5\boldsymbol{\varepsilon}_1}{\|\mathbf{v} - 5\boldsymbol{\varepsilon}_1\|} = \frac{1}{10}(-10, 0, 0, 0) = -\boldsymbol{\varepsilon}_1$$

You can see the disaster that would have happened if we had picked $a = -1$. We now write $\mathbf{u} = -\boldsymbol{\varepsilon}_1$ as a column vector and compute

$$Q = I - 2\mathbf{u}\mathbf{u}^T = I - 2\begin{bmatrix} -1 \\ 0 \\ 0 \\ 0 \end{bmatrix}[-1 \quad 0 \quad 0 \quad 0]$$

$$= \begin{bmatrix} 1 & & & \\ & 1 & & \\ & & 1 & \\ & & & 1 \end{bmatrix} - 2\begin{bmatrix} 1 & & & \\ & 0 & & \\ & & 0 & \\ & & & 0 \end{bmatrix} = \begin{bmatrix} -1 & & & \\ & 1 & & \\ & & 1 & \\ & & & 1 \end{bmatrix}$$ ■

NOTE The answer we obtained for part (b) is the one we always get if $\mathbf{u}$ is a multiple of $\boldsymbol{\varepsilon}_1$. See Exercise 32.

The Algorithm

We now proceed with the description of the algorithm that will compute the decomposition $A = QR$. The idea is very much like Gaussian elimination in that at each stage we make all the entries below the diagonal in a fixed column equal to zero. However, here we obtain the zeros using reflection matrices instead of row operations.

To keep the description simple A must be an $m \times n$ matrix

$$A = \begin{bmatrix} a_{11} & \cdots & a_{1n} \\ \vdots & & \vdots \\ a_{m1} & \cdots & a_{mn} \end{bmatrix}$$

with linearly independent columns $\mathbf{c}_1, \ldots, \mathbf{c}_n$. Thus $A = [\mathbf{c}_1 \vdots \cdots \vdots \mathbf{c}_n]$ and $m \geq n$.

Because c_1 is nonzero (since the columns are linearly independent), we can apply Theorem (4.75) with $\mathbf{c}_1 = \mathbf{v}$ and obtain an

(4.76) m-dimensional vector $\mathbf{u}_1$ and an $m \times m$ orthogonal matrix Q_1

with the property $Q_1\mathbf{c}_1 = \pm\|\mathbf{c}_1\|\boldsymbol{\varepsilon}_1$. Let $A_1 = Q_1 A$ so that

$$A_1 = Q_1[\mathbf{c}_1 \vdots \mathbf{c}_2 \vdots \cdots \vdots \mathbf{c}_n] = [Q_1\mathbf{c}_1 \vdots Q_1\mathbf{c}_1 \vdots \cdots \vdots Q_1\mathbf{c}_n]$$
$$= \begin{bmatrix} \pm\|\mathbf{c}_1\| & a'_{12} & \cdots & a'_{1n} \\ 0 & a'_{22} & \cdots & a'_{2n} \\ \vdots & \vdots & & \vdots \\ 0 & a'_{m2} & \cdots & a'_{mn} \end{bmatrix}$$

For the second step, we first work with the last $m - 1$ entries of the second column, $\mathbf{c}'_2 = [a'_{22} \quad a'_{32} \quad \cdots \quad a'_{m2}]^T$. Since $\mathbf{c}'_2$ is nonzero (since otherwise $\mathbf{c}_2$ is a multiple of $\mathbf{c}_1$), we can again apply Theorem (4.75) and obtain an

(4.77) $(m - 1)$-dimensional $\mathbf{u}_2$ and an $(m - 1) \times (m - 1)$ orthogonal matrix Q'_2

with the property $Q'_2\mathbf{c}'_2 = \pm\|\mathbf{c}'_2\|\boldsymbol{\varepsilon}_1$. (Note that this $\boldsymbol{\varepsilon}_1$ is an $(m - 1)$-dimensional column vector.) If we form the $m \times m$ matrix Q_2 by

$$Q_2 = \left[\begin{array}{c|ccc} 1 & 0 & \cdots & 0 \\ \hline 0 & & & \\ \vdots & & Q'_2 & \\ 0 & & & \end{array}\right]$$

then you can check Q_2 is orthogonal. If $A_3 = Q_2 A_2$, then

$$A_3 = Q_2 A_2 = \begin{bmatrix} \pm\|\mathbf{c}_1\| & a'_{12} & a'_{13} & \cdots & a'_{1n} \\ 0 & \pm\|\mathbf{c}'_2\| & a''_{23} & \cdots & a''_{2n} \\ 0 & 0 & a''_{33} & \cdots & a''_{3n} \\ \vdots & \vdots & \vdots & & \vdots \\ 0 & 0 & a''_{m3} & \cdots & a''_{mn} \end{bmatrix}$$

For the third step, we work with the last $m - 2$ entries of the third column $\mathbf{c}''_3 = [a''_{33} \quad \cdots \quad a''_{m3}]^T$, obtain an

(4.78) $(m - 2)$-dimensional vector $\mathbf{u}_3$ and $(m - 2) \times (m - 2)$ orthogonal matrix Q'_3

and form Q_3 by

$$Q_3 = \left[\begin{array}{cc|c} 1 & & \\ & 1 & \\ \hline & & Q'_3 \end{array}\right]$$

We keep going in this manner, obtaining

$$\text{vectors } \mathbf{u}_1, \ldots, \mathbf{u}_n \qquad \text{of length} \qquad m, \ldots, m-n+1 \text{ respectively} \tag{4.79}$$

and orthogonal matrices $Q_1, \ldots, Q_n$ with the property

$$Q_n \cdots Q_1 A = R, \qquad \text{where } R \text{ is upper triangular} \tag{4.80}$$

as promised in Equation (4.69).* Thus since the Q_i's are symmetric, we end up back where we started, with Equation (4.70):

$$A = Q_1 \cdots Q_n R \qquad \text{or} \qquad A = QR$$

where $Q = Q_1 \cdots Q_n$ is orthogonal.

Programming the Algorithm

We now turn to considerations when programming the algorithm just discussed. We must present the mathematics behind the algorithm in such a way that we can both efficiently encode and store the results and then efficiently access the results later. One of the keys is the fact that $Q = I - 2\mathbf{u}\mathbf{u}^T$; how this is used is illustrated in the next example.

Example 3 Let Q and $\mathbf{u}$ be as computed in Example 2(a), and let $\mathbf{x} = [3 \quad -1 \quad 2]^T$. Find $Q\mathbf{x}$ in two ways

(a) Compute $Q\mathbf{x}$ directly.
(b) Compute $(I - 2\mathbf{u}\mathbf{u}^T)\mathbf{x}$.

Solution Using the computation in Example 2(a),

(a)
$$Q\mathbf{x} = \begin{bmatrix} -\frac{2}{3} & -\frac{1}{3} & \frac{2}{3} \\ -\frac{1}{3} & \frac{14}{15} & \frac{2}{15} \\ \frac{2}{3} & \frac{2}{15} & \frac{11}{15} \end{bmatrix} \begin{bmatrix} 3 \\ -1 \\ 2 \end{bmatrix} = \begin{bmatrix} -\frac{1}{3} \\ -\frac{5}{3} \\ \frac{10}{3} \end{bmatrix}$$

(b)
$$(I - 2\mathbf{u}\mathbf{u}^T)\mathbf{x} = \mathbf{x} - 2\mathbf{u}\mathbf{u}^T\mathbf{x} = \mathbf{x} - 2\frac{1}{\sqrt{30}}\begin{bmatrix} 5 \\ 1 \\ -2 \end{bmatrix}\frac{1}{\sqrt{30}}[5 \quad 1 \quad -2]\begin{bmatrix} 3 \\ -1 \\ 2 \end{bmatrix}$$

$$= \begin{bmatrix} 3 \\ -1 \\ 2 \end{bmatrix} - \frac{2}{35}\begin{bmatrix} 5 \\ 1 \\ -2 \end{bmatrix} 10 = \begin{bmatrix} 3 \\ -1 \\ 2 \end{bmatrix} - \begin{bmatrix} \frac{10}{3} \\ \frac{2}{3} \\ -\frac{4}{3} \end{bmatrix} = \begin{bmatrix} -\frac{1}{3} \\ -\frac{5}{3} \\ \frac{10}{3} \end{bmatrix} \qquad ■$$

---

* *If $n = m$, that is, if A is square, there seems to be nothing to do for the last step. However you might note the algorithm still gives us a $\mathbf{u}_n$, the 1×1 vector $\mathbf{u}_n = [\pm 1]$.*

This example illustrates that, of course, you get the same thing either way. But go back and examine how much work it took after we had $\mathbf{u}$ to compute Q and then to compute $Q\mathbf{x}$. Then compare this with how much it took just to compute $[I - 2\mathbf{u}\mathbf{u}^T]\mathbf{x}$. Not only is this less work, but if we had n Q's all of which were $m \times m$, the total work would be considerably less using $(I - 2\mathbf{u}\mathbf{u}^T)$'s.

Moreover, there is a question of storage. Do we compute the product $Q = Q_1 \cdots Q_n$ (more work) and store Q and R, or do we store R and all the Q_i's (much more storage)? If we are content to work with the $(I - 2\mathbf{u}\mathbf{u}^T)$'s, we need only store the $\mathbf{u}$'s. Furthermore, as shown in (4.76)–(4.79), the $\mathbf{u}$'s decrease in size. Thus we can store R, which is upper triangular, and $\mathbf{u}_1, \ldots, \mathbf{u}_n$ in an $(m + 1) \times n$ matrix

(4.81)

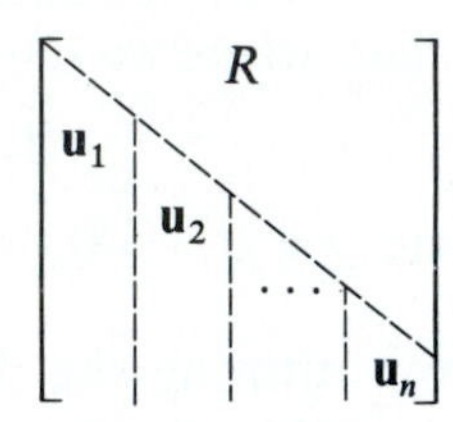

This is the essence of the idea, and a reasonable code could now be written. However, there are a few more technicalities to a very efficient code. If you are interested in more details, an excellent source is G. W. Stewart, *Introduction to Matrix Computations*, Academic Press, 1973, page 236, and associated material.

Exercise 4.6

1. In $\mathbb{R}^3$ let $\mathbf{v} = (0, 1, 0)$ and $\mathbf{w} = (0, 0, 1)$. Let Q be the function that reflects $\mathbf{v}$ to $\mathbf{w}$ and $\mathbf{w}$ to $\mathbf{v}$.

   (a) Find the hyperplane left fixed by Q.
   (b) Find $Q(1, 2, 3)$, $Q(4, 1, 1)$, $Q(0, -2, 2)$.

2. Same question as Exercise 1 assuming $\mathbf{v} = (1, 1, 0)$ and $\mathbf{w} = (-1, 1, 0)$.

In Exercises 3–10, (a) use (4.72) to find an algebraic description of the function Q that reflects the two given vectors, and (b) find $Q(\boldsymbol{\varepsilon}_1)$, $Q(\boldsymbol{\varepsilon}_1 + \boldsymbol{\varepsilon}_2)$, and $Q(\boldsymbol{\varepsilon}_1 - \boldsymbol{\varepsilon}_2)$, where $\boldsymbol{\varepsilon}_1, \ldots, \boldsymbol{\varepsilon}_n$ is the standard basis for $\mathbb{R}^n$.

3. $(0, 1), (-1, 0)$
4. $(3, 4), (3, -4)$
5. $(0, 5, 0), (0, 3, 4)$
6. $(1, 1, 1), (0, 0, \sqrt{3})$
7. $(2, 1, 2), (3, 0, 0)$
8. $(2, 1, 0), (0, 1, -2)$
9. $(1, 1, 1, 1), (0, 0, 0, 2)$
10. $(1, 1, 0, 0), (0, 0, 1, 1)$

In Exercises 11–16, find the $n \times n$ matrix $Q = I - 2\mathbf{u}\mathbf{u}^T$ for the given vector $\mathbf{u}$.

11. $(1, 1)$
12. $(-3, 4)$
13. $(-1, 1, -\sqrt{2})$
14. $(\sqrt{3}, 0, 1)$
15. $(1, 1, 1, 1)$
16. $(0, 0, 1, -1)$

In Exercises 17–22, find the vector $\mathbf{u}$ such that $Q = I - 2\mathbf{u}\mathbf{u}^T$ as described in Theorem (4.75). Then find $Q(\boldsymbol{\varepsilon}_1 + \boldsymbol{\varepsilon}_2)$.

17. $(3, -4)$
18. $(-4, 3)$
19. $(1, 2, -2)$
20. $(-2, -2, 1)$
21. $(-1, -1, 1, -1)$
22. $(2, -2, 4, 5)$

In Exercises 23–31, apply the alogrithm described in the text to obtain a QR decomposition. Give your answer (a) in the form indicated in (4.81) and (b) as matrices Q and R. If numbers become messy, round to three significant figures.

23. $\begin{bmatrix} 0 & 3 \\ 2 & 1 \end{bmatrix}$
24. $\begin{bmatrix} -3 & 2 \\ 4 & -3 \end{bmatrix}$
25. $\begin{bmatrix} -1 & 3 \\ 2 & 1 \\ -2 & 2 \end{bmatrix}$
26. $\begin{bmatrix} 1 & 2 \\ 1 & 1 \\ \sqrt{2} & 2 \end{bmatrix}$
27. $\begin{bmatrix} 0 & 1 & 1 \\ 1 & 0 & 1 \\ 1 & 1 & 0 \end{bmatrix}$
28. $\begin{bmatrix} -1 & 0 & 0 \\ 1 & 1 & 0 \\ 1 & 1 & 1 \end{bmatrix}$
29. $\begin{bmatrix} 1 & 1 \\ 1 & 2 \\ 1 & 0 \\ 1 & 1 \end{bmatrix}$
30. $\begin{bmatrix} -2 & 1 \\ 2 & 1 \\ -4 & 1 \\ 5 & 1 \end{bmatrix}$
31. $\begin{bmatrix} 1 & 1 & 2 \\ \sqrt{2} & \sqrt{2} & 0 \\ 1 & 1 & 0 \\ 0 & 1 & 1 \end{bmatrix}$

32. Let $\mathbf{v} = k\boldsymbol{\varepsilon}_1$, k a nonzero constant. Show that Equation (4.74) gives $\mathbf{u} = \pm\boldsymbol{\varepsilon}_1$ and that the resulting matrix Q given by Theorem (4.75) is

$$Q = \begin{bmatrix} -1 & & & \\ & 1 & & \\ & & \ddots & \\ & & & 1 \end{bmatrix}$$

33. Suppose $\mathbf{v} \neq \mathbf{w}$ but $\|\mathbf{v}\| = \|\mathbf{w}\|$. Show that

$$\frac{\mathbf{v} \cdot (\mathbf{v} - \mathbf{w})}{(\mathbf{v} - \mathbf{w}) \cdot (\mathbf{v} - \mathbf{w})} = \frac{\|\mathbf{v}\|^2 - \mathbf{v} \cdot \mathbf{w}}{2(\|\mathbf{v}\|^2 - \mathbf{v} \cdot \mathbf{w})} = \frac{1}{2}$$

34. In Equation (4.72), use Exercise 33 to show $Q(\mathbf{v}) = \mathbf{w}$, and in a similar way show $Q(\mathbf{w}) = \mathbf{v}$.

4.7 General Matrices of Linear Transformations; Similarity*

In Section 4.1, we saw how matrices correspond to linear transformations, and vice versa. The standard basis in $\mathbb{R}^n$ plays a very important role in this correspondence. If we were to continue to work solely with the standard basis, most linear transformations would appear to be very complicated. However, if we use other bases, linear transformations (and hence the way matrices act) can be greatly simplified. Diagonalization in Chapter 5 and the singular value decomposition in Chapter 6 provide two examples, but others can be given, depending on the kind of action to be explained. Essentially, if we look at the behavior restricted to a particular basis, then the action can be broken up into very simple forms. This section establishes the language and machinery to make this procedure rigorous. We begin with an introductory example.

Example 1 Let $T: \mathbb{R}^2 \to \mathbb{R}^2$ be given by $T(x, y) = (2x + 3y, 4x + 3y)$. Following the constructions in the proof of Theorem (4.4), Section 4.1, we find

$$T(\boldsymbol{\varepsilon}_1) = T(1, 0) = (2, 4) = 2\boldsymbol{\varepsilon}_1 + 4\boldsymbol{\varepsilon}_2$$
$$T(\boldsymbol{\varepsilon}_2) = T(0, 1) = (3, 3) = 3\boldsymbol{\varepsilon}_1 + 3\boldsymbol{\varepsilon}_2$$

If we put these coordinates into columns, we obtain

$$A = [T] = \begin{bmatrix} 2 & 3 \\ 4 & 3 \end{bmatrix}$$

For reasons that will be explained in Chapter 5, consider the vectors $\mathbf{v}_1 = [1 \quad -1]^T$ and $\mathbf{v}_2 = [3 \quad 4]^T$. Then $B = \{\mathbf{v}_1, \mathbf{v}_2\}$ is a basis for $\mathbb{R}^2$ and

$$T(\mathbf{v}_1) = T(1, -1) = (-1, 1) = -\mathbf{v}_1 = -\mathbf{v}_1 + 0\mathbf{v}_2$$
$$T(\mathbf{v}_2) = T(3, 4) = (18, 24) = 6\mathbf{v}_2 = 0\mathbf{v}_1 + 6\mathbf{v}_2$$

From Section 3.8, we see that

$$[T(\mathbf{v}_1)]_B = \begin{bmatrix} -1 \\ 0 \end{bmatrix}, \qquad [T(\mathbf{v}_2)]_B = \begin{bmatrix} 0 \\ 6 \end{bmatrix}$$

Forming a matrix with these vectors as columns, we obtain

$$\begin{bmatrix} -1 & 0 \\ 0 & 6 \end{bmatrix}$$

This matrix, denoted by $[T]_B$, is a very simple one. As we shall see in the following subsection, $[T]_B$ has properties similar to $[T]$ when the standard basis is replaced with B. ■

* *This section is optional and is not a prerequisite for any other part of the text. It will, of course, enrich the understanding of diagonalization in Chapter 5. Some instructors may prefer to introduce Section 4.7 earlier; it may be presented anywhere after Section 4.1.*

General Matrices of Linear Transformations

Before diving into the theory, we must emphasize that what we are about to discuss is not just used to simplify matrices. In Section 4.1, we dealt with $\mathbb{R}^n$; we now find out how to extend the concept of $[T]$ to all finite-dimensional nonzero vector spaces. You might quickly review Definition (4.2) and Theorem (4.4) and its proof in Section 4.1 in preparation for our main theorem, which follows.

(4.82)

Let V and W be finite-dimensional nonzero vector spaces with bases $B = \{\mathbf{v}_1, \ldots, \mathbf{v}_n\}$ and $C = \{\mathbf{w}_1, \ldots, \mathbf{w}_m\}$, respectively. Let $T: V \to W$ be a linear transformation.

DEFINITION The **matrix induced by T relative to the bases B and C**, denoted by $[T]_{BC}$, is the $m \times n$ matrix formed by placing the coordinate matrices $[T(\mathbf{v}_i)]_C$ as columns:

$$[T]_{BC} = \left[[T(\mathbf{v}_1)]_C \,\vdots\, [T(\mathbf{v}_2)]_C \,\vdots \cdots \vdots\, [T(\mathbf{v}_n)]_C\right]$$

THEOREM The matrix $[T]_{BC}$ has the very important property

$$[T]_{BC}[\mathbf{x}]_B = [T(\mathbf{x})]_C$$

for all vectors $\mathbf{x}$ in V.

NOTATION In the special case $W = V$ and $C = B$, so that $T: V \to V$, we can simplify $[T]_{BB}$ to $[T]_B$ so that Theorem (4.82) reads

$$[T]_B[\mathbf{x}]_B = [T(\mathbf{x})]_B$$

NOTE We also say the matrix $A = [T]_{BC}$ **represents** the linear transformation T in the bases B and C.

First we give some examples and then the proof.

Example 2 Let $T: P_2 \to P_3$ be given by $T[p(x)] = xp(x)$, and let $B = \{1, x, x^2\}$ and $C = \{1, x, x^2, x^3\}$ be the standard bases for P_2 and P_3, respectively. (You can check that T is linear.) Then

$$\begin{aligned} T(1) &= x & &\text{so} & [T(1)]_C &= [0 \;\; 1 \;\; 0 \;\; 0]^T \\ T(x) &= x^2 & &\text{so} & [T(x)]_C &= [0 \;\; 0 \;\; 1 \;\; 0]^T \\ T(x^2) &= x^3 & &\text{so} & [T(x^2)]_C &= [0 \;\; 0 \;\; 0 \;\; 1]^T \end{aligned}$$

Thus

$$[T]_{BC} = \begin{bmatrix} 0 & 0 & 0 \\ 1 & 0 & 0 \\ 0 & 1 & 0 \\ 0 & 0 & 1 \end{bmatrix}$$

Further, if p is in P_2, so that $p(x) = a + bx + cx^2$, then

$$T(p) = ax + bx^2 + cx^3, \qquad [p]_B = \begin{bmatrix} a \\ b \\ c \end{bmatrix}, \qquad [T(p)]_C = \begin{bmatrix} 0 \\ a \\ b \\ c \end{bmatrix}$$

and

$$[T]_{BC}[p]_B = \begin{bmatrix} 0 & 0 & 0 \\ 1 & 0 & 0 \\ 0 & 1 & 0 \\ 0 & 0 & 1 \end{bmatrix} \begin{bmatrix} a \\ b \\ c \end{bmatrix} = \begin{bmatrix} 0 \\ a \\ b \\ c \end{bmatrix} = [T(p)]_C$$

■

Example 3 *(from calculus)* Let $D: P_3 \to P_2$ be the derivative transformation (see Example 8, Section 4.1), and let C and B be the bases in Example 2. (Note that the domain and range are reversed.) Then

$$\begin{aligned} D(1) &= 0, && \text{so} && [D(1)]_B = [0 \quad 0 \quad 0]^T \\ D(x) &= 1, && \text{so} && [D(x)]_B = [1 \quad 0 \quad 0]^T \\ D(x^2) &= 2x, && \text{so} && [D(x^2)]_B = [0 \quad 2 \quad 0]^T \\ D(x^3) &= 3x^2, && \text{so} && [D(x^3)]_B = [0 \quad 0 \quad 3]^T \end{aligned}$$

Thus,

$$[D]_{CB} = \begin{bmatrix} 0 & 1 & 0 & 0 \\ 0 & 0 & 2 & 0 \\ 0 & 0 & 0 & 3 \end{bmatrix}$$

Further, if p is in P_3, so that $p(x) = a + bx + cx^2 + dx^3$, then

$$D[p(x)] = b + 2cx + 3dx^2, \qquad [p]_C = \begin{bmatrix} a \\ b \\ c \\ d \end{bmatrix}, \qquad [D(p)]_B = \begin{bmatrix} b \\ 2c \\ 3d \end{bmatrix}$$

and

$$[D]_{CB}[p]_C = \begin{bmatrix} 0 & 1 & 0 & 0 \\ 0 & 0 & 2 & 0 \\ 0 & 0 & 0 & 3 \end{bmatrix} \begin{bmatrix} a \\ b \\ c \\ d \end{bmatrix} = \begin{bmatrix} b \\ 2c \\ 3d \end{bmatrix} = [D(p)]_B$$

NOTE If we consider $D: P_3 \to P_3$, then

$$[D]_C = [D]_{CC} = \begin{bmatrix} 0 & 1 & 0 & 0 \\ 0 & 0 & 2 & 0 \\ 0 & 0 & 0 & 3 \\ 0 & 0 & 0 & 0 \end{bmatrix}$$

■

Example 4 Let $T:\mathbb{R}^2 \to \mathbb{R}^2$ be given by $T(x, y) = (\sqrt{2}x + y, \sqrt{2}y)$. Then if $S = \{\boldsymbol{\varepsilon}_1, \boldsymbol{\varepsilon}_2\}$ is the standard basis, you can check

$$[T]_S = [T]_{SS} = \begin{bmatrix} \sqrt{2} & 1 \\ 0 & \sqrt{2} \end{bmatrix}$$

In Chapter 5 (see Example 3, Section 5.3), we shall see that T does not have a nice basis like the transformation in Example 1. But for now let

$$B = \{\mathbf{v}_1, \mathbf{v}_2\} = \left\{ \begin{bmatrix} \dfrac{1}{\sqrt{3}} \\ \dfrac{\sqrt{2}}{\sqrt{3}} \end{bmatrix}, \begin{bmatrix} \dfrac{\sqrt{2}}{\sqrt{3}} \\ -\dfrac{1}{\sqrt{3}} \end{bmatrix} \right\}$$

$$C = \{\mathbf{w}_1, \mathbf{w}_2\} = \left\{ \begin{bmatrix} \dfrac{\sqrt{2}}{\sqrt{3}} \\ \dfrac{1}{\sqrt{3}} \end{bmatrix}, \begin{bmatrix} \dfrac{1}{\sqrt{3}} \\ -\dfrac{\sqrt{2}}{\sqrt{3}} \end{bmatrix} \right\}$$

These are two *orthonormal* bases for $\mathbb{R}^2$, and we can check that

$$T(\mathbf{v}_1) = 2\mathbf{w}_1, \qquad T(\mathbf{v}_2) = \mathbf{w}_2, \qquad \text{so} \qquad [T]_{BC} = \begin{bmatrix} 2 & 0 \\ 0 & 1 \end{bmatrix}$$

This *is* a nice matrix (and you can see, from looking ahead to Section 6.2, that it comes from the singular value decomposition). ■

We now turn to the proof of Theorem (4.82), which is parallel to the proof of Theorem (4.4), Section 4.1.

Proof of Theorem (4.82) Let $B = \{\mathbf{v}_1, \ldots, \mathbf{v}_n\}$ and $C = \{\mathbf{w}_1, \ldots, \mathbf{w}_m\}$ be as described in the hypotheses. Let $\mathbf{x}$ be any vector in V. Let $[\mathbf{x}]_B = [x_1 \cdots x_n]^T$, so that $\mathbf{x} = x_1\mathbf{v}_1 + \cdots + x_n\mathbf{v}_n$. Since T is linear,

$$T(\mathbf{x}) = T(x_1\mathbf{v}_1 + \cdots + x_n\mathbf{v}_n) = x_1 T(\mathbf{v}_1) + \cdots + x_n T(\mathbf{v}_n)$$

Then

$$\begin{aligned} [T(\mathbf{x})]_C &= [T(x_1\mathbf{v}_1 + \cdots + x_n\mathbf{v}_n)]_C \\ &= x_1[T(\mathbf{v}_1)]_C + \cdots + x_n[T(\mathbf{v}_n)]_C && \text{by Theorem (3.88), Section 3.8} \\ &= \left[[T(\mathbf{v}_1)]_C \,\vdots\, \cdots \,\vdots\, [T(\mathbf{v}_n)]_C \right] \begin{bmatrix} x_1 \\ \vdots \\ x_n \end{bmatrix} && \text{by Theorem (1.32), Section 1.3} \\ &= [T]_{BC}[\mathbf{x}]_B \end{aligned}$$

This is what we wanted to show. ■

Before turning to similarity, we should complete the analogy to Section 4.1.

(4.83)

Let V and W be finite-dimensional nonzero vector spaces with bases $B = \{\mathbf{v}_1, \ldots, \mathbf{v}_n\}$ and $C = \{\mathbf{w}_1, \ldots, \mathbf{w}_m\}$, respectively. Let A be an $m \times n$ matrix and $T: V \to W$ be a linear transformation.

DEFINITION The **induced function** $f_A: V \to W$ **relative to the bases** B **and** C is given by

$$f_A(\mathbf{x}) = \mathbf{y} \qquad \text{if} \qquad A[\mathbf{x}]_B = [\mathbf{y}]_C$$

THEOREM The function f_A is a linear transformation from V to W, and it satisfies the identities

$$[f_A]_{BC} = A \qquad \text{and} \qquad f_{[T]_{BC}} = T$$

Usually several exercises must be worked through before these statements make sense. A sufficient number of problems appear in the exercises, with sketches of the proofs.

Similarity

We now turn to the question: What happens if we change bases? The general case $T: V \to W$ is important, but to keep the presentation less complicated, we shall consider only the special case $T: V \to V$ with the same basis in both domain and range. This turns out to be a very important special case, and the answer to our question follows.

(4.84)

THEOREM Let V be a finite-dimensional nonzero vector space with bases $B = \{\mathbf{v}_1, \ldots, \mathbf{v}_n\}$ and $C = \{\mathbf{w}_1, \ldots, \mathbf{w}_n\}$. Let $T: V \to V$ be a linear transformation. Then

$$[T]_C = P^{-1}[T]_B P$$

where P is the transition matrix from C to B [see Definition (3.89), Section 3.8].

WARNING Be sure to keep the order of B and C straight. P is the transition matrix from C to B (*not* from B to C); by Theorem (3.94), Section 3.8, P^{-1} is the transition matrix from B to C. Thus we have

T relative to C ↘   *Transition from B to C* ↘   *T relative to B* ↘   *Transition from C to B* ↘

(4.85)
$$[T]_C \quad = \quad P^{-1} \quad [T]_B \quad P$$

■

First we give some examples and then the proof. The first example illustrates that it is easy to find P if B is the standard basis.

Example 5 From Example 1, let $T(x, y) = (2x + 3y, 4x + 3y)$, let $S = \{\boldsymbol{\varepsilon}_1, \boldsymbol{\varepsilon}_2\}$ be the standard basis, and let $C = \{\mathbf{v}_1, \mathbf{v}_2\}$, where $\mathbf{v}_1 = [1 \quad -1]^T$, $\mathbf{v}_2 = [3 \quad 4]^T$. First

$$[T]_S = \begin{bmatrix} 2 & 3 \\ 4 & 3 \end{bmatrix}$$

Next, by Theorem (3.91), Section 3.8, the way to find a transition matrix P from C to B is to perform row operations on an $n \times 2n$ matrix:

$$[B \,\vdots\, C] \xrightarrow[\text{operations}]{\text{row}} [I \,\vdots\, P]$$

(Recall that B and C are reversed.) Now in our case, S is the standard basis, so

$$[S \,\vdots\, C] \text{ “=” } [I \,\vdots\, C] \xrightarrow[\text{operations}]{\text{row}} [I \,\vdots\, P]$$

Thus P is simply the matrix that has the vectors of C as its columns:

$$P = \begin{bmatrix} 1 & 3 \\ -1 & 4 \end{bmatrix}$$

To find P^{-1}, we compute $[P \,\vdots\, I] \to [I \,\vdots\, P^{-1}]$:

$$\left[\begin{array}{cc:cc} 1 & 3 & 1 & 0 \\ -1 & 4 & 0 & 1 \end{array}\right] \to \left[\begin{array}{cc:cc} 1 & 3 & 1 & 0 \\ 0 & 7 & 1 & 1 \end{array}\right] \to \left[\begin{array}{cc:cc} 1 & 3 & 1 & 0 \\ 0 & 1 & \frac{1}{7} & \frac{1}{7} \end{array}\right] \to \left[\begin{array}{cc:cc} 1 & 0 & \frac{4}{7} & -\frac{3}{7} \\ 0 & 1 & \frac{1}{7} & \frac{1}{7} \end{array}\right]$$

Thus

$$P^{-1} = \begin{bmatrix} \frac{4}{7} & -\frac{3}{7} \\ \frac{1}{7} & \frac{1}{7} \end{bmatrix}$$

We now compute

$$\begin{aligned} [T]_C = P^{-1}[T]_S P &= \begin{bmatrix} \frac{4}{7} & -\frac{3}{7} \\ \frac{1}{7} & \frac{1}{7} \end{bmatrix} \begin{bmatrix} 2 & 3 \\ 4 & 3 \end{bmatrix} \begin{bmatrix} 1 & 3 \\ -1 & 4 \end{bmatrix} \\ &= \begin{bmatrix} \frac{4}{7} & -\frac{3}{7} \\ \frac{1}{7} & \frac{1}{7} \end{bmatrix} \begin{bmatrix} -1 & 18 \\ 1 & 24 \end{bmatrix} = \begin{bmatrix} -1 & 0 \\ 0 & 6 \end{bmatrix} \end{aligned}$$

which is what we obtained in Example 1. ■

The next example illustrates one common application. We need to determine $[T]_C$ for some reason, but the “straightforward” computation is cumbersome. So we pick a nice basis (usually the standard basis) for B and use Theorem (4.84).

Example 6 For P_2, let $S = \{1, x, x^2\}$ be the standard basis and let $C = \{1 - x, 1 + x, 1 + x + x^2\}$, which (as we easily can check) is also a basis. Let $T: P_2 \to P_2$ be given by

$$T(a + bx + cx^2) = (5a + b + c) + (b - a - c)x + (a + b)x^2$$

which (as we can check) is linear. By computing $T(1)$, $T(x)$, and $T(x^2)$ and putting the coordinates as columns, we obtain

$$[T]_S = \begin{bmatrix} 5 & 1 & 1 \\ -1 & 1 & -1 \\ 1 & 1 & 0 \end{bmatrix}$$

As in Example 5, the transition matrix from B to C is found by $[C \,\vdots\, B] \to [I \,\vdots\, P]$. Again C is the standard basis, so P simply has the coordinates of B as its columns:

$$P = \begin{bmatrix} 1 & 1 & 1 \\ -1 & 1 & 1 \\ 0 & 0 & 1 \end{bmatrix}$$

The inverse P^{-1} is found by $[P \,\vdots\, I] \to [I \,\vdots\, P^{-1}]$, with the result

$$P^{-1} = \begin{bmatrix} \frac{1}{2} & -\frac{1}{2} & 0 \\ \frac{1}{2} & \frac{1}{2} & -1 \\ 0 & 0 & 1 \end{bmatrix}$$

which we can readily check. We now compute

$$[T]_C = P^{-1}[T]_S P = \begin{bmatrix} \frac{1}{2} & -\frac{1}{2} & 0 \\ \frac{1}{2} & \frac{1}{2} & -1 \\ 0 & 0 & 1 \end{bmatrix} \begin{bmatrix} 5 & 1 & 1 \\ -1 & 1 & -1 \\ 1 & 1 & 0 \end{bmatrix} \begin{bmatrix} 1 & 1 & 1 \\ -1 & 1 & 1 \\ 0 & 0 & 1 \end{bmatrix}$$

$$= \begin{bmatrix} \frac{1}{2} & -\frac{1}{2} & 0 \\ \frac{1}{2} & \frac{1}{2} & -1 \\ 0 & 0 & 1 \end{bmatrix} \begin{bmatrix} 4 & 6 & 7 \\ -2 & 0 & -1 \\ 0 & 2 & 2 \end{bmatrix} = \begin{bmatrix} 3 & 3 & 4 \\ 1 & 1 & 1 \\ 0 & 2 & 2 \end{bmatrix}$$

The usefulness of such a computation depends on if (and for what purpose) we need $[T]_C$. ■

Proof of Theorem (4.84) Suppose V is a nonzero vector space with bases $B = \{\mathbf{v}_1, \ldots, \mathbf{v}_n\}$ and $C = \{\mathbf{w}_1, \ldots, \mathbf{w}_n\}$. Let P be the transition matrix from C to B, and let $T: V \to V$ be linear. To show that $[T]_C = P^{-1}[T]_B P$, it is sufficient to show that the ith column of $[T]_C$ is equal to the ith column of $P^{-1}[T]_B P$. Now for any matrix A, the ith column of A is $A\boldsymbol{\varepsilon}_i$. By Definition (4.82),

$$[T]_C \varepsilon_i = i\text{th column of } [T]_C = [T(\mathbf{w}_i)]_C \tag{4.86}$$

We now show that this result is the same as $P^{-1}[T]_B P\boldsymbol{\varepsilon}_i$.

Since P is the transition matrix from C to B, we know by Theorem (3.92), Section 3.8, that

(4.87) $$P\varepsilon_i = i\text{th column of } P = [\mathbf{w}_i]_B$$

By Theorem (4.82), we know that $[T]_B[\mathbf{x}]_B = [T(\mathbf{x})]_B$, so

(4.88) $$[T]_B[\mathbf{w}_i]_B = [T(\mathbf{w}_i)]_B$$

We know that P^{-1} is the transition matrix from B to C, so by Theorem (3.89), Section 3.8, this matrix has the property $P^{-1}[\mathbf{x}]_B = [\mathbf{x}]_C$. Thus,

(4.89) $$P^{-1}[T(\mathbf{w}_i)]_B = [T(\mathbf{w}_i)]_C$$

Putting Equations (4.86)–(4.89) together, we have

$$P^{-1}[T]_B P\varepsilon_i = P^{-1}[T]_B[\mathbf{w}_i]_B = P^{-1}[T(\mathbf{w}_i)]_B = [T(\mathbf{w}_i)]_C = [T]_C\varepsilon_i$$

Thus the two ith columns are equal, so we are done. ■

With this background, we now turn to our main concept and explain why it is useful.

(4.90) DEFINITION Two $n \times n$ matrices A and D are **similar** if there is an $n \times n$ invertible matrix P such that $D = P^{-1}AP$.

Why is such a concept important? By Theorem (4.84), A and D are similar if they represent the same linear transformation in two different bases. But the converse of this is also true. Suppose A and D are similar, so that $D = P^{-1}AP$. Let V be an n-dimensional vector space, and let B be a basis for V. (Perhaps $V = \mathbb{R}^n$, and B is the standard basis). Let $T: V \to V$ be the linear transformation with $[T]_B = A$ [given by Theorem (4.83)]. Then, using P, we can find another basis C for V such that $[T]_C = D$. (See Exercises 45, 46.) The result, which follows, ties together the geometry and algebra of similarity.

(4.91) THEOREM Two $n \times n$ matrices are similar if and only if they represent the same linear transformation on an n-dimensional vector space, but possibly in different bases.

WARNING There is really an awful lot going on here. If you feel somewhat befuddled, that is very reasonable! Many mathematical concepts must be seen several times (often in different courses) before they can be understood. Similarity is such a concept, because it involves algebra, geometry, and a relationship between them. The point to keep in mind (now and when you run into it later) is that if two matrices are similar, they have many of the same properties (for example, they have the same rank), because they represent the same function.

Examples are given in the exercises, in Chapter 5, and in almost all ensuing applications involving linear algebra.

Exercise 4.7 In Exercises 1–10, for the given $T: V \to W$ and $\mathbf{v}$ in V, find (a) $[T]_{SS}$, where S is the standard basis for V or W; (b) $[T]_{BC}$; (c) $[\mathbf{v}]_B$; (d) $T(\mathbf{v})$ in two ways, using (a) and (b) and Theorem (4.82).

1. $T: \mathbb{R}^2 \to \mathbb{R}^3$ by $T(x, y) = (x + y, x - y, x)$; $\mathbf{v} = (2, -4)$; $B = \{(-1, 1), (0, -1)\}$; $C = \{(0, 1, 1), (1, 0, 1), (1, 1, 0)\}$

2. $T: \mathbb{R}^2 \to \mathbb{R}^3$ by $T(x, y) = (x - y, 0, 2x)$; $\mathbf{v} = (-1, 5)$; $B = \{(1, -1), (2, 1)\}$; $C = \{(0, 0, 1), (0, 1, 1), (1, 1, 1)\}$

3. $T: \mathbb{R}^3 \to \mathbb{R}^2$ by $T(x, y, z) = (z - x, 2y - x)$; $\mathbf{v} = (2, -1, -3)$; $B = \{(0, 0, 1), (0, 1, 1), (1, 1, 1)\}$; $C = \{(1, -1), (2, 1)\}$

4. $T: \mathbb{R}^3 \to \mathbb{R}^2$ by $T(x, y, z) = (2z + x, y - 2x)$; $\mathbf{v} = (-4, -1, 1)$; $B = \{(0, 1, 1), (1, 0, 1), (1, 1, 0)\}$; $C = \{(-1, 1), (0, -1)\}$

5. $T: \mathbb{R}^2 \to \mathbb{R}^4$ by $T(x, y) = (x - y, 3x, -2y, x + y)$; $\mathbf{v} = (1, -6)$; $B = \{(1, 2), (2, 0)\}$; $C = \{(1, 1, 1, 1), (0, 1, 1, 1), (0, 0, 1, 1), (0, 0, 0, 1)\}$

6. $T: \mathbb{R}^4 \to \mathbb{R}^2$ by $T(x, y, z, w) = (2x - y + 3z - w, x - y - z + w)$; $\mathbf{v} = (2, -3, 0, -2)$; $B = \{(1, 1, 1, 1), (1, 1, 1, 0), (1, 1, 0, 0), (1, 0, 0, 0)\}$; $C = \{(1, 2), (0, 1)\}$

NOTE Exercises 7 and 8 tend to be confusing but are very instructive.

7. $T: \mathbb{R}^2 \to M_{22}$ by $T(x, y) = \begin{bmatrix} x - y & x + y \\ 2x & 3y \end{bmatrix}$; $\mathbf{v} = (5, -1)$; $B = \{(1, 1), (1, -1)\}$;
$C = \left\{\begin{bmatrix} 1 & 1 \\ 1 & 1 \end{bmatrix}, \begin{bmatrix} 1 & 1 \\ 1 & 0 \end{bmatrix}, \begin{bmatrix} 1 & 1 \\ 0 & 0 \end{bmatrix}, \begin{bmatrix} 1 & 0 \\ 0 & 0 \end{bmatrix}\right\}$

8. $T: P_2 \to M_{22}$ by $T(a + bx + cx^2) = \begin{bmatrix} a + b & a + c \\ c & b \end{bmatrix}$; $\mathbf{v} = 5 + x + 3x^2$; $B = \{1 + x, 1 + x^2, x\}$; $C = C$ in Exercise 7.

9. $T: P_1 \to P_3$ by $T[p(x)] = x^2 p(x)$; $\mathbf{v} = 3 - 2x$; $B = \{1 + x, 2x\}$; $C = \{1 + x + x^2 + x^3, 1 + x + x^2, 1 + x, 1\}$

10. $T: P_3 \to P_2$ by $T(a + bx + cx^2 + dx^3) = b - c + (a + d)x^2$; $\mathbf{v} = 2 - 3x^2 + x^3$; $B = C$ in Exercise 9; $C = \{1 + x + x^2, x + x^2, x^2\}$

11. *(calculus)* Let V be the subspace of $C^\infty(\mathbb{R})$ with basis $B = \{1, x, e^x, xe^x\}$. Let $D: V \to V$ be the derivative transformation, and let $f(x) = 2 + 3x - e^x - 2xe^x$. Find (a) $[f]_B$; (b) $[D]_B$; and (c) $[D(f)]_B$.

12. *(calculus)* Repeat Exercise 11 for $B = \{1, e^{2x}, xe^{2x}, x^2e^{2x}\}$ and $f(x) = 3 + (3 - 5x + x^2)e^{2x}$.

In Exercises 13–18, for the given T, A, B, and C, find $[T]_{BC}$, f_A, and verify that Theorem (4.83) holds.

13. T, B, C as in Exercise 1; $A = \begin{bmatrix} 2 & 1 \\ -1 & 0 \\ 0 & -3 \end{bmatrix}$.

14. T, B, C as in Exercise 2; $A = \begin{bmatrix} -1 & 2 \\ 2 & -2 \\ 3 & 0 \end{bmatrix}$.

15. T, B, C as in Exercise 3; $A = \begin{bmatrix} -1 & 0 & 1 \\ 1 & 1 & 0 \end{bmatrix}$.

16. T, B, C as in Exercise 4; $A = \begin{bmatrix} 2 & 1 & 0 \\ 1 & 0 & -1 \end{bmatrix}$.

17. T, B, C as in Exercise 5; $A = \begin{bmatrix} 3 & -1 & 0 & 0 \\ 0 & 1 & 2 & -1 \end{bmatrix}^T$.

18. T, B, C as in Exercise 6; $A = \begin{bmatrix} 0 & 1 & 0 & 2 \\ -1 & 1 & -1 & 0 \end{bmatrix}$.

In Exercises 19–22, refer to Theorem (4.83).

19. Explain why f_A is linear. [HINT Use Theorem (3.88), Section 3.8.]

20. Let M be the $1 \times m$ matrix with the vectors $\mathbf{w}_1, \ldots, \mathbf{w}_m$ as its entries. Explain why $f_A(x) = MA[x]_B$.

21. Prove that $[f_A]_{BC} = A$. [HINT Show that the ith column of both sides of the equation are the same. See the proof of Theorem (4.84), and note that $\boldsymbol{\varepsilon}_i = [\mathbf{v}_i]_B$.]

22. Prove that $f_{[T]_{BC}} = T$. [HINT First show that $f_{[T]_{BC}}(\mathbf{v}_i) = T(\mathbf{v}_i)$, and then use linearity.]

In Exercises 23–30, let B be the standard basis, C be as given, and P be the transition matrix from C to B. Find $[T]_B$, $[T]_C$, and verify that $[T]_C = P^{-1}[T]_BP$.

23. $T: \mathbb{R}^2 \to \mathbb{R}^2$ by $T(x, y) = (x - y, x + y)$; $C = \{(1, 1), (0, 1)\}$.

24. $T: \mathbb{R}^2 \to \mathbb{R}^2$ by $T(x, y) = (x - 2y, 2x + y)$; $C = \{(1, -2), (0, 2)\}$.

25. $T: \mathbb{R}^2 \to \mathbb{R}^2$ by $T(x, y) = (2y - x, x - y)$; $C = \{(2, 1), (1, 1)\}$.

26. $T:\mathbb{R}^3 \to \mathbb{R}^3$ by $T(x, y, z) = (x + y, y + z, x - z)$;
$C = \{(0, 0, 1), (0, 1, 1), (1, 1, 1)\}$.

27. $T:\mathbb{R}^3 \to \mathbb{R}^3$ by $T(x, y, z) = (x - z, x - y + z, 2y)$;
$C = \{(1, 0, 0), (1, 1, 0), (1, 1, 1)\}$.

28. $T:\mathbb{R}^3 \to \mathbb{R}^3$ by $T(x, y, z) = (x + y + 2z, x - 2y - z, 2x + y)$;
$C = \{(1, 0, -1), (0, 1, 1), (0, 0, 1)\}$.

29. $T:P_2 \to P_2$ by $T(a + bx + cx^2) = a - b + (b - c)x + (a + c)x^2$;
$C = \{x^2, 1 + x^2, x + x^2\}$.

30. $T:P_2 \to P_2$ by $T(a + bx + cx^2) = a + (a - 2b)x + (b + 3c)x^2$;
$C = \{1, 1 + x, 1 + x^2\}$.

31. Show that any $n \times n$ matrix is similar to itself.

32. Show that if A is similar to D, then D is similar to A.

33. Show that if A is similar to D and D is similar to E, then A is similar to E.

NOTE Exercises 31–33 show that similarity is an **equivalence relation**, an important concept that arises often in higher mathematics.

34. Find all $n \times n$ matrices that are similar to I_n.

35. Prove that if A is similar to B, then A^2 is similar to B^2.

36. Prove that if A is similar to B, then A^k is similar to B^k for all positive integers k.

37. Prove that if A is similar to B, then $rk(A) = rk(B)$.

38. Prove that if A is similar to B, then A is invertible if and only if B is invertible. If A and B are similar and invertible, show that A^{-1} is similar to B^{-1}.

39. Prove that if A is similar to B, then A is idempotent if and only if B is idempotent. (DEFINITION A is **idempotent** if $A^2 = A$.)

In Exercises 40–44, prove the statement or give a counterexample.

40. If A is similar to B, then $NS(A) = NS(B)$.

41. If $A = BC$ and B is invertible, then A is similar to CB.

42. Repeat Exercise 41 for arbitrary B.

43. If A is similar to a diagonal matrix, then A is similar to A^T.

44. $A = \begin{bmatrix} a & 1 & & & \\ & a & 1 & & \\ & & \ddots & \ddots & \\ & & & a & 1 \\ & & & & a \end{bmatrix}$ is similar to A^T.

(HINT Try the 2×2 and 3×3 cases first.)

45. Let V be a vector space with $\{\mathbf{v}_1, \ldots, \mathbf{v}_n\}$ for a basis. Let P be an invertible matrix, and let $\mathbf{w}_i$ be the vector in V such that $[\mathbf{w}_i]_B = P\mathbf{v}_i$. Show that $C = \{\mathbf{w}_1, \ldots, \mathbf{w}_n\}$ is a basis for V and that P is the transition matrix from C to B.

46. Let V, B, C, and P be as described in Exercise 45. Suppose that A is an $n \times n$ matrix and $D = P^{-1}AP$. Also suppose that $T: P \to P$ is linear and $[T]_B = A$. Show that $[T]_C = D$.

Review Exercises

1. Is $T: \mathbb{R}^2 \to \mathbb{R}^3$ by $T(x, y) = (2x - y, 3x, x + 4y)$ a linear transformation? If so, find the matrix A such that $T = f_A$.

2. Suppose $\mathbf{v}_1, \ldots, \mathbf{v}_m$ are fixed vectors in $\mathbb{R}^n$. Define $T: \mathbb{R}^n \to \mathbb{R}^m$ by $T(\mathbf{x}) = (\mathbf{x} \cdot \mathbf{v}_1, \mathbf{x} \cdot \mathbf{v}_2, \ldots, \mathbf{x} \cdot \mathbf{v}_m)$. Is T a linear transformation? If so, find a matrix A such that $T = f_A$.

3. Verify the triangle inequality and the Cauchy–Schwarz inequality for the vectors $\mathbf{u} = (2, 3, -1, 0, -2)$ and $\mathbf{v} = (4, -1, 3, 2, -2)$.

4. Verify that $A\mathbf{x} \cdot \mathbf{y} = \mathbf{x} \cdot A^T\mathbf{y}$ for

$$A = \begin{bmatrix} 1 & 2 \\ -3 & 0 \\ 2 & -2 \end{bmatrix}, \qquad \mathbf{x} = \begin{bmatrix} 3 \\ -1 \end{bmatrix}, \qquad \mathbf{y} = \begin{bmatrix} -2 \\ -1 \\ 3 \end{bmatrix}$$

5. Use the Cauchy-Schwarz inequality to prove that

$$(a_1 + a_2 + \cdots + a_n)\left(\frac{1}{a_1} + \frac{1}{a_2} + \cdots + \frac{1}{a_n}\right) \geq n^2$$

if $a_1, \ldots, a_n$ are *positive* real numbers.

6. Suppose that $\mathbf{y}$ and $\mathbf{z}$ are in $\mathbb{R}^n$. Prove that if $\mathbf{x} \cdot \mathbf{y} = \mathbf{x} \cdot \mathbf{z}$ for all $\mathbf{x}$ in $\mathbb{R}^n$, then $\mathbf{y} = \mathbf{z}$.

7. Let $\mathbf{x} = (1, -2, 3)$. Find the projection of $\mathbf{x}$ into (a) the line spanned by $(1, 1, 1)$, and (b) the plane spanned by $(1, 1, 1)$ and $(1, 0, -2)$.

8. Find the equation of the line that best fits the points $(0, 0)$, $(1, 0)$ and $(3, 2)$ in the least-squares sense.

9. What multiple of $(1, 1)$ should be subtracted from $(3, 0)$ to make the resulting vector perpendicular to $(1, 1)$? Sketch a figure.

10. What multiples of $\mathbf{v}_1 = (0, 1, 1)$ and $\mathbf{v}_2 = (0, -1, 1)$ should be subtracted from $\mathbf{w} = (1, 2, 3)$ to make the resulting vector perpendicular to $\mathbf{v}_1$ and $\mathbf{v}_2$? Note that $\mathbf{v}_1 \cdot \mathbf{v}_2 = 0$.

11. Find a unit vector, $\mathbf{u}$, in $\mathbb{R}^4$ that is perpendicular to ε_1 and ε_2 and that forms equal angles with ε_3 and ε_4. Here $\varepsilon_1, \ldots, \varepsilon_4$ form the standard basis for $\mathbb{R}^4$.

12. Show that the vectors ε_1, ε_2, $\mathbf{u}$ from Exercise 11 are orthonormal and extend them to an orthonormal basis for $\mathbb{R}^4$.

13. Factor

$$A = \begin{bmatrix} 3 & 2 \\ 4 & 0 \end{bmatrix}$$

into QR.

14. Suppose A is $m \times n$ and $A^T A$ is a diagonal matrix. Show that the columns of A are orthogonal.

15. (a) Find the 4×4 matrix $A = [T]$ that represents a cyclic permutation: $T(x_1, x_2, x_3, x_4) = (x_2, x_3, x_4, x_1)$.
    (b) What is the effect of A^2?
    (c) Show that $A^3 = A^{-1}$.

16. (a) Find the 4×3 matrix A that represents a *right shift*: $(x_1, x_2, x_3) \to (0, x_1, x_2, x_3)$.
    (b) Find the 3×4 matrix B that represents a *left shift*: $(x_1, x_2, x_3, x_4) \to (x_2, x_3, x_4)$.
    (c) Compute the products AB and BA.

17. Find the length of $\mathbf{v} = (1, -2, 2)$, and find two linearly independent vectors that are perpendicular to $\mathbf{v}$.

18. Find all vectors that are perpendicular to $\mathbf{v} = (2, 3, -1)$ and $\mathbf{w} = (-4, 1, 2)$ by placing them as rows of A and finding $NS(A)$.

19. If Q is orthogonal, is Q^4 also orthogonal? Why or why not?

20. How many 3×3 orthogonal matrices are there with only 0 and 1 for entries?

21. If Q is an orthogonal matrix with only $\frac{1}{4}$ or $-\frac{1}{4}$ for entries, what are the dimensions of Q?

22. In $\mathbb{R}^2$, give an example of linearly independent vectors that are not orthogonal. Then give an example of orthogonal vectors that are not linearly independent.

23. Find a matrix with a row space containing $(1, 2, -3)$ and a null space containing $(1, 1, 1)$, or prove that no such matrix exists. Do the same for a row space containing $(1, 2, -3)$ and a null space containing $(2, 1, 1)$.

24. Show that $\|A\mathbf{x}\| = \|A^T\mathbf{x}\|$ if $AA^T = A^TA$.

25. If the vectors $\mathbf{u}_1$, $\mathbf{u}_2$, and $\mathbf{u}_3$ are orthonormal, what linear combination of $\mathbf{u}_1$ and $\mathbf{u}_2$ is closest to $\mathbf{u}_3$?

26. Show that a matrix that is both orthogonal and lower triangular must be diagonal.

27. Find a Householder matrix H that interchanges (1, 2, 2) and (−2, −1, 2).

28. Let

$$A = \begin{bmatrix} 2 & 4 & 2 \\ -2 & 1 & 1 \\ -1 & 3 & -1 \end{bmatrix}$$

Find a Householder matrix H such that

$$HA = \begin{bmatrix} * & * & * \\ 0 & * & * \\ 0 & * & * \end{bmatrix}$$

29. Show that $CS(A) = CS(AA^T)$.

30. Show that if $\mathbf{x}$ is in both $NS(A)$ and $RS(A)$, then $\mathbf{x} = \mathbf{0}$.

31. What is minimized by the least square solution to $A\mathbf{x} = \mathbf{b}$?

32. Under what conditions is A^TA nonsingular?

33. Let A be an $m \times n$ matrix, let $\mathbf{b}$ be a vector in $\mathbb{R}^m$, and suppose $A\mathbf{x} = \mathbf{b}$ is consistent.

    (a) Show there is at least one solution $\mathbf{x}$ (to the equation $A\mathbf{x} = \mathbf{b}$) that is also in the row space of A.
    (b) Explain why there is at most one such $\mathbf{x}$.

34. If we want to project a vector onto a subspace V of $\mathbb{R}^n$, why is it advantageous to have an orthonormal basis for V rather than any basis?

Cumulative Review Exercises

NOTE Answers to *all* Cumulative Review Exercises appear in the *Instructor's Manual*. These problems are not answered in the back of the textbook.

1. Let A be a square matrix. Then A is invertible $\Leftrightarrow$ (make 12 equivalent statements, including at least two involving material from this chapter).

2. Let A be the 4×4 matrix with entries a_{ij} = the smaller of i and j.

    (a) Find A.
    (b) Find the LU decomposition of A.
    (c) Use (b) to solve $A\mathbf{x} = [2 \quad 2 \quad 3 \quad 0]^T$.
    (d) Find $\det(A)$ [from (b), if you like].

3. Give three different reasons why the matrix

$$A = \begin{bmatrix} * & * & 0 & 0 & 0 \\ * & * & 0 & 0 & 0 \\ * & * & 0 & 0 & 0 \\ * & * & * & * & * \\ * & * & * & * & * \end{bmatrix}$$

must be singular. At least two reasons should involve $RS(A)$, $CS(A)$, $NS(A)$, or $rk(A)$.

4. Let $\mathbf{v}_1 = (2, -2, 5, -4)$ and $\mathbf{v}_2 = (2, -1, 0, 0)$.

   (a) Use the Gram–Schmidt process to construct an orthonormal pair $\mathbf{u}_1$ and $\mathbf{u}_2$ from $\mathbf{v}_1$ and $\mathbf{v}_2$.
   (b) Express $\mathbf{v}_1$ and $\mathbf{v}_2$ as linear combinations of $\mathbf{u}_1$ and $\mathbf{u}_2$.
   (c) If A is a 4×2 matrix with $\mathbf{v}_1$ and $\mathbf{v}_2$ as columns, find its QR-decomposition $A = QR$.

5. (a) Find an orthonormal basis for the plane $x - y - z = 0$ in $\mathbb{R}^3$.
   (b) What point on this plane is closest to $(1, 1, 1)$?

6. Let $\mathbf{v} = (1, 2, -3, 1)$ and $\mathbf{w} = (-2, -4, 6, 3)$.

   (a) What does the Cauchy–Schwarz inequality say about $\mathbf{v}$ and $\mathbf{w}$?
   (b) Change only the last coordinate of $\mathbf{w}$, so that the Cauchy–Schwarz inequality becomes an equality.
   (c) Repeat (a) for the triangle inequality.
   (d) Repeat (b) for the triangle inequality.

7. Let T be the reflection in the xy-plane across the main diagonal (the line $y = x$).

   (a) Find the matrix for T in the standard basis S and the matrix for T in the basis $B = \left\{\frac{1}{\sqrt{2}}(1, 1), \frac{1}{\sqrt{2}}(1, -1)\right\}$.
   (b) Find an explicit P that shows that $[T]_S$ and $[T]_B$ are similar.
   (c) What special property does P have?

8. Let E be an elementary matrix. Are the following statements true or false? Provide a counterexample for each false statement.

   (a) $RS(EA) = RS(A)$ (b) $CS(EA) = CS(A)$
   (c) $NS(EA) = NS(A)$ (d) $\det(EA) = \det(A)$
   (e) A is invertible $\Leftrightarrow$ EA is invertible

9. Let A be square.

   (a) Show that $\|A\mathbf{x}\| = \|A^T\mathbf{x}\|$ if $AA^T = A^TA$.
   (b) Find a 2×2 matrix A such that $AA^T \neq A^TA$. For this A, find a vector $\mathbf{x}$ in $\mathbb{R}^2$ such that $\|A\mathbf{x}\| \neq \|A^T\mathbf{x}\|$.

10. Consider the system $\begin{aligned} ax + y &= 2 \\ x + ay &= 2 \end{aligned}$.

    (a) For which values of a does this system have no solution? One solution? Infinitely many solutions?
    (b) For which values of a is the system ill-conditioned?

11. Let V be an inner product space and S be a nonempty subset of V. Let $S^\perp = \{\mathbf{v} \text{ in } V \mid \mathbf{v} \cdot \mathbf{s} = 0, \text{ all } \mathbf{s} \text{ in } S\}$.

    (a) In $\mathbb{R}^3$, find $\{\boldsymbol{\varepsilon}_1\}^\perp$ and $\{\boldsymbol{\varepsilon}_1, \boldsymbol{\varepsilon}_2\}^\perp$.
    (b) Show that $S^\perp$ is a subspace of V.
    (c) Show that $S \subset (S^\perp)^\perp$.

Chapter 5

EIGENVECTORS AND EIGENVALUES

The first three chapters of this book introduced the first main part of linear algebra—the study of linear systems and related problems. We now embark upon the second main part of linear algebra—the study of eigenvectors, eigenvalues, and related problems.

Anyone who has been on a swing has experienced an eigenvalue problem. There is a natural place in the oscillation where additional force will make the swing go higher. This place corresponds to an eigenvalue of the dynamical system, and in fact all dynamical systems have eigenvalues. They can either be sought for good effects (as in a child's swing or a radio tuning circuit) or avoided for bad effects. Here are some bad effects to be avoided: When soldiers come to a bridge, traditionally they stop marching and just walk across, since otherwise they might march at an eigenvalue frequency of the bridge and cause it to swing, and even collapse. It is hard to believe, but a bridge did collapse this way in Manchester, England, in 1831. Moreover, bridge designers must carefully compute eigenvalues of a proposed bridge to avoid such a disaster caused by natural forces.*

In this chapter we shall see how eigenvalues also play a role with Fibonacci sequences, Markov processes, and diagonalizing bilinear forms (which have a broad spectrum of applications including analyzing reconnaissance photographs, analysing geologic stress in rocks, and matching automobile colors).

Except for simple or special systems, computing eigenvalues requires some fairly sophisticated mathematics. An introduction to this is given in the last section.

---

* *Nevertheless in 1940 the Tacoma Narrows Bridge oscillated badly and then collapsed, as a result of the steady wind producing previously unsuspected forces that were in tune with an eigenvalue of the bridge. See a delightful description in Braun, "Differential Equations and Their Applications," Springer-Verlag, pp. 248–252.*

5.1 A Brief Introduction to Determinants*

Determinants play an important role in both theoretical and computational aspects of eigenvalues, so some mention of determinants must be made before presenting eigenvalues. However, because some people wish to get to eigenvalues as quickly as possible, the purpose of this section is to present the minimum amount of information about determinants needed to begin the study of eigenvalues. Most proofs and a more complete development of determinants are given in Chapter 2.

The **determinant function** is denoted by **det**. It is a function that associates with every *square* matrix A a number, denoted by "$\det(A)$." For our purposes the following property is the main reason we study determinants.

(5.1) THEOREM If A is any square matrix, then

$$A \text{ is singular} \quad \text{if and only if} \quad \det(A) = 0$$

Equivalently,

$$A \text{ is nonsingular} \quad \text{if and only if} \quad \det(A) \neq 0$$

You probably can guess that this is a very important property. In the next sections, we shall see just how important, and useful, it is.

Definition of Determinant

We now discuss how to compute determinants.

(5.2) DEFINITION For a 1×1 matrix $A = [a]$,

$$\det[a] = a$$

Example 1 (a) $\det[3] = 3$, (b) $\det[-7.1] = -7.1$, (c) $\det[0] = 0$. ■

You can easily see Theorem (5.1) holds for 1×1 matrices.

(5.3) DEFINITION For a 2×2 matrix,

$$\text{If} \quad A = \begin{bmatrix} a & c \\ b & d \end{bmatrix} \quad \text{then} \quad \det(A) = ad - bc$$

Example 2 (a) $$\det\begin{bmatrix} 1 & 2 \\ 3 & 4 \end{bmatrix} = 1(4) - 3(2) = 4 - 6 = -2$$

* *Section 5.1 offers an alternative coverage of determinants to instructors who prefer just a quick overview. Chapter 2 provides a full coverage of determinants.*

(b) $\det\begin{bmatrix} 6 & -2 \\ -3 & 1 \end{bmatrix} = 6(1) - (-3)(-2) = 6 - 6 = 0$ ■

With a little thought, you can see

$$\begin{aligned} &\text{The determinant of a } 2 \times 2 \text{ matrix is } 0 \\ \Leftrightarrow \quad &\text{one row is a multiple of the other} \\ \Leftrightarrow \quad &\text{the matrix is singular} \end{aligned}$$

Thus we can see that Theorem (5.1) holds for 2×2 matrices.

NOTATION An alternative notation for the determinant of a matrix A is $|A|$ rather than $\det(A)$. Using this notation, the determinants in Example 2 can be written

$$\begin{vmatrix} 1 & 2 \\ 3 & 4 \end{vmatrix} = -2, \qquad \begin{vmatrix} 6 & -2 \\ -3 & 1 \end{vmatrix} = 0$$

Easy Formulas

The determinant of a 2×2 matrix can be memorized easily by thinking the product of one diagonal minus the product of the other diagonal, as illustrated in the following. There is a similar way to compute the determinant of a 3×3 matrix, and this is illustrated in the diagram:

$$\begin{bmatrix} a_{11} & a_{12} \\ a_{21} & a_{22} \end{bmatrix}$$

(a) Down arrow means $+a_{11}a_{22}$. Up arrow means $-a_{21}a_{12}$.

$$\begin{bmatrix} a_{11} & a_{12} & a_{13} \\ a_{21} & a_{22} & a_{23} \\ a_{31} & a_{32} & a_{33} \end{bmatrix} \begin{matrix} a_{11} & a_{12} \\ a_{21} & a_{22} \\ a_{31} & a_{32} \end{matrix}$$

(b) Down arrows mean $+a_{11}a_{22}a_{33}$, $+a_{12}a_{23}a_{31}$, $a_{13}a_{21}a_{32}$.
Up arrows mean $-a_{31}a_{22}a_{13}$, $-a_{32}a_{23}a_{11}$, $-a_{33}a_{21}a_{12}$.

Example 3 (a) $\det\begin{bmatrix} 2 & 3 \\ 4 & 5 \end{bmatrix} = \begin{vmatrix} 2 & 3 \\ 4 & 5 \end{vmatrix} = 2(5) - 4(3) = 10 - 12 = -2$

(b) $\det\begin{bmatrix} 2 & -3 & 1 \\ 4 & 0 & -2 \\ 3 & -1 & -3 \end{bmatrix} = \begin{vmatrix} 2 & -3 & 1 \\ 4 & 0 & -2 \\ 3 & -1 & -3 \end{vmatrix} \begin{matrix} 2 & -3 \\ 4 & 0 \\ 3 & -1 \end{matrix}$

$$\begin{aligned} &= 2(0)(-3) + (-3)(-2)(3) + 1(4)(-1) \\ &\quad - 3(0)(1) - (-1)(-2)(2) - (-3)(4)(-3) \\ &= 0 + 18 - 4 - 0 - 4 - 36 = -26 \end{aligned}$$ ■

WARNING Although these diagrams work for 2×2 and 3×3 matrices, there are no corresponding diagrams for $n \times n$ matrices when $n > 3$. Thus we now turn to the general definition of determinants.

General Definition of Determinants

We now set out to describe the pattern for the determinant of 3×3 matrices in such a way that will lead into a description of the general pattern for the determinant of $n \times n$ matrices.

(5.4)

DEFINITION Suppose in a 3×3 matrix A we delete the ith row and jth column to obtain a 2×2 submatrix. The determinant of this submatrix is called the (i, j)th **minor** of A and is denoted by M_{ij}. The number $(-1)^{i+j}M_{ij}$ is called the (i, j)th **cofactor** of A and is denoted by C_{ij}.

Example 4 Let

$$A = \begin{bmatrix} 2 & -3 & 1 \\ 4 & 0 & -2 \\ 3 & -1 & -3 \end{bmatrix}.$$

Then

(a) $$M_{11} = \begin{vmatrix} 2 & -3 & 1 \\ 4 & 0 & -2 \\ 3 & -1 & -3 \end{vmatrix} = \begin{vmatrix} 0 & -2 \\ -1 & -3 \end{vmatrix} = 0(-3) - (-1)(-2) = -2$$

$$C_{11} = (-1)^{1+1}M_{11} = (+1)M_{11} = -2$$

(b) $$M_{12} = \begin{vmatrix} 2 & -3 & 1 \\ 4 & 0 & -2 \\ 3 & -1 & -3 \end{vmatrix} = \begin{vmatrix} 4 & -2 \\ 3 & -3 \end{vmatrix} = 4(-3) - 3(-2) = -6$$

$$C_{12} = (-1)^{1+2}M_{12} = (-1)M_{12} = 6$$

(c) $$M_{23} = \begin{vmatrix} 2 & -3 & 1 \\ 4 & 0 & 2 \\ 3 & -1 & -3 \end{vmatrix} = \begin{vmatrix} 2 & -3 \\ 3 & -1 \end{vmatrix} = 2(-1) - 3(-3) = 7$$

$$C_{23} = (-1)^{2+3}M_{23} = (-1)M_{23} = -7$$

■

The following way of defining the determinant function is called **expansion by cofactors**.

(5.5)

DEFINITION Let $A = (a_{ij})$ be a 3×3 matrix. Then the **determinant** of A, $|A|$, is defined by

$$|A| = a_{11}C_{11} + a_{12}C_{12} + a_{13}C_{13}$$

THEOREM Let i be the index of any row of A and j be the index of any column of A. Then

$$|A| = a_{i1}C_{i1} + a_{i2}C_{i2} + a_{i3}C_{i3}$$
$$|A| = a_{1j}C_{1j} + a_{2j}C_{2j} + a_{3j}C_{3j}$$

Thus the determinant of A can be evaluated by expanding along *any* row or column.

This theorem is proved in Chapter 2.

Example 5* Let

$$A = \begin{bmatrix} 2 & -3 & 1 \\ 4 & 0 & -2 \\ 3 & -1 & -3 \end{bmatrix}$$

be as in Example 3. Then by the definition of a determinant given in (5.5),

$$\begin{aligned} |A| &= 2C_{11} + (-3)C_{12} + 1C_{13} \\ &= 2(-1)^{1+1}\begin{vmatrix} 0 & -2 \\ -1 & -3 \end{vmatrix} - 3(-1)^{1+2}\begin{vmatrix} 4 & -2 \\ 3 & -3 \end{vmatrix} + 1(-1)^{1+3}\begin{vmatrix} 4 & 0 \\ 3 & -1 \end{vmatrix} \\ &= 2[0(-3) - (-1)(-2)] + 3[4(-3) - 3(-2)] + [4(-1) - 3(0)] \\ &= 2(-2) + 3(-6) + (-4) = -26 \end{aligned}$$

If we expand along the second row, we obtain

$$\begin{aligned} 4C_{21} + 0C_{22} - 2C_{23} &= 4(-1)^{2+1}\begin{vmatrix} -3 & 1 \\ -1 & -3 \end{vmatrix} + 0 + (-2)(-1)^{2+3}\begin{vmatrix} 2 & -3 \\ 3 & -1 \end{vmatrix} \\ &= -4[-3(-3) - (-1)(1)] + 2[2(-1) - 3(-3)] \\ &= -4(10) + 2(7) = -26 = |A| \end{aligned}$$

If we expand along the third column, we obtain

$$\begin{aligned} &1C_{13} + (-2)C_{23} + (-3)C_{33} \\ &\quad = 1(-1)^{1+3}\begin{vmatrix} 4 & 0 \\ 3 & -1 \end{vmatrix} - 2(-1)^{2+3}\begin{vmatrix} 2 & -3 \\ 3 & -1 \end{vmatrix} - 3(-1)^{3+3}\begin{vmatrix} 2 & -3 \\ 4 & 0 \end{vmatrix} \\ &\quad = 1[4(-1) - 3(0)] + 2[2(-1) - 3(-3)] - 3[2(0) - 4(-3)] \\ &\quad = -4 + 2(7) - 3[12] = -26 = |A| \end{aligned}$$

These second two calculations illustrate Theorem (5.5). ■

* *Compare the computations here with Example 3.*

The definitions and theorems in (5.4) and (5.5) illustrate exactly what concepts we need to find the determinants of $n \times n$ matrices. Following are some formal definitions of these concepts.

(5.6)

Let A be an $n \times n$ matrix where $n \geq 2$.

(a) DEFINITION The (i, j)th **minor** of A, M_{ij}, is the determinant of the $(n-1) \times (n-1)$ submatrix of A obtained by deleting the ith row and jth column from A.

(b) DEFINITION The (i, j)th **cofactor** of A, C_{ij}, is the number $C_{ij} = (-1)^{i+j} M_{ij}$.

(c) DEFINITION The **determinant** of A is defined by expansion by cofactors along the first row,

$$|A| = a_{11}C_{11} + a_{12}C_{12} + \cdots + a_{1n}C_{1n}$$

(d) THEOREM We can compute the determinant of A by expanding by cofactor along any row or along any column. That is, if $1 \leq i, j \leq n$, then

$$|A| = a_{i1}C_{i1} + a_{i2}C_{i2} + \cdots + a_{in}C_{in}$$
$$|A| = a_{1j}C_{1j} + a_{2j}C_{2j} + \cdots + a_{nj}C_{nj}$$

We leave it to an exercise (Exercise 39) to show that for $n = 2$, Definition (5.6c) is equivalent to Definition (5.3).

Example 6 If

$$A = \begin{bmatrix} 3 & 4 & -1 & 0 \\ 4 & -1 & 0 & 3 \\ -6 & 4 & 8 & -2 \\ -1 & 1 & 2 & 7 \end{bmatrix}$$

then

$$|A| = 3(-1)^{1+1}\begin{vmatrix} -1 & 0 & 3 \\ 4 & 8 & -2 \\ 1 & 2 & 7 \end{vmatrix} + 4(-1)^{1+2}\begin{vmatrix} 4 & 0 & 3 \\ -6 & 8 & -2 \\ -1 & 2 & 7 \end{vmatrix}$$
$$+ (-1)(-1)^{1+3}\begin{vmatrix} 4 & -1 & 3 \\ -6 & 4 & -2 \\ -1 & 1 & 7 \end{vmatrix} + 0(-1)^{1+4}\begin{vmatrix} 4 & -1 & 0 \\ -6 & 4 & 8 \\ -1 & 1 & 2 \end{vmatrix}$$

We now expand the four 3×3 determinants as we have done before and eventually calculate that $|A| = -1162$, as you can check. ■

As you contemplate working through all the calculations indicated in Example 6, you might feel some joy that this is not a 10×10, or even higher dimensional matrix. In fact it is very natural to feel that there ought to be a better way of doing this! Indeed there are better ways, and we shall begin to introduce some of them here. (More details are given in Chapter 2.) However, even though it is our purpose in this section to present a minimal amount of material necessary to compute eigenvalues, the following important fact should also be considered when we decide how much computation to learn.

IMPORTANT FACT Determinants are very important for theoretical purposes, but it is seldom we actually have to calculate them to solve real-world problems. We solve those problems by more computationally efficient procedures.

In particular you should read the discussion following Cramer's rule in Section 2.4.

Facts about Determinants

It is surprising how often we have to compute the determinant of an upper or lower triangular matrix. Fortunately this is very easy.

(5.7)

THEOREM If A is either upper or lower triangular

$$A = \begin{bmatrix} a_{11} & & & \\ a_{21} & a_{22} & & \\ \vdots & \vdots & \ddots & \\ a_{n1} & a_{n2} & \cdots & a_{nn} \end{bmatrix}$$

or

$$A = \begin{bmatrix} a_{11} & a_{12} & \cdots & a_{1n} \\ & a_{22} & \cdots & a_{2n} \\ & & \ddots & \vdots \\ & & & a_{nn} \end{bmatrix}$$

then the det(A) is the product of the diagonal entries,

$$\det(A) = a_{11}a_{22}\cdots a_{nn}$$

Proof Suppose A is lower triangular. Then expand by cofactors along the first row, obtaining

$$|A| = a_{11}C_{11} + 0C_{12} + \cdots + 0C_{1n} = a_{11}(-1)^2 M_{11}$$
$$= a_{11}\begin{vmatrix} a_{22} & & & \\ a_{32} & a_{33} & & \\ \vdots & \vdots & \ddots & \\ a_{n2} & a_{n3} & \cdots & a_{nn} \end{vmatrix}$$

We expand this new determinant by cofactors along the first row again, obtaining

$$|A| = a_{11}\left[a_{22}\begin{vmatrix} a_{33} & & \\ \vdots & \ddots & \\ a_{n3} & \cdots & a_{nn} \end{vmatrix}\right] = a_{11}a_{22}\begin{vmatrix} a_{33} & & \\ \vdots & \ddots & \\ a_{n3} & \cdots & a_{nn} \end{vmatrix}$$

You can see that we can keep going like this, and we eventually get

$$|A| = a_{11}a_{22}\cdots a_{nn}$$

If A is upper triangular, we proceed similarly, except we expand along the first column each time to obtain the result. ■

Example 7 (a) $\begin{vmatrix} 3 & 1 & 2 & 7 \\ & -2 & 8 & 1 \\ & & 4 & 7 \\ & & & \frac{1}{6} \end{vmatrix} = 3(-2)(4)(\tfrac{1}{6}) = -4$

(b) $\begin{vmatrix} 1 & & & \\ 5 & 1 & & \\ -2 & 0 & 1 & \\ 3 & 0 & 0 & 1 \end{vmatrix} = 1(1)(1)(1) = 1$

(c) $\begin{vmatrix} 2 & & \\ & -7 & \\ & & \frac{1}{3} \end{vmatrix} = 2(-7)(\tfrac{1}{3}) = -\tfrac{14}{3}$ ■

The following corollary should now be obvious.

(5.8)

COROLLARY (a) Suppose L is of the form

$$L = \begin{bmatrix} 1 & & & & & \\ & \ddots & & & & \\ & & 1 & & & \\ & & m_{i+1} & & & \\ & & \vdots & & \ddots & \\ & & m_n & & & 1 \end{bmatrix}$$

Then $\det(L) = 1$.

(b) If D is a diagonal matrix (so D is both upper and lower triangular), then $\det D = d_{11}d_{22}\cdots d_{nn}$.

(c) In particular, $\det(I) = 1$.

There is just one more important fact about determinants that we shall need to know before we study eigenvalues.

(5.9)

THEOREM If A and B are both $n \times n$ matrices, then

$$\det(AB) = \det(A)\det(B)$$

That is, the determinant of a product is the product of the determinants.

See Chapter 2 for the proof.

Example 8 Suppose

$$A = \begin{bmatrix} 3 & -1 \\ 2 & 4 \end{bmatrix} \quad \text{and} \quad B = \begin{bmatrix} 4 & -2 \\ -5 & 3 \end{bmatrix}$$

Then

$$AB = \begin{bmatrix} 17 & -9 \\ -12 & 8 \end{bmatrix}$$

and

$$|A| = 12 - (-2) = 14, \qquad |B| = 12 - 10 = 2$$

and

$$|AB| = 136 - 108 = 28 = |A|\,|B|$$

■

Example 9 If A is a 5×5 matrix, then

$$\det(2A) = \det((2I)A) = \det(2I)\det(A) = 2^5\det(A)$$

since $\det(2I) = 2^5$ by Corollary (5.8b). ■

There is a very important corollary to Theorem (5.9).

(5.10)

COROLLARY If A is an invertible $n \times n$ matrix, then

$$\det(A^{-1}) = [\det(A)]^{-1}$$

Proof Since $AA^{-1} = I$ and $\det(I) = 1$ [by Corollary (5.8c)], we have, by Theorem (5.9),

$$[\det(A)][\det(A^{-1})] = \det(AA^{-1}) = \det(I) = 1$$

Thus $\det(A^{-1}) = \dfrac{1}{\det(A)} = [\det(A)]^{-1}$. ■

Example 10 Suppose

$$A = \begin{bmatrix} -2 & 4 \\ -3 & 5 \end{bmatrix}$$

Then

$$A^{-1} = \tfrac{1}{2}\begin{bmatrix} 5 & -4 \\ 3 & -2 \end{bmatrix} = \begin{bmatrix} \frac{5}{2} & -2 \\ \frac{3}{2} & -1 \end{bmatrix}$$

and

$$\det(A) = -10 - (-12) = 2, \qquad \det(A^{-1}) = -\tfrac{5}{2} + 3 = \tfrac{1}{2}$$ ■

As a final remark, up to this point the entries in a matrix have been numbers, so their determinants have also been numbers. However, in the computation of eigenvalues (and other situations as well), entries of matrices are functions. For such cases their determinants are also functions.

Example 11 Suppose

$$A = \begin{bmatrix} x-1 & 3 & x \\ 4 & x+2 & -1 \\ -2 & x & x+1 \end{bmatrix}$$

Then

$$\begin{aligned} \det(A) &= (x-1)\begin{vmatrix} x+2 & -1 \\ x & x+1 \end{vmatrix} - 3\begin{vmatrix} 4 & -1 \\ -2 & x+1 \end{vmatrix} + x\begin{vmatrix} 4 & x+2 \\ -2 & x \end{vmatrix} \\ &= (x-1)[(x+2)(x+1) + x] - 3[4(x+1) - 2] + x[4x + 2(x+2)] \\ &= (x-1)(x^2 + 4x + 2) - 3(4x + 2) + x(6x + 4) \\ &= x^3 + 9x^2 - 10x - 8 \end{aligned}$$ ■

Exercise 5.1 In Exercises 1–20, (a) evaluate the determinant of the given matrix (be judicious in your choice of columns or rows along which to expand) and (b) determine if the matrix is invertible (Y or N).

1. $[3.1]$ 2. $[-5.7]$ 3. $[0]$

4. $[\pi]$

5. $\begin{bmatrix} 1 & 2 \\ -2 & 3 \end{bmatrix}$

6. $\begin{bmatrix} 7 & -11 \\ 2 & -4 \end{bmatrix}$

7. $\begin{bmatrix} -2 & 3 \\ 4 & -6 \end{bmatrix}$

8. $\begin{bmatrix} -1.1 & 2.9 \\ -2.2 & 5.8 \end{bmatrix}$

9. $\begin{bmatrix} 3 & -1 \\ 4 & 0 \end{bmatrix}$

10. $\begin{bmatrix} 8 & 2 \\ 0 & -3 \end{bmatrix}$

11. $\begin{bmatrix} 2 & 1 & 3 \\ 4 & -4 & 1 \\ 1 & 0 & -1 \end{bmatrix}$

12. $\begin{bmatrix} 3 & -1 & -2 \\ 3 & 7 & 0 \\ 8 & 1 & 1 \end{bmatrix}$

13. $\begin{bmatrix} 2 & 5 & 1 \\ 3 & -3 & 2 \\ 7 & 7 & 4 \end{bmatrix}$

14. $\begin{bmatrix} 3 & -1 & -2 \\ -2 & 2 & 1 \\ -1 & 3 & 0 \end{bmatrix}$

15. $\begin{bmatrix} 2 & 3 & -4 \\ 0 & 4 & 1 \\ 0 & 0 & -1 \end{bmatrix}$

16. $\begin{bmatrix} 8 & 1 & 2 \\ -1 & 3 & 0 \\ 4 & 0 & 0 \end{bmatrix}$

17. $\begin{bmatrix} 4 & 1 & 2 & 1 \\ 3 & 0 & 0 & 0 \\ -1 & 0 & 2 & 1 \\ 4 & 5 & 0 & 2 \end{bmatrix}$

18. $\begin{bmatrix} 1 & 2 & 0 & 3 \\ -2 & 1 & 0 & 2 \\ 3 & 2 & 2 & -4 \\ 0 & 5 & 0 & 2 \end{bmatrix}$

19. $\begin{bmatrix} 4 & 2 & 7 & 1 & 3 \\ -2 & 1 & 3 & -2 & 1 \\ -4 & 2 & 5 & 3 & 2 \\ 0 & 0 & 0 & 0 & 0 \\ 3 & 4 & 1 & 2 & 7 \end{bmatrix}$

20. $\begin{bmatrix} 0 & 1 & 0 & 0 & 0 \\ 0 & 0 & 1 & 0 & 0 \\ 1 & 0 & 0 & 0 & 0 \\ 0 & 0 & 0 & 0 & 1 \\ 0 & 0 & 0 & 1 & 0 \end{bmatrix}$

In Exercises 21–31, (a) evaluate the determinant of the given matrix, and (b) determine for which values of x, if any, the matrix is singular.

21. $[x]$

22. $[x - 5]$

23. $[x^2 - 2x - 3]$

24. $[x^2 + 1]$

25. $\begin{bmatrix} x+1 & x \\ 2 & 3 \end{bmatrix}$

26. $\begin{bmatrix} 4x & 5 \\ x & x-1 \end{bmatrix}$

27. $\begin{bmatrix} x+3 & 3 \\ 2 & x-2 \end{bmatrix}$

28. $\begin{bmatrix} x-1 & 2 \\ 2 & x-4 \end{bmatrix}$

29. $\begin{bmatrix} x & x+1 & x-1 \\ 2 & 3 & -4 \\ 1 & 5 & 1 \end{bmatrix}$

30. $\begin{bmatrix} x-1 & 0 & 4 \\ 2 & x+1 & 2 \\ 3 & 3 & x-1 \end{bmatrix}$

31. $\begin{bmatrix} x-2 & 3 & 4 & 1 \\ 0 & x+3 & 2 & -4 \\ 0 & 0 & x+1 & 2 \\ 0 & 0 & 0 & x-5 \end{bmatrix}$

32. Find matrices A and B such that $\det(A + B) \neq \det(A) + \det(B)$.

In Exercises 33–38, assume A, B, and C are invertible 3×3 matrices, $\det(A) = a$, $\det(B) = b$, and $\det(C) = c$. Evaluate the given expressions.

33. $\det(AB)$
34. $\det(AB^{-1}C^2)$
35. $\det(5A)$
36. $\det(-B)$
37. $\det(C^T)$
38. $\det(A^TB^2C^{-1})$
39. Let $A = \begin{bmatrix} a & b \\ c & d \end{bmatrix}$. Show directly that the definitions given in Definitions (5.3) and (5.6c) of $\det(A)$ agree.

5.2 Eigenvalues and Eigenvectors

There are many problems in engineering, mathematics, and science where an $n \times n$ matrix A arises, and the person involved has to determine if there is a number λ and a nonzero vector $\mathbf{x}$ such that $A\mathbf{x} = \lambda\mathbf{x}$. In this section we discuss some of the theory behind determining this. Over the next several sections we discuss special aspects of the problem and then investigate a few applications.

Definition of Eigenvalue

(5.11)

DEFINITION Let A be an $n \times n$ matrix. Suppose $\mathbf{x}$ is a *nonzero* vector in $\mathbb{R}^n$ and λ is a number (possibly zero) such that

$$A\mathbf{x} = \lambda\mathbf{x}$$

that is, $A\mathbf{x}$ is a scalar multiple of $\mathbf{x}$. Then $\mathbf{x}$ is called an **eigenvector** of A, and λ is called an **eigenvalue** of A. We say that λ is the eigenvalue **associated with** or **corresponding to** $\mathbf{x}$, and $\mathbf{x}$ is an eigenvector **associated with** or **corresponding to** λ.

The word "eigen" in German means "proper." Historically eigenvalues are also called **proper values**, **characteristic values**, or **latent roots**, and the same terminology holds for eigenvectors.

NOTE Eigenvalues and eigenvectors are defined *only* for *square* matrices.

Example 1 Suppose $A = \begin{bmatrix} -1 & 2 \\ 0 & 3 \end{bmatrix}$. Then $\mathbf{v} = \begin{bmatrix} 1 \\ 2 \end{bmatrix}$ is an eigenvector corresponding to the eigenvalue 3 since

$$\begin{bmatrix} -1 & 2 \\ 0 & 3 \end{bmatrix}\begin{bmatrix} 1 \\ 2 \end{bmatrix} = \begin{bmatrix} 3 \\ 6 \end{bmatrix} = 3\begin{bmatrix} 1 \\ 2 \end{bmatrix} \qquad \text{or} \qquad A\mathbf{v} = 3\mathbf{v}$$

Also, $\mathbf{w} = \begin{bmatrix} 1 \\ 0 \end{bmatrix}$ is an eigenvector corresponding to the eigenvalue -1 since

$$\begin{bmatrix} -1 & 2 \\ 0 & 3 \end{bmatrix}\begin{bmatrix} 1 \\ 0 \end{bmatrix} = \begin{bmatrix} -1 \\ 0 \end{bmatrix} = (-1)\begin{bmatrix} 1 \\ 0 \end{bmatrix} \quad \text{or} \quad A\mathbf{w} = (-1)\mathbf{w}$$ ■

Suppose $A\mathbf{x} = \lambda\mathbf{x}$ and r is a number. Then

$$A(r\mathbf{x}) = r(A\mathbf{x}) = r(\lambda\mathbf{x}) = \lambda(r\mathbf{x})$$

Thus if $\mathbf{x}$ is an eigenvector of A associated with an eigenvalue λ, then any nonzero multiple of $\mathbf{x}$ is also an eigenvector of A associated with λ. Geometrically this says that if $\mathbf{x}$ is an eigenvector of A, then A multiplies every vector on the line through $\mathbf{x}$ (and the origin) by λ.

Example 2 Let A, $\mathbf{v}$, and $\mathbf{w}$ be as in Example 1. Then A multiplies every vector on the line through $\mathbf{v}$ by 3 and A multiplies every vector on the line through $\mathbf{w}$ by -1. See Figure 5.1. Thus in this case vectors on the line through $\mathbf{v}$ are expanded or stretched out, and vectors on the line through $\mathbf{w}$ are reflected through the origin. ■

Figure 5.1

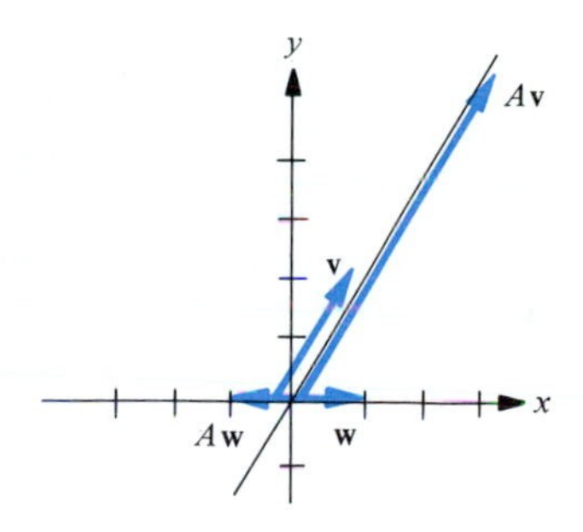

Finding Eigenvalues

You will now see why we had to discuss determinants before eigenvalues. Suppose we want to find all the eigenvalues of an $n \times n$ matrix A. We start by rewriting the equation $A\mathbf{x} = \lambda\mathbf{x}$:

$$\begin{aligned} A\mathbf{x} &= \lambda\mathbf{x} \\ A\mathbf{x} &= \lambda I\mathbf{x} \\ \lambda I\mathbf{x} - A\mathbf{x} &= \mathbf{0} \\ (\lambda I - A)\mathbf{x} &= \mathbf{0} \end{aligned}$$

We now see that

$$\begin{aligned} &\lambda \text{ is an eigenvalue of } A \\ \Leftrightarrow\quad & (\lambda I - A)\mathbf{x} = \mathbf{0} \text{ has a nontrivial solution} \\ \Leftrightarrow\quad & \lambda I - A \text{ is singular} \\ \Leftrightarrow\quad & \det(\lambda I - A) = 0 \end{aligned}$$

We have now proved the following.

(5.12)

Let A be an $n \times n$ matrix.

THEOREM The number λ is an eigenvalue of A if and only if $\det(\lambda I - A) = 0$.

DEFINITION When expanded, the determinant $\det(\lambda I - A)$ is a polynomial in λ of degree n that is called the **characteristic polynomial** of A. The equation $\det(\lambda I - A) = 0$ is called the **characteristic equation** of A.

Example 3 Find all the eigenvalues of the matrix

$$A = \begin{bmatrix} 2 & 3 \\ 4 & 3 \end{bmatrix}$$

Solution First find

$$\lambda I - A = \begin{bmatrix} \lambda & 0 \\ 0 & \lambda \end{bmatrix} - \begin{bmatrix} 2 & 3 \\ 4 & 3 \end{bmatrix} = \begin{bmatrix} \lambda - 2 & -3 \\ -4 & \lambda - 3 \end{bmatrix}$$

The characteristic polynomial of A is then

$$\det(\lambda I - A) = \begin{vmatrix} \lambda - 2 & -3 \\ -4 & \lambda - 3 \end{vmatrix} = (\lambda - 2)(\lambda - 3) - 12 = \lambda^2 - 5\lambda - 6$$

The solutions to the characteristic equation

$$\lambda^2 - 5\lambda - 6 = 0$$

are $\lambda = 6$ and $\lambda = -1$. These are the eigenvalues of A. ■

Example 4 Find all of the eigenvalues of the matrix

$$A = \begin{bmatrix} 5 & -7 & 7 \\ 4 & -3 & 4 \\ 4 & -1 & 2 \end{bmatrix}$$

Solution First find

$$\lambda I - A = \begin{bmatrix} \lambda & & \\ & \lambda & \\ & & \lambda \end{bmatrix} - \begin{bmatrix} 5 & -7 & 7 \\ 4 & -3 & 4 \\ 4 & -1 & 2 \end{bmatrix} = \begin{bmatrix} \lambda - 5 & 7 & -7 \\ -4 & \lambda + 3 & -4 \\ -4 & 1 & \lambda - 2 \end{bmatrix}$$

The characteristic polynomial of A is then

$$\det(\lambda I - A) = \begin{vmatrix} \lambda - 5 & 7 & -7 \\ -4 & \lambda + 3 & -4 \\ -4 & 1 & \lambda - 2 \end{vmatrix}$$

$$= (\lambda - 5)\begin{vmatrix} \lambda + 3 & -4 \\ 1 & \lambda - 2 \end{vmatrix} - 7\begin{vmatrix} -4 & -4 \\ -4 & \lambda - 2 \end{vmatrix} - 7\begin{vmatrix} -4 & \lambda + 3 \\ -4 & 1 \end{vmatrix}$$

$$= (\lambda - 5)(\lambda^2 + \lambda - 2) - 7(-4\lambda - 8) - 7(4\lambda + 8)$$

$$= \lambda^3 - 4\lambda^2 - 7\lambda + 10$$

Thus the characteristic equation is

$$\lambda^3 - 4\lambda^2 - 7\lambda + 10 = 0$$

We hope there are nice roots to this equation. Any root to this equation that is an integer must be a factor of 10, that is, the only nice candidates are ± 1, ± 2, ± 5, ± 10. We try $\lambda = 1$ and see $1^3 - 4(1^2) - 7(1) + 10 = 0$. Hence $\lambda = 1$ is a root of the characteristic equation, so $\lambda - 1$ is a factor of the characteristic polynomial. We divide this polynomial by $\lambda - 1$ and obtain

$$\lambda^3 - 4\lambda^2 - 7\lambda + 10 = (\lambda - 1)(\lambda^2 - 3\lambda - 10)$$

We can easily determine that the zeros of $\lambda^2 - 3\lambda - 10$ are $\lambda = 5, -2$. Thus altogether the roots of the characteristic equation are $\lambda = 1, -2, 5$. These are the eigenvalues of A. ■

After looking at this example, you can probably guess that it is usually very difficult to find eigenvalues by solving the characteristic equation, except in very special circumstances. However, we shall see later in this chapter that the mathematical theory leads to some very interesting and surprising mathematical techniques for computing eigenvalues.

But first there is a technical detail we must address. To find eigenvalues we solve the characteristic equation $\det(\lambda I - A) = 0$. However, this does not always have real solutions, so the question is: What do we do then? The next example illustrates the problem and its resolution.

Example 5 Find all eigenvalues of

$$A = \begin{bmatrix} 0 & -1 \\ 1 & 0 \end{bmatrix}$$

Solution First find

$$\lambda I - A = \begin{bmatrix} \lambda & \\ & \lambda \end{bmatrix} - \begin{bmatrix} 0 & -1 \\ 1 & 0 \end{bmatrix} = \begin{bmatrix} \lambda & 1 \\ -1 & \lambda \end{bmatrix}$$

The characteristic polynomial is then $\lambda^2 + 1$, so the characteristic equation is

$$\lambda^2 + 1 = 0$$

This equation has no real roots so this matrix A has no real eigenvalues. You can actually see why geometrically. By Example 5, Section 4.1, this matrix A rotates the plane, $\mathbb{R}^2$, by 90°. Thus every nonzero vector is rotated by A and hence is not a multiple of itself. Therefore, there are no real eigenvectors, and correspondingly there are no real eigenvalues.

However, the equation $\lambda^2 + 1 = 0$ *does* have two complex roots, namely $\pm i$. Recall that the set of **complex numbers**, $\mathbb{C}$, is the set of all numbers of the form $a + bi$ where a and b are real numbers and $i^2 = -1$. We can add, subtract, multiply, and divide complex numbers, and most important in this case, both $\lambda = i$ and $\lambda = -i$ satisfy $\lambda^2 + 1 = 0$. Thus $\pm i$ are candidates for eigenvalues, and if we allow vectors to have complex entries, you can check that

$$\begin{bmatrix} 0 & -1 \\ 1 & 0 \end{bmatrix}\begin{bmatrix} 1 \\ i \end{bmatrix} = \begin{bmatrix} -i \\ 1 \end{bmatrix} = (-i)\begin{bmatrix} 1 \\ i \end{bmatrix} \text{ and } \begin{bmatrix} 0 & -1 \\ 1 & 0 \end{bmatrix}\begin{bmatrix} 1 \\ -i \end{bmatrix} = \begin{bmatrix} i \\ 1 \end{bmatrix} = i\begin{bmatrix} 1 \\ -i \end{bmatrix}$$

Therefore, there are nonzero eigenvectors associated with $\pm i$. Even though complex numbers are historically called "imaginary," they have many real applications, ranging from electrical engineering to nuclear physics. Hence there are many real applications that require complex eigenvalues and eigenvectors. However, there are many applications that do not allow complex numbers (just as many applications do not allow negative real numbers), and developing them here would unnecessarily complicate this introductory material. Hence, unless specifically stated otherwise, we will assume all eigenvalues we seek are real. ■

We now summarize our results so far.

(5.13)

THEOREM Let A be an $n \times n$ matrix. Then the following are equivalent.

(a) λ is a (real) eigenvalue of A.
(b) There is a nonzero vector $\mathbf{x}$ in $\mathbb{R}^n$ such that $A\mathbf{x} = \lambda\mathbf{x}$.
(c) The system of equations $(\lambda I - A)\mathbf{x} = \mathbf{0}$ has a nontrivial solution.
(d) λ is a real solution of the characteristic equation $\det(\lambda I - A) = 0$.

Finding Eigenvectors and Eigenspaces

Once we know how to find eigenvalues, we next turn to the problem of associated eigenvectors. So suppose we know λ is an eigenvalue of the $n \times n$ matrix A. We wish to find all (nonzero) $\mathbf{x}$ such that $A\mathbf{x} = \lambda\mathbf{x}$. However,

$$A\mathbf{x} = \lambda\mathbf{x} \quad \Leftrightarrow \quad (\lambda I - A)\mathbf{x} = \mathbf{0} \quad \Leftrightarrow \quad \mathbf{x} \text{ is in } NS(\lambda I - A)$$

We have just shown the following.

(5.14)

Let λ be an eigenvalue of the $n \times n$ matrix A.

THEOREM The eigenvectors of A associated with λ are the nonzero vectors in $NS(\lambda I - A)$ and hence, together with the zero vector, form a subspace of $\mathbb{R}^n$.

DEFINITION $NS(\lambda I - A)$ is called the **eigenspace** of A associated with λ.

NOTATION We shall denote this eigenspace by W_λ.

Therefore, to find a basis for an eigenspace, we find a basis for a null space: $NS(\lambda I - A)$. We can sometimes do this by inspection, but otherwise we follow the method described in Chapter 3. This is illustrated in the next two examples.

Example 6 Find bases for the eigenspaces of the matrix

$$A = \begin{bmatrix} 4 & 2 \\ -1 & 1 \end{bmatrix}$$

Solution The characteristic equation of A is

$$(\lambda - 4)(\lambda - 1) + 2 = \lambda^2 - 5\lambda + 6 = (\lambda - 2)(\lambda - 3).$$

Thus $\lambda = 2$ and 3 are the eigenvalues. Theory tells us that the matrices $2I - A$ and $3I - A$ are singular, and their null spaces are the corresponding eigenspaces. Thus we are asked to find a basis for the null space of

$$\lambda I - A = \begin{bmatrix} \lambda & \\ & \lambda \end{bmatrix} - \begin{bmatrix} 4 & 2 \\ -1 & 1 \end{bmatrix} = \begin{bmatrix} \lambda - 4 & -2 \\ 1 & \lambda - 1 \end{bmatrix}$$

when $\lambda = 2$ or 3.

Case $\lambda = 2$ We wish to find a basis for the null space of

$$2I - A = \begin{bmatrix} -2 & -2 \\ 1 & 1 \end{bmatrix}$$

By inspection we can see $rk(2I - A) = 1$, so $\dim NS(2I - A) = 1$. Furthermore by inspection we can see $\mathbf{u}_1 = [1 \quad -1]^T$ is a nonzero vector in this null space. (If we did not see this by inspection, we would use the methods of Section 3.6.) Hence $\{\mathbf{u}_1\}$ is a basis for this null space and so for eigenspace W_2. Thus theory tells us that $A\mathbf{u}_1 = 2\mathbf{u}_1$, which we can verify directly.

$$A\mathbf{u}_1 = \begin{bmatrix} 4 & 2 \\ -1 & 1 \end{bmatrix} \begin{bmatrix} 1 \\ -1 \end{bmatrix} = \begin{bmatrix} 2 \\ -2 \end{bmatrix} = 2 \begin{bmatrix} 1 \\ -1 \end{bmatrix} = 2\mathbf{u}_1$$

Case $\lambda = 3$ We wish to find a basis for the null space of

$$3I - A = \begin{bmatrix} -1 & -2 \\ 1 & 2 \end{bmatrix}$$

By inspection we can see that the rank of $3I - A$, denoted by $rk(3I - A)$, is 1, so dim $NS(3I - A) = 1$. Furthermore, by inspection, we can see that $\mathbf{u}_2 = [2 \quad -1]^T$ is a nonzero vector in this null space. Hence $\{\mathbf{u}_2\}$ is a basis for this null space and so for the eigenspace W_3. Thus theory tells us that $A\mathbf{u}_2 = 3\mathbf{u}_2$, which we can verify directly.

$$A\mathbf{u}_2 = \begin{bmatrix} 4 & 2 \\ -1 & 1 \end{bmatrix}\begin{bmatrix} 2 \\ -1 \end{bmatrix} = \begin{bmatrix} 6 \\ -3 \end{bmatrix} = 3\begin{bmatrix} 2 \\ -1 \end{bmatrix} = 3\mathbf{u}_2$$ ■

Example 7 Find bases for the eigenspaces of the matrix

$$A = \begin{bmatrix} 2 & 3 & 0 \\ 4 & 3 & 0 \\ 0 & 0 & 6 \end{bmatrix}$$

Solution The characteristic equation of A is $(\lambda - 6)^2(\lambda + 1) = 0$, as you can readily verify. Thus the eigenvalues are $\lambda = -1, 6$. The theory tells us that the matrices $(-1)I - A$ and $6I - A$ are singular and their null spaces are the corresponding eigenspaces. Therefore, we are asked to find a basis for the null space of

$$\lambda I - A = \begin{bmatrix} \lambda & & \\ & \lambda & \\ & & \lambda \end{bmatrix} - \begin{bmatrix} 2 & 3 & 0 \\ 4 & 3 & 0 \\ 0 & 0 & 6 \end{bmatrix} = \begin{bmatrix} \lambda - 2 & -3 & 0 \\ -4 & \lambda - 3 & 0 \\ 0 & 0 & \lambda - 6 \end{bmatrix} \tag{5.15}$$

when $\lambda = -1$ or 6.

Case $\lambda = -1$ We first substitute $\lambda = -1$ into Equation (5.15), obtaining

$$(-1)I - A = \begin{bmatrix} -3 & -3 & 0 \\ -4 & -4 & 0 \\ 0 & 0 & -7 \end{bmatrix}$$

We next reduce this to echelon form, obtaining

$$U = \begin{bmatrix} -3 & -3 & 0 \\ 0 & 0 & -7 \\ 0 & 0 & 0 \end{bmatrix}$$

We know $NS(U) = NS[(-1)I - A]$, and using the methods of Section 3.6, we obtain

$$\mathbf{u}_1 = [1 \quad -1 \quad 0]^T$$

for the basis of $NS(U)$, as you should check. Thus $\{\mathbf{u}_1\}$ is a basis for W_{-1}. You can also verify directly that $A\mathbf{u}_1 = (-1)\mathbf{u}_1$.

Case $\lambda = 6$ We substitute $\lambda = 6$ into Equation (5.15) and then reduce to echelon form, obtaining

$$6I - A = \begin{bmatrix} 4 & -3 & 0 \\ -4 & 3 & 0 \\ 0 & 0 & 0 \end{bmatrix} \to U = \begin{bmatrix} 4 & -3 & 0 \\ 0 & 0 & 0 \\ 0 & 0 & 0 \end{bmatrix}$$

We see $rk(U) = 1$, so we know that $\dim NS(U) = \dim W_6 = 2$. By the methods of Section 3.6 (or by inspection), we determine that $\{\mathbf{u}_2, \mathbf{u}_3\}$ is a basis for $NS(U) = W_6$, where

$$\mathbf{u}_2 = [\tfrac{3}{4} \quad 1 \quad 0]^T, \qquad \mathbf{u}_3 = [0 \quad 0 \quad 1]^T$$

Thus we obtained bases for the two eigenspaces, so we are done. ■

NOTE In Example 6 the matrix A is 3×3, but it has only two distinct eigenvalues. More precisely, it has three eigenvalues, but one of them, $\lambda = 6$, is a multiple root of the characteristic equation, which we call a repeated eigenvalue. Nevertheless we have found three linearly independent eigenvectors. This does not always happen, that is, an $n \times n$ matrix does not always have n linearly independent eigenvectors. Whether it does or does not has important consequences for the matrix and what we can do with it, and this is the topic of the next section.

Two Important Properties of Eigenvalues (Optional)

We now present two elementary but important properties of eigenvalues. This subsection is optional because these properties are not used elsewhere in the text; however, they are very important in the theory of abstract and computational linear algebra. First we provide a definition.

(5.16)

DEFINITION Let $A = [a_{ij}]$ be an $n \times n$ matrix. Then the **trace** of A, denoted by $\text{tr}(A)$, is the sum of the diagonal entries:

$$\text{tr}(A) = a_{11} + a_{22} + \cdots + a_{nn}$$

Theorem (5.17b), which is the main result, hints at one reason why $\text{tr}(A)$ is important.

(5.17)

THEOREM Let A be an $n \times n$ matrix with eigenvalues $\lambda_1, \lambda_2, \ldots, \lambda_n$. Then

(a) $\det(A) = \lambda_1 \lambda_2 \cdots \lambda_n$
(b) $\text{tr}(A) = \lambda_1 + \lambda_2 + \cdots + \lambda_n$

Proof The eigenvalues are the n roots of the characteristic polynomial of A, $\det(\lambda I - A)$. Thus this polynomial can be factored completely as

$$\det(\lambda I - A) = (\lambda - \lambda_1)(\lambda - \lambda_2)\cdots(\lambda - \lambda_n) \tag{5.18}$$

For (a), set $\lambda = 0$ in Equation (5.18) and compare the resulting equation with Theorem (5.17a). [Remember that $\det(-A) = (-1)^n \det(A)$.]
For (b), first show that

$$\det(\lambda I - A) = (\lambda - a_{11})(\lambda - a_{22})\cdots(\lambda - a_{nn}) + \text{a polynomial (in } \lambda\text{) of degree} \leq n - 2 \tag{5.19}$$

Substitute Equation (5.19) into Equation (5.18), and compare the coefficients of λ^{n-1} in the resulting equation. See Exercises 36, 37. ■

Example 8 (a) Let $A = \begin{bmatrix} 2 & 3 \\ 4 & 3 \end{bmatrix}$ from Example 2, so that $\lambda = -1, 6$ are the eigenvalues. Then we can see that

$$\det(A) = 2(3) - 4(3) = -6 = \lambda_1\lambda_2, \qquad \text{tr}(A) = 2 + 3 = 5 = \lambda_1 + \lambda_2$$

(b) Let $A = \begin{bmatrix} 5 & -7 & 7 \\ 4 & -3 & 4 \\ 4 & -1 & 2 \end{bmatrix}$ from Example 4, so that $\lambda = 1, -2, 5$ are the eigenvalues. Then you can check that $\det(A) = -10$, so we can see that

$$\det(A) = \lambda_1\lambda_2\lambda_3, \qquad \text{tr}(A) = 5 - 3 + 2 = 4 = \lambda_1 + \lambda_2 + \lambda_3$$

(c) Let $A = \begin{bmatrix} 0 & -1 \\ 1 & 0 \end{bmatrix}$ from Example 5, so that $\lambda = i, -i$ are the eigenvalues. Then we can see that

$$\det A = 0(0) - 1(-1) = 1 = \lambda_1\lambda_2$$
$$\text{tr}(A) = 0 + 0 = 0 = \lambda_1 + \lambda_2$$
■

Indeed, Theorem 5.17 remains true (with the same proof) when we allow complex eigenvalues (and eigenvectors), as Example 8c illustrates.

Exercise 5.2

In Exercises 1–6, show directly that the given vectors are eigenvectors of the given matrix. What are the corresponding eigenvalues?

1. $\mathbf{v}_1 = \begin{bmatrix} 1 \\ 3 \end{bmatrix}, \quad \mathbf{v}_2 = \begin{bmatrix} 1 \\ 1 \end{bmatrix}, \quad A = \begin{bmatrix} 3 & 1 \\ -3 & 7 \end{bmatrix}$

2. $\mathbf{v}_1 = \begin{bmatrix} -3 \\ 1 \end{bmatrix}, \quad \mathbf{v}_2 = \begin{bmatrix} 1 \\ -2 \end{bmatrix}, \quad A = \begin{bmatrix} 2 & -3 \\ 2 & 9 \end{bmatrix}$

3. $\mathbf{v}_1 = \begin{bmatrix} -1 \\ 2 \end{bmatrix}, \quad \mathbf{v}_2 = \begin{bmatrix} -1 \\ 4 \end{bmatrix}, \quad A = \begin{bmatrix} 4 & 1 \\ -8 & -2 \end{bmatrix}$

4. $\mathbf{v}_1 = \begin{bmatrix} 1 \\ -3 \end{bmatrix}$, $\mathbf{v}_2 = \begin{bmatrix} 2 \\ 1 \end{bmatrix}$, $A = \begin{bmatrix} 1 & -2 \\ -3 & 6 \end{bmatrix}$

5. $\boldsymbol{\varepsilon}_1, \ldots, \boldsymbol{\varepsilon}_n, I_n$

6. $\boldsymbol{\varepsilon}_1, \ldots, \boldsymbol{\varepsilon}_n, \mathbf{0} = n \times n$ matrix of all zeros

7. Suppose A is an $n \times n$ matrix, $\mathbf{v}$ is a nonzero vector in $NS(A)$. Show directly that $\mathbf{v}$ is an eigenvector of A. What is its corresponding eigenvalue?

In Exercises 8–10, let

$$A_2 = \begin{bmatrix} 1 & 1 \\ 1 & 1 \end{bmatrix}, \quad A_3 = \begin{bmatrix} 1 & 1 & 1 \\ 1 & 1 & 1 \\ 1 & 1 & 1 \end{bmatrix}, \ldots, \quad A_n = \begin{bmatrix} 1 & 1 & \cdots & 1 \\ \vdots & \vdots & & \vdots \\ 1 & 1 & \cdots & 1 \end{bmatrix}, \ldots$$

8. Show directly the $n \times 1$ vector $[1 \quad 1 \quad \cdots \quad 1]^T$ is an eigenvector of A_n. What is its associated eigenvalue?

9. Show directly that

$$\mathbf{v}_2 = [1 \quad -1 \quad 0 \quad \cdots \quad 0]^T, \quad \mathbf{v}_3 = [0 \quad 1 \quad -1 \quad 0 \quad \cdots \quad 0]^T, \quad \ldots,$$
$$\mathbf{v}_n = [0 \quad \cdots \quad 0 \quad 1 \quad -1]^T$$

are $n - 1$ linearly independent vectors in $NS(A_n)$.

10. Use Exercises 7–9 to find n linearly independent eigenvectors of A_n. What are their associated eigenvalues?

In Exercises 11–22, for the given matrix:

(a) Find its characteristic polynomial (in factored form).
(b) Find its eigenvalues.
(c) Find bases for its eigenspaces.
(d) Graph the eigenspaces in Exercises 11–20 only.
(e) Verify directly that $A\mathbf{v} = \lambda\mathbf{v}$, for all associated eigenvectors and eigenvalues you found.
(f) Verify that Theorem (5.17) holds.

11. $\begin{bmatrix} 2 & 0 \\ 0 & -3 \end{bmatrix}$ 12. $\begin{bmatrix} 0 & -1 \\ -4 & 0 \end{bmatrix}$ 13. $\begin{bmatrix} 5 & -8 \\ 1 & -1 \end{bmatrix}$

14. $\begin{bmatrix} -1 & -3 \\ -3 & 7 \end{bmatrix}$ 15. $\begin{bmatrix} 1 & 3 \\ 4 & -3 \end{bmatrix}$ 16. $\begin{bmatrix} 5 & 6 \\ 2 & 1 \end{bmatrix}$

17. $\begin{bmatrix} 1 & -1 & 0 \\ -1 & 2 & -1 \\ 0 & -1 & 1 \end{bmatrix}$ 18. $\begin{bmatrix} 2 & 0 & -1 \\ 2 & -1 & 0 \\ -12 & 0 & 1 \end{bmatrix}$

19. $\begin{bmatrix} 2 & -5 & 5 \\ 0 & 3 & -1 \\ 0 & -1 & 3 \end{bmatrix}$ 20. $\begin{bmatrix} 5 & 0 & -4 \\ 0 & -3 & 0 \\ -4 & 0 & -1 \end{bmatrix}$

21. $\begin{bmatrix} -3 & 2 & 0 & 0 \\ -3 & 4 & 0 & 0 \\ 0 & 0 & -5 & -4 \\ 0 & 0 & -2 & 2 \end{bmatrix}$

22. $\begin{bmatrix} -4 & 0 & 6 & 0 \\ 0 & -5 & 0 & -4 \\ -1 & 0 & 1 & 0 \\ 0 & 3 & 0 & 2 \end{bmatrix}$

In Exercises 23–28, $\mathbf{v}$ and λ are associated eigenvectors, eigenvalues of an $n \times n$ matrix A.

23. Show directly that $\mathbf{v}$ is also an eigenvector of A^2. What is its associated eigenvalue?

24. Same question as Exercise 23 for A^k, k a positive integer.

25. Same question as Exercise 23 for A^{-1}, if A is invertible.

26. Same question as Exercise 24, k any integer.

27. Same question as Exercise 23 for $A - 7I$.

28. Same question as Exercise 23 for $A - aI$.

29. A square matrix B is called nilpotent if $B^k = 0$ for some integer $k > 1$. Show that 0 is the only eigenvalue of a nilpotent matrix.

30. A square matrix C is called idempotent if $C^2 = C$. What are the possible eigenvalues of an idempotent matrix?

31. Suppose that A is a 3×3 matrix with eigenvalues 0, 2, 4 and corresponding eigenvectors $\mathbf{u}_1$, $\mathbf{u}_2$, $\mathbf{u}_3$.
    (a) Find bases for $NS(A)$ and $CS(A)$.
    (b) Solve $A\mathbf{x} = \mathbf{u}_2 + \mathbf{u}_3$.
    (c) Show that $A\mathbf{x} = \mathbf{u}_1$ has no solution.

32. For each of the three types of elementary row operations, give a 2×2 example to show that the eigenvalues might be changed when that operation is performed on a matrix.

33. Prove that A and A^T have the same eigenvalues. (HINT Compare their characteristic polynomials.)

34. Find the eigenvalues and eigenvectors of $A = \begin{bmatrix} 3 & 4 \\ 4 & -3 \end{bmatrix}$ and of $B = \begin{bmatrix} a & b \\ b & -a \end{bmatrix}$.

35. (a) Find 2×2 matrices A and B such that the eigenvalues of the product AB are not the product of the eigenvalues of A and B.
    (b) Repeat (a) for sum instead of product.
    (c) Verify in (a) and (b) that the sum of the eigenvalues of $A + B$ equals the sum of the individual eigenvalues of A and B. Then verify the same for products.
    (d) Explain why (c) is true for any $n \times n$ matrices A and B.

36. In $(x - a_1)(x - a_2) \cdots (x - a_n)$, show that the coefficient of x^{n-1} is $-(a_1 + a_2 + \cdots + a_n)$.

37. Go through all the details of the proof of Theorem (5.17b). In particular:

    (a) Verify Equation (5.19) by expanding $\det(\lambda I - A)$ by minors along the first row, then expanding the first determinant again, and so on.

    (b) Substitute Equation (5.19) into Equation (5.18), and use Exercise 36.

5.3 Diagonalization

We begin with a fact that is essential to both theory and applications, and it will be used repeatedly throughout the chapter. The computation that verifies this fact is surprisingly simple.

(5.20) THEOREM Suppose the $n \times n$ matrix A has n *linearly independent* eigenvectors $\mathbf{v}_1, \ldots, \mathbf{v}_n$. Form an $n \times n$ matrix S whose ith column is $\mathbf{v}_i$. Then S is invertible and $S^{-1}AS$ is a diagonal matrix Λ:

$$S^{-1}AS = \Lambda = \begin{bmatrix} \lambda_1 & & & \\ & \lambda_2 & & \\ & & \ddots & \\ & & & \lambda_n \end{bmatrix}$$

where λ_i is the eigenvalue of A associated with $\mathbf{v}_i$.

Proof Suppose $\mathbf{v}_1, \ldots, \mathbf{v}_n$ are n linearly independent eigenvectors of A and $A\mathbf{v}_i = \lambda_i \mathbf{v}_i$, $1 \le i \le n$. Construct

$$S = [\mathbf{v}_1 \vdots \mathbf{v}_2 \vdots \cdots \vdots \mathbf{v}_n]$$

Now $rk(S) = n$, since its columns are linearly independent, so S is invertible.

Compute the product AS one column at a time,

$$AS = A[\mathbf{v}_1 \vdots \cdots \vdots \mathbf{v}_n] = [A\mathbf{v}_1 \vdots \cdots \vdots A\mathbf{v}_n] = [\lambda_1\mathbf{v}_1 \vdots \cdots \vdots \lambda_n\mathbf{v}_n]$$

Next factor the last matrix as indicated.

$$[\lambda_1\mathbf{v}_1 \vdots \cdots \vdots \lambda_n\mathbf{v}_n] = [\mathbf{v}_1 \vdots \cdots \vdots \mathbf{v}_n]\begin{bmatrix} \lambda_1 & & \\ & \ddots & \\ & & \lambda_n \end{bmatrix}$$

IMPORTANT You should check this last multiplication to see why we get $S\Lambda$ and *not* ΛS (i.e., see ΛS multiples the ith *row* by λ_i and $S\Lambda$ multiples the

ith *column* by λ_i). Therefore, we have shown

(5.21) $$AS = S\Lambda \qquad \text{which implies} \qquad S^{-1}AS = \Lambda$$

(by multiplying on the left by S^{-1}), so we are done. ■

Note that if we had multiplied by S^{-1} on the right in (5.21), we would have obtained

(5.22) COROLLARY If A, Λ, and S are as in Theorem (4.16), then

$$A = S\Lambda S^{-1}$$

Example 1 Let A be the matrix of Example 7 in Section 5.2, and let $\mathbf{v}_1$, $\mathbf{v}_2$, and $\mathbf{v}_3$ be the eigenvectors of A computed there. Thus

$$A = \begin{bmatrix} 2 & 3 & 0 \\ 4 & 3 & 0 \\ 0 & 0 & 6 \end{bmatrix}, \qquad \mathbf{v}_1 = \begin{bmatrix} 1 \\ -1 \\ 0 \end{bmatrix}, \qquad \mathbf{v}_2 = \begin{bmatrix} \frac{3}{4} \\ 1 \\ 0 \end{bmatrix}, \qquad \mathbf{v}_3 = \begin{bmatrix} 0 \\ 0 \\ 1 \end{bmatrix}$$

and $A\mathbf{v}_1 = -\mathbf{v}_1$, $A\mathbf{v}_2 = 6\mathbf{v}_2$, $A\mathbf{v}_3 = 6\mathbf{v}_3$. If we let

$$S = \begin{bmatrix} 1 & \frac{3}{4} & 0 \\ -1 & 1 & 0 \\ 0 & 0 & 1 \end{bmatrix} \qquad \text{and compute} \qquad S^{-1} = \begin{bmatrix} \frac{4}{7} & -\frac{3}{7} & 0 \\ \frac{4}{7} & \frac{4}{7} & 0 \\ 0 & 0 & 1 \end{bmatrix}$$

then we can check directly that

$$S^{-1}AS = \begin{bmatrix} \frac{4}{7} & -\frac{3}{7} & 0 \\ \frac{4}{7} & \frac{4}{7} & 0 \\ 0 & 0 & 1 \end{bmatrix} \begin{bmatrix} 2 & 3 & 0 \\ 4 & 3 & 0 \\ 0 & 0 & 6 \end{bmatrix} \begin{bmatrix} 1 & \frac{3}{4} & 0 \\ -1 & 1 & 0 \\ 0 & 0 & 1 \end{bmatrix} = \begin{bmatrix} -1 & & \\ & 6 & \\ & & 6 \end{bmatrix} = \Lambda$$

■

NOTE Given an A, then S and Λ are *not* unique. If we use different eigenvectors or change their order, we will get a different S and possibly a different Λ.

Example 2 For the A and S in Example 1, let $S_1 = [\mathbf{v}_3 \,\vdots\, \mathbf{v}_2 \,\vdots\, \mathbf{v}_1]$. Then $S_1 \neq S$ and

$$S_1^{-1}AS_1 = \Lambda_1 \qquad \text{where} \qquad \Lambda_1 = \begin{bmatrix} 6 & & \\ & 6 & \\ & & -1 \end{bmatrix}$$ ■

(5.23) DEFINITION An $n \times n$ matrix A is **diagonalizable** if there is an invertible matrix S such that $S^{-1}AS$ is a diagonal matrix, Λ. We say such a matrix S **diagonalizes** A. We also say the *process* of finding such an S and Λ **diagonalizes** A.

We can now reinterpret Theorem (5.20) as saying that if an $n \times n$ matrix A has n linearly independent eigenvectors, then A is diagonalizable. The converse is also true; see Exercise 31. Therefore, altogether we have

(5.24) **THEOREM** An $n \times n$ matrix A is diagonalizable if and only if A has n linearly independent eigenvectors.

Suppose A is diagonalizable so that A has n linearly independent eigenvectors. These eigenvectors are in $\mathbb{R}^n$, and they form a basis for $\mathbb{R}^n$. Thus we have

(5.25) **THEOREM** An $n \times n$ matrix A is diagonalizable if and only if there is a basis of $\mathbb{R}^n$ consisting of eigenvectors of A.

This will be very important for both theory and applications.

NOTE If you have covered Section 4.7, you should now interpret diagonalization in terms of similarity. In particular, by applying the definitions, we obtain

(5.26) **THEOREM** An $n \times n$ matrix A is diagonalizable if and only if it is similar to a diagonal matrix.

See Exercises 35–37 for more details.

The next theorem says a matrix S that diagonalizes A must be very special.

(5.27) **THEOREM** If A is $n \times n$ and the equation $AS = S\Lambda$ holds, where Λ is a diagonal matrix, then the columns of S must be eigenvectors of A and the diagonal entries of Λ must be the corresponding eigenvalues.

Proof We essentially do the proof of Theorem (5.20) in reverse. Suppose $AS = S\Lambda$ and $\mathbf{v}_1, \ldots, \mathbf{v}_n$ are the columns of S. Then

$$A[\mathbf{v}_1 \vdots \cdots \vdots \mathbf{v}_n] = [\mathbf{v}_1 \vdots \cdots \vdots \mathbf{v}_n]\begin{bmatrix} \lambda_1 & & \\ & \ddots & \\ & & \lambda_n \end{bmatrix}$$

$$[A\mathbf{v}_1 \vdots \cdots \vdots A\mathbf{v}_n] = [\lambda_1\mathbf{v}_1 \vdots \cdots \vdots \lambda_n\mathbf{v}_n]$$

from which we conclude, by definition of matrix equality, that $A\mathbf{v}_1 = \lambda\mathbf{v}_1, \ldots, A\mathbf{v}_n = \lambda_n\mathbf{v}_n$, so we are done. ■

NOTE The S in Theorem (5.27) does not have to be invertible.

It is very important to note that:

(5.28) Not all $n \times n$ matrices are diagonalizable.

The next example gives the classic example of a matrix that is not diagonalizable. We give two different solutions, as they are both instructive.

Example 3 Show the matrix

$$\begin{bmatrix} a & 1 \\ & a \end{bmatrix}$$

is not diagonalizable.

Solution The characteristic polynomial of A is

$$\det[\lambda I - A] = (\lambda - a)^2$$

Thus A has only one distinct eigenvalue, $\lambda = a$. So if A is diagonalizable, by Theorem (4.16) we must have the diagonal matrix $\Lambda = S^{-1}AS$ where

$$\Lambda = \begin{bmatrix} a & \\ & a \end{bmatrix} \tag{5.29}$$

Method 1 Suppose A is diagonalizable. Then by Corollary (5.22) $A = S\Lambda S^{-1}$, where Λ is given in Equation (5.29). However, Λ is aI, so $\Lambda S = S\Lambda$. Then we must have

$$A = S\Lambda S^{-1} = \Lambda SS^{-1} = \Lambda$$

Since $A \neq \Lambda$, A is not diagonalizable.

Method 2 Suppose A is diagonalizable. Then by Theorem (5.24) there are two linearly independent eigenvectors of A associated with the eigenvalue a. Hence $\dim NS(aI - A) \geq 2$. However,

$$U = aI - A = \begin{bmatrix} 0 & -1 \\ 0 & 0 \end{bmatrix}$$

By inspection, $rk(U) = 1$, so $\dim NS(U) = n - rk(U) = 1 < 2$. Thus A is not diagonalizable. ■

Now that we know not all square matrices are diagonalizable, the natural question is: What matrices are diagonalizable. Unfortunately there is no simple answer to this. There are, however, interesting and useful criteria which imply diagonalizability. We shall discuss two such special criteria. One of them is symmetry, and we discuss this in the next section. We now give the other.

(5.30)

> **THEOREM** Let A be an $n \times n$ matrix. If A has n real, distinct eigenvalues, then A is diagonalizable.

The proof of Theorem (5.30) requires the results of the following useful theorem.

(5.31)

> **THEOREM** If $\mathbf{v}_1, \ldots, \mathbf{v}_k$ are eigenvectors of an $n \times n$ matrix A and the associated eigenvalues $\lambda_1, \ldots, \lambda_k$ are *distinct*, then the $\mathbf{v}_i$'s are linearly independent.

Of course, since an $n \times n$ matrix can have at most n eigenvalues, we must have $k \leq n$. We defer the proof of Theorem (5.31) until the end of this section.

Proof of Theorem (5.30) Let A be an $n \times n$ matrix and suppose A has n distinct real eigenvalues, $\lambda_1, \ldots, \lambda_n$. By the definition of eigenvector [Definition (5.11), Section 5.2], each λ_i has an associated eigenvector, $\mathbf{v}_i$, $1 \leq i \leq n$. By Theorem (5.31), the n eigenvectors $\mathbf{v}_1, \ldots, \mathbf{v}_n$ are linearly independent since the n λ_i's are distinct. By Theorem (5.20), A is diagonalizable, so we are done. ■

WARNING The hypothesis of Theorem (5.30) is *not* necessary. For example, the 3×3 matrix A of Example 1 has only two distinct eigenvalues, yet we showed it is diagonalizable in that example. For another example, which is extreme, the $n \times n$ identity matrix I_n has only a single eigenvalue, namely $\lambda = 1$, but it is clearly diagonalizable. To be diagonalizable, a matrix must have n linearly independent eigen*vectors*; it does not *have* to have n distinct eigen*values*.

Altogether then, we have the following situation.

(5.32)

> If an $n \times n$ matrix A has n distinct eigenvalues, then it also has n linearly independent eigenvectors and hence it is diagonalizable. If it does not have n distinct eigenvalues, then it may or may not be diagonalizable. If the characteristic polynomial has a multiple real zero, you must check further to determine if n linearly independent eigenvectors can be found.

Example 4 Determine if the following two matrices are diagonalizable.

$$\text{(a)}\quad A = \begin{bmatrix} 5 & -7 & 7 \\ 4 & -3 & 4 \\ 4 & -1 & 2 \end{bmatrix} \qquad \text{(b)}\quad B = \begin{bmatrix} -6 & 12 & -1 \\ -6 & 11 & 0 \\ 0 & 0 & 2 \end{bmatrix}$$

Solution (a) This is the matrix of Example 4 in Section 5.2. By the computations there, the eigenvalues of A are $\lambda = 1, -2, 5$. Since A is 3×3 and it has three distinct eigenvalues, A is diagonalizable by Theorem (5.30).

(b) The characteristic polynomial of B is

$$\det(\lambda I - B) = \det\begin{bmatrix} \lambda + 6 & -12 & 1 \\ 6 & \lambda - 11 & 0 \\ 0 & 0 & \lambda - 2 \end{bmatrix} = (\lambda - 2)\begin{vmatrix} \lambda + 6 & -12 \\ 6 & \lambda - 11 \end{vmatrix} \tag{5.33}$$

$$= (\lambda - 2)(\lambda^2 - 5\lambda + 6) = (\lambda - 2)^2(\lambda - 3)$$

(Note how it helped to expand the first determinant along the bottom row.) We see that B has only two distinct eigenvalues, namely, $\lambda = 2$ and $\lambda = 3$. We are not done yet, however; we need to know how many linearly independent eigenvectors there are (again, as Example 1 shows). To do this, we must determine $\dim[NS(\lambda I - A)]$ for $\lambda = 2, 3$.

Case $\lambda = 3$ Using the first step of Equation (5.33),

$$3I - B = \begin{bmatrix} 9 & -12 & 1 \\ 6 & -8 & 0 \\ 0 & 0 & 1 \end{bmatrix}$$

which reduces to

$$U = \begin{bmatrix} 6 & -8 & 0 \\ 0 & 0 & 1 \\ 0 & 0 & 0 \end{bmatrix}$$

Since $rk(U) = 2$, $1 = \dim NS(U) = \dim NS(3I - A)$, so there is at most one linearly independent eigenvector of A associated with $\lambda = 3$. For example, $\mathbf{v}_1 = (4, 3, 0)$ is such an eigenvector.

Case $\lambda = 2$ Using the first step of Equation (5.33),

$$2I - B = \begin{bmatrix} 8 & -12 & 1 \\ 6 & -9 & 0 \\ 0 & 0 & 0 \end{bmatrix}$$

which reduces to

$$U = \begin{bmatrix} 6 & -9 & 0 \\ 0 & 0 & 1 \\ 0 & 0 & 0 \end{bmatrix}$$

Again, for this U, $rk(U) = 2$ so dim $NS(2I - A) = 1$. There is at most one linearly independent eigenvector A associated with $\lambda = 2$. For example, $\mathbf{v}_2 = (3, 2, 0)$ is such an eigenvector.

Since we have run out of eigenvalues, altogether we can have at most two linearly independent eigenvectors for this B. Since B is 3×3 and it does not have three linearly independent eigenvalues, B is not diagonalizable. ■

You should note that Theorem (5.31) says $\mathbf{v}_1$ and $\mathbf{v}_2$ of Example 4 must be linearly independent, a fact you can easily check.

There are many reasons why it is important to know that a given matrix is diagonalizable. The following theorem is one of the reasons, for you will see that this theorem simplifies both theory and computations.

(5.34)

THEOREM If an $n \times n$ matrix is diagonalizable so that $A = S\Lambda S^{-1}$ by Corollary (5.22), then

$$A^k = S\Lambda^k S^{-1} = S\begin{bmatrix} \lambda_1^k & & \\ & \ddots & \\ & & \lambda_n^k \end{bmatrix} S^{-1}$$

for any positive integer k. If all the eigenvalues are nonzero, this holds for any negative integer k as well. In particular,

$$A^{-1} = S\Lambda^{-1}S^{-1} = S\begin{bmatrix} \lambda_1^{-1} & & \\ & \ddots & \\ & & \lambda_n^{-1} \end{bmatrix} S^{-1}$$

provided all the λ's are nonzero.

The straightforward proof is left to the exercises. See Exercises 27 and 28.

OPTIONAL ***Proof of Theorem (5.31)*** Let $\mathbf{v}_1, \mathbf{v}_2, \ldots, \mathbf{v}_k$ be eigenvectors of an $n \times n$ matrix A that are associated with distinct eigenvalues, $\lambda_1, \lambda_2, \ldots, \lambda_k$. We wish to show the $\mathbf{v}_i$'s are linearly independent.

Since $\mathbf{v}_1$ is an eigenvector, $\mathbf{v}_1 \neq \mathbf{0}$ by definition, so $\{\mathbf{v}_1\}$ is linearly independent. Let m be the largest integer such that $\{\mathbf{v}_1, \ldots, \mathbf{v}_m\}$ is linearly independent. We have $1 \leq m \leq k$. If we show $m = k$, we are done. So suppose $m < k$, and we shall derive a contradiction.

We now assume $\mathbf{v}_1, \ldots, \mathbf{v}_m$ are linearly independent but $\mathbf{v}_1, \ldots, \mathbf{v}_m, \mathbf{v}_{m+1}$ are linearly dependent. Thus there are numbers $c_1, \ldots, c_m, c_{m+1}$, at least one of which is nonzero, such that

$$c_1\mathbf{v}_1 + \cdots + c_m\mathbf{v}_m + c_{m+1}\mathbf{v}_{m+1} = \mathbf{0} \tag{5.35}$$

First multiply both sides of Equation (5.35) on the left by A, obtaining

$$c_1A\mathbf{v}_1 + \cdots + c_mA\mathbf{v}_m + c_{m+1}A\mathbf{v}_{m+1} = A\mathbf{0} = \mathbf{0} \tag{5.36}$$

Since $A\mathbf{v}_i = \lambda_i\mathbf{v}_i$, $1 \leq i \leq m$, Equation (5.36) reduces to

$$c_1\lambda_1\mathbf{v}_1 + \cdots + c_m\lambda_m\mathbf{v}_m + c_{m+1}\lambda_{m+1}\mathbf{v}_{m+1} = \mathbf{0} \tag{5.37}$$

Now multiply both sides of Equation (5.35) by λ_{m+1} and subtract the resulting equation from Equation (5.37) to obtain

$$c_1(\lambda_1 - \lambda_{m+1})\mathbf{v}_1 + \cdots + c_m(\lambda_m - \lambda_{m+1})\mathbf{v}_m + 0\mathbf{v}_{m+1} = \mathbf{0} \tag{5.38}$$

Because $\mathbf{v}_1, \ldots, \mathbf{v}_m$ are linearly independent, we must have

$$c_1(\lambda_1 - \lambda_{m+1}) = 0, \quad \ldots, \quad c_m(\lambda_m - \lambda_{m+1}) = 0$$

Since all the λ_i's are distinct, $\lambda_j - \lambda_{m+1} \neq 0$ for $1 \leq j \leq m$. Therefore, $c_1 = 0, \ldots, c_m = 0$, and Equation (5.35) reduces to

$$c_{m+1}\mathbf{v}_{m+1} = \mathbf{0}$$

Since at least one of the c_i's is nonzero, we must have $c_{m+1} \neq 0$. Therefore, $\mathbf{v}_{m+1} = \mathbf{0}$. However, $\mathbf{v}_{m+1}$ is an eigenvector and hence is nonzero by definition. We now have our contradiction. ■

Exercise 5.3 In Exercises 1–6, n linearly independent eigenvectors and their corresponding eigenvalues of an $n \times n$ matrix A are given.

(a) Find a matrix S and a diagonal matrix Λ such that $A = S\Lambda S^{-1}$.
(b) Find A.

1. (1, 1), 2; (1, 0), −3
2. (2, 1), −1; (3, 1), 4
3. (1, 0, 1), −2; (0, 1, 0), −1; (1, 0, −1), 0
4. (1, 1, 1), 3; (1, 1, −2), −3; (1, −2, 0), 3
5. (1, 0, 0, 0, 0), 2; (0, 1, 0, 0, 0), 3; (0, 0, 1, 0, 0), 4; (0, 0, 0, 1, 0), 5; (0, 0, 0, 0, 1), 6
6. (1, 1, 0, 0, 0, 0), 2; (1, −1, 0, 0, 0, 0), 2; (0, 0, 1, 1, 0, 0), 3; (0, 0, 1, −1, 0, 0), 3; (0, 0, 0, 0, 1, 1), 4; (0, 0, 0, 0, 1, −1), −4

In Exercises 7–12, show that the given matrix is *not* diagonalizable.

7. $\begin{bmatrix} 4 & 0 \\ 1 & 4 \end{bmatrix}$ 8. $\begin{bmatrix} 1 & 1 \\ -1 & -1 \end{bmatrix}$ 9. $\begin{bmatrix} 2 & 1 & 0 \\ 0 & 2 & 0 \\ 0 & 0 & -3 \end{bmatrix}$

10. $\begin{bmatrix} 5 & 1 & 0 \\ 0 & 5 & 1 \\ 0 & 0 & 5 \end{bmatrix}$ 11. $\begin{bmatrix} 7 & 1 & & & \\ & 7 & 1 & & \\ & & 7 & 1 & \\ & & & 7 & 1 \\ & & & & 7 \end{bmatrix}$ 12. $\begin{bmatrix} 7 & 1 & & & \\ & 7 & & & \\ & & 7 & & \\ & & & 7 & \\ & & & & 7 \end{bmatrix}$

In Exercises 13–16, a matrix S and a matrix Λ are given. Let $A = S\Lambda S^{-1}$, but *do not* calculate A. Find *different* S_1 and Λ_1 such that the same A satisfies $A = S_1\Lambda_1 S_1^{-1}$.

13. $S = \begin{bmatrix} 1 & 1 \\ 2 & -1 \end{bmatrix}$, $\Lambda = \begin{bmatrix} 2 & \\ & 3 \end{bmatrix}$ 14. $S = \begin{bmatrix} 2 & 0 \\ 0 & 4 \end{bmatrix}$, $\Lambda = \begin{bmatrix} 3 & \\ & 0 \end{bmatrix}$

15. $S = \begin{bmatrix} 1 & 1 & 1 \\ 1 & 1 & 0 \\ 1 & 0 & 0 \end{bmatrix}$, $\Lambda = \begin{bmatrix} -2 & & \\ & 3 & \\ & & 0 \end{bmatrix}$

16. $S = \begin{bmatrix} 2 & 1 & 1 \\ 1 & 2 & 0 \\ -1 & 4 & 0 \end{bmatrix}$, $\Lambda = \begin{bmatrix} -5 & & \\ & -5 & \\ & & 3 \end{bmatrix}$

In Exercises 17–27, determine if the given matrix is diagonalizable. If so, find matrices S and Λ such the given matrix equals $S\Lambda S^{-1}$.

17. $\begin{bmatrix} 2 & -4 \\ 1 & -2 \end{bmatrix}$ 19. $\begin{bmatrix} 1 & -1 \\ 1 & 3 \end{bmatrix}$ 18. $\begin{bmatrix} 1 & 4 \\ 1 & -2 \end{bmatrix}$

20. $\begin{bmatrix} 3 & 3 \\ 4 & 2 \end{bmatrix}$ 22. $\begin{bmatrix} 0 & 0 & 0 \\ 0 & 0 & 0 \\ 4 & 0 & 1 \end{bmatrix}$ 21. $\begin{bmatrix} 1 & 0 & 0 \\ 0 & 1 & 2 \\ 0 & 0 & 2 \end{bmatrix}$

23. $\begin{bmatrix} 1 & 0 & 0 \\ 1 & 1 & 0 \\ 0 & 1 & 1 \end{bmatrix}$ 24. $\begin{bmatrix} 1 & 0 & 1 \\ 0 & 2 & 0 \\ -1 & 0 & -1 \end{bmatrix}$

25. $\begin{bmatrix} 3 & 0 & 0 & 0 \\ 0 & 3 & 5 & -5 \\ 0 & 0 & -2 & 0 \\ 0 & 0 & 0 & -2 \end{bmatrix}$ 26. $\begin{bmatrix} 6 & 0 & 0 & 0 \\ 4 & 5 & 0 & 0 \\ 3 & 0 & 5 & 0 \\ 2 & 0 & 0 & 5 \end{bmatrix}$

27. Prove: If $A = S\Lambda S^{-1}$ and k is a positive integer, then $A^k = S\Lambda^k S^{-1}$. (HINT Write out $AA \cdots A$.)

28. Prove: If $A = S\Lambda S^{-1}$, then $A^{-1} = S\Lambda^{-1}S^{-1}$, if Λ^{-1} exists.

In Exercises 29 and 30, use Exercises 27 and 28 to compute A^3, A^{-1}.

29. The A described in Exercise 13.

30. The A described in Exercise 16.

31. Prove: If A is square, S is invertible, Λ is diagonal, and $A = S\Lambda S^{-1}$, then A has n linearly independent eigenvectors.

32. Suppose A and B are square, diagonalizable, and have the same eigenvectors (but not necessarily the same eigenvalues). Show that A and B commute, $AB = BA$.

33. Let T be a 3×3 upper triangular matrix with diagonal entries 1, 3, -6. Can T be diagonalized? Why or why not? If so, what is Λ?

34. Let $\mathbf{v}$ and $\mathbf{w}$ be nonzero $n \times 1$ vectors, and let $A = \mathbf{v}\mathbf{w}^T$, which is an $n \times n$ matrix.

    (a) Show directly that $\mathbf{v}$ is an eigenvector of A, and determine its associated eigenvalue.
    (b) Show that $\dim[NS(A)] = n - 1$, and interpret this in terms of eigenvalues and eigenvectors.
    (c) Use (a) and (b) to find the characteristic polynomial of A.
    (d) Compute $\operatorname{tr}(A)$ in two ways: from the sum of the diagonal, and from the sum of the eigenvalues. What does this say about the eigenvalues if $\mathbf{v} \cdot \mathbf{w} = 0$?
    (e) Show that A is diagonalizable if and only if $\mathbf{v} \cdot \mathbf{w} \neq 0$.

The remaining exercises are only for those who covered similarity in Section 4.7. First review the paragraph preceding Theorem (4.21).

35. Compare the construction of S in Theorem (5.20) with the construction of a transition matrix P in Theorem (3.92) on page 216. Use this comparison to show that S is a transition matrix, and determine from what to what.

36. Ignoring Exercise 35, compare $A = S\Lambda S^{-1}$ in Corollary (5.22) and $[T]_C = P^{-1}[T]_B P$ in Theorem (4.84) on page 302. Use this comparison to show that S is a transition matrix, and determine from what to what.

37. Suppose that the $n \times n$ matrix A is diagonalizable, so that $S^{-1}AS = \Lambda$. Thinking of A as $[f_A]$, look at Theorem (4.91) on page 305 and describe geometrically the action of f_A on the basis B of eigenvectors $\mathbf{v}_1, \ldots, \mathbf{v}_n$. (Consider separately the cases $\lambda_i > 1$, $\lambda_i = 1$, $0 < \lambda_i < 1$, $\lambda_i = 0$, and $\lambda_i < 0$.)

5.4 Symmetric Matrices

In Section 5.3, we mentioned that a symmetric matrix is diagonalizable. In fact something even stronger is true: an $n \times n$ symmetric matrix has n orthonormal eigenvectors. In this section we shall discuss these and other important facts about symmetric matrices and discuss some of their implications.

We begin with two surprising and fundamental facts about symmetric matrices.

(5.39) **THEOREM** Let A be an $n \times n$ symmetric matrix.

(a) All of the eigenvalues of A are real. That is, the n zeros (counting multiplicity) of the characteristic polynomial are all real.

(b) A has n linearly independent eigenvectors and hence is diagonalizable.

The proof of (a) is optional as it requires a little knowledge of complex numbers. You can find the minimal background necessary for the proof in the exercises. In particular we need to know two facts.

(5.40) $$\text{A complex number } z \text{ is real} \quad \text{if and only if} \quad z = \bar{z}$$

where if $z = a + bi$, $\bar{z} = a - bi$ is the complex conjugate of z. For the other fact, recall that inner products for real vectors and numbers have the property $(\lambda\mathbf{v}) \cdot \mathbf{w} = \lambda(\mathbf{v} \cdot \mathbf{w})$ and $(\mathbf{v} \cdot \lambda\mathbf{w}) = \lambda(\mathbf{v} \cdot \mathbf{w})$. [See Theorem (4.5), Section 4.2.] However, for complex numbers and vectors, the definition of dot product is slightly different. See Exercise 24. Under this definition,

(5.41) $$(\lambda\mathbf{v}) \cdot \mathbf{w} = \lambda(\mathbf{v} \cdot \mathbf{w}) \qquad \text{and} \qquad \mathbf{v} \cdot \lambda\mathbf{w} = \bar{\lambda}(\mathbf{v} \cdot \mathbf{w})$$

Proof of Theorem (5.39a) OPTIONAL Suppose A is a real symmetric matrix, λ is an eigenvalue of A (possibly a complex number but that we shall show is real), and $\mathbf{u}$ is an associated eigenvector (which may be an n-tuple of complex numbers in $\mathbb{C}^n$-see Example 5 in Section 5.2). By Theorem (4.15), Section 4.2, $A\mathbf{v} \cdot \mathbf{w} = \mathbf{v} \cdot A^T\mathbf{w}$. The crucial property is that since $A = A^T$, we have

$$A\mathbf{v} \cdot \mathbf{w} = \mathbf{v} \cdot A\mathbf{w}, \qquad \text{all } \mathbf{v} \text{ and } \mathbf{w}$$

Letting $\mathbf{v} = \mathbf{w} = \mathbf{u}$, we obtain

$$A\mathbf{u} \cdot \mathbf{u} = \mathbf{u} \cdot A\mathbf{u}$$

Since $A\mathbf{u} = \lambda\mathbf{u}$,

$$(\lambda\mathbf{u}) \cdot \mathbf{u} = \mathbf{u} \cdot (\lambda\mathbf{u})$$

or, by Equations (5.41),

$$\lambda(\mathbf{u} \cdot \mathbf{u}) = \bar{\lambda}(\mathbf{u} \cdot \mathbf{u}) \tag{5.42}$$

Since $\mathbf{u}$ is nonzero ($\mathbf{u}$ is an eigenvector), $\mathbf{u} \cdot \mathbf{u} \neq 0$. We can, therefore, divide Equation (5.42) by $\mathbf{u} \cdot \mathbf{u}$ to obtain

$$\lambda = \bar{\lambda} \qquad \text{or equivalently} \qquad \lambda \text{ is real}$$

by (5.40). Hence we are done. ■

The proof of (b), which uses (a), is more difficult. It requires that we know how to "factor off" an eigenvalue from A. This proof can be found in Hoffman and Kunze, *Linear Algebra*, page 314, Theorem 18.

We now turn to the next important fact concerning symmetric matrices.

(5.43) **THEOREM** Let A be a symmetric matrix. Eigenvectors of A associated with different eigenvalues are orthogonal.

You should compare this theorem with Theorem (5.31), Section 5.3, which says eigenvectors associated with distinct eigenvalues of a matrix A must be linearly independent. Theorem (5.43) says if A is symmetric, those eigenvectors must be orthogonal, not just independent. [Recall that orthogonality implies linear independence, by Theorem (4.46), Section 4.4.]

Proof of Theorem (5.43) Suppose A is a symmetric matrix, and $\mathbf{v}_1$ and $\mathbf{v}_2$ are eigenvectors of A associated with distinct eigenvalues. Thus

$$A\mathbf{v}_1 = \lambda_1\mathbf{v}_1, \qquad A\mathbf{v}_2 = \lambda_2\mathbf{v}_2, \qquad \lambda_1 \neq \lambda_2 \qquad (\text{and } \mathbf{v}_1, \mathbf{v}_2 \neq \mathbf{0})$$

We want to show $\mathbf{v}_1 \cdot \mathbf{v}_2 = 0$. Now, by Theorem (4.15), Section 4.2,

$$A\mathbf{v}_1 \cdot \mathbf{v}_2 = \mathbf{v}_1 \cdot A^T\mathbf{v}_2$$

Since A is symmetric,

$$A\mathbf{v}_1 \cdot \mathbf{v}_2 = \mathbf{v}_1 \cdot A\mathbf{v}_2 \qquad \text{or} \qquad (\lambda_1\mathbf{v}_1) \cdot \mathbf{v}_2 = \mathbf{v}_1 \cdot (\lambda_2\mathbf{v}_2)$$

Because λ_2 is real, by Theorem (5.39a), we have

$$\lambda_1(\mathbf{v}_1 \cdot \mathbf{v}_2) = \lambda_2(\mathbf{v}_1 \cdot \mathbf{v}_2) \qquad \text{or} \qquad (\lambda_1 - \lambda_2)(\mathbf{v}_1 \cdot \mathbf{v}_2) = 0$$

Since $\lambda_1 \neq \lambda_2$, we know that $\lambda_1 - \lambda_2 \neq 0$, so we conclude

$$\mathbf{v}_1 \cdot \mathbf{v}_2 = 0$$

and we are done. ■

Example 1 Find eigenvalues and eigenvectors of the matrix

$$A = \begin{bmatrix} 0 & 2 & 2 \\ 2 & 0 & -2 \\ 2 & -2 & 0 \end{bmatrix}$$

Solution The characteristic polynomial of A is

$$\det(\lambda I - A) = \begin{vmatrix} \lambda & -2 & -2 \\ -2 & \lambda & 2 \\ -2 & 2 & \lambda \end{vmatrix} = (\lambda - 2)^2(\lambda + 4)$$

as you can verify. Thus A has two eigenvalues, $\lambda = -4$ and $\lambda = 2$ (which has multiplicity two). To find associated eigenvectors, we reduce $\lambda I - A$ to echelon form and find a basis for $NS(\lambda I - A)$.

Case $\lambda = -4$

$$(-4)I - A = \begin{bmatrix} -4 & -2 & -2 \\ -2 & -4 & 2 \\ -2 & 2 & -4 \end{bmatrix} \quad \text{which reduces to} \quad U = \begin{bmatrix} -4 & -2 & -2 \\ 0 & -3 & 3 \\ 0 & 0 & 0 \end{bmatrix}$$

We see that $NS[(-4)I - A]$ has a basis consisting of

$$\mathbf{u}_1 = \begin{bmatrix} -1 \\ 1 \\ 1 \end{bmatrix}$$

We can check directly that $A\mathbf{u}_1 = -4\mathbf{u}_1$.

Case $\lambda = 2$

$$2I - A = \begin{bmatrix} 2 & -2 & -2 \\ -2 & 2 & 2 \\ -2 & 2 & 2 \end{bmatrix} \quad \text{which reduces to} \quad U = \begin{bmatrix} 2 & -2 & -2 \\ 0 & 0 & 0 \\ 0 & 0 & 0 \end{bmatrix}$$

We see that $NS(2I - A)$ has a basis consisting of

$$\mathbf{u}_2 = \begin{bmatrix} 1 \\ 1 \\ 0 \end{bmatrix} \quad \text{and} \quad \mathbf{u}_3 = \begin{bmatrix} 1 \\ 0 \\ 1 \end{bmatrix}$$

We can check directly that $A\mathbf{u}_2 = 2\mathbf{u}_2$ and $A\mathbf{u}_3 = 2\mathbf{u}_3$.

Most important, as Theorem (5.39b) promises, A has three linearly independent eigenvectors, $\mathbf{u}_1$, $\mathbf{u}_2$, and $\mathbf{u}_3$, and as Theorem (5.43) promises, $\mathbf{u}_1$ is orthogonal to $\mathbf{u}_2$ and $\mathbf{u}_3$, as you can check. ■

As you can easily see, $\mathbf{u}_2$ and $\mathbf{u}_3$ are *not* orthogonal. What happens if we apply the Gram–Schmidt process to them? We will then have orthogonal vectors, but will these new vectors still be eigenvectors? The answer is a resounding YES, because $\mathbf{u}_2$ and $\mathbf{u}_3$ are still associated with the *same* eigenvalue, $\lambda = 2$. The Gram–Schmidt process just takes particular linear combinations of vectors, so:

(5.44)

IMPORTANT FACT If the Gram–Schmidt process is applied to linearly independent vectors in a particular subspace, the resulting orthonormal vectors will remain in that subspace. Consequently, if the Gram–Schmidt process is applied to linearly independent eigenvectors associated with a single eigenvalue λ of a matrix A, the resulting orthonormal vectors will all still be eigenvectors of A associated with λ.

We have now almost proved the following theorem.

(5.45)

THEOREM Let A be an $n \times n$ symmetric matrix. Then A has n orthonormal eigenvectors. Hence, A can be expressed as

$$A = Q\Lambda Q^T$$

where Q is orthogonal and Λ is a diagonal matrix (with the eigenvalues of A on the diagonal).

Proof By Theorem (5.39b), A has n linearly independent eigenvectors. For each distinct λ, apply the Gram–Schmidt process to all of the eigenvectors associated with that λ. By (5.44), we then have n eigenvectors that are both normal and orthogonal to all the others associated with the same eigenvalue. But by Theorem (5.43), each eigenvector is automatically orthogonal to eigenvectors associated with different eigenvalues. Altogether, then, they are orthonormal. ■

In summary, we have the following.

PROCEDURE FOR FINDING N ORTHONORMAL EIGENVECTORS OF AN $N \times N$ SYMMETRIC MATRIX A

(a) Find the eigenvalues of A (they are all real).
(b) Find a basis for each eigenspace in the usual way.
(c) Apply the Gram–Schmidt process to obtain an orthonormal basis for each eigenspace. Collectively all of the vectors obtained this way form a set of n orthonormal eigenvectors.

Example 2 Find three orthonormal eigenvectors for the matrix of Example 1,

$$A = \begin{bmatrix} 0 & 2 & 2 \\ 2 & 0 & -2 \\ 2 & -2 & 0 \end{bmatrix}$$

Solution From Example 1 we know A has eigenvalues $\lambda = -4, 2$ and eigenvectors $\mathbf{u}_1$ associated with -4 and $\mathbf{u}_2$ and $\mathbf{u}_3$ associated with 2, where

$$\mathbf{u}_1 = \begin{bmatrix} -1 \\ 1 \\ 1 \end{bmatrix}, \qquad \mathbf{u}_2 = \begin{bmatrix} 1 \\ 1 \\ 0 \end{bmatrix}, \qquad \mathbf{u}_3 = \begin{bmatrix} 1 \\ 0 \\ 1 \end{bmatrix}$$

This completes steps (a) and (b) of the procedure in the box. For step (c), $\lambda = -4$ has only $\mathbf{u}_1$ as an associated (linearly independent) eigenvector. When we apply the Gram–Schmidt process to it, we merely make it normal, and we get

$$\mathbf{v}_1 = \frac{\mathbf{u}_1}{\|\mathbf{u}_1\|} = \frac{1}{\sqrt{3}}[-1 \quad 1 \quad 1]^T = \left[-\frac{1}{\sqrt{3}} \quad \frac{1}{\sqrt{3}} \quad \frac{1}{\sqrt{3}}\right]^T$$

The other eigenvalue, $\lambda = 2$, has two associated (linearly independent) eigenvectors, $\mathbf{u}_2$ and $\mathbf{u}_3$. We apply the Gram–Schmidt process to them, obtaining

$$\mathbf{v}_2 = \begin{bmatrix} \frac{1}{\sqrt{2}} \\ \frac{1}{\sqrt{2}} \\ 0 \end{bmatrix} \quad \text{and} \quad \mathbf{v}_3 = \begin{bmatrix} \frac{1}{\sqrt{6}} \\ -\frac{1}{\sqrt{6}} \\ \frac{2}{\sqrt{6}} \end{bmatrix}$$

as you can check. Altogether, $\mathbf{v}_1$, $\mathbf{v}_2$, and $\mathbf{v}_3$ are three orthonormal eigenvectors for A. ■

Let A be a symmetric $n \times n$ matrix. We now know that A has n orthonormal eigenvectors, $\mathbf{v}_1, \ldots, \mathbf{v}_n$. Let Q be the $n \times n$ matrix with columns $\mathbf{v}_1, \ldots, \mathbf{v}_n$. Then Q has two properties:

1. Q is orthogonal [by Definition (4.58), Section 4.5].
2. Q diagonalizes A [by Theorem (5.20), Section 5.3].

Therefore, A is not simply diagonalizable; it is diagonalizable by an orthogonal matrix. Amazingly this property characterizes symmetric matrices.

(5.46) An $n \times n$ matrix A is symmetric if and only if there is an orthogonal matrix Q that diagonalizes A.

We just proved the "only if" part. The "if" part is left to an exercise with a hint. (See Exercise 21.)

Example 3 Find an orthogonal matrix Q that diagonalizes the matrix A of Example 1.

Solution In Example 2 we found three orthonormal eigenvectors of A, $\mathbf{v}_1$, $\mathbf{v}_2$, and $\mathbf{v}_3$. If we form $Q = [\mathbf{v}_1 \vdots \mathbf{v}_2 \vdots \mathbf{v}_3]$, we obtain

$$Q = \begin{bmatrix} -\frac{1}{\sqrt{3}} & \frac{1}{\sqrt{2}} & \frac{1}{\sqrt{6}} \\ \frac{1}{\sqrt{3}} & \frac{1}{\sqrt{2}} & -\frac{1}{\sqrt{6}} \\ \frac{1}{\sqrt{3}} & 0 & \frac{2}{\sqrt{6}} \end{bmatrix}$$

Then we can check that $Q^T = Q^{-1}$ and that $Q^{-1}AQ = \Lambda$, where Λ is the diagonal matrix whose diagonal entries are the eigenvalues that correspond to the $\mathbf{v}_i$'s, that is,

$$\begin{bmatrix} -\frac{1}{\sqrt{3}} & \frac{1}{\sqrt{3}} & \frac{1}{\sqrt{3}} \\ \frac{1}{\sqrt{2}} & \frac{1}{\sqrt{2}} & 0 \\ \frac{1}{\sqrt{6}} & -\frac{1}{\sqrt{6}} & \frac{2}{\sqrt{6}} \end{bmatrix} \begin{bmatrix} 0 & 2 & 2 \\ 2 & 0 & -2 \\ 2 & -2 & 0 \end{bmatrix} \begin{bmatrix} -\frac{1}{\sqrt{3}} & \frac{1}{\sqrt{2}} & \frac{1}{\sqrt{6}} \\ \frac{1}{\sqrt{3}} & \frac{1}{\sqrt{2}} & -\frac{1}{\sqrt{6}} \\ \frac{1}{\sqrt{3}} & 0 & \frac{2}{\sqrt{6}} \end{bmatrix}$$

$$= \begin{bmatrix} -4 & & \\ & 2 & \\ & & 2 \end{bmatrix}$$ ■

Exercise 5.4

In Exercises 1–14, for the given $n \times n$ symmetric matrix A,

(a) Find any n linearly independent eigenvectors and verify those associated with distinct eigenvalues are orthogonal.

(b) Find an orthogonal matrix Q and a diagonal matrix Λ such that $Q^{-1}AQ = \Lambda$.

1. $\begin{bmatrix} 1 & 2 \\ 2 & 4 \end{bmatrix}$

2. $\begin{bmatrix} 2 & 3 \\ 3 & 2 \end{bmatrix}$

3. $\begin{bmatrix} 5 & -2 \\ -2 & 2 \end{bmatrix}$

4. $\begin{bmatrix} -1 & -6 \\ -6 & 4 \end{bmatrix}$

5. $\begin{bmatrix} 1 & 1 \\ 1 & 1 \end{bmatrix}$

6. $\begin{bmatrix} 0 & 1 \\ 1 & 0 \end{bmatrix}$

7. $\begin{bmatrix} 1 & 1 & 1 \\ 1 & 1 & 1 \\ 1 & 1 & 1 \end{bmatrix}$ 8. $\begin{bmatrix} 0 & 1 & 0 \\ 1 & 0 & 1 \\ 0 & 1 & 0 \end{bmatrix}$ 9. $\begin{bmatrix} -2 & -2 & 2 \\ -2 & 1 & -4 \\ 2 & -4 & 1 \end{bmatrix}$

10. $\begin{bmatrix} 2 & 1 & 1 \\ 1 & 2 & 1 \\ 1 & 1 & 2 \end{bmatrix}$ 11. $\begin{bmatrix} 3 & -2 & \\ -2 & 0 & \\ & & 1 \end{bmatrix}$ 12. $\begin{bmatrix} 4 & & \\ & 1 & 3 \\ & 3 & 1 \end{bmatrix}$

13. $\begin{bmatrix} 1 & 1 & 1 & 1 \\ 1 & 1 & 1 & 1 \\ 1 & 1 & 1 & 1 \\ 1 & 1 & 1 & 1 \end{bmatrix}$ 14. $\begin{bmatrix} 0 & 1 & 0 & 1 \\ 1 & 0 & 1 & 0 \\ 0 & 1 & 0 & 1 \\ 1 & 0 & 1 & 0 \end{bmatrix}$

(HINT By inspection the eigenvalues are $\lambda = 0, 0, 0, 4$ for Exercise 13 and $\lambda = 0, 0, 2, -2$ for Exercise 14. See Exercises 8–10, Section 5.2.)

In Exercises 15–20, eigenvectors and corresponding eigenvalues of a symmetric matrix A are given. Find a diagonal matrix Λ and an orthogonal matrix Q such that $A = Q\Lambda Q^{-1}$. You do not need to find A.

15. $2, [2 \quad 1]^T; -3, [1 \quad -2]^T$ 16. $3, [1 \quad -3]^T; 0, [3 \quad 1]^T$

17. $3, [1 \quad 2 \quad -2]^T; -4, [2 \quad 1 \quad 2]^T; -1, [-2 \quad 2 \quad 1]^T$

18. $5, [3 \quad 4 \quad 5]^T; -1, [4 \quad -3 \quad 0]^T; 0, [3 \quad 4 \quad -5]^T$

19. $2, [1 \quad 1 \quad 1]^T; 3, [1 \quad -1 \quad 0]^T; 3, [1 \quad 1 \quad -2]^T$

20. $-4, [3 \quad 2 \quad 1]^T; 5, [1 \quad -1 \quad -1]^T; 5, [1 \quad 4 \quad 5]^T$

21. Show that if there is an orthogonal matrix Q that diagonalizes A, then A is symmetric. (HINT If $Q^{-1}AQ = \Lambda$, then $A = Q\Lambda Q^{-1}$. Now find A^T and use the facts that $\Lambda^T = \Lambda$ and $Q^{-1} = Q^T$.)

22. Recall that two complex numbers $z = a + bi$ and $w = c + di$ are equal if and only if $a = c$ and $b = d$ (where a, b, c, and d are real). Also $z = a + bi$ is real if and only if $b = 0$. Use this to prove (5.40): A complex number $z = a + bi$ is real if and only if $z = \bar{z}$.

23. For a complex number $z = a + bi$, the length or absolute value of z is $|z| = \sqrt{a^2 + b^2}$. Show that $|z| = \sqrt{z\bar{z}}$.

24. For $\mathbf{z} = (z_1, \ldots, z_n)$ and $\mathbf{w} = (w_1, \ldots, w_n)$ in $\mathbb{C}^n$, define the length or norm of $\mathbf{z}$ by $\|\mathbf{z}\| = \sqrt{|z_1|^2 + \cdots + |z_n|^2}$ and the inner (or dot) product by $\mathbf{z} \cdot \mathbf{w} = z_1\bar{w}_1 + \cdots + z_n\bar{w}_n$. Use Exercise 23 to show that $\|\mathbf{z}\| = \sqrt{\mathbf{z} \cdot \mathbf{z}}$ (and the conjugates over the second factor in $\mathbf{z} \cdot \mathbf{w}$ are necessary for this to be true).

25. Use Exercise 24 to show (5.41): For $\mathbf{z}$ and $\mathbf{w}$ in $\mathbb{C}^n$ and a complex number λ, $(\lambda\mathbf{z}) \cdot \mathbf{w} = \lambda(\mathbf{z} \cdot \mathbf{w})$ and $\mathbf{z} \cdot (\lambda\mathbf{w}) = \bar{\lambda}(\mathbf{z} \cdot \mathbf{w})$.

5.5 An Application—Difference Equations: Fibonacci Sequences and Markov Processes

If a sequence of steps is taken and the question, "What happens on the nth step?" is asked, the resulting mathematical formulation is usually what is called a *difference equation*. Applications of difference equations arise in many situations. We shall study two such situations: the first originally arising in biology, and the second arising in demographics. However, the techniques derived in each problem have broad application elsewhere.

Fibonacci Sequences

These sequences are named after the Italian Leonardo Fibonacci (1170?–1250?). He discovered these sequences while studying a problem in rabbit population growth not unlike the problem illustrated in Exercise 1. A typical Fibonacci sequence is

$$0, 1, 1, 2, 3, 5, 8, 13, 21, \ldots \tag{5.47}$$

You probably see the pattern: Every number is the sum of its two predecessors,

$$F_{k+2} = F_{k+1} + F_k \tag{5.48}$$

This is a difference equation, and it turns up in a fantastic variety of applications both in nature and in pure mathematics.* The sequence given in (5.47) starts with $F_0 = 0$ and $F_1 = 1$, and a typical problem is to determine F_{1000}. Of course, one way to answer this is to compute $F_2, \ldots, F_{999}$; fortunately we can avoid such thankless and thoughtless computations with some of the tools in this chapter. For example, we add to Equation (5.48) the simple identity $F_{k+1} = F_{k+1}$ and form the system

$$\begin{aligned} F_{k+2} &= F_{k+1} + F_k \\ F_{k+1} &= F_{k+1} \end{aligned} \qquad \text{or} \qquad \begin{bmatrix} F_{k+2} \\ F_{k+1} \end{bmatrix} = \begin{bmatrix} 1 & 1 \\ 1 & 0 \end{bmatrix} \begin{bmatrix} F_{k+1} \\ F_k \end{bmatrix} \tag{5.49}$$

We define A and $\mathbf{x}_k$, $k \geq 0$ by

$$A = \begin{bmatrix} 1 & 1 \\ 1 & 0 \end{bmatrix} \quad \text{and} \quad \mathbf{x}_k = \begin{bmatrix} F_{k+1} \\ F_k \end{bmatrix}, \quad k \geq 0, \quad \mathbf{x}_0 = \begin{bmatrix} 1 \\ 0 \end{bmatrix}$$

so that the system in (5.49) reduces to the simple equation $\mathbf{x}_{k+1} = A\mathbf{x}_k$. We are given $\mathbf{x}_0$, and our problem is to determine $\mathbf{x}_{999}$. We observe

$$\mathbf{x}_1 = A\mathbf{x}_0, \qquad \mathbf{x}_2 = A\mathbf{x}_1 = A^2\mathbf{x}_0, \qquad \mathbf{x}_3 = A\mathbf{x}_2 = A^3\mathbf{x}_0, \ldots$$

* *For example, count the scales on the spiral-like paths on pinecones or observe the pattern in which many leaves, thorns, or stems grow. You might examine* Patterns in Nature, *by Peter Stevens (Little, Brown, 1974) or any issue of the journal* Fibonacci Quarterly.

and conclude (or prove by induction) that

$$\mathbf{x}_k = A^k\mathbf{x}_0 \quad \text{and in particular} \quad \mathbf{x}_{999} = A^{999}\mathbf{x}_0 \tag{5.50}$$

If we can diagonalize A, $A = S\Lambda S^{-1}$, then by Theorem (5.34), Section 5.3, $A^{999} = S\Lambda^{999}S^{-1}$. In this case,

$$\mathbf{x}_k = S\Lambda^k S^{-1}\mathbf{x}_0 \quad \text{and in particular} \quad \mathbf{x}_{999} = S\Lambda^{999}S^{-1}\mathbf{x}_0 \tag{5.51}$$

which is a fairly simple computation. So instead of computing $F_2, \ldots, F_{999}$, we enthusiastically set out to diagonalize A. Its characteristic polynomial is

$$\det(\lambda I - A) = \det\begin{bmatrix} \lambda - 1 & -1 \\ -1 & \lambda \end{bmatrix} = \lambda^2 - \lambda - 1$$

which has zeros

$$\lambda_1 = \frac{1 + \sqrt{5}}{2} \quad \text{and} \quad \lambda_2 = \frac{1 - \sqrt{5}}{2}$$

Using a little algebra along with the facts that $\lambda_1 + \lambda_2 = 1$ and $\lambda_1\lambda_2 = -1$, we can show that $\begin{bmatrix} \lambda_1 \\ 1 \end{bmatrix}$ and $\begin{bmatrix} \lambda_2 \\ 1 \end{bmatrix}$ are eigenvectors of A associated with λ_1 and λ_2 respectively. Therefore,

$$S = \begin{bmatrix} \lambda_1 & \lambda_2 \\ 1 & 1 \end{bmatrix} \quad \text{and} \quad S^{-1} = \frac{1}{\lambda_1 - \lambda_2}\begin{bmatrix} 1 & -\lambda_2 \\ -1 & \lambda_1 \end{bmatrix}$$

using Theorem (5.20), Section 5.3, and Example 3 in Section 1.4. Therefore, by Equations (5.51) and since $\lambda_1 - \lambda_2 = \sqrt{5}$,

$$\begin{aligned}
\begin{bmatrix} F_{k+1} \\ F_k \end{bmatrix} &= \mathbf{x}_k = S\Lambda^k S^{-1}\mathbf{x}_0 \\
&= \frac{1}{\lambda_1 - \lambda_2}\begin{bmatrix} \lambda_1 & \lambda_2 \\ 1 & 1 \end{bmatrix}\begin{bmatrix} \lambda_1 & \\ & \lambda_2 \end{bmatrix}^k\begin{bmatrix} 1 & -\lambda_2 \\ -1 & \lambda_1 \end{bmatrix}\begin{bmatrix} 1 \\ 0 \end{bmatrix} \\
&= \frac{1}{\sqrt{5}}\begin{bmatrix} \lambda_1 & \lambda_2 \\ 1 & 1 \end{bmatrix}\begin{bmatrix} \lambda_1^k & \\ & \lambda_2^k \end{bmatrix}\begin{bmatrix} 1 \\ -1 \end{bmatrix} \\
&= \frac{1}{\sqrt{5}}\begin{bmatrix} \lambda_1 & \lambda_2 \\ 1 & 1 \end{bmatrix}\begin{bmatrix} \lambda_1^k \\ -\lambda_2^k \end{bmatrix} \\
&= \frac{1}{\sqrt{5}}\begin{bmatrix} \lambda_1^{k+1} & -\lambda_2^{k+1} \\ \lambda_1^k & -\lambda_2^k \end{bmatrix}
\end{aligned}$$

Hence we conclude that

$$F_k = \frac{\lambda_1^k - \lambda_2^k}{\sqrt{5}} \tag{5.52}$$

and in particular

$$F_{1000} = \frac{1}{\sqrt{5}}\left[\left(\frac{1+\sqrt{5}}{2}\right)^{1000} - \left(\frac{1-\sqrt{5}}{2}\right)^{1000}\right] \tag{5.53}$$

This equation brings to mind several thoughts. First, from the difference equation F_{1000} is an integer, so it is fascinating to contemplate how the $\sqrt{5}$'s must cancel in Equation (5.53) and how everything else must add up to an integer. Second, $\sqrt{5} \approx 2.2$, so $\left|\frac{1-\sqrt{5}}{2}\right| < 1$. Hence $\left[\frac{1-\sqrt{5}}{2}\right]^{1000}$ is an extremely small number. Therefore,

$$F_{1000} \approx \frac{1}{\sqrt{5}}\left[\frac{1+\sqrt{5}}{2}\right]^{1000} \approx 4.3467 \times 10^{208}$$

which is an enormous number.

Third and finally, this approximation suggests that if we look at the ratio $\frac{F_{k+1}}{F_k}$ for large k, we obtain

$$\frac{F_{k+1}}{F_k} \approx \frac{[(1+\sqrt{5})/2]^{k+1}/\sqrt{5}}{[(1+\sqrt{5})/2]^{k}/\sqrt{5}} = \frac{1+\sqrt{5}}{2} \approx 1.6$$

This ratio, $\frac{1+\sqrt{5}}{2}$, was called the **golden mean** by the Greeks. It arises not only here but also in such diverse areas as geometry (the most pleasing looking rectangles have their sides in the ratio of 1.6:1 as all artists and carpenters know) and in search theory (with ratios of step sizes to take, no other information being available).

You can see why Fibonacci sequences have almost a mystical appeal for many people.

Markov Processes

You may recall that Example 4 of Section 1.1 was concerned with people moving in and out of Texas. This problem served as an introduction to linear algebra, and there was one part of that question we could not answer then. We are now ready to tackle that part.

Here was the situation. We assumed that each year 20% of the people inside of Texas moved out and 10% of the people outside of Texas moved in. This situation lends itself to difference equations. If we let

x_i = the number of people who are inside of Texas at the beginning of i years from now

y_i = the number of people who are outside of Texas (but in the United States) at the beginning of i years from now

Then we see

(5.54)
$$x_{i+1} = 0.8x_i + 0.1y_i \qquad \qquad \begin{bmatrix} x_{i+1} \\ y_{i+1} \end{bmatrix} = \begin{bmatrix} 0.8 & 0.1 \\ 0.2 & 0.9 \end{bmatrix} \begin{bmatrix} x_i \\ y_i \end{bmatrix}$$
$$y_{i+1} = 0.2x_i + 0.9y_i \qquad \text{or}$$

We let $\mathbf{u}_i = [x_i \quad y_i]^T$, so that Equations (5.54) simplify to $\mathbf{u}_{i+1} = A\mathbf{u}_i$.

This problem has several important features that also hold in many chemical, physical, biological, and economic processes:

> The total population stays fixed, and the individual populations are never negative. This is reflected in the matrix in that *each column adds up to 1*, and the *entries are nonnegative* and less than or equal to one.

Such a process is called a **Markov process*** and the associated matrix, a **Markov matrix** or **transition matrix**. We shall use this example to illustrate some of the special properties of Markov matrices. In particular we are interested in the following two questions.

(a) Is there a starting population, $\mathbf{u}_0$, that would remain fixed by the process?

(b) Given an arbitrary initial population, $\mathbf{u}_0$, do the resulting populations $\mathbf{u}_k$ tend to a limit as k becomes large?

It turns out these two questions are related. We approach question (a) first, which is equivalent to asking: Is there a nonzero vector $\mathbf{u}_0$ such that $A\mathbf{u}_0 = \mathbf{u}_0$?

Simply, this is asking if 1 is an eigenvalue of A, so we compute

$$\det(\lambda I - A) = \det\begin{bmatrix} \lambda - 0.8 & -0.1 \\ -0.2 & \lambda - 0.9 \end{bmatrix} = (\lambda - 0.8)(\lambda - 0.9) - 0.02$$
$$= \lambda^2 - 1.7\lambda + 0.7 = (\lambda - 1)(\lambda - 0.7)$$

Thus we see that $\lambda = 1$ is an eigenvalue. The two eigenvalues of this example, $\lambda_1 = 1$ and $\lambda_2 = 0.7$, illustrate the general situation.

(5.55)

> THEOREM If A is a Markov matrix, then
>
> (a) $\lambda = 1$ is an eigenvalue of A.
>
> (b) All the eigenvalues of A are ≤ 1 in absolute value.

Part of the proof will be outlined in the exercises. (See Exercise 16.)

We now consider question (b): Does the population tend to a fixed limit? The answer is no in certain special cases (also indicated in the exercises) but yes in most cases. To see why, we diagonalize our Markov matrix A.

* *An important philosophical point to note about a Markov process is that history is completely disregarded. Thus each new* $\mathbf{u}_{k+1}$ *dependents only upon the present* $\mathbf{u}_k$, *and not upon preceding ones.*

We know the eigenvalues are 1 and 0.7, and you can check

$$\mathbf{v}_1 = \begin{bmatrix} \frac{1}{3} \\ \frac{2}{3} \end{bmatrix} \qquad \text{is an eigenvector associated with } \lambda_1 = 1$$

$$\mathbf{v}_2 = \begin{bmatrix} 1 \\ -1 \end{bmatrix} \qquad \text{is an eigenvector associated with } \lambda_2 = 0.7$$

The fact that $A\mathbf{v}_1 = \mathbf{v}_1$ means that if a third of the total population of the United States started off in Texas and two-thirds started off outside Texas, then at the end of the year there would still be one-third inside and two-thirds outside, even though the moving companies had been quite busy.

The fact that $\mathbf{v}_2$ has both positive and negative coordinates means that there is no realistic population distribution having 0.7 as an eigenvalue. (After all, if the total population is fixed and positive, it cannot shrink by 0.7.)

Now let $\mathbf{u}_0 = [x_0 \quad y_0]^T$ be an arbitrary starting population. The eigenvectors $\mathbf{v}_1$, $\mathbf{v}_2$ form a basis for $\mathbb{R}^2$, so we can write $\mathbf{u}_0 = a_1\mathbf{v}_1 + a_2\mathbf{v}_2$. Since $A\mathbf{v}_i = \lambda_i\mathbf{v}_i$, we see that

$$\begin{aligned}
\mathbf{u}_1 &= A\mathbf{u}_0 = A(a_1\mathbf{v}_1 + a_2\mathbf{v}_2) = a_1A\mathbf{v}_1 + a_2A\mathbf{v}_2 = a_1\lambda_1\mathbf{v}_1 + a_2\lambda_2\mathbf{v}_2 \\
\mathbf{u}_2 &= A\mathbf{u}_1 = A(a_1\lambda_1\mathbf{v}_1 + a_2\lambda_2\mathbf{v}_2) = a_1\lambda_1A\mathbf{v}_1 + a_2\lambda_2A\mathbf{v}_2 = a_1\lambda_1^2\mathbf{v}_1 + a_2\lambda_2^2\mathbf{v}_2 \\
\mathbf{u}_3 &= A\mathbf{u}_2 = A(a_1\lambda_1^2\mathbf{v}_1 + a_2\lambda_2^2\mathbf{v}_2) = a_1\lambda_1^2A\mathbf{v}_1 + a_2\lambda_2^2A\mathbf{v}_2 = a_1\lambda_1^3\mathbf{v}_1 + a_2\lambda_2^3\mathbf{v}_2
\end{aligned}$$

and in general

$$\mathbf{u}_k = a_1\lambda_1^k\mathbf{v}_1 + a_2\lambda_2^k\mathbf{v}_2$$

Since $\lambda_1 = 1$ and $\lambda_2 = 0.7$, $\lambda_1^k = 1$ and $\lambda_2^k \to 0$ as $k \to \infty$. Thus

$$\mathbf{u}_k \to \mathbf{u}_\infty = a_1\mathbf{v}_1 \qquad \text{as} \qquad k \to \infty$$

In other words, the population approaches a multiple of the eigenvector associated with $\lambda_1 = 1$.

NOTE There are, of course, infinitely many eigenvectors associated with $\lambda_1 = 1$. We picked the eigenvector $\mathbf{v}_1 = [\frac{1}{3} \quad \frac{2}{3}]^T$ because its coordinates $[\frac{1}{3} \quad \frac{2}{3}]^T$ add up to 1—a technical detail that facilitates computation. This way, since the total population is preserved at each step, we know that

$$\begin{aligned}
x_0 + y_0 &= \text{the sum of the coordinates of } \mathbf{u}_0 \\
&= \text{the sum of the coordinates of } \mathbf{u}_\infty \\
&= \tfrac{1}{3}a_1 + \tfrac{2}{3}a_1 = a_1
\end{aligned}$$

What this says about our problem is that the population distribution given by $\mathbf{u}_k$ approaches one-third of the total population inside Texas and two-thirds of the total population outside Texas. This illustrates the general situation, which follows.

(5.56)

THEOREM For *most* Markov matrices A

(a) The eigenvalue $\lambda = 1$ of A has multiplicity 1, and all other eigenvalues are strictly less than 1 in absolute value.

(b) Eigenvectors corresponding to eigenvalues strictly less than 1 do not represent realistic distributions.

(c) No matter what the starting population $\mathbf{u}_0$ is, the population $\mathbf{u}_k = A^k\mathbf{u}_0$ tends to a limit as $k \to \infty$, and that limit is an eigenvector associated with the eigenvalue $\lambda = 1$.

(d) The sum of the coordinates of the limit eigenvector is equal to the sum of the coordinates of the initial vector $\mathbf{u}_0$.

More examples and other properties will be treated in the exercises.

Exercise 5.5

In Exercises 1 and 2, suppose we are studying a certain species of rabbits in which a pair (one female and one male) takes one month to mature. When mature, the pair gives birth to another pair at the end of each month. Suppose at the beginning of one month, you are given a pair of newborn rabbits.

1. Write down the number of pairs of rabbits you will have at the beginning of each of the next six months. Compare your results with the sequence in (5.47).

2. Use the techniques of this section to determine how many pairs of rabbits you will have at the end of 36 months.

In Exercises 3–5, assume a Fibonacci sequence starts with 1 and 3, but still $F_{k+2} = F_{k+1} + F_k$.

3. Write down the first six terms of the resulting sequence.

4. Compute F_{100}.

5. Show that the ratios $\dfrac{F_{k+1}}{F_k}$ of this new sequence still approach the golden mean.

In Exercises 6–8, suppose that with a different species of rabbits, newborns take two months to mature, but each mature pair gives birth to a pair at the end of each month. Suppose you start with one pair of newborns.

6. Write down the first 10 terms of the resulting sequence.

7. Explain why the difference equation is now $F_{k+3} = F_{k+2} + F_k$.

8. Using the equations $F_{k+3} = F_{k+2} + F_k$, $F_{k+2} = F_{k+2}$, $F_{k+1} = F_{k+1}$, find the corresponding transition matrix.

In Exercises 9–11, suppose we are studying a certain species of beetles that live for at most three years and that propagate in the third year. Assume that one-half of the newly hatched survive the first year, one-third of the 1-year-olds survive the second year, and each female beetle produces six female infants at the end of the third year. Let F_k, S_k, and T_k be the number of female beetles in year k that are in their first, second, and third year of life respectively.

9. Write down the difference equations that express these relationships.

10. Let $\mathbf{u}_k = [F_k \quad S_k \quad T_k]^T$. Find the transition matrix A that expresses the difference equation as $\mathbf{u}_{k+1} = A\mathbf{u}_k$.

11. (a) Show that $A^3 = I$, and (b) follow the distribution of beetles for six years starting with 6000 beetles in each age group.

In Exercises 12–15, suppose we are studying a certain fish population. Of the eggs that hatch each year, $\frac{1}{20}$ survive; of the 1-year-olds, $\frac{1}{5}$ survive. All of the two-year-old fish spawn and then die, producing on the average 100 eggs that hatch. Assume we start a given year with 1000 fish in each age group.

12. Write down the difference equations that express these relationships.

13. Find the transition matrix A.

14. Follow the distribution of the fish for the next three years.

15. Show that $A^3 = I$. What are the implications for future years?

In Exercises 16–18, let A be a square matrix such that the sum of the entries in each column is 1.

16. Show that $\lambda = 1$ must be an eigenvalue of A. (HINT Show that the rows of $\lambda I - A$, for $\lambda = 1$, are linearly dependent, so that $1I - A$ is singular.)

17. Show that if $A\mathbf{x} = \lambda\mathbf{x}$, $\lambda \neq 1$, then the components of $\mathbf{x}$ add to zero. (HINT $A\mathbf{x} = \lambda\mathbf{x}$ is n equations. Add these equations together and then use the hypotheses on A and on λ.)

18. Use Exercise 17 to show that if $\mathbf{x}$ is an eigenvector of A associated with $\lambda \neq 1$, then $\mathbf{x}$ cannot represent a realistic distribution if all the items in the distribution are required to be nonnegative (such as in populations).

In Exercises 19–21, assume for several years everyone inside of Texas moves out and everyone outside moves in.

19. Find the transition matrix A.

20. Compute the eigenvalues of A and corresponding eigenvectors.

21. What is the population distribution that remains fixed? Under what circumstances does an initial distribution tend toward this fixed distribution?

In Exercises 22–25, assume a certain truck rental company has centers in New York, Chicago, and Dallas. Every month half those in New York and Chicago go to Dallas and the other half stay where they are. Half of the trucks in Dallas go to New York and half to Chicago.

22. Write down the difference equations that express these relationships.

23. Find the transition matrix A and show that it is a Markov matrix.

24. Find the steady-state distribution.

25. Assume the company starts with 32 trucks each in New York and Chicago and 128 trucks in Dallas.
    (a) What does Exercise 24 say the distribution should tend toward?
    (b) Follow the distribution for five months, and compare the results.

5.6 An Application—Differential Equations

There are many relationships in biology, physics, economics, chemistry, and other fields that are expressed as **differential equations**, that is, expressed in terms of equations involving functions and their derivatives. This section is an introduction to the methods of solving such equations using eigenvalues and eigenvectors.

We start with an application from ecology and use a simplified model of how two species might interact to indicate how differential equations arise. Although this model is given in its most simple form for the sake of this presentation, generalizations of it accurately describe many other diverse interactions, for example, how certain chemicals interact and why the fish population plummeted in the oceans off Chile.

Suppose there is a reasonably large island on which live foxes and hares. We assume that both foxes and hares will reproduce if they find mates. Also, if there are more rabbits around, foxes will eat more and reproduce more; if there are fewer rabbits, foxes will eat less and reproduce less. Similarly, if there are fewer foxes around, more rabbits will survive to produce even more; if there are more foxes, more rabbits will be eaten and they will reproduce less. Thus the *rate* at which the rabbits reproduce, that is, the rate at which their number increases or decreases, depends upon both the number of rabbits and the number of foxes on the island. A similar statement is true for the foxes. Putting these relationships in mathematical form, we let

$u = u(t)$ be the number of rabbits on the island at time t

$v = v(t)$ be the number of foxes on the island at time t

then

(5.57)
$$\frac{du}{dt} \quad \text{is the rate at which the number of rabbits is changing}$$
$$\frac{dv}{dt} \quad \text{is the rate at which the number of foxes is changing}$$

This discussion shows that $\frac{du}{dt}$ and $\frac{dv}{dt}$ depend upon both u and v. Under certain hypotheses this relationship can be expressed as

$$\frac{du}{dt} = au + bv \qquad (5.58)$$
$$\frac{dv}{dt} = cu + dv$$

This is a system of differential equations. It is systems in this form we shall use linear algebra to solve. To solve this problem, we first define

$$\frac{d}{dt}\begin{bmatrix} u \\ v \end{bmatrix} = \begin{bmatrix} \frac{du}{dt} \\ \frac{dv}{dt} \end{bmatrix}$$

and then rewrite (5.58) as

$$\frac{d}{dt}\begin{bmatrix} u \\ v \end{bmatrix} = \begin{bmatrix} a & b \\ c & d \end{bmatrix}\begin{bmatrix} u \\ v \end{bmatrix} \qquad (5.59)$$

This equation is of the form

$$\frac{d\mathbf{x}}{dt} = A\mathbf{x} \qquad \text{where} \qquad \mathbf{x} = \begin{bmatrix} u \\ v \end{bmatrix}, \quad A = \begin{bmatrix} a & b \\ c & d \end{bmatrix} \qquad (5.60)$$

You may have seen an equation like this before,

$$\frac{dx}{dt} = ax \qquad \text{where } a \text{ and } x \text{ are real numbers} \qquad (5.61)$$

You may know the solution is

(5.62)
$$x = Ce^{at}, \qquad C \text{ is a constant}$$

The form of the system in (5.60) is similar to the form of Equation (5.61). Hence we hope the solution to (5.60) is similar to the solution to Equation (5.61) that is given in (5.62). In particular we hope the solution has the form

(5.63)
$$u = Ce^{\lambda t} \qquad \text{and} \qquad v = De^{\lambda t}$$

This is one of several approaches to solving (5.60), and it works quite well. For such u and v,

(5.64)
$$\frac{du}{dt} = C\lambda e^{\lambda t} \qquad \text{and} \qquad \frac{dv}{dt} = D\lambda e^{\lambda t}$$

so that

(5.65)
$$\begin{bmatrix} u \\ v \end{bmatrix} = \begin{bmatrix} Ce^{\lambda t} \\ De^{\lambda t} \end{bmatrix} = e^{\lambda t}\begin{bmatrix} C \\ D \end{bmatrix}$$

and

(5.66)
$$\frac{d}{dt}\begin{bmatrix} u \\ v \end{bmatrix} = \begin{bmatrix} C\lambda e^{\lambda t} \\ D\lambda e^{\lambda t} \end{bmatrix} = \lambda\begin{bmatrix} Ce^{\lambda t} \\ De^{\lambda t} \end{bmatrix} = \lambda e^{\lambda t}\begin{bmatrix} C \\ D \end{bmatrix}$$

Substituting these into (5.60), we see that

(5.67)
$$\lambda e^{\lambda t}\begin{bmatrix} C \\ D \end{bmatrix} = Ae^{\lambda t}\begin{bmatrix} C \\ D \end{bmatrix} \qquad \text{or} \qquad \lambda\begin{bmatrix} C \\ D \end{bmatrix} = A\begin{bmatrix} C \\ D \end{bmatrix}$$

(since we can divide by $e^{\lambda t}$, which is never zero). This leads us to the following remarkable fact.

(5.68)

THEOREM A system of differential equations

$$\frac{d}{dt}\begin{bmatrix} u \\ v \end{bmatrix} = A\begin{bmatrix} u \\ v \end{bmatrix}$$

has solutions of the form

$$u = Ce^{\lambda t} \qquad \text{and} \qquad v = De^{\lambda t}$$

where λ and $\begin{bmatrix} C \\ D \end{bmatrix}$ are an associated eigenvalue and eigenvector of A.

If the eigenvalues are real and distinct, any solution has this form. If there are repeated eigenvalues or if the eigenvalues are complex, there is a bit more to the story, which we shall not discuss.

Example 1 Solve the system

$$\frac{du}{dt} = 3u - 2v$$

$$\frac{dv}{dt} = 5u - 4v$$

Solution This system is equivalent to

(5.69)
$$\frac{d}{dt}\begin{bmatrix} u \\ v \end{bmatrix} = A\begin{bmatrix} u \\ v \end{bmatrix} \qquad \text{where} \qquad A = \begin{bmatrix} 3 & -2 \\ 5 & -4 \end{bmatrix}$$

By the previous discussion, we assume solutions are of the form $u = Ce^{\lambda t}$ and $v = De^{\lambda t}$, and we set out to find eigenvalues and eigenvectors of A. The characteristic polynomial is

$$\det(\lambda I - A) = \det\begin{bmatrix} \lambda - 3 & 2 \\ -5 & \lambda + 4 \end{bmatrix} = (\lambda - 3)(\lambda + 4) + 10$$
$$= \lambda^2 + \lambda - 2 = (\lambda + 2)(\lambda - 1)$$

Thus the eigenvalues are $\lambda = 1, -2$. So now we need to find the corresponding eigenvectors.

Case $\lambda = 1$ We need to find the null space of

$$1I - A = \begin{bmatrix} -2 & 2 \\ -5 & 5 \end{bmatrix} \qquad \text{which reduces to} \qquad U = \begin{bmatrix} -2 & 2 \\ 0 & 0 \end{bmatrix}$$

Easily, $[1 \quad 1]^T$ forms a basis for this null space, so

(5.70)
$$\begin{bmatrix} 1 \\ 1 \end{bmatrix} \quad \text{is an eigenvector of } A \text{ associated with } \lambda = 1$$

Case $\lambda = -2$ We need to find a basis for the null space of

$$-2I - A = \begin{bmatrix} -5 & 2 \\ -5 & 2 \end{bmatrix} \qquad \text{which reduces to} \qquad U = \begin{bmatrix} -5 & 2 \\ 0 & 0 \end{bmatrix}$$

Easily $[2 \quad 5]^T$ forms a basis for this null space so that

(5.71)
$$\begin{bmatrix} 2 \\ 5 \end{bmatrix} \quad \text{is an eigenvector of } A \text{ associated with } \lambda = -2$$

Using (5.70) and (5.71) and Theorem (5.68), we know that

$$\begin{bmatrix} u \\ v \end{bmatrix} = \begin{bmatrix} e^t \\ e^t \end{bmatrix} = e^t\begin{bmatrix} 1 \\ 1 \end{bmatrix}$$

and

$$\begin{bmatrix} u \\ v \end{bmatrix} = \begin{bmatrix} 2e^{-2t} \\ 5e^{-2t} \end{bmatrix} = e^{-2t}\begin{bmatrix} 2 \\ 5 \end{bmatrix}$$

are solutions to the system in (5.69). Any linear combination of solutions to the system in (5.69) is again a solution. Thus these equations can be combined into one equation to give us

$$\begin{bmatrix} u \\ v \end{bmatrix} = ae^{t}\begin{bmatrix} 1 \\ 1 \end{bmatrix} + be^{-2t}\begin{bmatrix} 2 \\ 5 \end{bmatrix} \tag{5.72}$$

■

This equation is called the **general solution** to the system. Very often, when a system such as (5.69) is given, side conditions are also given. For example, if the values of u and v are given when $t = 0$, then altogether the problem is called an **initial-value problem**. In the example at the beginning of this section, we might be given the present population of foxes and hares on the island and be asked to determine the population trends. To solve such problems, we usually find the general solution first, if possible, and then find the **particular solution** that satisfies the given conditions.

Example 2 Solve the initial value problem

$$\frac{du}{dt} = 3u - 2v \qquad u = 13 \quad \text{when} \quad t = 0$$

$$\frac{dv}{dt} = 5u - 4v \qquad v = 22 \quad \text{when} \quad t = 0$$

Solution By Example 1 the general solution is

$$\begin{bmatrix} u \\ v \end{bmatrix} = ae^{t}\begin{bmatrix} 1 \\ 1 \end{bmatrix} + be^{-2t}\begin{bmatrix} 2 \\ 5 \end{bmatrix}$$

We substitute the initial conditions, $u = 13$, $v = 22$, $t = 0$, into this equation, obtaining

$$\begin{bmatrix} 13 \\ 22 \end{bmatrix} = ae^{0}\begin{bmatrix} 1 \\ 1 \end{bmatrix} + be^{-2(0)}\begin{bmatrix} 2 \\ 5 \end{bmatrix}$$

which, since $e^0 = 1$, reduces to

$$\begin{bmatrix} 13 \\ 22 \end{bmatrix} = a\begin{bmatrix} 1 \\ 1 \end{bmatrix} + b\begin{bmatrix} 2 \\ 5 \end{bmatrix} \quad \text{or} \quad \begin{bmatrix} 13 \\ 22 \end{bmatrix} = \begin{bmatrix} 1 & 2 \\ 1 & 5 \end{bmatrix}\begin{bmatrix} a \\ b \end{bmatrix}$$

This we solve in the usual manner, obtaining $a = 7$ and $b = 3$. Therefore, the particular solution to our problem is

$$\begin{bmatrix} \mathbf{u} \\ \mathbf{v} \end{bmatrix} = 7e^{t}\begin{bmatrix} 1 \\ 1 \end{bmatrix} + 3e^{-2t}\begin{bmatrix} 2 \\ 5 \end{bmatrix}$$

■

These procedures can be used to solve larger systems as well.

Example 3 Solve the system

$$\frac{du}{dt} = 5u - 7v + 7w$$
$$\frac{dv}{dt} = 4u - 3v + 4w$$
$$\frac{dw}{dt} = 4u - v + 2w$$

Solution We rewrite the system as

$$\frac{d}{dt}\begin{bmatrix} u \\ v \\ w \end{bmatrix} = \begin{bmatrix} 5 & -7 & 7 \\ 4 & -3 & 4 \\ 4 & -1 & 2 \end{bmatrix}\begin{bmatrix} u \\ v \\ w \end{bmatrix} \tag{5.73}$$

Because the form of this system is similar to (5.60) and (5.61), we hope the solutions are of the form

$$u = Ce^{\lambda \tau}, \qquad v = De^{\lambda t}, \qquad w = Ee^{\lambda t} \tag{5.74}$$

We substitute these expressions into (5.73), simplify using the procedures introduced in (5.63)–(5.67), and obtain

$$\lambda\begin{bmatrix} C \\ D \\ E \end{bmatrix} = \begin{bmatrix} 5 & -7 & 7 \\ 4 & -3 & 4 \\ 4 & -1 & 2 \end{bmatrix}\begin{bmatrix} C \\ D \\ E \end{bmatrix} \qquad \text{or} \qquad \lambda\mathbf{x} = A\mathbf{x}$$

as you can check. So we have an eigenvalue problem as before. By Example 4 in Section 5.2, the eigenvalues are

$$\lambda_1 = 5, \qquad \lambda_2 = -2, \qquad \lambda_3 = 1$$

We can compute in the usual way the eigenvectors corresponding to these eigenvalues and obtain

$$\begin{bmatrix} 1 \\ 1 \\ 1 \end{bmatrix}, \quad \begin{bmatrix} -1 \\ 0 \\ 1 \end{bmatrix}, \quad \begin{bmatrix} 0 \\ 1 \\ 1 \end{bmatrix}$$

as you can check. Along the lines of (5.72), we obtain the general solution to the system in (5.73).

$$\begin{bmatrix} u \\ v \\ w \end{bmatrix} = ae^{5t}\begin{bmatrix} 1 \\ 1 \\ 1 \end{bmatrix} + be^{-2t}\begin{bmatrix} -1 \\ 0 \\ 1 \end{bmatrix} + ce^{t}\begin{bmatrix} 0 \\ 1 \\ 1 \end{bmatrix}$$

■

Exercise 5.6 In Exercises 1–10, (a) find the general solution to the system of differential equations and (b) find the particular solution that satisfies the given initial conditions.

1. $$\frac{du}{dt} = 2u - 3v$$
$$\frac{dv}{dt} = 4u - 5v$$
$u = 7, v = 13$ when $t = 0$

2. $$\frac{du}{dt} = 2u + 4v$$
$$\frac{dv}{dt} = u - v$$
$u = -7, v = 5$ when $t = 0$

3. $$\frac{du}{dt} = 5u - 8v$$
$$\frac{dv}{dt} = u - v$$
$u = 3, v = 2$ when $t = 0$

4. $$\frac{du}{dt} = u + 3v$$
$$\frac{dv}{dt} = 4u - 3v$$
$u = 5, v = 1$ when $t = 0$

5. $$\frac{du}{dt} = 5u + 6v$$
$$\frac{dv}{dt} = 2u + v$$
$u = 7, v = 5$ when $t = 0$

6. $$\frac{du}{dt} = -u - 3v$$
$$\frac{dv}{dt} = -3u + 7v$$
$u = 4, v = 7$ when $t = 0$

7. $$\frac{du}{dt} = u - v$$
$$\frac{dv}{dt} = -u + 2v - w$$
$$\frac{dw}{dt} = -v + w$$
$u = 3, v = 4, w = 5$ when $t = 0$

8. $$\frac{du}{dt} = 2u - w$$
$$\frac{dv}{dt} = 2u - v$$
$$\frac{dw}{dt} = -12u + w$$
$u = 7, v = 4, w = 3$ when $t = 0$

9. $$\frac{du}{dt} = -6u - 5v + 5w$$
$$\frac{dv}{dt} = 3v - w$$
$$\frac{dw}{dt} = -v + 3w$$
$u = 1, v = 4, w = 2$ when $t = 0$

10. $$\frac{du}{dt} = -2u + w$$
$$\frac{dv}{dt} = -2u + v$$
$$\frac{dw}{dt} = 4u + w$$
$u = -1, v = 1, w = 0$ when $t = 0$

In Exercises 11–14, use the hint given in Exercise 11 to (a) find the general solution to the differential equation and (b) find the particular solution that satisfies the given initial conditions.

11. $y'' - y' - 2y = 0$, $y = 5$, $y' = 4$ when $t = 0$. (HINT Let $u = y$ and $v = y'$. Then $u' = v$ and $v' = y'' = y' + 2y = v + 2u$. Thus we have the system

$$\frac{du}{dt} = v$$
$$\frac{dv}{dt} = 2u + v$$

and initial conditions $u = y = 5$, $v = y' = 4$, when $t = 0$.)

12. $y'' - 5y' + 6y = 0$, $y = 2$, $y' = 3$ when $t = 0$

13. $y''' - 6y'' + 11y' - 6y = 0$, $y = 1$, $y' = -1$, $y'' = 2$ when $t = 0$

14. $y''' - 4y'' + y' + 6y = 0$, $y = 2$, $y' = 1$, $y'' = -1$ when $t = 0$

5.7 An Application—Quadratic Forms

Expressions such as

$$ax^2 + bxy + cy^2 + dx + ey + f \tag{5.75}$$

and

$$ax^2 + bxy + cy^2 + dyz + ez^2 + fxz + gx + hy + kz + l \tag{5.76}$$

are called **quadratic forms**. The associated quadratic forms that contain all the degree two terms, for example,

$$ax^2 + bxy + cy^2 \tag{5.77}$$

and

$$ax^2 + bxy + cy^2 + dyz + ez^2 + fxz \tag{5.78}$$

are called their **principal parts**. Quadratic forms arise throughout mathematics and in many diverse applications as well. For example, in geology the results of stress on rock strata are sometimes described using quadratic forms, and in the automobile industry, when matching colors in plastic and metal parts of automobile bodies, quadratic forms are used to express relationships between shades, hues, and brightness. This section is an introduction to quadratic forms and how eigenvalues and eigenvectors are used to describe and simplify them. We shall work only with principal parts, as this simplifies our work and the principal part contains most of the important information anyway.

We begin the discussion with conic sections and a brief review of the basic facts about them. A more thorough discussion can be found in almost any college algebra or calculus textbook.

Example 1 Describe and graph the following

(a) $9x^2 + 4y^2 = 1$ (b) $4x^2 + 9y^2 = 1$ (c) $9x^2 - 4y^2 = 1$
(d) $4y^2 - 9x^2 = 1$ (e) $y - 4x^2 = 0$ (f) $4y^2 + x = 0$

Solution (a) and (b) are **ellipses**. Their graphs are in Figure 5.2a.

(c) and (d) are **hyperbolas**. Their graphs are in Figure 5.2b. Note that they each have the same asymptotes, $y = \pm\frac{3}{2}x$.

(e) and (f) are **parabolas**. Their graphs are in Figure 5.2c.

Figure 5.2

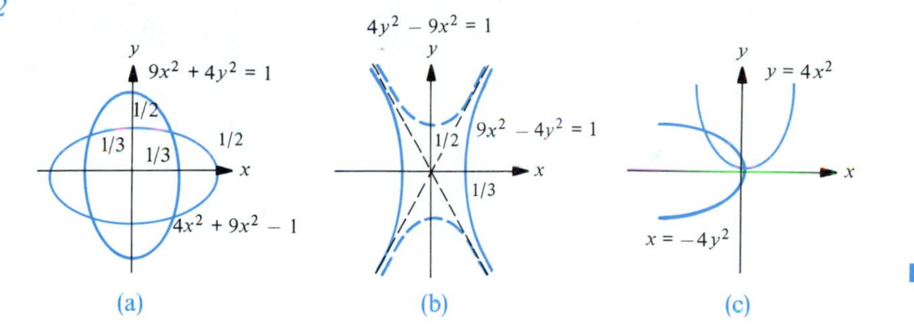

(a) (b) (c)

Notice that the equations of parts (a)–(d) of Example 1 involve principal parts of quadratic forms, as in (5.77), but they have no xy, or **crossed**, terms. Our goal is to use linear algebra to describe expressions like these that do have crossed terms. Our first step is to rewrite principal parts of quadratic forms using matrices.

(5.79) THEOREM The principal part of any quadratic form can be rewritten as

$$\mathbf{x}^T A \mathbf{x}$$

where A is a square symmetric matrix.

Proof The quadratic form

$$ax^2 + bxy + cy^2$$

can be rewritten as

$$[x \quad y]\begin{bmatrix} a & \frac{1}{2}b \\ \frac{1}{2}b & c \end{bmatrix}\begin{bmatrix} x \\ y \end{bmatrix} \tag{5.80}$$

since

(5.81)
$$[x \quad y]\begin{bmatrix} a & \frac{1}{2}b \\ \frac{1}{2}b & c \end{bmatrix}\begin{bmatrix} x \\ y \end{bmatrix} = [x \quad y]\begin{bmatrix} ax + \frac{1}{2}by \\ \frac{1}{2}bx + cy \end{bmatrix}$$
$$= [ax^2 + \tfrac{1}{2}bxy + \tfrac{1}{2}bxy + cy^2] = ax^2 + bxy + cy^2$$

NOTE This requires the identification of a number r with the 1×1 matrix $[r]$, just as we did for Theorem (4.16), Section 4.2.

The quadratic form

$$ax^2 + bxy + cy^2 + dyz + ez^2 + fxz$$

can be rewritten as

(5.82)
$$[x \quad y \quad z]\begin{bmatrix} a & \frac{1}{2}b & \frac{1}{2}f \\ \frac{1}{2}b & c & \frac{1}{2}d \\ \frac{1}{2}f & \frac{1}{2}d & e \end{bmatrix}\begin{bmatrix} x \\ y \\ z \end{bmatrix}$$

as can be readily verified by a computation similar to (5.81).

The general pattern should now be clear, but we shall not be working with case where $n > 3$. ■

You might note that other A's give the same quadratic form, for example,

$$ax^2 + bxy + cy^2 = [x \quad y]\begin{bmatrix} a & b \\ 0 & c \end{bmatrix}\begin{bmatrix} x \\ y \end{bmatrix}$$

But the matrices given in (5.80) and (5.82) are symmetric. This is *very* important for our purposes since symmetric matrices are always diagonalizable and have orthonormal eigenvectors, as we know from Theorems (5.39) and (5.45), Section 5.4. These eigenvectors, in turn, can be used to rewrite the quadratic form in a simplified form without any crossed terms. Hence in this form, the quadratic form is easy to graph and to understand, as we illustrate in the next examples.

Example 2 For the quadratic form $q(x, y) = 3x^2 - 4xy + 3y^2$,

(a) Find a symmetric matrix A such that $q = \mathbf{x}^T A\mathbf{x}$.
(b) Find an orthonormal Q that diagonalizes A, that is, such that $A = Q\Lambda Q^{-1}$, where Λ is a diagonal matrix.

NOTE Hereafter we shall call this process **diagonalizing** the quadratic form.

Solution (a) As described by the matrices in (5.80),

$$A = \begin{bmatrix} 3 & -2 \\ -2 & 3 \end{bmatrix} \quad \text{so} \quad 3x^2 - 4xy + 3y^2 = [x \quad y]\begin{bmatrix} 3 & -2 \\ -2 & 3 \end{bmatrix}\begin{bmatrix} x \\ y \end{bmatrix}$$

(b) First compute

$$\det(\lambda I - A) = \det\begin{bmatrix} \lambda - 3 & 2 \\ 2 & \lambda - 3 \end{bmatrix} = \lambda^2 - 6\lambda + 5 = (\lambda - 5)(\lambda - 1)$$

Thus the eigenvalues are $\lambda = 5, 1$. We compute the eigenvectors in the usual way, obtaining $\mathbf{v}_1 = [1 \quad -1]^T$ and $\mathbf{v}_2 = [1 \quad 1]^T$, which correspond to $\lambda = 5$ and 1, respectively, as you can check. As theory promises, $\mathbf{v}_1$ and $\mathbf{v}_2$ are orthogonal (since A is symmetric). We normalize them, obtaining $\mathbf{u}_1 = \begin{bmatrix} \frac{1}{\sqrt{2}} & -\frac{1}{\sqrt{2}} \end{bmatrix}^T$, $\mathbf{u}_2 = \begin{bmatrix} \frac{1}{\sqrt{2}} & \frac{1}{\sqrt{2}} \end{bmatrix}^T$. We construct Q with $\mathbf{u}_1$ and $\mathbf{u}_2$ as columns, obtaining

$$Q = \begin{bmatrix} \frac{1}{\sqrt{2}} & \frac{1}{\sqrt{2}} \\ -\frac{1}{\sqrt{2}} & \frac{1}{\sqrt{2}} \end{bmatrix} \quad \text{such that} \quad A = Q\begin{bmatrix} 5 & \\ & 1 \end{bmatrix}Q^{-1}$$ ■

We are now in the position of knowing

$$q = \mathbf{x}^T A\mathbf{x} \quad \text{and} \quad A = Q\Lambda Q^T$$

This is of the very simple form

$$q = \mathbf{x}'^T\Lambda\mathbf{x}' \quad \text{for} \quad \mathbf{x}' = \begin{bmatrix} x' \\ y' \end{bmatrix} = Q^T\begin{bmatrix} x \\ y \end{bmatrix}$$

In particular the equation $q(x', y') = 1$ is

$$[x' \quad y']\begin{bmatrix} 5 & \\ & 1 \end{bmatrix}\begin{bmatrix} x' \\ y' \end{bmatrix} = 1 \quad \text{or} \quad 5x'^2 + y'^2 = 1$$

which is an ellipse in x', y' coordinates. To graph this ellipse, we need to know which directions correspond to $[1 \quad 0]^T$ and $[0 \quad 1]^T$ in the prime coordinates. So if

$$\begin{bmatrix} 1 \\ 0 \end{bmatrix} = Q^T\begin{bmatrix} x \\ y \end{bmatrix} \quad \text{then} \quad \begin{bmatrix} x \\ y \end{bmatrix} = Q\begin{bmatrix} 1 \\ 0 \end{bmatrix} = \mathbf{u}_1$$

since $Q^T = Q^{-1}$. Similarly if

$$\begin{bmatrix} 0 \\ 1 \end{bmatrix} = Q^T \begin{bmatrix} x \\ y \end{bmatrix} \quad \text{then} \quad \begin{bmatrix} x \\ y \end{bmatrix} = Q \begin{bmatrix} 0 \\ 1 \end{bmatrix} = \mathbf{u}_2$$

Thus if we rotate the axes so that they point in the direction of the columns of Q, which are the eigenvectors of A, then the graph is easy to find.

Example 3 Graph the equation $q(x, y) = 1$, where

$$q(x, y) = 3x^2 - 4xy + 3y^2$$

is from Example 2.

Solution We rotate so that the x- and y-axes point in the direction of

$$\mathbf{u}_1 = \begin{bmatrix} \frac{1}{\sqrt{2}} & -\frac{1}{\sqrt{2}} \end{bmatrix}^T \quad \text{and} \quad \mathbf{u}_2 = \begin{bmatrix} \frac{1}{\sqrt{2}} & \frac{1}{\sqrt{2}} \end{bmatrix}^T$$

respectively, where $\mathbf{u}_1$ and $\mathbf{u}_2$ were found in Example 2. As we discussed, in this direction the equation is $5x'^2 + y'^2 = 1$. See Figure 5.3.

Figure 5.3

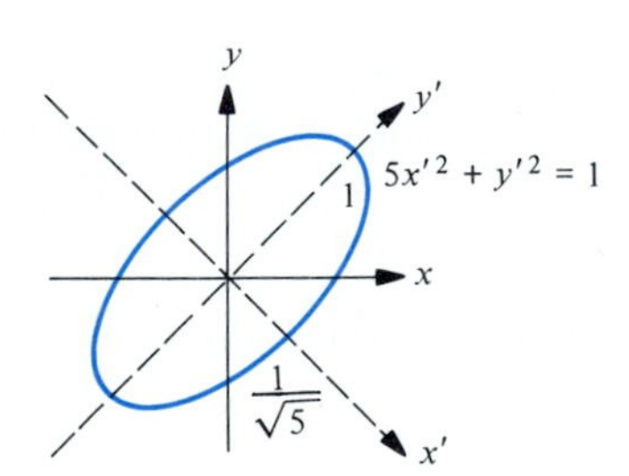

■

Example 4 Diagonalize the quadratic form

$$q(x, y) = xy$$

rewrite the equation in the new coordinate system, and use this to graph $xy = 1$.

Solution We rewrite the quadratic form as

$$xy = [x \quad y] A \begin{bmatrix} x \\ y \end{bmatrix} = [x \quad y] \begin{bmatrix} 0 & \frac{1}{2} \\ \frac{1}{2} & 0 \end{bmatrix} \begin{bmatrix} x \\ y \end{bmatrix}$$

Next we diagonalize this A, obtaining an orthonormal Q such that $A = Q\Lambda Q^{-1}$.

$$\det(\lambda I - A) = \det\begin{bmatrix} \lambda & -\frac{1}{2} \\ -\frac{1}{2} & \lambda \end{bmatrix} = \lambda^2 - \tfrac{1}{4} = (\lambda - \tfrac{1}{2})(\lambda + \tfrac{1}{2})$$

Thus the eigenvalues are $\frac{1}{2}$ and $-\frac{1}{2}$ and their corresponding orthonormal eigenvectors are $\begin{bmatrix} \frac{1}{\sqrt{2}} & \frac{1}{\sqrt{2}} \end{bmatrix}^T$ and $\begin{bmatrix} -\frac{1}{\sqrt{2}} & \frac{1}{\sqrt{2}} \end{bmatrix}^T$, as you can check. Thus if

$$Q = \begin{bmatrix} \frac{1}{\sqrt{2}} & -\frac{1}{\sqrt{2}} \\ \frac{1}{\sqrt{2}} & \frac{1}{\sqrt{2}} \end{bmatrix} \qquad \text{then} \qquad A = Q\begin{bmatrix} \frac{1}{2} & \\ & -\frac{1}{2} \end{bmatrix}Q^T$$

We can now write the quadratic form as

$$\begin{aligned} xy &= [x \quad y]A\begin{bmatrix} x \\ y \end{bmatrix} = [x \quad y]Q\Lambda Q^T\begin{bmatrix} x \\ y \end{bmatrix} \\ &= [x' \quad y']\begin{bmatrix} \frac{1}{2} & \\ & -\frac{1}{2} \end{bmatrix}\begin{bmatrix} x' \\ y' \end{bmatrix} = \tfrac{1}{2}x'^2 - \tfrac{1}{2}y'^2 \end{aligned}$$

Thus we have an hyperbola, and the graph of $xy = 1$ or $\frac{1}{2}x'^2 - \frac{1}{2}y'^2 = 1$ is in Figure 5.4.

Figure 5.4

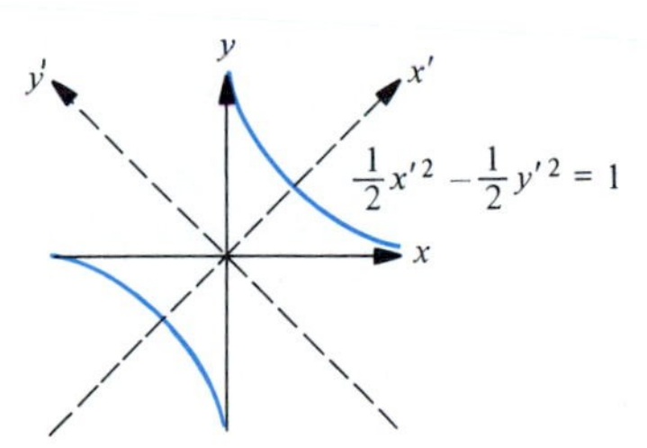

■

All of these examples generalize to three and higher dimensions. However, to keep the exposition unencumbered, we shall restrict our attention to ellipsoids, which have the following basic equation

$$\frac{x^2}{a^2} + \frac{y^2}{b^2} + \frac{z^2}{c^2} = 1$$

and the basic graph is in Figure 5.5.

Figure 5.5

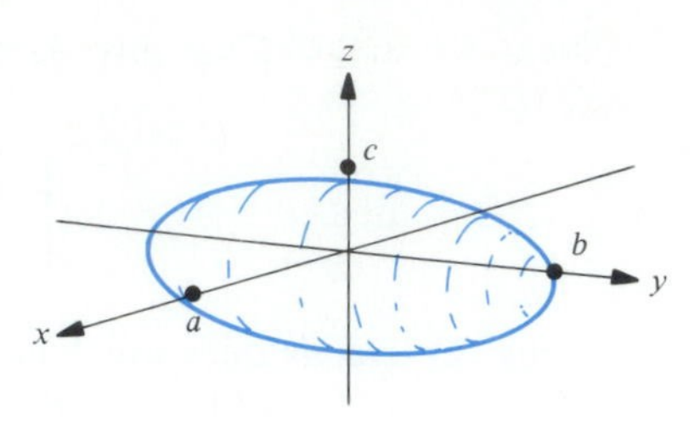

The procedures for handling quadratic forms in three variables are essentially the same as we have discussed. This is illustrated in the next example.

Example 5 For the quadratic form

$$q(x, y, z) = \tfrac{1}{3}(9x^2 - 4xy + 8y^2 + 10z^2 - 4xz)$$

(a) Find a symmetric matrix A such that $q = x^T Ax$.
(b) Diagonalize A.
(c) Graph the equation $q(x, y, z) = 12$.

Solution (a) We proceed as described in the proof of Theorem (5.79), obtaining

$$A = \tfrac{1}{3}\begin{bmatrix} 9 & -2 & -2 \\ -2 & 8 & 0 \\ -2 & 0 & 10 \end{bmatrix}$$

You can check that $q = \mathbf{x}^T A\mathbf{x}$, where $\mathbf{x} = [x \quad y \quad z]^T$.

(b) To diagonalize A, we first find

$$\det(\lambda I - A) = \det\begin{bmatrix} \lambda - 3 & \frac{2}{3} & \frac{2}{3} \\ \frac{2}{3} & \lambda - \frac{8}{3} & 0 \\ \frac{2}{3} & 0 & \lambda - \frac{10}{3} \end{bmatrix}$$
$$= \lambda^3 - 9\lambda^2 + 26\lambda - 24 = (\lambda - 2)(\lambda - 3)(\lambda - 4)$$

as you can check. Therefore the eigenvalues are $\lambda_1 = 2$, $\lambda_2 = 3$, and $\lambda_3 = 4$. We compute the corresponding eigenvectors in the usual way, normalize them, and obtain

$$\mathbf{u}_1 = \begin{bmatrix} \frac{2}{3} \\ \frac{2}{3} \\ \frac{1}{3} \end{bmatrix}, \qquad \mathbf{u}_2 = \begin{bmatrix} -\frac{1}{3} \\ \frac{2}{3} \\ -\frac{2}{3} \end{bmatrix}, \qquad \mathbf{u}_3 = \begin{bmatrix} -\frac{2}{3} \\ \frac{1}{3} \\ \frac{2}{3} \end{bmatrix}$$

You can check directly that $A\mathbf{u}_i = \lambda_i \mathbf{u}_i$ and that the $\mathbf{u}_i$'s are orthonormal. Therefore, if

$$S = \begin{bmatrix} \frac{2}{3} & -\frac{1}{3} & -\frac{2}{3} \\ \frac{2}{3} & \frac{2}{3} & \frac{1}{3} \\ \frac{1}{3} & -\frac{2}{3} & \frac{2}{3} \end{bmatrix} \quad \text{and} \quad \Lambda = \begin{bmatrix} 2 & & \\ & 3 & \\ & & 4 \end{bmatrix}$$

then $A = S\Lambda S^T$. Thus if we let $S^T\mathbf{x} = \mathbf{x}' = [x' \quad y' \quad z']^T$, then

$$\begin{aligned} q(x, y, z) &= \mathbf{x}^T A\mathbf{x} = \mathbf{x}^T S\Lambda S^T\mathbf{x} \\ &= x'^T \Lambda x' = 2x'^2 + 3y'^2 + 4z'^2 \end{aligned}$$

so that

$$q(x', y', z') = 2x'^2 + 3y'^2 + 4z'^2$$

(c) If we rotate the x-, y-, z-axes obtaining x'-, y'-, z'-axes in the direction of $\mathbf{u}_1$, $\mathbf{u}_2$, and $\mathbf{u}_3$ respectively, then the graph of $q(x, y, z) = 12$ becomes

$$2x'^2 + 3y'^2 + 4z'^2 = 12 \qquad \text{or} \qquad \frac{x'^2}{6} + \frac{y'^2}{4} + \frac{z'^2}{3} = 1$$

The graph is in Figure 5.6.

Figure 5.6

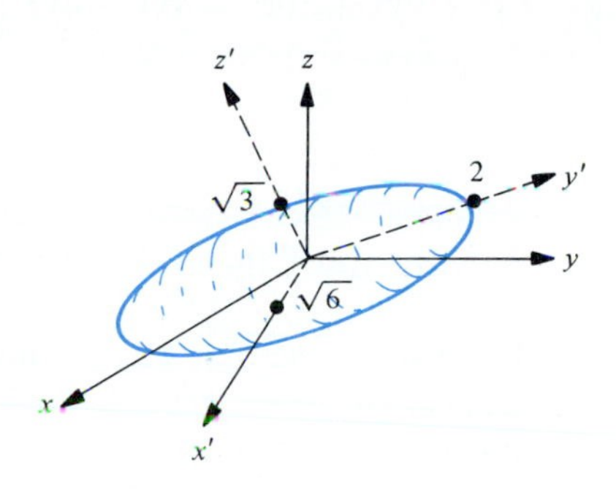

Exercise 5.7 In Exercises 1–4, find the principal parts of the given quadratic form.

1. $2x^2 + 3x + 4xy - 5y + y^2 - 6$
2. $3y^2 - 2y - yz - 8z^2 + 3z + 2$
3. $3x^2 - 8x + xy + 4y - 2y^2 - 4z - 8yz + 2z^2 - 5xz - 4$
4. $5x^2 - 4y^2 - 3z^2 - 8x + 4y - 2z + xz - 4xy + 2yz - 1$

In Exercises 5–10, describe and graph the given conic section

5. $4x^2 + 9y^2 = 36$
6. $25y^2 - 9x^2 = 16$
7. $9x^2 - y = 0$
8. $25x^2 + 4y^2 = 9$
9. $8x^2 - 18y^2 = 2$
10. $4x - 5y^2 = 0$

In Exercises 11–18, rewrite the quadratic form as $\mathbf{x}^T A\mathbf{x}$, where A is symmetric.

11. $3x^2 + 4xy - 5y^2$
12. $x^2 + 6xy + 2y^2$
13. $5x^2 - 2xy + 2y^2$
14. $3x^2 + 3xy + 4y^2$

15. $2x^2 - 8xy - 5y^2 - 6yz - 7z^2 + xz$
16. $9x^2 - 7xy - 5y^2 + 8yz - z^2 + 3xz$
17. $2x^2 - xz + 9y^2 + 3yz$
18. $2xy + 4xz + 6yz$

In Exercises 19–30, (a) diagonalize the quadratic form q and (b) graph the equation $q(x, y) = 1$ or $q(x, y, z) = 1$.

19. $4x^2 + 4xy + 4y^2$
20. $x^2 + 4xy + y^2$
21. $3x^2 + 2\sqrt{2}xy + 2y^2$
22. $2x^2 + 6xy + 2y^2$
23. $2x^2 - 3xy - 2y^2$
24. $4x^2 + 7y^2 + 4xy$
25. $-2x^2 + 4xy + y^2$
26. $2x^2 - 2xy + 2y^2$
27. $xy - \frac{3}{4}x^2$
28. $y^2 + xy$
29. $8x^2 + 9y^2 + 10z^2 - 4xy - 4yz$
30. $2x^2 + 2xy + 3y^2 + 2yz + 2z^2$

5.8 Solving the Eigenvalue Problem Numerically

There is no one best way to solve the eigenvalue problem numerically. For some applications you need only the eigenvalue that is the largest in absolute value, or the largest several. For other problems you need them all. For some problems you need associated eigenvectors, and for others you do not. Which method is "best" depends not only upon which of these types of problems you have, but also upon the matrix itself—is it large or small, sparse or full, symmetric or not, . . . ? To complicate matters more, there are some methods which, although they are fine in theory, are terrible numerically and should never be tried.

The purpose of this section is to present a brief introduction to a few of the methods and some of the numerical considerations that are involved. The bibliography contains several references with more complete treatments; the bible of this subject is Wilkenson, *The Algebraic Eigenvalue Problem*, but this is quite advanced. Well-written subroutines are available in IMSL and EISPACK, as described in an appendix.

Preparation of Matrices

When solving an eigenvalue problem numerically outside the mathematics classroom, the matrix A almost always should be "preprocessed" first. This preprocessing should bring the matrix to **tridiagonal form** if A is symmetric or to **upper-Hessenberg form** if the matrix is not symmetric. Both forms are illustrated in Figure 5.7. Without this preprocessing, convergence is usually quite slow, and consequently the solution is unnecessarily expensive and less accurate.

Figure 5.7

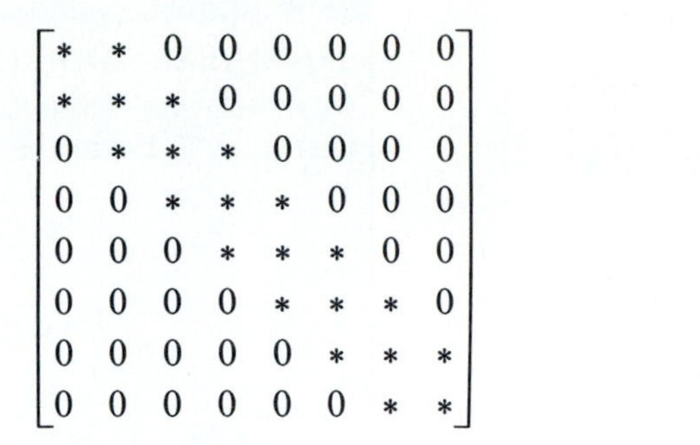

(a) *Tridiagonal form*

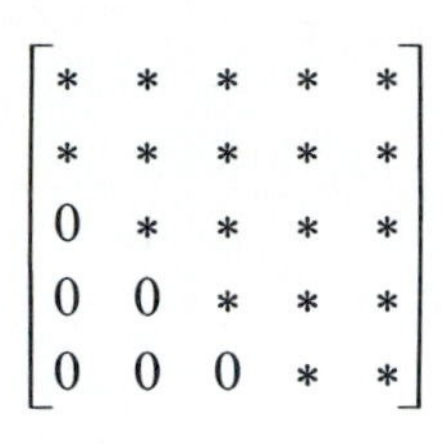

(b) *Upper-Hessenberg form*

The preprocessing consists of finding orthogonal matrices $Q_1, \ldots, Q_k$ such that

$$Q_k \cdots Q_1 A Q_1^T \cdots Q_k^T$$

is in the desired form. This does not affect the eigenvalues; see Exercise 1. How to find such Q_i's is outlined in the exercises. What is important at this point is not that you know how to do it but just that you know it should almost always be done when you work with such problems.

An Assumption

To simplify the exposition for this section, we shall assume the following about A.

(5.83)

> ASSUMPTION Assume the $n \times n$ matrix A is diagonalizable with eigenvalues $\lambda_1, \ldots, \lambda_n$, and assume that
>
> $$|\lambda_1| > |\lambda_2| > \cdots > |\lambda_n| > 0$$
>
> Also assume that $\mathbf{v}_1, \ldots, \mathbf{v}_n$ are *unit* eigenvectors of A associated with $\lambda_1, \ldots, \lambda_n$ respectively. As a consequence, $\mathbf{v}_1, \ldots, \mathbf{v}_n$ are linearly independent, by Theorem (5.32), Section 5.3, so they form a basis for $\mathbb{R}^n$. In particular, any vector $\mathbf{w}$ in $\mathbb{R}^n$ can be expressed as $\mathbf{w} = a_1\mathbf{v}_1 + \cdots + a_n\mathbf{v}_n$.

This assumption holds for a great many of the matrices you run into in practice. In any case, most of what we say is true in general, but the proofs are considerably more involved.

Direct Iteration

The first method of finding numerical answers to eigenvalue problems we discuss is called **direct iteration**. This method is a simple way of finding λ_1 and $\mathbf{v}_1$, and it forms a basis for other more preferred methods.

(5.84)

THEOREM. DIRECT ITERATION Let $\mathbf{w}_0 = a_1\mathbf{v}_1 + \cdots + a_n\mathbf{v}_n$ be any vector in $\mathbb{R}^n$ such that $a_1 \neq 0$. Form the sequences

$$\begin{array}{lll} A\mathbf{w}_0, & \|A\mathbf{w}_0\|, & \mathbf{w}_1 = \dfrac{A\mathbf{w}_0}{\|A\mathbf{w}_0\|} \\ A\mathbf{w}_1, & \|A\mathbf{w}_1\|, & \mathbf{w}_2 = \dfrac{A\mathbf{w}_1}{\|A\mathbf{w}_1\|} \\ A\mathbf{w}_2, & \|A\mathbf{w}_2\|, & \mathbf{w}_3 = \dfrac{A\mathbf{w}_2}{\|A\mathbf{w}_2\|} \\ \vdots & \vdots & \vdots \end{array}$$

Then the sequence of vectors

$$\mathbf{w}_1, \mathbf{w}_2, \mathbf{w}_3, \ldots \quad \text{approaches} \quad \pm\mathbf{v}_1$$

and the sequence of numbers

$$\|A\mathbf{w}_0\|,\ \|A\mathbf{w}_1\|,\ \|A\mathbf{w}_2\|, \ldots \quad \text{approaches} \quad |\lambda_1|$$

The proof is straightforward, but it is optional and will be put off until the end of the section.

Example 1 Let $A = \begin{bmatrix} 4 & -2 \\ 3 & -1 \end{bmatrix}$. You can check that $\lambda = 2, 1$ are eigenvalues with associated unit eigenvectors $\mathbf{v}_1 = \begin{bmatrix} \frac{1}{\sqrt{2}} & \frac{1}{\sqrt{2}} \end{bmatrix}^T \approx [0.707 \quad 0.707]^T$, $\mathbf{v}_2 = \begin{bmatrix} \frac{2}{\sqrt{13}} & \frac{3}{\sqrt{13}} \end{bmatrix}^T \approx [0.555 \quad 0.832]^T$, respectively. Let $\mathbf{w}_0 = [1 \quad 0]^T$. Then $\mathbf{w}_0 = 3\sqrt{2}\mathbf{v}_1 - \sqrt{13}\mathbf{v}_2$, so $a_1 \neq 0$. Now compute $\mathbf{w}_1, \mathbf{w}_2, \ldots$

$$A\mathbf{w}_0 = \begin{bmatrix} 4 & -2 \\ 3 & -1 \end{bmatrix}\begin{bmatrix} 1 \\ 0 \end{bmatrix} = \begin{bmatrix} 4 \\ 3 \end{bmatrix}, \qquad \|A\mathbf{w}_0\| = 5, \qquad \mathbf{w}_1 = \tfrac{1}{5}\begin{bmatrix} 4 \\ 3 \end{bmatrix} = \begin{bmatrix} 0.8 \\ 0.6 \end{bmatrix}$$

$$A\mathbf{w}_1 = \begin{bmatrix} 4 & -2 \\ 3 & -1 \end{bmatrix}\begin{bmatrix} 0.8 \\ 0.6 \end{bmatrix} = \begin{bmatrix} 2.0 \\ 1.8 \end{bmatrix}, \qquad \|A\mathbf{w}_1\| \approx 2.69, \qquad \mathbf{w}_2 \approx \frac{1}{2.69}\begin{bmatrix} 2.0 \\ 1.8 \end{bmatrix} \approx \begin{bmatrix} 0.743 \\ 0.669 \end{bmatrix}$$

$$A\mathbf{w}_2 = \begin{bmatrix} 4 & -2 \\ 3 & -1 \end{bmatrix}\begin{bmatrix} 0.743 \\ 0.669 \end{bmatrix} \approx \begin{bmatrix} 1.63 \\ 1.56 \end{bmatrix}, \qquad \|A\mathbf{w}_2\| \approx 2.26, \qquad \mathbf{w}_3 \approx \frac{1}{2.26}\begin{bmatrix} 1.63 \\ 1.56 \end{bmatrix} \approx \begin{bmatrix} 0.721 \\ 0.690 \end{bmatrix}$$

$$A\mathbf{w}_3 = \begin{bmatrix} 4 & -2 \\ 3 & -1 \end{bmatrix}\begin{bmatrix} 0.721 \\ 0.690 \end{bmatrix} \approx \begin{bmatrix} 1.50 \\ 1.47 \end{bmatrix}, \qquad \|A\mathbf{w}_3\| \approx 2.10, \qquad \mathbf{w}_4 \approx \frac{1}{2.10}\begin{bmatrix} 1.50 \\ 1.47 \end{bmatrix} \approx \begin{bmatrix} 0.714 \\ 0.700 \end{bmatrix}$$

$$A\mathbf{w}_4 = \begin{bmatrix} 4 & -2 \\ 3 & -1 \end{bmatrix}\begin{bmatrix} 0.714 \\ 0.700 \end{bmatrix} = \begin{bmatrix} 1.456 \\ 1.442 \end{bmatrix}, \qquad \|A\mathbf{w}_4\| \approx 2.049, \qquad \mathbf{w}_5 \approx \frac{1}{2.049}\begin{bmatrix} 1.456 \\ 1.442 \end{bmatrix} \approx \begin{bmatrix} 0.7106 \\ 0.7038 \end{bmatrix}$$

You can observe the convergence

$$\|A\mathbf{w}_1\|, \|A\mathbf{w}_2\|, \|A\mathbf{w}_3\|, \ldots \to \lambda_1 = 2$$

and

$$\mathbf{w}_1, \mathbf{w}_2, \mathbf{w}_3, \ldots \to \mathbf{v}_1 \approx \begin{bmatrix} 0.707 \\ 0.707 \end{bmatrix}$$ ■

But also observe the convergence is fairly slow. It is very important when working numerically to know not only if a sequence converges but also how fast, or slowly, it converges. Fortunately the answer is easy to state in this case.

(5.85)

> THEOREM Under Assumption (5.83), direct iteration converges in such a way that the error is multiplied by a factor of about $\left|\frac{\lambda_2}{\lambda_1}\right|$ on each iteration.

This estimate works only when the estimates are "close" to the answer. We shall also discuss this at the end of this section.

Example 2 In Example 1 we know $\lambda_1 = 2$ and $\lambda_2 = 1$. Therefore, with each iteration the error is multiplied by $\frac{\lambda_2}{\lambda_1} = \frac{1}{2} = 0.5$. So in three iterations the errors are multiplied by $(\frac{1}{2})^3 = \frac{1}{8} = 0.125$. Since the error has been decreased by about 0.1 in three iterations, our estimates improve by about one significant figure in three iterations! You can see this in Example 1. Since $\lambda_1 = 2$, we know the sequence

$$\|A\mathbf{w}_0\| = 5, \quad \|A\mathbf{w}_1\| = 2.69, \quad \|A\mathbf{w}_2\| = 2.26,$$
$$\|A\mathbf{w}_3\| = 2.10, \quad \|A\mathbf{w}_4\| = 2.049, \ldots$$

is converging to 2.000 .... Ignoring the first step (since 5 is not very close to 2), you can see the three steps from $\|A\mathbf{w}_1\|$ to $\|A\mathbf{w}_4\|$ pick up about one significant figure. ■

This is not very fast, and usually it is much worse! For example, if $\lambda_1 = 50$ and $\lambda_2 = 40$, then $\frac{\lambda_2}{\lambda_1} = \frac{4}{5} = 0.8$. Here it would take about 10 iterations to pick up one significant figure. If $\lambda_1 = 500$ and $\lambda_2 = 490$, then it would take over 110 iterations. Hence you can readily believe that:

(5.86)

> Except when seeking a very crude estimate for the largest eigenvalue, you should almost never use direct iteration.

Although the method by itself usually does not work well, it forms the basis for a method that does. We now take a second step toward that method.

Inverse Iteration

We know if $A\mathbf{v} = \lambda\mathbf{v}$, $\lambda \neq 0$, and if A^{-1} exists, then $A^{-1}\mathbf{v} = \left(\frac{1}{\lambda}\right)\mathbf{v}$. See Exercise 25, Section 5.2. Therefore, since A has eigenvalues $|\lambda_1| > |\lambda_2| > \cdots > |\lambda_n|$ and associated eigenvectors, $\mathbf{v}_1, \ldots, \mathbf{v}_n$, A^{-1} has eigenvalues $\frac{1}{|\lambda_1|} < \frac{1}{|\lambda_2|} < \cdots < \frac{1}{|\lambda_n|}$ with the same associated eigenvectors $\mathbf{v}_1, \mathbf{v}_2, \ldots, \mathbf{v}_n$. Consequently direct iteration with A^{-1} converges to the reciprocal of the smallest eigenvalue of A and its associated eigenvector. Of course, as we mentioned in Chapter 1, we usually do not compute $\mathbf{y}_{n+1} = A^{-1}\mathbf{w}_n$ since we seldom want to compute A^{-1}. At each stage we know $\mathbf{w}_n$ and solve $A\mathbf{y}_{n+1} = \mathbf{w}_n$ for $\mathbf{y}_{n+1}$. This is one of the many situations that arises quite naturally where we need to solve the equation $A\mathbf{x} = \mathbf{b}$ many times with the same A but with different $\mathbf{b}$'s.

Altogether, then, we have shown the following.

(5.87)

THEOREM. INVERSE ITERATION Let A be as in Assumption (5.83). Let $\mathbf{w}_0 = a_1\mathbf{v}_1 + \cdots + a_n\mathbf{v}_n$ be any vector in $\mathbb{R}^n$ such that $a_n \neq 0$. Form the sequences $\mathbf{y}_1, \mathbf{y}_2, \ldots$ and $\mathbf{w}_1, \mathbf{w}_2, \ldots$ as follows:

$$\text{Solve } A\mathbf{y}_1 = \mathbf{w}_0, \quad \text{find } \|\mathbf{y}_1\|, \quad \text{let } \mathbf{w}_1 = \frac{\mathbf{y}_1}{\|\mathbf{y}_1\|}$$

$$\text{Solve } A\mathbf{y}_2 = \mathbf{w}_1, \quad \text{find } \|\mathbf{y}_2\|, \quad \text{let } \mathbf{w}_2 = \frac{\mathbf{y}_2}{\|\mathbf{y}_2\|}$$

$$\vdots \qquad\qquad \vdots \qquad\qquad \vdots$$

Then the sequence of numbers

$$\|\mathbf{y}_1\|, \|\mathbf{y}_2\|, \ldots \quad \text{approaches} \quad \frac{1}{|\lambda_n|} \quad \text{so}$$

$$\frac{1}{\|\mathbf{y}_1\|}, \frac{1}{\|\mathbf{y}_2\|}, \ldots \quad \text{approaches} \quad |\lambda_n|$$

and the sequence of vectors

$$\mathbf{w}_1, \mathbf{w}_2, \ldots \quad \text{approaches} \quad \pm\mathbf{v}_n$$

Example 3 Let $A = \begin{bmatrix} 4 & -2 \\ 3 & -1 \end{bmatrix}$ be as in Example 1. Thus $\lambda_1 = 2$, $\lambda_2 = 1$, $\mathbf{v}_1 = \begin{bmatrix} \frac{1}{\sqrt{2}} & \frac{1}{\sqrt{2}} \end{bmatrix}^T \approx [0.707 \quad 0.707]^T$, and $\mathbf{v}_2 = \begin{bmatrix} \frac{2}{\sqrt{13}} & \frac{3}{\sqrt{13}} \end{bmatrix}^T \approx [0.555 \quad 0.832]^T$ are its associated eigenvalues and eigenvectors. Again let $\mathbf{w}_0 = [1 \quad 0]^T$ so that $\mathbf{w}_0 = 3\sqrt{2}\mathbf{v}_1 - \sqrt{13}\mathbf{v}_2$ and $a_2 \neq 0$. We now find the first several $\mathbf{y}_i$'s and $\mathbf{w}_i$'s to three significant figures.

$$\text{Solve } A\mathbf{y}_1 = \mathbf{w}_0, \quad \mathbf{y}_1 = \begin{bmatrix} -0.5 \\ -1.5 \end{bmatrix}, \quad \|\mathbf{y}_1\| \approx 1.58, \quad \mathbf{w}_1 \approx \begin{bmatrix} -0.316 \\ -0.949 \end{bmatrix}$$

$$\text{Solve } A\mathbf{y}_2 = \mathbf{w}_1, \quad \mathbf{y}_2 = \begin{bmatrix} -0.791 \\ -1.42 \end{bmatrix}, \quad \|\mathbf{y}_2\| \approx 1.63, \quad \mathbf{w}_2 \approx \begin{bmatrix} -0.486 \\ -0.872 \end{bmatrix}$$

$$\text{Solve } A\mathbf{y}_3 = \mathbf{w}_2, \quad \mathbf{y}_3 = \begin{bmatrix} -0.629 \\ -1.02 \end{bmatrix}, \quad \|\mathbf{y}_3\| \approx 1.19, \quad \mathbf{w}_3 \approx \begin{bmatrix} -0.527 \\ -0.854 \end{bmatrix}$$

$$\text{Solve } A\mathbf{y}_4 = \mathbf{w}_3, \quad \mathbf{y}_4 = \begin{bmatrix} -0.591 \\ -0.918 \end{bmatrix}, \quad \|\mathbf{y}_4\| \approx 1.09, \quad \mathbf{w}_4 \approx \begin{bmatrix} -0.541 \\ -0.841 \end{bmatrix}$$

Again you can see the convergence

$$\|\mathbf{y}_1\|, \|\mathbf{y}_2\|, \|\mathbf{y}_3\|, \|\mathbf{y}_4\|, \ldots \to \frac{1}{\lambda_2} = 1$$

and

$$\mathbf{w}_1, \mathbf{w}_2, \mathbf{w}_3, \mathbf{w}_4, \ldots \to -\mathbf{v}_2$$

And again you can see the convergence is slow. The rate of convergence is given by the ratio

$$\frac{1/\lambda_{n-1}}{1/\lambda_n} = \frac{\lambda_n}{\lambda_{n-1}} = \frac{1}{2}$$

Thus again we would pick up one significant figure about every three iterations. ■

Although, just as in direct iteration, inverse iteration is usually not very fast, the next modification allows us to find a variation that is quite fast. Furthermore we shall be able to use it to find any eigenvalue.

Shifted Inverse Iteration

Let a be any number. It is easy to see that if $A\mathbf{v} = \lambda\mathbf{v}$, then $(A - aI)\mathbf{v} = (\lambda - a)\mathbf{v}$. See Exercise 2. Therefore, if A has eigenvalues $\lambda_1, \ldots, \lambda_n$ with associated unit eigenvectors $\mathbf{v}_1, \ldots, \mathbf{v}_n$, then $A - aI$ has eigenvalues $\lambda_1 - a, \ldots, \lambda_n - a$ with the same eigenvectors, $\mathbf{v}_1, \ldots, \mathbf{v}_n$.

Suppose we wish to find λ_k fairly accurately and we know $a \approx \lambda_k$. Then most likely $|\lambda_k - a|$ is the smallest of all the $|\lambda_i - a|$'s. Therefore, inverse iteration with $A - aI$ will converge to $|\lambda_k - a|^{-1}$ and $\pm\mathbf{v}_k$. Furthermore, if a is a "good" approximation to λ_k so that $|\lambda_k - a|$ is small relative to $|\lambda_i - a|$, $i \neq k$, then the ratios $\dfrac{|\lambda_k - a|}{|\lambda_i - a|}$ will all be small. Consequently the convergence will be quite rapid. We now summarize this.

(5.88)

THEOREM. SHIFTED INVERSE ITERATION Let A be as in Assumption (5.83), let a be any number, and let λ_k be the eigenvalue of A closest to a. Then inverse iteration with $A - aI$ will converge to $|\lambda_k - a|^{-1}$ and $\pm\mathbf{v}_k$. The rate of convergence will be the largest of the ratios $\dfrac{|\lambda_k - a|}{|\lambda_i - a|}$, $i \neq k$.

Example 4 Suppose in Example 1 we had stopped direct iteration after computing $\|A\mathbf{w}_2\| \approx 2.26$ and $\mathbf{w}_3 = [0.721 \quad 0.690]^T$. Now let $a = 2.26$ and proceed using inverse iteration with $A - aI$. The process will then converge to $|\lambda_2 - a|^{-1} = 0.26^{-1} \approx 3.846$ and $\mathbf{v}_1 \approx [0.7071 \quad 0.7071]^T$. The rate of convergence is $\dfrac{|2 - a|}{|1 - a|} = \dfrac{0.26}{1.26} \approx 0.2$. Since $(0.2)^3 = 0.008 \approx 0.01$, this means every three iterations we should pick up better than two significant digits. To see this happen,

$$A - aI = \begin{bmatrix} 4 & -2 \\ 3 & -1 \end{bmatrix} - \begin{bmatrix} 2.26 & \\ & 2.26 \end{bmatrix} = \begin{bmatrix} 1.74 & -2 \\ 3 & -3.26 \end{bmatrix}$$

Then

$$(A - aI)^{-1} \approx \frac{1}{0.3276}\begin{bmatrix} -3.26 & 2 \\ -3 & 1.74 \end{bmatrix} \approx \begin{bmatrix} -9.951 & 6.105 \\ -9.158 & 5.311 \end{bmatrix}$$

We let $\mathbf{w}_0 = [0.721 \quad 0.690]^T$ and proceed as in Example 3.

$$\text{Solve } (A - aI)\mathbf{y}_1 = \mathbf{w}_0, \quad \mathbf{y}_1 \approx \begin{bmatrix} -2.962 \\ -2.938 \end{bmatrix}, \quad \|\mathbf{y}_1\| \approx 4.172, \quad \mathbf{w}_1 \approx \begin{bmatrix} -0.7100 \\ -0.7042 \end{bmatrix}$$

$$\text{Solve } (A - aI)\mathbf{y}_2 = \mathbf{w}_1, \quad \mathbf{y}_2 \approx \begin{bmatrix} 2.766 \\ 2.762 \end{bmatrix}, \quad \|\mathbf{y}_2\| \approx 3.909, \quad \mathbf{w}_2 \approx \begin{bmatrix} 0.7076 \\ 0.7066 \end{bmatrix}$$

$$\text{Solve } (A - aI)\mathbf{y}_3 = \mathbf{w}_2, \quad \mathbf{y}_3 \approx \begin{bmatrix} -2.7275 \\ -2.7274 \end{bmatrix}, \quad \|\mathbf{y}_3\| \approx 3.857, \quad \mathbf{w}_3 \approx \begin{bmatrix} -0.70716 \\ -0.70713 \end{bmatrix}$$

You can see how quickly this is converging as compared with Example 1. However, we could have done better! If we had let $a = \|A\mathbf{w}_3\| = 2.10$, then the rate of convergence would have been determined by $\dfrac{|2-a|}{|1-a|} = \dfrac{0.1}{1.1} \approx 0.09$. In this case each iteration improves the results by more than a significant figure! ■

But what about the next question? How do we obtain estimates of the eigenvalues, especially of the ones between the smallest and largest? One answer is provided by the final method we shall discuss.

The *QR* Algorithm

The *QR* algorithm is deceptively easy to describe, and it is almost magical the way it works.

(5.89)

THEOREM (THE (UNSHIFTED) QR ALGORITHM) Let $A = A_0$ be almost any nonsingular square matrix. Form the following sequence, where at each step $A_k = Q_k R_k$ is the QR decomposition of orthogonal times upper triangular described in Section 4.5.

$$\begin{array}{ll} \text{Factor } A_0 = Q_0 R_0 & \text{Let } A_1 = R_0 Q_0 \\ \text{Factor } A_1 = Q_1 R_1 & \text{Let } A_2 = R_1 Q_1 \\ \text{Factor } A_2 = Q_2 R_2 & \text{Let } A_3 = R_2 Q_2 \\ \vdots & \vdots \end{array}$$

Then

(a) Each matrix A_k has the same eigenvalues as A_{k-1} (since $A_k = R_{k-1} Q_{k-1} = Q_{k-1}^T Q_{k-1} R_{k-1} Q_{k-1} = Q_{k-1}^T A_{k-1} Q_{k-1}$; see Exercise 1). Similarly, A_{k-1} has the same eigenvalues as A_{k-2}, and so forth. Putting these all together, it follows that A_k has the same eigenvalues as A_0.

(b) The sequence of matrices $A_0, A_1, A_2, \ldots$ tends toward an upper triangular matrix with the eigenvalues of A down the diagonal in descending order.

Example 5 Let $A = \begin{bmatrix} 4 & -2 \\ 3 & -1 \end{bmatrix}$ be as in Example 1. We apply the *QR* algorithm as described in Theorem (5.89).

$$A_0 = \begin{bmatrix} 4 & -2 \\ 3 & -1 \end{bmatrix} = \begin{bmatrix} \frac{4}{5} & -\frac{3}{5} \\ \frac{3}{5} & \frac{4}{5} \end{bmatrix} \begin{bmatrix} 5 & -\frac{11}{5} \\ & \frac{2}{5} \end{bmatrix} = Q_0 R_0$$

$$R_0 Q_0 = \begin{bmatrix} 5 & -\frac{11}{5} \\ & \frac{2}{5} \end{bmatrix} \begin{bmatrix} \frac{4}{5} & -\frac{3}{5} \\ \frac{3}{5} & \frac{4}{5} \end{bmatrix} = \begin{bmatrix} \frac{67}{25} & -\frac{119}{25} \\ \frac{6}{25} & \frac{8}{25} \end{bmatrix} = A_1$$

$$A_1 = \begin{bmatrix} 2.68 & -4.76 \\ 0.24 & 0.32 \end{bmatrix} \approx \begin{bmatrix} 0.9960 & -0.0892 \\ 0.0892 & 0.9960 \end{bmatrix} \begin{bmatrix} 2.691 & -4.712 \\ 0 & 0.7433 \end{bmatrix} = Q_1 R_1$$

$$R_1 Q_1 = \begin{bmatrix} 2.691 & -4.712 \\ & 0.7433 \end{bmatrix} \begin{bmatrix} 0.9960 & -0.0892 \\ 0.0892 & 0.9960 \end{bmatrix} \approx \begin{bmatrix} 2.260 & -4.934 \\ 0.0663 & 0.7403 \end{bmatrix} = A_2$$

$$A_2 \approx \begin{bmatrix} 0.99957 & -0.02933 \\ 0.02933 & 0.99957 \end{bmatrix} \begin{bmatrix} 2.2606 & -4.9099 \\ & 0.88470 \end{bmatrix} = Q_2 R_2$$

$$R_2 Q_2 = \begin{bmatrix} 2.2606 & -4.9099 \\ & 0.88470 \end{bmatrix} \begin{bmatrix} 0.99957 & -0.02933 \\ 0.02933 & 0.99957 \end{bmatrix} \approx \begin{bmatrix} 2.1157 & -4.9741 \\ 0.02595 & 0.88432 \end{bmatrix} = A_3$$

$$A_3 \approx \begin{bmatrix} 0.99992 & -0.01226 \\ 0.01226 & 0.99992 \end{bmatrix} \begin{bmatrix} 2.1158 & -4.9628 \\ & 0.94525 \end{bmatrix} = Q_3 R_3$$

$$R_3 Q_3 = \begin{bmatrix} 2.1158 & -4.9628 \\ & 0.94525 \end{bmatrix} \begin{bmatrix} 0.99992 & -0.01226 \\ 0.01226 & 0.99992 \end{bmatrix} \approx \begin{bmatrix} 2.0548 & -4.9884 \\ 0.01159 & 0.94518 \end{bmatrix} = A_4$$

■

You can see that the sequence of A_i's is converging as described in Theorem (5.89), but that this convergence is slow. For some applications this is sufficient, but for others we wish the process speeded up. Here is a brief idea of how to speed the algorithm up.

The Shifted QR Algorithm

1. "Preprocess" the matrix, as described in the first subsection of this section, obtaining a matrix with zeros below the first subdiagonal,

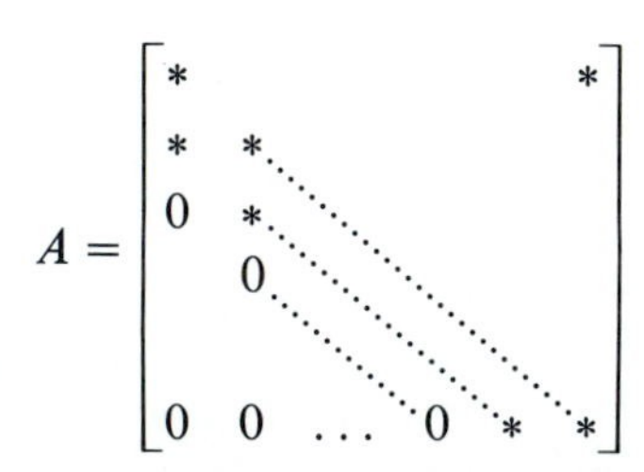

2. Theory will show that two or three iterations of the unshifted QR algorithm will bring a_{nn} moderately close to the smallest eigenvalue λ_n.
3. We now turn to the heart of the *shifted* QR algorithm. This is factoring $A_k - aI = Q_k R_k$ and then defining $A_{k+1} = R_k Q_k + aI$. The matrix A_{k+1}

has the same eigenvalues as A_k (see Exercise 23) and if $a = a_{nn}$ is close to λ_n, then this sequence converges very quickly to a matrix that looks like

$$\left[\begin{array}{cccc|c} * & * & * & * & * \\ * & * & * & * & * \\ 0 & * & * & * & * \\ 0 & 0 & * & * & * \\ \hline 0 & 0 & 0 & \varepsilon & a_{nn} \end{array}\right]$$

where this new a_{nn} is very close to λ_n and ε is very close to zero. We then say $\lambda_n \approx a_{nn}$ and repeat the process with the smaller matrix (the 4×4 matrix in the above illustration).

Steps 1–3 essentially describe the complete QR algorithm as it is usually used in practice to compute all the eigenvalues of a given matrix. If the eigenvectors are also needed, a single-step shifted inverse iteration with each $\lambda_i = a$ yields each v_i in turn.

This discussion is intended to give you just a basic background in case you should run into such problems further on. For more information see Wilkenson, *The Algebraic Eigenvalue Problem.*

OPTIONAL ***Proof of Theorems (5.84) and (5.85)*** Let $\mathbf{w}_0 = a_1\mathbf{v}_1 + \cdots + a_n\mathbf{v}_n$ be any vector in $\mathbb{R}^n$ such that $a_1 \neq 0$. Form the sequence

$$\begin{array}{lll} A\mathbf{w}_0, & \|A\mathbf{w}_0\|, & \mathbf{w}_1 = \dfrac{A\mathbf{w}_0}{\|A\mathbf{w}_0\|} \\ A\mathbf{w}_1, & \|A\mathbf{w}_1\|, & \mathbf{w}_2 = \dfrac{A\mathbf{w}_1}{\|A\mathbf{w}_1\|} \\ \vdots & \vdots & \vdots \end{array}$$

We wish to show the sequence of vectors $\mathbf{w}_1, \mathbf{w}_2, \ldots$ approaches $\pm\mathbf{v}_1$ and the sequence of numbers $\|A\mathbf{w}_0\|, \|A\mathbf{w}_1\|, \ldots$ approaches $|\lambda_1|$.

First consider the sequence of vectors $\mathbf{x}_0 = \mathbf{w}_0$, $\mathbf{x}_1 = A\mathbf{x}_0$, $\mathbf{x}_2 = A\mathbf{x}_1, \ldots$. Then

$$\begin{aligned} \mathbf{x}_0 &= a_1\mathbf{v}_1 + \cdots + a_1\mathbf{v}_n \\ \mathbf{x}_1 &= A\mathbf{x}_0 = a_1A\mathbf{v}_1 + \cdots + a_1A\mathbf{v}_n = a_1\lambda_1\mathbf{v}_1 + \cdots + a_1\lambda_n\mathbf{v}_n \\ \mathbf{x}_2 &= A\mathbf{x}_1 = a_1\lambda_1A\mathbf{v}_1 + \cdots + a_n\lambda_nA\mathbf{v}_n = a_1\lambda_1^2\mathbf{v}_1 + \cdots + a_n\lambda_n^2\mathbf{v}_n \\ \mathbf{x}_3 &= A\mathbf{x}_2 = a_1\lambda_1^2A\mathbf{v}_1 + \cdots + a_n\lambda_n^2A\mathbf{v}_n = a_1\lambda_1^3\mathbf{v}_1 + \cdots + a_n\lambda_n^3\mathbf{v}_n \\ &\vdots \end{aligned}$$

and in general we obtain

$$\mathbf{x}_k = a_1\lambda_1^k\mathbf{v}_1 + \cdots + a_n\lambda_n^k\mathbf{v}_n$$

We now compute $\mathbf{w}_1, \mathbf{w}_2, \ldots$

$$\mathbf{w}_1 = \frac{A\mathbf{w}_0}{\|A\mathbf{w}_0\|} = \frac{\mathbf{x}_1}{\|\mathbf{x}_1\|}$$

$$\mathbf{w}_2 = \frac{A\mathbf{w}_1}{\|A\mathbf{w}_1\|} = \frac{A\left(\dfrac{\mathbf{x}_1}{\|\mathbf{x}_1\|}\right)}{\left\|A\left(\dfrac{\mathbf{x}_1}{\|\mathbf{x}_1\|}\right)\right\|}$$

$$= \frac{\dfrac{1}{\|\mathbf{x}_1\|} A\mathbf{x}_1}{\dfrac{1}{\|\mathbf{x}_1\|} \|A\mathbf{x}_1\|} = \frac{A\mathbf{x}_1}{\|A\mathbf{x}_1\|}$$

$$\mathbf{w}_3 = \frac{A\mathbf{w}_2}{\|A\mathbf{w}_2\|} = \frac{A\left(\dfrac{\mathbf{x}_2}{\|\mathbf{x}_2\|}\right)}{\left\|A\left(\dfrac{\mathbf{x}_2}{\|\mathbf{x}_2\|}\right)\right\|}$$

$$= \frac{\dfrac{1}{\|\mathbf{x}_2\|} A\mathbf{x}_2}{\dfrac{1}{\|\mathbf{x}_2\|} \|A\mathbf{x}_2\|} = \frac{A\mathbf{x}_2}{\|A\mathbf{x}_2\|}$$

and in general we obtain

(5.90)
$$\mathbf{w}_k = \frac{A\mathbf{x}_k}{\|A\mathbf{x}_k\|} = \frac{a_1\lambda_1^k\mathbf{v}_1 + a_2\lambda_2^k\mathbf{v}_2 + \cdots + a_n\lambda_n^k\mathbf{v}_n}{\|a_1\lambda_1^k\mathbf{v}_1 + a_2\lambda_2^k\mathbf{v}_2 + \cdots + a_n\lambda_n^k\mathbf{v}_n\|}$$
$$= \frac{\lambda_1^k(a_1\mathbf{v}_1 + a_2(\lambda_2/\lambda_1)^k\mathbf{v}_2 + \cdots + a_n(\lambda_n/\lambda_1)^k\mathbf{v}_n)}{|\lambda_1^k|\,\|a_1\mathbf{v}_1 + a_2(\lambda_2/\lambda_1)^k\mathbf{v}_2 + \cdots + a_n(\lambda_n/\lambda_1)^k\mathbf{v}_n\|}$$

Now $\dfrac{\lambda_1^k}{|\lambda_1^k|} = \pm 1$. Also, by Assumption (5.83), $|\lambda_1| > |\lambda_i|$, $i > 1$, so $\left|\dfrac{\lambda_i}{\lambda_1}\right| < 1$ for $i > 1$. Therefore, as $k \to \infty$, $\left|\dfrac{\lambda_i}{\lambda_1}\right|^k \to 0$ for $i > 1$. Also,

(5.91)
$$\frac{a_1\mathbf{v}_1}{\|a_1\mathbf{v}_1\|} = \frac{a_1}{|a_1|}\frac{\mathbf{v}_1}{\|\mathbf{v}_1\|} = \pm\mathbf{v}_1$$

since $\|\mathbf{v}_1\| = 1$ and $a_1 \neq 0$. Altogether, as $k \to \infty$, we have

(5.92)
$$\mathbf{w}_k \to \pm\frac{a_1\mathbf{v}_1}{\|a_1\mathbf{v}_1\|} = \pm\mathbf{v}_1$$

Now that we know $\mathbf{w}_k \to \pm\mathbf{v}_1$, we also know

$$\|A\mathbf{w}_k\| \to \|\pm A\mathbf{v}_1\| = \|\pm\lambda_1\mathbf{v}_1\| = |\lambda_1|\,\|\pm\mathbf{v}_1\| = |\lambda_1|$$

By Equation (5.90), the rate of convergence is determined by how quickly the various $\left(\frac{\lambda_i}{\lambda_1}\right)^k \to 0$. Since $|\lambda_2| > |\lambda_i|$, for $i > 2$, $\left(\frac{\lambda_2}{\lambda_1}\right)^k \to 0$ converges the slowest of all. This is why the fraction λ_2/λ_1 determines the rate of convergence. The actual conclusion of Theorem (5.85) follows from some technical algebra, which we shall not do. ■

Exercise 5.8

1. Let A be an $n \times n$ matrix, let S be any invertible $n \times n$ matrix, and let $B = S^{-1}AS$. Show that A and B have the same eigenvalues. (HINT If $A\mathbf{v} = \lambda\mathbf{v}$, show that $B\mathbf{w} = \lambda\mathbf{w}$ where $\mathbf{w} = S^{-1}\mathbf{v}$.)

2. Let A be an $n \times n$ matrix, let a be any number, and let $B = A - aI$. Show that if $A\mathbf{v} = \lambda\mathbf{v}$, then $B\mathbf{v} = (\lambda - a)\mathbf{v}$. Thus A and B have the same eigenvectors and shifted eigenvalues. (HINT Compute $B\mathbf{v}$ directly and see what happens.)

In Exercises 3–8, starting with $x_0 = [1 \quad 1]^T$,

(a) Proceed through several steps of direct iteration.
(b) Proceed through several steps of inverse iteration.
(c) Solve the eigenvalue-eigenvector problem by hand and compare yours answers.

NOTE Everyone should work through at least one problem by hand, perhaps with a hand-held calculator. After that it might be reasonable to use a computer, if one is available (provided you write your own program).

3. $\begin{bmatrix} 5 & -8 \\ 1 & -1 \end{bmatrix}$
4. $\begin{bmatrix} -1 & -3 \\ -3 & 7 \end{bmatrix}$
5. $\begin{bmatrix} 1 & 3 \\ 4 & -3 \end{bmatrix}$
6. $\begin{bmatrix} 5 & 6 \\ 2 & 1 \end{bmatrix}$
7. $\begin{bmatrix} 8 & -8 \\ 1 & 2 \end{bmatrix}$
8. $\begin{bmatrix} -3 & -3 \\ -3 & 5 \end{bmatrix}$

In Exercises 9–14, assume a 4×4 matrix has the four given numbers as eigenvalues. What would be the (approximate) rate of convergence of direct iteration? Of inverse iteration? (Give your answer in the form of how many iterations it takes to improve by one significant figure, when close.)

9. 10, 6, 5, 1
10. $-0.1, -2, -10, -20$
11. $20, 5, -1, -2$
12. $50, 1, -0.001, -5$
13. $100, 0.0001, -0.01, -1$
14. $8, 2, -1, -6$

In Exercises 15–18, assume a 4×4 matrix has the given numbers as eigenvalues. Assume shifted inverse iteration is used with each of the given values of a. To what eigenvalue of the original matrix would it converge and what would be its rate of convergence?

15. Same as Exercise 9: $a = 10.1, 2$

16. Same as Exercise 10: $a = -10.1, -1$

17. Same as Exercise 11: $a = 4.9, -1.9$

18. Same as Exercise 12: $a = 49, -1$

In Exercises 19–22, proceed through several steps of the QR algorithm. The note that accompanies the instructions for Exercises 3–8 applies here as well.

19. $\begin{bmatrix} 3 & -8 \\ 1 & -3 \end{bmatrix}$

20. $\begin{bmatrix} 0 & -3 \\ -3 & 8 \end{bmatrix}$

21. $\begin{bmatrix} 3 & 3 \\ 4 & -1 \end{bmatrix}$

22. $\begin{bmatrix} 3 & 6 \\ 2 & -1 \end{bmatrix}$

23. Suppose $A - aI = QR$ and $B = RQ + aI$. Show that A and B have the same eigenvalues. (HINT Show $QBQ^{-1} = A$ and apply Exercise 1.)

In Exercises 24–26, we show that for an $n \times n$ matrix A there are orthogonal matrices $Q_1, \ldots, Q_{n-1}$ such that $Q_{n-1} \cdots Q_1 A Q_1^T \cdots Q_{n-1}^T$ is in upper-Hessenberg form. For this we shall need Theorem (4.75), Section 4.6, and the material leading up to it.

24. Let $A = [a_{ij}]$, let $\mathbf{v} = [a_{21} \quad \cdots \quad a_{n1}]^T$ and assume $\mathbf{v} \neq [a_{21} \quad 0 \quad \cdots \quad 0]^T$. (If it is, let $Q_1 = I$.) Let Q be the resulting $(n-1) \times (n-1)$ orthogonal matrix given in Theorem (4.75), Section 4.6. Let $Q_1 = \left[\begin{array}{c|c} 1 & \\ \hline & Q \end{array}\right]$ and let $A_1 = Q_1 A Q_1^T$. Show that Q_1 is orthogonal and that A_1 has the form

$$A_1 = \begin{bmatrix} a'_{11} & a'_{12} & \cdots & a'_{1n} \\ a'_{21} & a'_{22} & \cdots & a'_{2n} \\ 0 & a'_{23} & \cdots & a'_{3n} \\ \vdots & \vdots & & \vdots \\ 0 & a'_{n3} & \cdots & a'_{nn} \end{bmatrix}$$

25. From the A_1 obtained in Exercise 24, let $\mathbf{v} = [a'_{23} \quad \cdots \quad a'_{n3}]^T$ and assume $\mathbf{v} \neq [a'_{23} \quad 0 \quad \cdots \quad 0]$. Let Q be the resulting $(n-2) \times (n-2)$ orthogonal matrix given in Theorem (4.75), Section 4.6. Let $Q_2 = \left[\begin{array}{cc|c} 1 & & \\ & 1 & \\ \hline & & Q \end{array}\right]$ and let

$A_2 = Q_2 A_1 Q_2^T$. Show that Q_2 is orthogonal and that A_2 has the form

$$A_2'' = \begin{bmatrix} a_{11}'' & a_{12}'' & a_{13}'' & \cdots & a_{1n}'' \\ a_{21}'' & a_{22}'' & a_{23}'' & \cdots & a_{2n}'' \\ 0 & a_{32}'' & a_{33}'' & \cdots & a_{3n}'' \\ 0 & 0 & a_{43}'' & \cdots & a_{4n}'' \\ \vdots & \vdots & \vdots & & \vdots \\ 0 & 0 & a_{n3}'' & \cdots & a_{nn}'' \end{bmatrix}$$

26. Describe the next step in the process described in Exercises 24 and 25 and explain why the whole process takes $n - 1$ steps, that is, why A_{n-1} is in upper-Hessenberg form.
27. Show that if A is symmetric, each A_k is also symmetric and hence A_{n-1} is in fact tridiagonal and symmetric.

Review Exercises

In Exercises 1–4, find the determinants of the given matrix.

1. $\begin{bmatrix} 2 & 3 \\ -3 & -4 \end{bmatrix}$
2. $\begin{bmatrix} 4 & 3 & -1 \\ -1 & 2 & -1 \\ 4 & 1 & 3 \end{bmatrix}$
3. $\begin{bmatrix} 3 - x & 4 \\ -2 & 1 - x \end{bmatrix}$
4. $\begin{bmatrix} x & x^3 \\ x^2 & x^4 \end{bmatrix}$
5. Choose the second row of the *companion matrix* $A = \begin{bmatrix} 0 & 1 \\ \cdot & \cdot \end{bmatrix}$ so that its characteristic polynomial is $\lambda^2 + 2\lambda - 3$.
6. Choose the third row of the companion matrix $A = \begin{bmatrix} 0 & 1 & 0 \\ 0 & 0 & 1 \\ \cdot & \cdot & \cdot \end{bmatrix}$ so that its characteristic polynomial is $\lambda^3 - 3\lambda^2 + 2\lambda - 5$.

In Exercises 7 and 8, diagonalize the given matrix.

7. $\begin{bmatrix} 1 & 4 \\ 3 & 5 \end{bmatrix}$
8. $\begin{bmatrix} 1 & 2 & 3 \\ & 4 & 5 \\ & & 6 \end{bmatrix}$
9. Show that if $b \neq 0$, the matrix $\begin{bmatrix} a & b \\ & a \end{bmatrix}$ is not diagonalizable.

10. Find a real matrix B such that $B^2 = A$ if

$$A = \begin{bmatrix} 1 & 3 & 2 \\ & 4 & 5 \\ & & 9 \end{bmatrix}$$

How many such matrices B are there?

11. Suppose A is a 2×2 matrix with eigenvalues 0 and 1 and corresponding eigenvectors $[1 \quad 3]^T$ and $[3 \quad -1]^T$.

(a) Why is A symmetric?
(b) What is the determinant of A?
(c) How many such A are there? Find one.

12. Let A be the matrix of Exercise 11.

(a) What are the eigenvalues and eigenvectors of A^2?
(b) What is the relationship between A and A^2?

In Exercises 11 and 12, prove or find a counterexample.

If B is obtained from A by interchanging two rows, then B is similar to A.

If A and B are diagonalizable $n \times n$ matrices, then $A + B$ is diagonalizable.

In Exercises 13 and 14, prove or find a counterexample.

13. If B is obtained from A by interchanging two rows, then B is similar to A.

14. If A and B are diagonalizable $n \times n$ matrices, then $A + B$ is diagonalizable.

15. Suppose A is an $n \times n$ matrix and $A^2 = I$.

(a) What are the possible eigenvalues of A?
(b) If $n = 2$ and $A \neq I, -I$, find the trace and determinant of A.
(c) If, in addition to (b), the first row of A is $[3 \quad -1]$, find the second row of A.

16. Suppose A has eigenvalues 2, -3, and 4.

(a) Find $\det(A^2)$ and $\operatorname{tr}(A^2)$.
(b) Find $\det[(A^T)^{-1}]$.

17. Suppose A is a 3×3 matrix and has eigenvalues 2, 2, and -3. Which of the following statements are true and which may or may not be true? Give a reason or counterexample for each answer.

(a) A is invertible.
(b) A is symmetric.
(c) A is diagonalizable.

18. Suppose A is a 3×3 matrix and the only eigenvectors of A are multiples of $\boldsymbol{\varepsilon}_1 = [1 \quad 0 \quad 0]^T$. Are the following statements true or false?

    (a) A is invertible.
    (b) A is diagonalizable.
    (c) $\det(\lambda I - A) = (\lambda - a)^3$

19. Let $\{F_k\}$ be the Fibonacci sequence with $F_0 = 0$, $F_1 = 1$, and $F_{k+1} = F_k + F_{k-1}$. Suppose we allow negative subscripts. Find F_{-1}, F_{-2} and determine how F_{-k} is related to F_k.

20. Suppose there are three major brands of beer, brand A with 40% of the market and brands B and C with 30% of the market each. Suppose market research over the past years indicates that regular beer drinkers tend to switch brands in the following way.

    Of those who drink brand A, 25% switch to brand B, 25% switch to brand C, and the remainder stay. Of those who drink brands B and C, 50% switch to brand A and the remainder stay with what they have.

    (a) Find the transition matrix for this relationship.
    (b) Toward what distribution will the regular beer drinkers tend (if there is no major change in advertizing)?

21. Solve the initial value problem:

$$\frac{du}{dt} = u + v \qquad u = 1 \quad \text{when} \quad t = 0$$
$$\frac{dv}{dt} = 4u - 2v \qquad v = 6 \quad \text{when} \quad t = 0$$

22. Diagonalize the quadratic form $q(x, y) = x^2 + 2xy - 2y^2$ and graph the equation $q(x, y) = 1$.

23. For the matrix $A = \begin{bmatrix} 3 & 1 \\ 4 & 3 \end{bmatrix}$, start with the vector $x_0 = \begin{bmatrix} 1 \\ 0 \end{bmatrix}$ and proceed through three steps of (a) direct iteration and then (b) inverse iteration. Round calculations off to three significant figures. Then diagonalize A and compare the results.

Cumulative Review Exercises

NOTE Answers to *all* Cumulative Review Exercises appear in the *Instructor's Manual.* These problems are not answered in the back of the textbook.

1. Let A be a square matrix. Then A is invertible $\Leftrightarrow$ (make 15 equivalent statements, including at least two involving material from this chapter).

2. (a) Show that if B is invertible, then AB is similar to BA.
   (b) Show that A is never similar to $A + I$. (HINT Look at the eigenvalues.)

(c) Find a diagonal matrix Q with only ± 1's on the diagonal to show that

$$\begin{bmatrix} a & 1 & & \\ 1 & b & 1 & \\ & 1 & c & 1 \\ & & 1 & d \end{bmatrix} \text{ is similar to } \begin{bmatrix} a & -1 & & \\ -1 & b & -1 & \\ & -1 & c & -1 \\ & & -1 & d \end{bmatrix}$$

3. (a) Rewrite $2x^2 - 2\sqrt{2}xy + 3y^2 = 1$ as $\mathbf{x}^T A\mathbf{x} = 1$. Then diagonalize A and graph $\mathbf{x}^T A\mathbf{x} = 1$.
   (b) For the A in (a), write $A = LU$ and then $A = LDL^T$. Compare this D with the diagonal matrix you found in (a).

4. Note that $M = \begin{bmatrix} 0.4 & 0.3 \\ 0.6 & 0.7 \end{bmatrix}$ is an invertible Markov matrix.
   (a) Find the limit M_∞ of M^k as $k \to \infty$.
   (b) Are M^k, $k \geq 1$, and M_∞ all invertible?
   (c) For any $\begin{bmatrix} a \\ b \end{bmatrix}$ in $\mathbb{R}^2$, find the limit of $M^k\begin{bmatrix} a \\ b \end{bmatrix}$ as $k \to \infty$.
   (d) If $\mathbf{x}_k$ solves $M^k\mathbf{x}_k = \mathbf{b}$, find the limit of $\mathbf{x}_k$ as $k \to \infty$ if:
   (1) $\mathbf{b} = \begin{bmatrix} 1 \\ 2 \end{bmatrix}$, (2) $\mathbf{b} = \begin{bmatrix} 1 \\ 1 \end{bmatrix}$

5. Let $A = \begin{bmatrix} 1 & 2 & -3 & 2 \\ 0 & 3 & -2 & -5 \\ 0 & 0 & 4 & -3 \end{bmatrix} = [\mathbf{c}_1 \quad \mathbf{c}_2 \quad \mathbf{c}_3 \quad \mathbf{c}_4]$.
   (a) Find a nonzero $\mathbf{x}$ in $NS(A)$.
   (b) Use $\mathbf{x}$ to show explicitly that the columns of A are linearly dependent.
   (c) Give a simple reason why the rows of A are linearly independent. Then find $rk(A)$.
   (d) Let B be the 4×4 matrix obtained from A by adding a fourth row $\mathbf{r}$ perpendicular to $\mathbf{x}$.
   (1) Explain why $\mathbf{x}$ is an eigenvector of B. What is its corresponding eigenvalue?
   (2) Explain why $\mathbf{r}$ must be a linear combination of the rows of A. [HINT What is the rank of B? Hence, what is the relationship between $RS(A)$ and $RS(B)$?]

6. Consider each of the three types of 2×2 elementary matrices:

$$E_1 = \begin{bmatrix} m & 0 \\ 0 & 1 \end{bmatrix}, \quad E_2 = \begin{bmatrix} 0 & 1 \\ 1 & 0 \end{bmatrix}, \quad E_3 = \begin{bmatrix} 1 & 0 \\ m & 1 \end{bmatrix}, \quad m \neq 0$$

What are the eigenvalues of each matrix? Which matrices, if any, are diagonalizable?

7. Let $A = \begin{bmatrix} 1 & a & 0 \\ a & 1 & a \\ 0 & a & 1 \end{bmatrix}$.

(a) For what values of a is $\det(A) < 0$?

(b) For what values of a is A singular, and what is the row-echelon form in these cases?

Chapter 6

FURTHER DIRECTIONS

The elementary foundations of linear algebra have now been laid. From here linear algebra expands in many directions. In addition to specializing in solving purely linear algebraic problems, it intertwines with all the other branches of mathematics.

The purpose of this chapter is to give a taste of some of the diversity. Section 6.1 introduces the linear algebra in function theory, the broad interaction between linear algebra and analysis, and some of the applications. Section 6.2 examines the singular value decomposition and generalized inverses, an area of pure linear algebra growing out of attempts to deal with numerical difficulties. Section 6.3 shows how the equation $A\mathbf{x} = \mathbf{b}$ can be solved by iterative methods, an approach often needed when A is very large and sparse. Section 6.4 introduces matrix norms and provides the theory needed for Section 6.3, for defining and understanding condition numbers, and for taking the first step into Banach spaces. Section 6.5 provides an introduction to abstract linear algebra over an arbitrary field. This is the proper context for abstract matrix theory, for the interaction between linear algebra and abstract algebra, and for understanding many applications such as coding theory.

These five sections are independent of one another and may be presented individually.

6.1 Function Spaces

The purpose of this section is to bring together and extend some of the facts already discussed about the most "famous" infinite-dimensional vector space, $C[a, b]$. This space has many applications, and we shall see how it is applied to several topics we have already discussed. In particular we shall:

1. Discuss Fourier series in terms of orthonormal bases.
2. Apply the Gram–Schmidt orthogonalization to the polynomials 1, $x, x^2, \ldots$.
3. Find the least-squares approximation of a function f by a straight line.

These three topics are just the beginning of a vast array of theory and applications that revolve around $C[a, b]$.

Review

Recall from Examples 9 and 10 in Section 3.3 that $C[a, b]$ is the set of all continuous real-valued functions defined on the interval $[a, b]$. It not only is a vector space, it is also an inner product space, with $f \cdot g$ defined by

(6.1) $$f \cdot g = \int_a^b f(x)g(x)\,dx$$

From this the norm of a function f is given by

(6.2) $$\|f\| = (f \cdot f)^{1/2} = \left(\int_a^b f(x)^2\,dx\right)^{1/2}$$

Fourier Series

For this subsection we restrict our attention to $[0, 2\pi]$ and discuss how to express certain functions f defined on this interval as

(6.3) $$f(x) = a_0 + a_1 \cos x + b_1 \sin x + a_2 \cos 2x + b_2 \sin 2x + \cdots$$

This is called the **Fourier series** of f.

For certain technical reasons, which require some sophisticated analysis to explain adequately, we must require more of such f's than continuity. We shall require f to be in $C^1[0, 2\pi]$, the set of all functions with continuous first derivatives on $[0, 2\pi]$.

First let

(6.4) $$g_0(x) = 1, \quad g_n(x) = \cos nx, \quad h_n(x) = \sin nx, \quad n \geq 1, \quad n \text{ a positive integer}$$

We have the following interesting fact.

(6.5) THEOREM The g's and h's are orthogonal, in the sense that

$$g_n \cdot g_m = 0, \quad h_n \cdot h_m = 0 \qquad n \neq m$$
$$g_n \cdot h_m = 0 \qquad \text{all } m, n$$

Proof We shall show $g_n \cdot g_m = 0$ for $n \neq m$ and leave the remainder to the exercises. (See Exercises 1 and 2.) We shall use the trigonometric identity

$$\cos u \cos v = \tfrac{1}{2}[\cos(u+v) + \cos(u-v)] \tag{6.6}$$

For $n, m \geq 0$, $n \neq m$,

$$\begin{aligned} g_n \cdot g_m &= \int_0^{2\pi} \cos nx \cos mx \, dx \\ &= \int_0^{2\pi} \frac{1}{2}[\cos(n+m)x + \cos(n-m)x] \, dx \qquad \text{by (6.6)} \\ &= \frac{1}{2}\int_0^{2\pi} \cos(n+m)x \, dx + \frac{1}{2}\int_0^{2\pi} \cos(n-m)x \, dx \\ &= \frac{1}{2}\left(\frac{\sin(n+m)x}{n+m}\right)_0^{2\pi} + \frac{1}{2}\left(\frac{\sin(n-m)x}{n-m}\right)_0^{2\pi} \\ &= \frac{1}{2}\left(\frac{0-0}{n+m}\right) + \frac{1}{2}\left(\frac{0-0}{n-m}\right) = 0 \end{aligned}$$

since $\sin 2\pi k = 0$ for k any integer. ■

Since the g's and h's are orthogonal and nonzero, they are linearly independent. [Theorem (4.46), Section 4.4, says this for a finite number of vectors, and we can extend this proof to infinitely many vectors.] If $\mathbf{v}_1, \ldots, \mathbf{v}_n$ is a finite number of orthogonal nonzero vectors and $\mathbf{y}$ is in their span, then

$$\mathbf{y} = c_1\mathbf{v}_1 + \cdots + c_n\mathbf{v}_n \qquad \text{where} \qquad c_i = \frac{\mathbf{y} \cdot \mathbf{v}_i}{\mathbf{v}_i \cdot \mathbf{v}_i} \tag{6.7}$$

This is exactly what we proved in Theorem (4.49), Section 4.4. What we would like to know is that this also holds in this case, where there are infinitely many $\mathbf{v}_i$'s and we are taking an infinite sum. Infinite sums are **series**, and they do not always converge (or add up). This is where analysis comes in and interacts with linear algebra. However, there are very powerful theorems that say that the series discussed here do converge, and hence all this makes sense and works.

Thus using the identities

$$\cos^2 u = \tfrac{1}{2}(1 + \cos 2u) \qquad \text{and} \qquad \sin^2 u = \tfrac{1}{2}(1 - \cos 2u) \tag{6.8}$$

we compute

$$\begin{aligned} g_0 \cdot g_0 &= \int_0^{2\pi} 1 \cdot 1 \, dx = 2\pi \\ g_n \cdot g_n &= \int_0^{2\pi} \cos^2 nx \, dx = \int_0^{2\pi} \tfrac{1}{2}[1 + \cos 2nx] \, dx \\ &= \pi \qquad \text{if} \qquad n > 0 \\ h_n \cdot h_n &= \pi \qquad \text{if} \qquad n > 0\text{, by a similar argument.} \end{aligned}$$

Then keeping in mind Equation (6.7) as an analogy,

(6.9)

> THEOREM Let $f \in C^1[0, 2\pi]$,
>
> $$a_0 = \frac{f \cdot g_0}{g_0 \cdot g_0} = \frac{1}{2\pi} \int_0^{2\pi} f(x) \cdot 1 \, dx$$
>
> $$a_n = \frac{f \cdot g_n}{g_n \cdot g_n} = \frac{1}{\pi} \int_0^{2\pi} f(x) \cos nx \, dx, n > 0$$
>
> $$b_n = \frac{f \cdot h_n}{h_n \cdot h_n} = \frac{1}{\pi} \int_0^{2\pi} f(x) \sin nx \, dx, n > 0$$
>
> Then
>
> $$f(x) = a_0 + a_1 \cos x + b_1 \sin x + a_2 \cos 2x + b_2 \sin 2x + \cdots$$
>
> for all x in $[0, 2\pi]$.

Examples with hints are given in the exercises. (See Exercises 5–14.)

Orthogonalizing Polynomials

Polynomials are very useful, and the functions $f_0(x) = 1, f_1(x) = x, f_2(x) = x^2, \ldots$ are the simplest to use as a basis. Unfortunately they are not orthogonal over any interval $[a, b]$. For example,

$$f_2 \cdot f_0 = \int_a^b f_2(x) f_0(x) = \int_a^b x^2(1) \, dx > 0$$

since $x^2 > 0$ for $x \neq 0$. Thus f_2 and f_0 are not orthogonal. However, there are some applications, as we shall see in the next subsection, for which it would be useful to have orthogonal polynomials. One natural way of constructing them is to apply the Gram–Schmidt process to the functions $f_n(x) = x^n$, $n \geq 0$. We then obtain functions $q_0, q_1, q_2, \ldots$ that are orthogonal by following the procedure described in Section 4.4. Following is a review of the first three steps.

We restrict our attention to the interval $[-1, 1]$. This simplifies our computations since

$$f_{\text{odd}} \cdot f_{\text{even}} = \int_{-1}^{1} x^{\text{odd}} x^{\text{even}} \, dx = \int_{-1}^{1} x^{\text{odd}} \, dx = \left(\frac{x^{\text{even}}}{\text{even}} \right)_{-1}^{1} = 0 \tag{6.10}$$

Thus over the interval $[-1, 1]$, x^{odd} and x^{even} are orthogonal.

If we let $q_0(x) = f_0(x) = 1$ and $q_1(x) = f_1(x) = x$, then we have $q_0 \cdot q_1 = f_0 \cdot f_1 = 0$ by (6.10). To construct q_2,

$$q_2 = f_2 - \frac{f_2 \cdot q_0}{q_0 \cdot q_0} q_0 - \frac{f_2 \cdot q_1}{q_1 \cdot q_1} q_1 \tag{6.11}$$

Since $f_2 \cdot q_1 = f_2 \cdot f_1 = 0$, by (6.10), we need only compute

$$f_2 \cdot q_0 = \int_{-1}^{1} x^2(1)\, dx = \tfrac{1}{3}x^3 \Big]_{-1}^{1} = \tfrac{2}{3}$$

$$q_0 \cdot q_0 = \int_{-1}^{1} 1^2\, dx = x \Big]_{-1}^{1} = 2$$

Substituting these computations into (6.11), we obtain

$$q_2(x) = x^2 - \frac{\frac{2}{3}}{2} 1 = x^2 - \frac{1}{3}$$

These are the first three **Legendre polynomials**, polynomials that are obtained with this procedure and are orthogonal to each other over the interval $[-1, 1]$. If we wanted orthonormal functions, we would divide each q_i by its length $\|q_i\|$ and obtain the **normalized Legendre polynomials**.

Least-Squares Approximation

Let $f(x) = \sqrt{x}$, $0 \le x \le 1$. Suppose we are given the problem of finding the straight line $y = mx + b$ that best fits f over the interval $[0, 1]$, using least-squares approximations. Of course, the first question to ask is: What is meant by that? To answer this, suppose instead of all of f we had just a finite number of points $(x_1, y_1), \ldots, (x_n, y_n)$. Then by Section 4.4,

(6.12)

DISCRETE LEAST-SQUARES PROBLEM
Find m and b that minimize

$$E^2 = \varepsilon_1^2 + \cdots + \varepsilon_n^2 \qquad \text{where} \qquad \varepsilon_k = y_k - (mx + b)$$

See Figure 6.1a.

The natural generalization of this is to let

$$\varepsilon(x) = y - (mx + b) \qquad \text{or} \qquad \varepsilon(x) = f(x) - (mx + b)$$

Figure 6.1

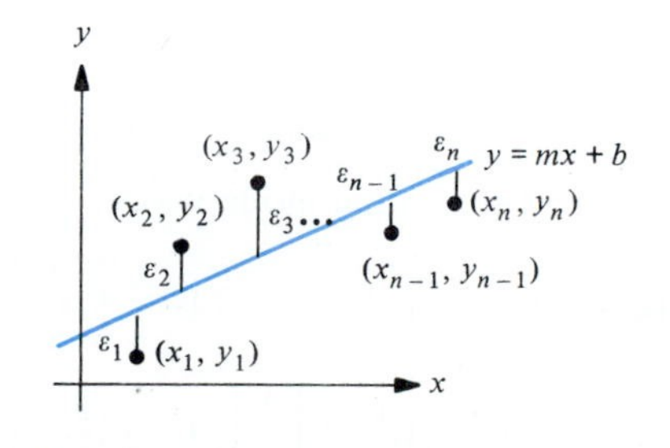

(a) $\varepsilon_k = y_k - (mx_k + b)$;

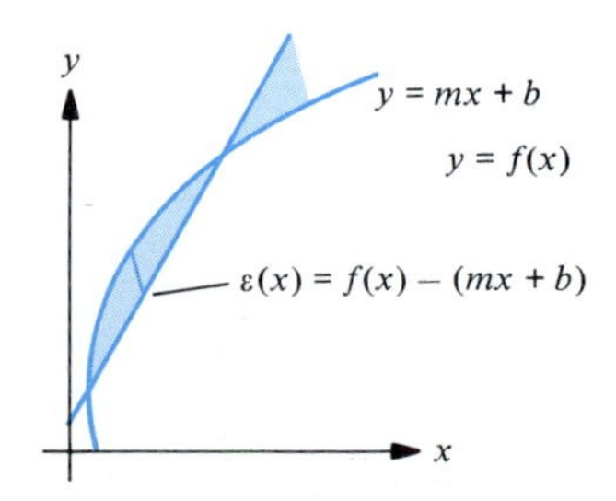

(b) $\varepsilon(x) = f(x) - (mx + b)$.

and to minimize $\int \varepsilon(x)\,dx$. Thus the problem now is

(6.13) **CONTINUOUS LEAST-SQUARES PROBLEM OVER THE INTERVAL [0, 1]**
Find m and b that minimize

$$E^2 = \int_0^1 \varepsilon(x)^2\,dx = \int_0^1 [f(x) - (mx + b)]^2\,dx$$

See Figure 6.1b.

There are several different ways of solving this problem. We shall briefly discuss the two that correspond to the methods we used in Chapter 4 to solve the discrete problem.

In Chapter 4, we expressed the least-squares problem as

$$A\mathbf{x} = \mathbf{y} \quad \text{where} \quad A = \begin{bmatrix} 1 & x_1 \\ 1 & x_2 \\ \vdots & \vdots \\ 1 & x_n \end{bmatrix}, \quad \mathbf{x} = \begin{bmatrix} b \\ m \end{bmatrix}, \quad \mathbf{y} = \begin{bmatrix} y_1 \\ \vdots \\ y_n \end{bmatrix} \tag{6.14}$$

This generalizes nicely to

$$A\mathbf{x} = \mathbf{y} \quad \text{where here} \quad A = [1 \quad x], \quad \mathbf{x} = \begin{bmatrix} b \\ m \end{bmatrix}, \quad \mathbf{y} = [f] \tag{6.15}$$

The first method we used to solve (6.14) was to derive the normal equations $A^T A\mathbf{x} = A^T\mathbf{y}$ [see Theorem (4.35), Section 4.3], or

$$\begin{bmatrix} 1 & \cdots & 1 \\ x_1 & \cdots & x_n \end{bmatrix} \begin{bmatrix} 1 & x_1 \\ 1 & x_2 \\ \vdots & \vdots \\ 1 & x_n \end{bmatrix} \begin{bmatrix} b \\ m \end{bmatrix} = \begin{bmatrix} 1 & \cdots & 1 \\ x_1 & \cdots & x_n \end{bmatrix} \begin{bmatrix} y_1 \\ \vdots \\ y_n \end{bmatrix} \tag{6.16}$$

and solve them directly. When we compute the products $A^T A$ and $A^T\mathbf{y}$ in (6.16), we obtain entries such as

$$1x_1 + 1x_2 + \cdots + 1x_n \qquad \text{and} \qquad x_1 y_1 + \cdots + x_n y_n$$

If we recognize these as *inner products*,

$$(1, \ldots, 1) \cdot (x_1, \ldots, x_n) \qquad \text{and} \qquad (x_1, \ldots, x_n) \cdot (y_1, \ldots, y_n)$$

then we can solve (6.15) in an analogous way:

$$A\mathbf{x} = \mathbf{y} \quad \text{is} \quad [1 \quad x] \begin{bmatrix} b \\ m \end{bmatrix} = [f]$$

$$A^T A\mathbf{x} = A^T\mathbf{y} \quad \text{is} \quad \begin{bmatrix} 1 \\ x \end{bmatrix} [1 \quad x] \begin{bmatrix} b \\ m \end{bmatrix} = \begin{bmatrix} 1 \\ x \end{bmatrix} [f] \tag{6.17}$$

Multiplying out (6.17) using dot products, we obtain

$$\begin{bmatrix} 1\cdot 1 & 1\cdot x \\ 1\cdot x & x\cdot x \end{bmatrix}\begin{bmatrix} b \\ m \end{bmatrix} = \begin{bmatrix} 1\cdot f \\ x\cdot f \end{bmatrix} \tag{6.18}$$

Next we evaluate the dot products, using $f(x) = \sqrt{x}$,

$$\begin{gathered} 1\cdot 1 = \int_0^1 1^2\,dx = 1, \qquad 1\cdot x = \int_0^1 x\,dx = \tfrac{1}{2} \\ x\cdot x = \int_0^1 x^2\,dx = \tfrac{1}{3}, \qquad 1\cdot f = \int_0^1 1\sqrt{x}\,dx = \tfrac{2}{3} \\ x\cdot f = \int_0^1 x\sqrt{x}\,dx = \int_0^1 x^{3/2}\,dx = \tfrac{2}{5} \end{gathered} \tag{6.19}$$

Substituting these in (6.18) yields

$$\begin{bmatrix} 1 & \frac{1}{2} \\ \frac{1}{2} & \frac{1}{3} \end{bmatrix}\begin{bmatrix} b \\ m \end{bmatrix} = \begin{bmatrix} \frac{2}{3} \\ \frac{2}{5} \end{bmatrix} \tag{6.20}$$

We solve this by Gaussian elimination, obtaining $(b, m) = (\frac{4}{15}, \frac{4}{5})$. So the line that best fits $y = \sqrt{x}$ on $[0, 1]$ is $y = \frac{4}{5}x + \frac{4}{15}$. See Figure 6.2.

Figure 6.2

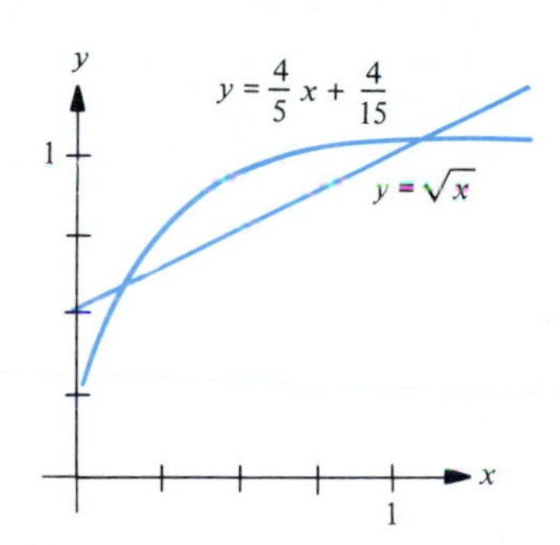

The second method of solving (6.14) is to make the columns of A orthogonal first and then solve the resulting normal equations. (See Section 4.5.) Here we apply the Gram-Schmidt process to 1 and x over the interval $[0, 1]$ and obtain 1 and $x - \frac{1}{2}$. You can check that

$$1\cdot(x - \tfrac{1}{2}) = \int_0^1 1(x - \tfrac{1}{2})\,dx = 0 \tag{6.21}$$

To simplify the presentation, we approach the problem slightly differently from the way we did in the discrete case. We try to find the best fit of $B(1) + M(x - \frac{1}{2})$ to $\sqrt{x}$ on $[0, 1]$; that is, we solve

$$\begin{bmatrix} 1 & x - \frac{1}{2} \end{bmatrix}\begin{bmatrix} B \\ M \end{bmatrix} = [\sqrt{x}]$$

To do this we form the normal equations

$$\begin{bmatrix} 1 \\ x - \frac{1}{2} \end{bmatrix} [1 \quad x - \tfrac{1}{2}] \begin{bmatrix} B \\ M \end{bmatrix} = \begin{bmatrix} 1 \\ x - \frac{1}{2} \end{bmatrix} [\sqrt{x}]$$

which gives us

$$\begin{bmatrix} 1 \cdot 1 & 0 \\ 0 & (x - \frac{1}{2}) \cdot (x - \frac{1}{2}) \end{bmatrix} \begin{bmatrix} B \\ M \end{bmatrix} = \begin{bmatrix} 1 \cdot \sqrt{x} \\ (x - \frac{1}{2}) \cdot \sqrt{x} \end{bmatrix} \tag{6.22}$$

since $1 \cdot (x - \frac{1}{2}) = 0$ by (6.21). Solving this we obtain

$$B = \frac{1 \cdot \sqrt{x}}{1 \cdot 1} \qquad \text{and} \qquad M = \frac{(x - \frac{1}{2}) \cdot \sqrt{x}}{(x - \frac{1}{2}) \cdot (x - \frac{1}{2})}$$

Thus we observe the solution is

$$y = B(1) + M(x - \tfrac{1}{2}) = \underbrace{\frac{1 \cdot \sqrt{x}}{1 \cdot 1} 1}_{\substack{\text{The projection} \\ \text{of } \sqrt{x} \text{ on } 1}} + \underbrace{\frac{(x - \frac{1}{2}) \cdot \sqrt{x}}{(x - \frac{1}{2}) \cdot (x - \frac{1}{2})} (x - \tfrac{1}{2})}_{\substack{\text{The projection} \\ \text{of } \sqrt{x} \text{ on } x - \frac{1}{2}}} \tag{6.23}$$

Observe that the answer is the sum of the projections of $\sqrt{x}$ on 1 and $x - \frac{1}{2}$. This happens because 1 and $x - \frac{1}{2}$ are orthogonal. A different argument would have given us this answer more directly.

We know $1 \cdot 1 = 1$ and $1 \cdot \sqrt{x} = \frac{2}{3}$ from (6.19), so we compute

$$(x - \tfrac{1}{2}) \cdot \sqrt{x} = \int_0^1 (x - \tfrac{1}{2})\sqrt{x}\, dx = \int_0^1 (x^{3/2} - \tfrac{1}{2}x^{1/2})\, dx = \tfrac{2}{5} - \tfrac{1}{3} = \tfrac{1}{15}$$

$$(x - \tfrac{1}{2}) \cdot (x - \tfrac{1}{2}) = \int_0^1 (x - \tfrac{1}{2})^2\, dx = \tfrac{1}{3}(x - \tfrac{1}{2})^3 \Big]_0^1 = \tfrac{1}{12}$$

Substituting these computations into (6.23), we obtain

$$y = B(1) + M(x - \tfrac{1}{2}) = \frac{\frac{2}{3}}{1} + \frac{\frac{1}{15}}{\frac{1}{12}}(x - \tfrac{1}{2}) = \tfrac{2}{3} + \tfrac{4}{5}(x - \tfrac{1}{2})$$

Rewriting this in the form $y = mx + b$ gives us $y = \frac{4}{5}x + \frac{4}{15}$ as before.

Exercise 6.1

In Exercises 1–2, use the identities $\sin u \sin v = \frac{1}{2}[\cos(u - v) - \cos(u + v)]$ and $\sin u \cos v = \frac{1}{2}[\sin(u + v) + \sin(u - v)]$.

1. Show $\sin nx \cdot \sin mx = 0$ for $n \neq m$.
2. Show $\sin nx \cdot \cos mx = 0$, all m, n.
3. Work through all the details to show $\sin nx \cdot \sin nx = \pi$, $n \geq 1$.
4. Same question as Exercise 3, for $\cos nx \cdot \cos nx$.

5. Let $f(x) = 1$, $0 \leq x \leq 2\pi$. Explain why for this f, $a_n = 0$, $b_n = 0$, $n \geq 1$, and $a_0 = 1$ so that the Fourier series for f is $f(x) = 1$.

6. Let $f(x) = 0$, $0 \leq x \leq 2\pi$. Show that $a_n = 0$, $b_n = 0$, for all n so that the Fourier series for this f is $f(x) = 0$.

For Exercises 7–9, let $f(x) = x$, $0 \leq x \leq 2\pi$.

7. Use the formula $\int x \sin nx\, dx = \dfrac{x \cos nx}{n} + \dfrac{\sin nx}{n^2} + C$ to compute $b_n = \dfrac{f \cdot h_n}{h_n \cdot h_n}$, $n \geq 1$.

8. Use the formula $\int x \cos nx\, dx = \dfrac{x \sin nx}{n} + \dfrac{\cos nx}{n^2} + C$ to compute $a_n = \dfrac{f \cdot g_n}{g_n \cdot g_n}$, $n \geq 1$.

9. Use Exercises 7 and 8 to find the Fourier series for f. (Do not forget a_0.)

For Exercises 10 and 11 let $f(x) = x^2$, $0 \leq x \leq 2\pi$.

10. Use the formulas $\int x^2 \sin nx\, dx = \dfrac{-x^2 \cos nx}{n} + \dfrac{2x \sin nx}{n^2} + \dfrac{2 \cos nx}{n^3} + C$ and $\int x^2 \cos nx\, dx = \dfrac{x^2 \sin nx}{n} + \dfrac{2x \cos nx}{n^2} - \dfrac{2 \sin nx}{n^3} + C$ to compute the Fourier coefficients a_n, b_n, $n \geq 1$.

11. Find the Fourier series for f.

In Exercises 12–14 let f be the **square-wave** function

$$f(x) = \begin{Bmatrix} 0, & x = 0, \pi, 2\pi \\ 1, & 0 < x < \pi \\ -1, & \pi < x < 2\pi \end{Bmatrix}$$

12. Use a geometric argument to show $f \cdot \cos nx = 0$, $n \geq 0$.

13. Find

$$b_n = \frac{f(x) \cdot h_n(x)}{h_n(x) \cdot h_n(x)}, \qquad n \geq 1$$

and the Fourier series

$$b_1 \sin x + b_2 \sin 2x + b_3 \sin 3x + \cdots$$

14. Graph the first three "partial sums" $b_1 \sin x$, $b_1 \sin x + b_2 \sin 2x$, $b_1 \sin x + b_2 \sin 2x + b_3 \sin 3x$. It is fun to watch the Fourier series converge to this function, even though it is not continuous.

Legendre Polynomials

15. Find q_3 and q_4, the fourth and fifth Legendre polynomials.

16. Find the first four normalized Legendre polynomials.

Least-Squares Approximation

17. Let $f(x) = x^2$, $0 \leq x \leq 1$. Find the equation of the straight line that best fits f using least-squares approximations by solving (6.18).

18. Same question as Exercise 17, except using the interval $-1 \leq x \leq 1$.

19. Same question as Exercise 17, except using orthogonal projection and solving the equations corresponding to (6.22).

20. Same question as Exercise 19, except using the interval $-1 \leq x \leq 1$ and hence projecting on the first two Legendre polynomials.

21. Same as Exercise 17, except using $f(x) = x^3$.

22. Same as Exercise 18, except using $f(x) = x^3$.

23. Same as Exercise 19, except using $f(x) = x^3$.

24. Same as Exercise 20, except using $f(x) = x^3$.

6.2 The Singular Value Decomposition—Generalized Inverses, the General Least-Squares Problem, and an Approach to Ill-Conditioned Systems

Suppose we wanted to solve the equation $A\mathbf{x} = \mathbf{b}$, where A is square and nonsingular. Then the theoretical answer is $\mathbf{x} = A^{-1}\mathbf{b}$. Suppose instead A is $m \times n$ and the equation $A\mathbf{x} = \mathbf{b}$ is a least-squares problem. Then if A has linearly independent columns, by Theorem (4.35), Section 4.3, the theoretical answer is $\mathbf{x} = A^{+}\mathbf{b}$, where $A^{+} = (A^TA)^{-1}A^T$. Thus A^{+} plays the role of an inverse, and it is called a **generalized inverse** or **pseudoinverse** of A (when A has linearly independent columns). In this section we shall derive a pseudoinverse, A^{+}, for an arbitrary $m \times n$ matrix A. We shall then see that it has the property that $\bar{\mathbf{x}} = A^{+}\mathbf{b}$ is the least-squares solution to the system $A\mathbf{x} = \mathbf{b}$.

The Singular Value Decomposition

As a first step in our discussion of pseudoinverses, we need to define the *singular value decomposition* of a matrix. The singular value decomposition is a very important theoretical and computational tool, and it is nowhere near so famous as it ought to be. We shall use this tool first to construct the pseudoinverse, and later to solve ill-conditioned systems.

The singular value decomposition theorem is easy to state.

(6.24) **THEOREM (SINGULAR VALUE DECOMPOSITION)**
Any $m \times n$ matrix can be factored as

(6.25)
$$A = V\Sigma U^T$$

where V is $m \times m$ and orthogonal, U is $n \times n$ and orthogonal, and Σ is $m \times n$ of the special form

(6.26)
$$\Sigma = \begin{bmatrix} D & \vdots & 0 \\ \cdots & & \cdots \\ 0 & \vdots & 0 \end{bmatrix} \qquad \text{where} \qquad D = \begin{bmatrix} \sigma_1 & & \\ & \ddots & \\ & & \sigma_k \end{bmatrix}$$

The numbers $\sigma_1, \ldots, \sigma_k$ are all positive real numbers, $\sigma_1 \geq \sigma_2 \geq \cdots \geq \sigma_k > 0$, and are called the **singular values** of A.

The proof is surprisingly straightforward, if you do the right thing at the right time.

Proof Consider A^TA, which is $n \times n$. It has two essential properties.

1. A^TA is symmetric, so it has n real eigenvalues and n orthonormal eigenvectors, by Theorems (5.39) and (5.45), Section 5.4.
2. The eigenvalues are in fact all nonnegative. For if $A^TA\mathbf{x} = \lambda\mathbf{x}$, $\mathbf{x} \neq \mathbf{0}$, then

$$0 \leq A\mathbf{x} \cdot A\mathbf{x} = A^TA\mathbf{x} \cdot \mathbf{x} = \lambda\mathbf{x} \cdot \mathbf{x} = \lambda(\mathbf{x} \cdot \mathbf{x})$$

If we divide by $\mathbf{x} \cdot \mathbf{x} > 0$, we get $\lambda \geq 0$.

Now we order the eigenvalues of A^TA so that $\lambda_1 \geq \lambda_2 \geq \cdots \geq \lambda_n \geq 0$ and let $\mathbf{u}_1, \ldots, \mathbf{u}_n$ be corresponding orthonormal eigenvectors. Assume that λ_k is the smallest positive eigenvalue so that altogether we have

(6.27)
$$\lambda_1 \geq \cdots \geq \lambda_k > 0, \qquad \lambda_{k+1} = \cdots = \lambda_n = 0, \qquad A^TA\mathbf{u}_i = \lambda_i\mathbf{u}_i, \quad \mathbf{u}_i \cdot \mathbf{u}_j = \delta_{ij}{}^*$$

In particular, for $i > k$ we have $A^TA\mathbf{u}_i = \lambda_i\mathbf{u}_i = 0\mathbf{u}_i = \mathbf{0}$. Hence $A\mathbf{u}_i = \mathbf{0}$ for $i > k$, since $NS(A^TA) = NS(A)$ by Theorem (4.18a), Section 4.2.

Next let U be the orthogonal matrix whose columns are the $\mathbf{u}_i$'s

$$U = [\mathbf{u}_1 \vdots \mathbf{u}_2 \vdots \cdots \vdots \mathbf{u}_n]$$

* *This symbol δ_{ij} is a very handy notational device called the **Kronecker delta**.* It is defined by

$$\delta_{ij} = \begin{cases} 0, & \text{if} \quad i \neq j \\ 1, & \text{if} \quad i = j \end{cases}$$

For $1 \leq i \leq k$, let

$$\sigma_i = +\sqrt{\lambda_i}, \quad \mathbf{v}_i = \frac{1}{\sigma_i} A\mathbf{u}_i \tag{6.28}$$

Then the $\mathbf{v}_i$'s are orthonormal for

$$\begin{aligned}\mathbf{v}_i \cdot \mathbf{v}_j &= \frac{1}{\sigma_i} A\mathbf{u}_i \cdot \frac{1}{\sigma_j} A\mathbf{u}_j = \frac{1}{\sigma_i\sigma_j} A\mathbf{u}_i \cdot A\mathbf{u}_j \\ &= \frac{1}{\sigma_i\sigma_j} A^T A\mathbf{u}_i \cdot \mathbf{u}_j = \frac{1}{\sigma_i\sigma_j} \lambda_i \mathbf{u}_i \cdot \mathbf{u}_j \\ &= \frac{\lambda_i}{\sigma_i\sigma_j} \mathbf{u}_i \cdot \mathbf{u}_j\end{aligned}$$

If $i \neq j$, $\mathbf{u}_i \cdot \mathbf{u}_j = 0$, so $\mathbf{v}_i \cdot \mathbf{v}_j = 0$; if $i = j$, $\mathbf{u}_i \cdot \mathbf{u}_i = 1$ and $\sigma_i\sigma_i = \sqrt{\lambda_i}\sqrt{\lambda_i} = \lambda_i$, so $\mathbf{v}_i \cdot \mathbf{v}_i = 1$. Therefore, $\mathbf{v}_1, \ldots, \mathbf{v}_k$ are orthonormal vectors in $\mathbb{R}^m$. Next pick any $\mathbf{v}_{k+1}, \ldots, \mathbf{v}_m$ so that altogether $\mathbf{v}_1, \ldots, \mathbf{v}_m$ are orthonormal. Let V be the $m \times m$ orthogonal matrix with the $\mathbf{v}_i$'s as its columns

$$V = [\mathbf{v}_1 \vdots \mathbf{v}_2 \vdots \cdots \vdots \mathbf{v}_m] \tag{6.29}$$

and let $\Sigma = [c_{ij}] = V^T A U$. Then

$$\begin{aligned}c_{ij} &= (i\text{th row of } V^T) \cdot (j\text{th column of } AU) \\ &= \mathbf{v}_i^T A\mathbf{u}_j \qquad \text{since } AU = [A\mathbf{u}_1 \vdots \cdots \vdots A\mathbf{u}_n] \\ &= \mathbf{v}_i \cdot A\mathbf{u}_j\end{aligned}$$

since the ith row of V^T is the ith column of V. If $j > k$, $A\mathbf{u}_j = \mathbf{0}$, so $c_{ij} = 0$. If $j \leq k$, $A\mathbf{u}_j = \sigma_j \mathbf{v}_j$, so $c_{ij} = \mathbf{v}_i \cdot \sigma_j \mathbf{v}_j = \sigma_j(\mathbf{v}_i \cdot \mathbf{v}_j) = \sigma_j \delta_{ij}$. Altogether we have

$$c_{ij} = \begin{cases}\sigma_j, & \text{if } i = j \text{ and } j \leq k \\ 0, & \text{otherwise}\end{cases}$$

Therefore,

$$\Sigma = \begin{bmatrix} D & \vdots & 0 \\ \cdots & & \cdots \\ 0 & \vdots & 0 \end{bmatrix} \qquad \text{where} \qquad D = \begin{bmatrix} \sigma_1 & & \\ & \ddots & \\ & & \sigma_k \end{bmatrix}$$

as described in (6.26). Since $V^T A U = \Sigma$, $A = V\Sigma U^T$, because U and V are orthogonal. This is what we need to complete the proof of the singular value decomposition. ■

Pseudoinverse

By the singular value decomposition, any $m \times n$ matrix $A = V\Sigma U^T$. If A is square and Σ is invertible (i.e., $k = n$ and all $\sigma_i > 0$), then

$$A^{-1} = (V\Sigma U^T)^{-1} = (U^T)^{-1}\Sigma^{-1}V^{-1} = U\Sigma^{-1}V^T$$

since U and V are orthogonal. If Σ^{-1} does not exist, the following is the best we can do.

(6.30) DEFINITION If A is an $m \times n$ matrix and $A = V\Sigma U^T$ is its singular value decomposition, then the **pseudoinverse*** of A, A^+, is defined by

$$A^+ = U\Sigma^+ V^T$$

where Σ^+ is the $n \times m$ matrix

$$\Sigma^+ = \begin{bmatrix} D & \vdots & 0 \\ \cdots & & \cdots \\ 0 & \vdots & 0 \end{bmatrix}^+ = \begin{bmatrix} D^{-1} & \vdots & 0 \\ \cdots & & \cdots \\ 0 & \vdots & 0 \end{bmatrix}$$

Some of the properties of the pseudoinverse A^+ will be considered in the exercises.

Least-Squares Problems

In Chapter 4 we described the least-squares problem for $A\mathbf{x} = \mathbf{b}$ as

Find a vector $\bar{\mathbf{x}}$ so that $\|\boldsymbol{\varepsilon}\|$ is a minimum, where $\boldsymbol{\varepsilon} = \mathbf{b} - A\bar{\mathbf{x}}$

[See (4.33), Section 4.3.] If A has linearly independent columns, we saw in Section 4.3 that there is a unique $\bar{\mathbf{x}}$. However, if A does not have linearly independent columns, $NS(A) \neq \{\mathbf{0}\}$ and there are many such $\bar{\mathbf{x}}$. When this happens, we usually pick the $\bar{\mathbf{x}}$ that makes $\|\bar{\mathbf{x}}\|$ a minimum.

(6.31) DEFINITION The *least-squares problem* for the system $A\mathbf{x} = \mathbf{b}$ is to find a vector $\bar{\mathbf{x}}$ for which $\|\boldsymbol{\varepsilon}\| = \|\mathbf{b} - A\mathbf{x}\|$ is a minimum *and* for which $\|\bar{\mathbf{x}}\|$ is a minimum.

We can now easily give the solution.

(6.32) THEOREM The solution for the least-squares problem for $A\mathbf{x} = \mathbf{b}$ is given by

$$\bar{\mathbf{x}} = A^+\mathbf{b}$$

Proof We first prove the theorem for the special case $A = \Sigma$. We then use this special case to prove the general case.

SPECIAL CASE Solve the least-squares problem

$$\Sigma\mathbf{z} = \mathbf{y}, \quad \text{where } \Sigma = \begin{bmatrix} D & \vdots & 0 \\ \cdots & & \cdots \\ 0 & \vdots & 0 \end{bmatrix}, \quad D = \begin{bmatrix} \sigma_1 & & \\ & \ddots & \\ & & \sigma_k \end{bmatrix}, \quad \sigma_i > 0, 1 \le i \le k \tag{6.33}$$

* *This pseudoinverse is called the **Moore–Penrose inverse**, for its discoverers. It is one, and the best known, of several types of generalized inverses.*

Here Σ is $m \times n$, and $\mathbf{z} = [z_1 \ \cdots \ z_n]^T$, $\mathbf{y} = [y_1 \ \cdots \ y_m]^T$. Now $\Sigma\mathbf{z} = [\sigma_1\mathbf{z}_1 \ \cdots \ \sigma_k\mathbf{z}_k \ 0 \ \cdots \ 0]^T$, so that

$$\boldsymbol{\varepsilon} = \mathbf{y} - \Sigma\mathbf{z} = [y_1 - \sigma_1 z_1 \ \cdots \ y_k - \sigma_k z_k \ y_{k+1} \ \cdots \ y_m]$$

Keeping in mind that $\mathbf{y}$ is fixed but $\mathbf{z}$ varies, we see

$$\begin{aligned} \|\boldsymbol{\varepsilon}\| \text{ is a minimum} \quad &\Leftrightarrow \quad \text{the first } k \text{ components of } \boldsymbol{\varepsilon} \text{ are } 0 \\ &\Leftrightarrow \quad y_i - \sigma_i z_i = 0, \qquad 1 \le i \le k \\ &\Leftrightarrow \quad z_i = \frac{y_i}{\sigma_i}, \qquad 1 \le i \le k \\ &\Leftrightarrow \quad \begin{bmatrix} z_i \\ \vdots \\ z_k \end{bmatrix} = D^{-1}\begin{bmatrix} y_1 \\ \vdots \\ y_k \end{bmatrix} \end{aligned}$$

Thus for $\|\boldsymbol{\varepsilon}\|$ to be a minimum, the first k components of $\mathbf{z}$ are determined by $z_i = \dfrac{y_i}{\sigma_i}$, but the last $n - k$ components $z_{k+1}, \ldots, z_n$ are arbitrary. Therefore, the solution $\bar{\mathbf{z}}$ looks like

$$\bar{\mathbf{z}} = \begin{bmatrix} \dfrac{y_1}{\sigma_1} & \cdots & \dfrac{y_k}{\sigma_k} & z_{k+1} & \cdots & z_n \end{bmatrix}^T$$

However, since the $\dfrac{y_i}{\sigma_i}$'s are fixed but the z_j's are arbitrary, we can easily see that

$$\|\bar{\mathbf{z}}\| = \sqrt{\left(\frac{y_1}{\sigma_1}\right)^2 + \cdots + \left(\frac{y_k}{\sigma_k}\right)^2 + z_{k+1}^2 + \cdots + z_n^2}$$

is a minimum exactly when $z_{k+1} = \cdots = z_n = 0$. Therefore, the least-squares solution to $\Sigma\mathbf{z} = \mathbf{y}$ is

$$\bar{\mathbf{z}} = \begin{bmatrix} \dfrac{y_1}{\sigma_1} \\ \vdots \\ \dfrac{y_k}{\sigma_k} \\ 0 \\ \vdots \\ 0 \end{bmatrix} = \begin{bmatrix} D^{-1} & 0 \\ 0 & 0 \end{bmatrix}\begin{bmatrix} y_1 \\ \vdots \\ y_n \end{bmatrix} = \Sigma^+ y \tag{6.34}$$

GENERAL CASE Solve the least-squares problem $A\mathbf{x} = \mathbf{b}$, where A is $m \times n$. Let $A = V\Sigma U^T$ be the singular value decomposition for A. Then our problem is to

$$\text{minimize } \|\boldsymbol{\varepsilon}\|, \qquad \boldsymbol{\varepsilon} = \mathbf{b} - A\mathbf{x} = \mathbf{b} - V\Sigma U^T\mathbf{x}, \qquad \text{where } \mathbf{b} \text{ is fixed and } \mathbf{x} \text{ varies} \tag{6.35}$$

Now V is orthogonal, so V^T is orthogonal, which means

$$\|V^T\mathbf{v}\| = \|\mathbf{v}\|, \qquad \text{for all vectors } \mathbf{v} \text{ in } \mathbb{R}^m \tag{6.36}$$

by Theorem (4.62b), Section 4.5. Thus

$$\|\boldsymbol{\varepsilon}\| = \|V^T\boldsymbol{\varepsilon}\| = \|V^T(\mathbf{b} - V\Sigma U^T\mathbf{x})\| = \|V^T\mathbf{b} - \Sigma U^T\mathbf{x}\| \tag{6.37}$$

Combining (6.35) and (6.37), we see our problem is to

$$\text{minimize } \|V^T\mathbf{b} - \Sigma U^T\mathbf{x}\|, \qquad \text{where } \mathbf{b} \text{ is fixed and } \mathbf{x} \text{ varies} \tag{6.38}$$

Substituting $\mathbf{y} = V^T\mathbf{b}$ and $\mathbf{z} = U^T\mathbf{x}$ into (6.38) yields

$$\text{minimize } \|\mathbf{y} - \Sigma\mathbf{z}\|, \qquad \text{where } \mathbf{y} \text{ is fixed and } \mathbf{z} \text{ varies} \tag{6.39}$$

By (6.34) of the special case proof, the answer is

$$\bar{\mathbf{z}} = \Sigma^+\mathbf{y} \tag{6.40}$$

Since $\mathbf{y} = V^T\mathbf{b}$, we see

$$\bar{\mathbf{z}} = \Sigma^+V^T\mathbf{b} \tag{6.41}$$

Now $\mathbf{z} = U^T\mathbf{x}$, so $\mathbf{x} = (U^T)^{-1}\mathbf{z} = U\mathbf{z}$. Multiplying (6.41) by U yields

$$U\bar{\mathbf{z}} = U\Sigma^+V^T\mathbf{b}$$

or

$$\bar{\mathbf{x}} = U\Sigma^+V^T\mathbf{b} \tag{6.42}$$

where $\bar{\mathbf{x}} = U\bar{\mathbf{z}}$. Since $A^+ = U\Sigma^+V^T$ by the definition of pseudoinverse (6.30), the solution to the least squares problem $A\mathbf{x} = \mathbf{b}$ is $\bar{\mathbf{x}} = A^+\mathbf{b}$, by (6.42), so we are done. ■

Ill-Conditioned Problems

We conclude this section with a brief mention of how the singular value decomposition can help solve ill-conditioned problems.

Suppose we wish to solve the system $A\mathbf{x} = \mathbf{b}$, where A is $n \times n$ and invertible, but unfortunately the problem is ill conditioned. To get around this difficulty, first form the singular value decomposition for A,

$$A = V\Sigma U^T, \qquad \Sigma = \begin{bmatrix} \sigma_1 & & \\ & \ddots & \\ & & \sigma_n \end{bmatrix}$$

Since A is invertible, all the σ_i's are nonzero, and hence all are positive,

$$\sigma_1 \geq \sigma_2 \geq \cdots \geq \sigma_n > 0$$

It turns out that since A is ill conditioned, σ_1 will be *much* larger than σ_n. In other words, it is a large disparity in the magnitude of the singular values that causes the ill-conditioning. What we do is find a σ_k somewhere in the middle that is not significantly smaller than σ_1, but after which the remaining σ_i's are small. We then set $\sigma_{k+1} = \cdots = \sigma_n = 0$ and solve

$$A'\mathbf{x} = \mathbf{b}, \qquad \text{where } A' = V\Sigma' U^T, \qquad \Sigma' = \begin{bmatrix} \sigma_1 & & & & & \\ & \ddots & & & & \\ & & \sigma_k & & & \\ & & & 0 & & \\ & & & & \ddots & \\ & & & & & 0 \end{bmatrix}$$

Of course A' is singular, so $A'\mathbf{x} = \mathbf{b}$ is a least squares problem. But by Theorem (6.32), the answer is

$$\bar{\mathbf{x}} = (A')^+\mathbf{b} = U(\Sigma')^+ V^T\mathbf{b}, \qquad (\Sigma')^+ = \begin{bmatrix} \sigma_1^{-1} & & & & & \\ & \ddots & & & & \\ & & \sigma_k^{-1} & & & \\ & & & 0 & & \\ & & & & \ddots & \\ & & & & & 0 \end{bmatrix}$$

Thus the procedure for solving ill-conditioned problems is quite simple.

(6.43)

PROCEDURE FOR SOLVING ILL-CONDITIONED PROBLEMS $A\mathbf{x} = \mathbf{b}$ USING THE SINGULAR VALUE DECOMPOSITION

(a) Find the singular value decomposition, $A = V\Sigma U^T$.

(b) Set small singular values to zero, $\sigma_{k+1} = \cdots = \sigma_n = 0$, obtaining Σ'. The answer is

$$\bar{\mathbf{x}} = U(\Sigma')^+ V^T\mathbf{b},$$

$$(\Sigma')^+ = \begin{bmatrix} \sigma_1^{-1} & & & & & \\ & \ddots & & & & \\ & & \sigma_k^{-1} & & & \\ & & & 0 & & \\ & & & & \ddots & \\ & & & & & 0 \end{bmatrix}$$

Of course, the question is how good an answer is $\bar{\mathbf{x}}$? The answer depends on three things.

1. How much numerical error has crept into computing V, Σ, and U.
2. Whether or not the vector $\mathbf{b}$ is close to $CS(A')$. (In general, methods for dealing with ill-conditioned problems have to take $\mathbf{b}$ into consideration as well as A.)
3. The cut point taken in the singular values below which all values are set to zero. Some problems have an obvious separation between large and small values, but some do not.

   There is a trade-off between the consideration in 2 and the problem's being ill conditioned.

In general there is no one best way for dealing with all ill-conditioned problems. The singular value decomposition is just one of several approaches, and it works well in many, but not all, situations.

Exercise 6.2

In Exercises 1–8, find the singular values of the given matrix.

1. $\begin{bmatrix} -2 & \\ & 1 \end{bmatrix}$

2. $\begin{bmatrix} 9 & \\ & -4 \end{bmatrix}$

3. $\begin{bmatrix} -2 & \\ & 0 \end{bmatrix}$

4. $\begin{bmatrix} -2 & & \\ & 1 & \\ & & 0 \end{bmatrix}$

5. $\begin{bmatrix} -2 & -1 \\ & 0 \end{bmatrix}$

6. $\begin{bmatrix} -2 & 2 \\ 1 & -1 \end{bmatrix}$

7. $\begin{bmatrix} 2 & 1 & \\ & 2 & \\ & & 0 \end{bmatrix}$

8. $\begin{bmatrix} -3 & 1 & 0 \\ & -2 & 1 \\ & & -1 \end{bmatrix}$

In Exercises 9–20, find the singular value decomposition of the given matrix.

9. $\begin{bmatrix} -3 & \\ & 1 \end{bmatrix}$

10. $\begin{bmatrix} 0 & \\ & -4 \end{bmatrix}$

11. $\begin{bmatrix} 1 & -2 \end{bmatrix}$

12. $\begin{bmatrix} 1 \\ -2 \end{bmatrix}$

13. $\begin{bmatrix} 1 & -2 \\ -2 & 4 \end{bmatrix}$

14. $\begin{bmatrix} -3 & -1 \\ 6 & 2 \end{bmatrix}$

15. $\begin{bmatrix} \sqrt{2} & 1 & 0 \\ 0 & \sqrt{2} & 0 \end{bmatrix}$

16. $\begin{bmatrix} \sqrt{2} & 0 \\ 1 & \sqrt{2} \\ 0 & 0 \end{bmatrix}$

17. $\begin{bmatrix} 2 & -1 & \\ 2 & 2 & \\ & & -2 \end{bmatrix}$

18. $\begin{bmatrix} -3 & & \\ & -2 & 2 \\ & -1 & -2 \end{bmatrix}$

19. $\begin{bmatrix} 1 & & & & \\ & 2 & & & \\ & & 3 & & \\ & & & \ddots & \\ & & & & n \end{bmatrix}$

20. $\begin{bmatrix} 1 & & & & & \\ & 1 & & & & \\ & & -2 & & & \\ & & & -2 & & \\ & & & & -3 & \\ & & & & & 3 \end{bmatrix}$

In Exercises 21–28, find the pseudoinverse for the given matrix.

21. The matrix in Exercise 9.
22. The matrix in Exercise 10.
23. The matrix in Exercise 11.
24. The matrix in Exercise 12.
25. The matrix in Exercise 13.
26. The matrix in Exercise 14.
27. The matrix in Exercise 15.
28. The matrix in Exercise 16.

When Penrose originally defined A^+, he showed that for any A there was one and only one matrix A^+ satisfying

(i) $AA^+A = A$ (ii) $A^+AA^+ = A^+$ (iii) $(AA^+)^T = AA^+$
(iv) $(A^+A)^T = A^+A$

Our geometric approach is more intuitive, but his algebraic approach makes it simple to verify properties.

29. Show that our definition of A^+ satisfies Penrose's conditions (i) and (iii).

30. Show that our definition of A^+ satisfies Penrose's conditions (ii) and (iv).

31. Suppose A^+ satisfies Penrose's conditions (i)–(iv) for A. Show that $(A^+)^T$ satisfies Penrose's conditions (i)–(iv) for A^T. Thus $(A^T)^+ = (A^+)^T$ by Penrose's work.

32. Use Exercise 31 to show: If A is symmetric, then A^+ is symmetric.

33. Suppose A^+ satisfies Penrose's four conditions for A. Show that A satisfies the four conditions for A^+. Hence $(A^+)^+ = A$.

34. Suppose A is square, symmetric, and $A^2 = A$. Show that $A^+ = A$.

In Exercises 35–39, use the pseudoinverse to solve the least-squares problem.

35. $\begin{bmatrix} 1 & 1 & 1 \\ 1 & 1 & 1 \end{bmatrix}\begin{bmatrix} x \\ y \\ z \end{bmatrix} = \begin{bmatrix} 1 \\ 0 \end{bmatrix}$

36. $\begin{bmatrix} 1 & 1 & -1 \\ 1 & 1 & -1 \end{bmatrix}\begin{bmatrix} x \\ y \\ z \end{bmatrix} = \begin{bmatrix} 2 \\ -2 \end{bmatrix}$

37. $\begin{bmatrix} 1 & 2 \\ 2 & 4 \end{bmatrix}\begin{bmatrix} x \\ y \end{bmatrix} = \begin{bmatrix} 2 \\ 3 \end{bmatrix}$

38. $\begin{bmatrix} 1 & 3 \\ 3 & 9 \end{bmatrix}\begin{bmatrix} x \\ y \end{bmatrix} = \begin{bmatrix} 0 \\ 1 \end{bmatrix}$

39. Let $A = \begin{bmatrix} 1 & 1 \\ 1 & 1.001 \end{bmatrix}$, $\mathbf{b} = \begin{bmatrix} 1 \\ 1.1 \end{bmatrix}$, $\mathbf{c} = \begin{bmatrix} 0 \\ 1 \end{bmatrix}$.

(a) Solve $A\mathbf{x} = \mathbf{b}$ and $A\mathbf{x} = \mathbf{c}$.

(b) Find the singular value decomposition for A, set the small singular value to zero, and solve the corresponding least-squares problems as described in the text.

(c) Compare the answers in (a) and (b) and explain the differences.

6.3 Iterative Methods

Many matrices that arise in applications tend to be quite large. In fact, 1000×1000 now represents a medium-sized matrix. When they get this big, matrices can no longer fit into many computer memories. And even when they do fit, solving $A\mathbf{x} = \mathbf{b}$ by standard Gaussian elimination is quite slow.

Large matrices that arise in applications are usually *sparse*, meaning that lots of their entries are zero. A typical 1000×1000 matrix may have 5–20 nonzeros in each row. In such cases, only the nonzeros are stored and worked with; schemes for doing this are more involved, and this is a separate subject in itself. We shall not consider this interesting aspect of the problem here.*

When A is large and sparse, there are roughly two approaches to finding the solution to $A\mathbf{x} = \mathbf{b}$: direct methods, and iterative methods. In *direct methods*, Gaussian elimination is generalized to this context and the equivalent to a LU-decomposition is usually found. In *iterative methods*, a sequence of vectors is found that gives increasingly better approximations to the answer. Different approaches work better for different kinds of problems.

The purpose of this section is to provide an intuitive introduction to iterative methods. There are no proofs in this section, but there are examples. (Hageman and Young, *Applied Iterative Methods*, is one classic textbook on the subject in which proofs can be found.) We begin with an intuitive introduction to iterative methods in general.

Background

The iterative methods we shall be discussing later involve functions $F: \mathbb{R}^n \to \mathbb{R}^n$. To give you an intuitive feel for the way general iterative methods work, we begin by using simple functions $f: \mathbb{R} \to \mathbb{R}$. Roughly, here is the idea. Suppose that we want to solve $f(x) = x$ for some function $f: \mathbb{R} \to \mathbb{R}$. Further suppose that we have a general idea of the answer we are seeking (say, x_0). Then we compute $x_1 = f(x_0)$, $x_2 = f(x_1)$, $x_3 = f(x_2)$, ... If f has certain nice properties, then we can prove that $x_n \to$ the solution.

* *But see Duff, Erisman, and Reid,* Direct Methods for Sparse Matrices, *Clarendon Press, 1989.*

Example 1 Use iterative methods to solve $\cos x = x$.

Solution Let $f(x) = \cos x$. The graphs $y = \cos x$ and $y = x$ appear in Figure 6.3, and we can see that there is a unique solution r. Further, we might guess that $r \approx 0.7$. So starting with $x_0 = 0.7$, we compute (using a scientific hand-held calculator, in *radian* mode):

$$\begin{aligned} x_1 &= \cos x_0 \approx 0.7648422 \\ x_2 &= \cos x_1 \approx 0.7214916 \\ x_3 &= \cos x_2 \approx 0.7508213 \\ x_4 &= \cos x_3 \approx 0.7311288 \\ &\vdots \end{aligned} \tag{6.44}$$

Figure 6.3

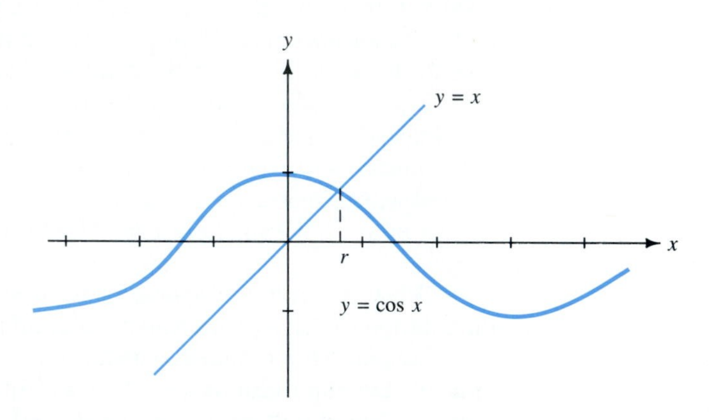

By pushing the cos button repeatedly, you eventually will get as close as your calculator can to the answer $r = 0.7390851\ldots$. You can see that the sequence in (6.44) does converge to the answer, but at a fairly slow rate of convergence.

In the next example, the rate of convergence is faster. This example uses calculus, but you can follow the main principle even if you have not had calculus.

Example 2 Find a good approximation to $\sqrt{2}$ by applying Newton's method to the function $f(x) = x^2 - 2$.

Solution The first (and main) point of this example is that sometimes we can solve a problem (in this case, $x^2 - 2 = 0$) by turning it into a different problem that can be solved iteratively. The solution to the changed problem is then a solution to the original problem as well. Applying Newton's method, the problem

$$\text{"Solve } f(x) = 0\text{"} \qquad \text{is changed into} \qquad \text{"Solve } x = x - \frac{f(x)}{f'(x)}\text{"}$$

It is easy to see that

$$\text{If } F(x) = x - \frac{f(x)}{f'(x)}, \qquad \text{then} \qquad f(r) = 0 \Leftrightarrow F(r) = r$$

It turns out that if f has certain nice properties (and most functions f that you run into normally do) and if you can start reasonably close to the answer, then the sequence $x_1 = F(x_0)$, $x_2 = F(x_1), \cdots$ converges to a zero of f quite quickly. We shall see this happen in our problem: $f(x) = x^2 - 2$. First, with a little algebra,

$$F(x) = x - \frac{f(x)}{f'(x)} = x - \frac{x^2 - 2}{2x} = \frac{x + (2/x)}{2} \tag{6.45}$$

See Exercise 4. If we start by guessing that $x_0 = 1.5 \approx \sqrt{2}$, then

$$\begin{aligned} x_1 &= F(x_0) \approx 1.4146667 \\ x_2 &= F(x_1) \approx 1.4142157 \\ x_3 &= F(x_2) \approx 1.4142136 \end{aligned} \tag{6.46}$$

Equations (6.46) provide the answer to eight significant figures in only three steps! ■

In general, iterative methods are seldom this fast, and much research is spent on speeding up the process.

In summary, the way iterative processes work is to rewrite the original problem in the form $x = F(x)$ and generate a sequence x_0, $x_1 = F(x_0)$, $x_2 = F(x_1), \ldots$. If the original problem has certain nice properties, then the sequence approaches the solution to the original problem. We now do this for linear systems $A\mathbf{x} = \mathbf{b}$.

Jacobi's Method

We begin with a simple description of Jacobi's method. We start with a linear system to solve:

$$\begin{aligned} a_{11}x_1 + a_{12}x_2 + \cdots + a_{1n}x_n &= b_1 \\ a_{21}x_1 + a_{22}x_2 + \cdots + a_{2n}x_n &= b_2 \\ &\vdots \\ a_{n1}x_1 + a_{n2}x_2 + \cdots + a_{nn}x_n &= b_n \end{aligned} \tag{6.47}$$

Keeping in mind our goal of rewriting this as $\mathbf{x} = F(\mathbf{x})$, where $\mathbf{x} = [x_1 \quad x_2 \quad \cdots \quad x_n]^T$, we solve the first equation for x_1, the second equation for x_2, and so on. First we solve for $a_{ii}x_i$:

$$\begin{aligned} a_{11}x_1 &= \qquad\qquad -a_{12}x_2 - a_{13}x_3 - \cdots - a_{1n}x_n + b_1 \\ a_{22}x_2 &= -a_{21}x_1 \qquad\qquad - a_{23}x_3 - \cdots - a_{2n}x_n + b_2 \\ &\vdots \\ a_{nn}x_n &= -a_{n1}x_1 - a_{n2}x_2 - \cdots - a_{n,n-1}x_{n-1} + b_n \end{aligned} \tag{6.48}$$

Then we divide the ith equation by a_{ii}:

$$\begin{aligned} x_1 &= (\qquad\qquad -a_{12}x_2 - a_{13}x_3 - \cdots - a_{1n}x_n + b_1)/a_{11} \\ x_2 &= (-a_{21}x_1 \qquad\qquad - a_{23}x_3 - \cdots - a_{2n}x_n + b_2)/a_{22} \\ &\vdots \\ x_n &= (-a_{n1}x_1 - a_{n2}x_2 - \quad \cdots \quad - a_{n,n-1}x_{n-1} + b_n)/a_{nn} \end{aligned} \tag{6.49}$$

For this last step to be legitimate, we need

(6.50)

HYPOTHESIS 1 The $n \times n$ matrix A has the property

$$a_{ii} \neq 0 \qquad \text{for} \qquad i = 1, 2, \ldots, n$$

We are now ready to define our first method.

(6.51)

DEFINITION Let $F(\mathbf{x})$ be given by the right-hand side of (6.49). Then **Jacobi's Method** is to generate the sequence $\mathbf{x}_1 = F(\mathbf{x}_0)$, $\mathbf{x}_2 = F(\mathbf{x}_1), \ldots$.

Of course, having $a_{ii} \neq 0$ for all i does not make A nonsingular, so we do not know if $A\mathbf{x} = \mathbf{b}$ has a solution. Further, even if A is nonsingular, Jacobi's method may not converge to the solution. The next hypotheses guarantees convergence; it is one (of several) that is satisfied by many problems arising in actual applications.

(6.52)

HYPOTHESIS 2 The $n \times n$ matrix A has the property that for each row,

$$|a_{ii}| > |a_{i1}| + \cdots + |a_{i,i-1}| + |a_{i,i+1}| + \cdots + |a_{in}|$$

Such a matrix is said to be **strictly diagonally dominant**.

Note that Hypothesis 2 implies Hypothesis 1. Hypothesis 2 can be weakened to $\geq$ if additional care is taken. We now have what we need.

(6.53)

THEOREM If the $n \times n$ matrix A satisfies Hypothesis 2, then A is invertible. Further, for *any* starting vector $\mathbf{x}_0$, Jacobi's method converges to the unique solution of $A\mathbf{x} = \mathbf{b}$.

Here is an example.

Example 3 Consider the matrix $A = \begin{bmatrix} 2 & -1 \\ -1 & 2 \end{bmatrix}$ (the 2×2 version of a matrix that arises in finite-difference approximations). If we rewrite system $A\mathbf{x} = \mathbf{b}$ following steps (6.48) and (6.49), we obtain

$$\begin{aligned} x_1 &= (\qquad x_2 + b_1)/2 \\ x_2 &= (x_1 \qquad + b_2)/2 \end{aligned} \tag{6.54}$$

Now let $\mathbf{b} = [3 \quad 6]^T$, so that the exact solution is $\mathbf{x} = [4 \quad 5]^T$, as you can check. Now pretend we do not know the answer and have no idea where to start. So we begin with $\mathbf{x}_0 = [0 \quad 0]^T$ and obtain

For $\mathbf{x}_1$, $$x_1 = \frac{0+3}{2} = 1.5$$

$$x_2 = \frac{0+6}{2} = 3.0$$

For $\mathbf{x}_2$, $$x_1 = \frac{3+3}{2} = 3.00$$

$$x_2 = \frac{1.3+6}{2} = 3.75$$

Putting the next five steps into tabular form, we obtain

	$\mathbf{x}_3$	$\mathbf{x}_4$	$\mathbf{x}_5$	$\mathbf{x}_6$	$\mathbf{x}_7$
x_1	3.375	3.75	3.84375	3.9375	3.9609375
x_2	4.5	4.6975	4.875	4.921875	4.96875

You can see that the $\mathbf{x}_i$'s are approaching $[4 \quad 5]^T$, but not too quickly ■

It would be nice to have a faster method, so we turn to the Gauss–Seidel method.

The Gauss–Seidel Method*

Take a quick look back at the equations in (6.49) that describe Jacobi's method. At each step, all of the old x_i's are on the right and all of the new x_i's are on the left:

$$\begin{aligned} x_1^{\text{NEW}} &= (-a_{12}x_2^{\text{OLD}} - a_{13}x_3^{\text{OLD}} - \cdots - a_{1n}x_n^{\text{OLD}} + b_1)/a_{11} \\ x_2^{\text{NEW}} &= (-a_{21}x_1^{\text{OLD}} - a_{23}x_3^{\text{OLD}} - \cdots - a_{2n}x_n^{\text{OLD}} + b_2)/a_{22} \\ &\vdots \\ x_n^{\text{NEW}} &= (-a_{n1}x_1^{\text{OLD}} - a_{n2}x_2^{\text{OLD}} - \cdots - a_{n,n-1}x_{n-1}^{\text{OLD}} + b_n)/a_{nn} \end{aligned} \tag{6.55}$$

* *This method is labeled "Gauss–Seidel," even though Gauss apparently did not know of it and Seidel did not recommend it. This is suprising, since the method is quite reasonable for many applications.*

Think of computing these equations one at a time from the top down. Thus we know x_1^{NEW} when we compute x_2^{NEW}, we know x_1^{NEW} and x_2^{NEW} when we compute x_3^{NEW}, and so on. The idea behind the Gauss–Seidel method is to start using new values as soon as they become available:

(6.56)
$$\begin{aligned}
x_1^{\text{NEW}} &= (-a_{12}x_2^{\text{OLD}} - a_{13}x_3^{\text{OLD}} - \cdots - a_{1n}x_n^{\text{OLD}} + b_1)/a_{11} \\
x_2^{\text{NEW}} &= (-a_{21}x_1^{\text{NEW}} - a_{23}x_3^{\text{OLD}} - \cdots - a_{2n}x_n^{\text{OLD}} + b_2)/a_{22} \\
x_3^{\text{NEW}} &= (-a_{31}x_1^{\text{NEW}} - a_{32}x_2^{\text{NEW}} - \cdots - a_{3n}x_n^{\text{OLD}} + b_3)/a_{33} \\
&\vdots \\
x_n^{\text{NEW}} &= (-a_{n1}x_1^{\text{NEW}} - a_{n2}x_2^{\text{NEW}} - \cdots - a_{n,n-1}x_{n-1}^{\text{NEW}} + b_n)/a_{nn}
\end{aligned}$$

Fortunately, the same condition implies convergence.

(6.57) THEOREM If the $n \times n$ matrix satisfies Hypothesis 2 (6.52), then for any starting vector $\mathbf{x}_0$, the Gauss–Seidel method converges to the unique solution of $A\mathbf{x} = \mathbf{b}$.

Example 4 If we apply the Gauss–Seidel method to A and $\mathbf{b}$ in Example 3, then the defining equations in (6.54) become

(6.58)
$$\begin{aligned}
x_1^{\text{NEW}} &= (\phantom{x_1^{\text{NEW}}} \quad x_2^{\text{OLD}} + b_1)/2 \\
x_2^{\text{NEW}} &= (x_1^{\text{NEW}} \quad\quad\quad + b_2)/2
\end{aligned}$$

Remembering that $\mathbf{b} = [3 \quad 6]^T$ and starting again with $\mathbf{x}_0 = [0 \quad 0]^T$, we obtain

For $\mathbf{x}_1$,
$$x_1 = \frac{0 + 3}{2} = 1.5$$
$$x_2 = \frac{1.5 + 6}{2} = 3.75$$

For $\mathbf{x}_2$,
$$x_1 = \frac{3.75 + 3}{2} = 3.375$$
$$x_2 = \frac{3.375 + 6}{2} = 4.6875$$

Reverting again to tabular form for the next three steps,

	$\mathbf{x}_3$	$\mathbf{x}_4$	$\mathbf{x}_5$
x_1	3.85375	3.9609375	3.9902344
x_2	4.921875	4.9804688	4.9951172

We can see that the $\mathbf{x}_i$'s are again approaching $[4 \quad 5]^T$, but this time the convergence rate is significantly faster. ■

Example 4 illustrates the fact that for many applications the Gauss–Seidel method converges about twice as fast as Jacobi's method. However, be forewarned that this is not always the case; it is possible to construct examples for which the Jacobi method converges and the Gauss–Seidel method fails (or conversely). Further, convergence rates can be significantly improved by more sophisticated methods, but this will have to wait for other courses. We now turn to a matrix description of the Jacobi and Gauss–Seidel methods; this ties the descriptions of these methods to the introduction and lays the foundation for Section 6.4 and further courses.

A Matrix Description of the Jacobi and Gauss–Seidel Methods

We begin by expressing A as the sum of three matrices, each of which contains an important part of A:

(6.59) $A = L + D + U$, where D contains the diagonal of A, L contains the lower part of A, and U contains the upper part of A

For example, if $A = \begin{bmatrix} 2 & -1 \\ -1 & 2 \end{bmatrix}$, then

$$L = \begin{bmatrix} 0 & 0 \\ -1 & 0 \end{bmatrix}, \qquad D = \begin{bmatrix} 2 & 0 \\ 0 & 2 \end{bmatrix}, \qquad U = \begin{bmatrix} 0 & -1 \\ 0 & 0 \end{bmatrix}$$

Now return to the derivation of Jacobi's method in Equations (6.47)–(6.49). Equation (6.47) is of the form

$$A\mathbf{x} = \mathbf{b} \quad \text{or} \quad (L + D + U)\mathbf{x} = \mathbf{b} \tag{6.60}$$

To obtain the next step (6.48), we keep the diagonal part on the left side of Equation (6.60) and everything else on the right side:

$$D\mathbf{x} = -(L + U)\mathbf{x} + \mathbf{b} \tag{6.61}$$

This is, in fact, the important step, and we should think of it as

(6.62)

> The matrix description of Jacobi's method is given by
>
> $$D\mathbf{x}^{\text{NEW}} = -(L + U)\mathbf{x}^{\text{OLD}} + \mathbf{b}$$

For completeness, we solve for $\mathbf{x}^{\text{NEW}}$ and obtain the matrix equivalent to

Equations (6.49): $\mathbf{x}^{\text{NEW}} = D^{-1}[-(L + U)\mathbf{x}^{\text{OLD}} + \mathbf{b}]$, or

(6.63)

The function description of Jacobi's method is given by

$$\mathbf{x}_{k+1} = F(\mathbf{x}_k) \qquad \text{where}$$
$$F(\mathbf{x}) = D^{-1}[-(L + U)\mathbf{x} + \mathbf{b}]$$

Of course, D^{-1} exists since $d_{ii} \neq 0$ by Hypothesis 2 (6.52).

We now turn to the derivation of the Gauss–Seidel method from the Jacobi method. By examining Equations (6.49) and (6.56), we see that the x_i^{NEW} entries on the right side of Equation (6.56) are multiplied by the lower entries of A. Thus if Equation (6.49) is $\mathbf{x}^{\text{NEW}} = D^{-1}[-(L + U)\mathbf{x}^{\text{OLD}} + \mathbf{b}]$, then Equation (6.56) gives us

(6.64)

The matrix description of the Gauss–Seidel method is given by

$$\mathbf{x}^{\text{NEW}} = D^{-1}[-L\mathbf{x}^{\text{NEW}} - U\mathbf{x}^{\text{OLD}} + \mathbf{b}]$$

We now find the function description of the Gauss–Seidel method. First we solve for $\mathbf{x}^{\text{NEW}}$ by multiplying Equation (6.64) by D to obtain $D\mathbf{x}^{\text{NEW}} = -L\mathbf{x}^{\text{NEW}} - U\mathbf{x}^{\text{OLD}} + \mathbf{b}$ and then

$$(L + D)\mathbf{x}^{\text{NEW}} = -U\mathbf{x}^{\text{OLD}} + \mathbf{b} \tag{6.65}$$

Now $L + D$ is invertible, since it is lower triangular with nonzero entries on the diagonal. We multiply Equation (6.65) by $(L + D)^{-1}$ to obtain $\mathbf{x}^{\text{NEW}} = (L + D)^{-1}[-U\mathbf{x}^{\text{OLD}} + \mathbf{b}]$, or

(6.66)

The function description of the Gauss–Seidel method is given by

$$\mathbf{x}_{k+1} = F(\mathbf{x}_k) \qquad \text{where}$$
$$F(\mathbf{x}) = (L + D)^{-1}[-U\mathbf{x} + \mathbf{b}]$$

Exercise 6.3

1. Show that $f(x) = x^2$ has two fixed points $r = 0, 1$. Then pick several starting values x_0 near each r (both larger and smaller than r) and see what happens to the sequence $x_0, x_1 = f(x_0), x_2 = f(x_1), \ldots$.
2. Show graphically that $f(x) = 1 + \sin x$ has a unique fixed point r. Can you find starting values x_0 for which the sequence $x_0, x_1 = f(x_0), x_2 = f(x_1), \ldots$ does not converge to r?
3. Let $F(x) = x - f(x)/f'(x)$. Show that if $f'(r) \neq 0$, then $f(r) = 0 \Leftrightarrow F(r) = r$.

4. If $f(x) = x^2 - a$ and $a > 0$, show that the F in Exercise 3 simplifies to $F(x) = (x + a/x)/2$. (Assume that $f'(x) = 2x$.) Next, pick an a with a square root you know (say, $a = 9$). Start with x_0 near the square root of a, and see how quickly the sequence x_0, $x_1 = f(x_0)$, $x_2 = f(x_1)$, ... converges.

5. For $f(x) = \tan^{-1} x$, assume that $f'(x) = 1/(1 + x^2)$. Find and simplify $F(x)$ from Exercise 3. Since $r = 0$ is the only zero of f, it follows that $r = 0$ is the only fixed point of F. Can you find an x_0 for which the sequence x_0, $x_1 = F(x_0)$, $x_2 = F(x_1)$, ... does not converge to $r = 0$?

6. Let $f(x) = \tan x$. Show graphically that in each interval $(-\pi/2 + k\pi, \pi/2 + k\pi)$, where k is any integer, $f(x)$ has a unique fixed point. Estimate the fixed point r graphically in the intervals where $k = -1, 0, 1$. Pick x_0's close to each of these three r's, and see if the sequence x_0, $x_1 = f(x_0)$, $x_2 = f(x_1)$, ... converges to anything. (Do not forget to use radian mode on your calculator.)

In Exercises 7–10, apply Jacobi's method to the given system, using $\mathbf{x}_0 = [0 \quad 0 \quad \cdots \quad 0]^T$. Continue the iterations until two successive approximations agree to three significant figures.

7. $$\begin{aligned} 4x_1 + x_2 &= 11 \\ -x_1 + 3x_2 &= 7 \end{aligned}$$

8. $$\begin{aligned} -5x_1 + 3x_2 &= 21 \\ 2x_1 + 4x_2 &= 12 \end{aligned}$$

9. $$\begin{aligned} 3x_1 + x_2 - x_3 &= 7 \\ x_1 - 4x_2 - 2x_3 &= 5 \\ -x_1 + 2x_2 + 5x_3 &= 0 \end{aligned}$$

10. $$\begin{aligned} 7x_1 + 2x_2 - 2x_3 &= 12 \\ x_1 + 2x_2 &= -4 \\ 3x_1 + x_2 - 5x_3 &= 13 \end{aligned}$$

11. Apply the Gauss–Seidel method to Exercise 7, and compare the rates of convergence.

12. Repeat Exercise 11 as it applies to Exercise 8.

13. Repeat Exercise 11 as it applies to Exercise 9.

14. Repeat Exercise 11 as it applies to Exercise 10.

In Exercises 15–20, determine if the given matrix is diagonally dominant.

15. $\begin{bmatrix} 3 & -2 \\ -4 & 5 \end{bmatrix}$

16. $\begin{bmatrix} -5 & 4 \\ 3 & -4 \end{bmatrix}$

17. $\begin{bmatrix} 4 & -5 \\ -3 & 4 \end{bmatrix}$

18. $\begin{bmatrix} -5 & 4 \\ 3 & -3 \end{bmatrix}$

19. $\begin{bmatrix} 14 & -7 & 4 \\ 8 & 15 & -8 \\ 4 & -5 & 10 \end{bmatrix}$

20. $\begin{bmatrix} -9 & -7 & 1 \\ 10 & 16 & -4 \\ -3 & -8 & 15 \end{bmatrix}$

In Exercises 21–24, determine that the Gauss–Seidel method diverges for the given system, starting with $\mathbf{x}_0 = [0 \quad 0 \quad \cdots \quad 0]^T$. Observe that the given system is *not* diagonally dominant.

21. $$\begin{aligned} 2x_1 - 7x_2 &= -34 \\ 5x_1 + x_2 &= -11 \end{aligned}$$

22. $$\begin{aligned} -2x_1 + 3x_2 &= 11 \\ 4x_1 - x_2 &= 3 \end{aligned}$$

23. $$\begin{aligned} 3x_1 + x_2 - 7x_3 &= -3 \\ 10x_1 - 4x_2 - 3x_3 &= 34 \\ -x_1 + 9x_2 + 5x_3 &= -3 \end{aligned}$$

24. $$\begin{aligned} 3x_1 + 9x_2 - 2x_3 &= -19 \\ -x_1 + 2x_2 - 7x_3 &= -44 \\ 12x_1 + x_2 - 4x_3 &= -14 \end{aligned}$$

In Exercises 25–28, interchange the equations of the given system to obtain a diagonally dominant system. Then use the Gauss–Seidel method to solve the new system, starting with $\mathbf{x}_0 = [0 \;\; 0 \;\; \cdots \;\; 0]^T$. Continue the iterations until two successive approximations agree to two significant figures. (NOTE This shows that the *order of equations can be extremely important when solving systems iteratively*.)

25. The system in Exercise 21.

26. The system in Exercise 22.

27. The system in Exercise 23.

28. The system in Exercise 24.

In Exercises 29–30, note that the given system is *not* diagonally dominant. Show that the Gauss–Seidel method converges anyway, starting with $\mathbf{x}_0 = [0 \;\; 0 \;\; \cdots \;\; 0]^T$.

29. $$\begin{aligned} 5x_1 - 6x_2 &= -17 \\ -2x_1 + 3x_2 &= 8 \end{aligned}$$

30. $$\begin{aligned} 5x_1 + 2x_2 - 3x_3 &= 23 \\ x_1 - 4x_2 - 4x_3 &= -7 \\ -2x_1 + 3x_2 + 5x_3 &= -13 \end{aligned}$$

6.4 Matrix Norms

In this section, we shall study the concept of a norm of a matrix, which at first may seem quite strange to you. You already know that a norm of a vector is its "size" or "length." But what should the "size" or "length" of a matrix be? And of what use could it be? It is the purpose of this section to begin to answer these questions.

For us, the norm of a matrix A will mean how much the size of $\mathbf{x}$ is changed by A; to be specific, the norm is the largest ratio $\|A\mathbf{x}\|/\|\mathbf{x}\|$, $\mathbf{x} \neq \mathbf{0}$. We shall also briefly discuss how to "measure" norms in different ways. Finally we shall examine two independent applications: the condition number (discussed briefly in Section 1.7), and the convergence rates of iterative methods (discussed in Section 6.3).

Definitions

We know that if an $n \times n$ matrix A is orthogonal, then $\|A\mathbf{x}\| = \|\mathbf{x}\|$ for all $\mathbf{x}$ in $\mathbb{R}^n$. When A is not orthogonal, then $\|A\mathbf{x}\| \neq \|\mathbf{x}\|$ for all $\mathbf{x}$, and hence A can "magnify" or "shrink" the size of $\mathbf{x}$. We are concerned with the largest amount by which A can increase the size of $\mathbf{x}$; that is, we would like to find the best bound B such that

(6.67) $$\|A\mathbf{x}\| \le B\|\mathbf{x}\| \qquad \text{for all } \mathbf{x} \text{ in } \mathbb{R}^n$$

It turns out that this bound B exists for $n \times n$ matrices and is what we call the *norm* of A.

(6.68)

DEFINITION If A is an $n \times n$ matrix, then the **norm** of A is denoted by $\|A\|$ and is the number defined by

$$\|A\| = \max_{\mathbf{x} \neq \mathbf{0}} \frac{\|A\mathbf{x}\|}{\|\mathbf{x}\|} \qquad \text{where } \mathbf{x} \text{ is in } \mathbb{R}^n$$

THEOREM This maximum exists and, since $\dfrac{\|A\mathbf{x}\|}{\|\mathbf{x}\|} \le \|A\|$ for any particular $\mathbf{x}$, we have

(6.69) $$\|A\mathbf{x}\| \le \|A\|\,\|\mathbf{x}\| \qquad \text{for all } \mathbf{x} \text{ in } \mathbb{R}^n$$

Example 1 Suppose that A is a diagonal matrix $A = \begin{bmatrix} \lambda_1 & & \\ & \ddots & \\ & & \lambda_n \end{bmatrix}$. Let λ_M be the eigenvalue that is largest in absolute value, so that $|\lambda_M| \ge |\lambda_i|$, $1 \le i \le n$. Then $\|A\| = |\lambda_M|$. To see this,

$$\begin{aligned} \|A\mathbf{x}\|^2 = \|[\lambda_1 x_1 \cdots \lambda_n x_n]^T\|^2 &= \lambda_1^2 x_1^2 + \cdots + \lambda_n^2 x_n^2 \le \lambda_M^2 x_1^2 + \cdots + \lambda_M^2 x_n^2 \\ &= \lambda_M^2(x_1^2 + \cdots + x_n^2) = \lambda_M^2\|\mathbf{x}\|^2 \end{aligned}$$

Thus, if $\mathbf{x} \neq \mathbf{0}$, then

(6.70) $$\|A\mathbf{x}\| \le |\lambda_M|\,\|\mathbf{x}\| \qquad \text{so} \qquad \frac{\|A\mathbf{x}\|}{\|\mathbf{x}\|} \le \lambda_M$$

Further, since λ_M is an eigenvalue, $A\mathbf{v} = \lambda_M\mathbf{v}$ for some nonzero $\mathbf{v}$ in $\mathbb{R}^n$. Thus

(6.71) $$\|A\mathbf{v}\| = |\lambda_M|\,\|\mathbf{v}\| \qquad \text{or} \qquad \frac{\|A\mathbf{v}\|}{\|\mathbf{v}\|} = |\lambda_M| \qquad \text{so} \qquad \max_{\mathbf{x} \neq \mathbf{0}} \frac{\|A\mathbf{x}\|}{\|\mathbf{x}\|} \ge |\lambda_M|$$

Together, Equations (6.70) and (6.71) imply $\|A\| = |\lambda_M|$. ■

Example 2 By Example 1,

$$\left\|\begin{bmatrix} 3 & \\ & -2 \end{bmatrix}\right\| = 3 \qquad \text{and} \qquad \left\|\begin{bmatrix} 3 & \\ & -5 \end{bmatrix}\right\| = 5.$$ ■

Example 3 Suppose that A is a symmetric $n \times n$ matrix and that λ_M is the eigenvalue of A that has the largest absolute value. Then again, $\|A\| = |\lambda_M|$. To see this, we know that since A is symmetric, $A = Q\Lambda Q^T$, where Λ is diagonal and Q and Q^T are orthogonal by Theorem (5.45), Section 5.4. Now pick an arbitrary $\mathbf{x}$ in $\mathbb{R}^n$, and let $\mathbf{y} = Q^T\mathbf{x}$. Then $\|\mathbf{y}\| = \|\mathbf{x}\|$ and

$$\frac{\|A\mathbf{x}\|}{\|\mathbf{x}\|} = \frac{\|Q\Lambda Q^T\mathbf{x}\|}{\|\mathbf{x}\|} = \frac{\|\Lambda Q^T\mathbf{x}\|}{\|\mathbf{x}\|} = \frac{\|\Lambda\mathbf{y}\|}{\|\mathbf{y}\|}$$

Thus $\|A\| = \|\Lambda\| = \lambda_M$ by Example 1. ■

Example 4 By Example 1, Section 5.4, the eigenvalues of $A = \begin{bmatrix} 0 & 2 & 2 \\ 2 & 0 & -2 \\ 2 & -2 & 0 \end{bmatrix}$ are 2, 2, -4. Thus $\|A\| = 4$. ■

However, if A is not symmetric, then probably $\|A\| \neq |\lambda_M|$. In fact, in general,

(6.72)

> THEOREM If an $n \times n$ matrix A is arbitrary, then $\|A\| = \sigma_1$, the largest singular value of A.

Singular values are discussed in Section 6.2.

Example 5 Let $A = \begin{bmatrix} 1 & 1 \\ 0 & 2 \end{bmatrix}$. Then the eigenvalues of A are 1 and 2. To see that $\|A\| > 2$, let $\mathbf{x} = [1 \quad 2]^T$. Then $\|\mathbf{x}\| = \sqrt{5}$, and

$$A\mathbf{x} = \begin{bmatrix} 1 & 1 \\ 0 & 2 \end{bmatrix}\begin{bmatrix} 1 \\ 2 \end{bmatrix} = \begin{bmatrix} 3 \\ 4 \end{bmatrix} \qquad \text{so} \qquad \|A\mathbf{x}\| = 5$$

Thus $\|A\| \geq \|A\mathbf{x}\|/\|\mathbf{x}\| = \sqrt{5} \approx 2.24$. In fact, if you have looked at Section 6.2, you can compute $\|A\| = \sigma_1 = \sqrt{3 + \sqrt{5}} \approx 2.28$. ■

We now briefly examine a different norm on $\mathbb{R}^n$ because we shall need it later and also because it introduces the concept of more general norms.

(6.73)

> DEFINITION The **infinity norm**, written ∞-norm, on $\mathbb{R}^n$ is denoted by $\|\mathbf{x}\|_\infty$ and is defined by
>
> $$\|\mathbf{x}\|_\infty = \max\{|x_1|, |x_2|, \ldots, |x_n|\}$$
>
> for $\mathbf{x} = [x_1 \cdots x_n]^T$ in $\mathbb{R}^n$.

This norm satisfies the following properties.

(6.74)

THEOREM If $\mathbf{x}$ and $\mathbf{y}$ are in $\mathbb{R}^n$ and r is a real number, then

(a) $\|\mathbf{x}\|_\infty \geq 0$ and $\mathbf{x} = \mathbf{0} \Leftrightarrow \|\mathbf{x}\|_\infty = 0$
(b) $\|r\mathbf{x}\|_\infty = |r|\,\|\mathbf{x}\|_\infty$
(c) $\|\mathbf{x} + \mathbf{y}\|_\infty \leq \|\mathbf{x}\|_\infty + \|\mathbf{y}\|_\infty$ (triangle inequality)

We know from Section 4.2 that $\|\mathbf{x}\| = \sqrt{\mathbf{x} \cdot \mathbf{x}}$ also satisfies all the properties of Theorem (6.74). These properties define a norm on a general vector space; you will see them again in higher mathematics courses.

We can now use $\|\mathbf{x}\|_\infty$ to define $\|A\|_\infty$ in a way similar to Definition (6.68).

(6.75)

DEFINITION If A is an $n \times n$ matrix, then the **infinity norm** of A, written ∞-norm of A, is denoted by $\|A\|_\infty$ and is the number defined by

$$\|A\|_\infty = \max_{\mathbf{x} \neq \mathbf{0}} \frac{\|A\mathbf{x}\|_\infty}{\|\mathbf{x}\|_\infty} \qquad \text{where } \mathbf{x} \text{ is in } \mathbb{R}^n$$

THEOREM This maximum exists and, since $\dfrac{\|A\mathbf{x}\|_\infty}{\|\mathbf{x}\|_\infty} \leq \|A\|_\infty$ for any particular $\mathbf{x}$, we have

(6.76)
$$\|A\mathbf{x}\|_\infty \leq \|A\|_\infty \|\mathbf{x}\|_\infty \qquad \text{for all } \mathbf{x} \text{ in } \mathbb{R}^n$$

There is an obvious similarity between properties (6.67) and (6.76), but exploiting it will have to wait for other courses. We should point out that $\|\mathbf{x}\|_\infty$ and $\|A\|_\infty$ are *considerably* easier to calculate on a computer than $\|\mathbf{x}\|$ and $\|A\|$; thus the ∞-norms play an important role in numerical mathematics. See Exercises 21–26.

We now turn to the application.

When Iterative Methods Converge*

If you look back at Section 6.3, you can see that the function descriptions of Jacobi's method given in (6.63) and of the Gauss–Seidel method given in (6.66) can be rewritten in the form

(6.77)
$$\mathbf{x}_{k+1} = F(\mathbf{x}_k) \qquad \text{where} \qquad F(\mathbf{x}) = B\mathbf{x} + \mathbf{c}$$

* *This subsection assumes you have covered Section 6.3.*

In this subsection, we shall sketch the ideas behind showing that

1. If B is any $n \times n$ matrix with $\|B\|_* < 1$ (where $\|\cdot\|_* = \|\cdot\|$ or $\|\cdot\|_\infty$), then starting at any vector $\mathbf{x}_0$ in $\mathbb{R}^n$, the sequence generated by $\mathbf{x}_0$, $\mathbf{x}_1 = F(\mathbf{x}_0)$, $\mathbf{x}_2 = F(\mathbf{x}_1), \ldots$ converges to a unique vector $\mathbf{r}$ such that $F(\mathbf{r}) = \mathbf{r}$ (called a **fixed point** of F).
2. If B is derived from Jacobi's method, then $\|B\|_\infty < 1$.

These two statements, together with the derivations of the function descriptions given in (6.63) and (6.66), Section 6.3, explain why Jacobi's method works. The convergence of the Gauss–Seidel method is similar in spirit, but it takes a bit more care and will not be covered here. (See Ortega, *Numerical Analysis*, p. 120.) We shall consider the second statement first.

The hypothesis underlying Jacobi's method and the Gauss–Seidel method is that A is diagonally dominant; that is,

$$|a_{ii}| > |a_{i1}| + \cdots + |a_{i,\,i-1}| + |a_{i,i+1}| + \cdots + |a_{in}| \tag{6.78}$$

for each row $i = 1, \ldots, n$. Let $A = L + D + U$, as described in Equation (6.59), Section 6.3. For Jacobi's method, $B = -D^{-1}(L + U)$ [see Equation (6.63), Section 6.3], so that

$$b_{ii} = 0 \qquad \text{and} \qquad b_{ij} = -\frac{a_{ij}}{a_{ii}} \qquad \text{for } i \neq j \tag{6.79}$$

See Exercise 27. From (6.78) and (6.79), we see that $|b_{i1}| + \cdots + |b_{in}| < 1$. We need to show that

$$\frac{\|B\mathbf{x}\|_\infty}{\|\mathbf{x}\|_\infty} < 1 \qquad \text{for all nonzero } \mathbf{x} \text{ in } \mathbb{R}^n \tag{6.80}$$

So we pick an arbitrary $\mathbf{x}$ in $\mathbb{R}^n$ and let $\mathbf{y} = B\mathbf{x}$. If we let i and j be indices such that $|y_i| = \|\mathbf{y}\|_\infty$ and $|x_j| = \|\mathbf{x}\|_\infty$, then

$$\begin{aligned}
\|B\mathbf{x}\|_\infty &= |y_i| = |b_{i1}x_1 + \cdots + b_{in}x_n| \\
&\leq |b_{i1}||x_1| + \cdots + |b_{in}||x_n| \\
&\leq |b_{i1}||x_j| + \cdots + |b_{in}||x_j| \\
&= [|b_{i1}| + \cdots + |b_{in}|]|x_j| \\
&< |x_j| = \|\mathbf{x}\|_\infty
\end{aligned}$$

Thus $\|B\mathbf{x}\|_\infty / \|\mathbf{x}\|_\infty < 1$, so we are done.

We now turn to the first statement and examine the concepts behind convergence. This topic is a little advanced compared to the rest of the material in this text. It is previewed here to give you a flavor for one important way in which vector norms and matrix norms are used.

First, in general, a function f may have many *fixed points* (points r such that $f(r) = r$). For example, if $f(x) = x^3$, then $f(r) = r$ for $r = -1, 0, 1$, so this f has three fixed points. However, if $F(\mathbf{x}) = B\mathbf{x} + \mathbf{c}$, as in Equation

(6.77), and $\|B\|_* < 1$, then F cannot have more than one fixed point. For if $F(\mathbf{r}) = \mathbf{r}$ and $F(\mathbf{s}) = \mathbf{s}$, then

$$\begin{aligned} \|\mathbf{r} - \mathbf{s}\|_* &= \|F(\mathbf{r}) - F(\mathbf{s})\|_* = \|B\mathbf{r} + \mathbf{c} - (B\mathbf{s} + \mathbf{c})\|_* \\ &= \|B(\mathbf{r} - \mathbf{s})\|_* \le \|B\|_* \|\mathbf{r} - \mathbf{s}\|_* \end{aligned}$$

Thus

$$\|\mathbf{r} - \mathbf{s}\|_* \le \|B\|_* \|\mathbf{r} - \mathbf{s}\|_* \tag{6.81}$$

If $\mathbf{r} = \mathbf{s}$, then $\mathbf{r} - \mathbf{s} = \mathbf{0}$, $\|\mathbf{r} - \mathbf{s}\|_* = 0$, and (6.81) reads $0 \le 0$, which is true. But if $\mathbf{r} \ne \mathbf{s}$, then $\mathbf{r} - \mathbf{s} \ne \mathbf{0}$ and $\|\mathbf{r} - \mathbf{s}\|_* > 0$. So we can divide (6.81) by $\|\mathbf{r} - \mathbf{s}\|_*$ to obtain $1 \le \|B\|_*$. This is a contradiction, since $\|B\|_* < 1$ by hypothesis. Thus we cannot have $\mathbf{r} \ne \mathbf{s}$, and F can have at most one fixed point.

In our case, we know F has at least one fixed point, because we started with the system $A\mathbf{x} = \mathbf{b}$, which has a solution $\mathbf{r}$. For this $\mathbf{r}$, we showed in Section 6.3 that $F(\mathbf{r}) = \mathbf{r}$. It happens that any F of the form $F(\mathbf{x}) = B\mathbf{x} + \mathbf{c}$, with $\|B\|_\infty < 1$, has a fixed point (whether or not it comes from a system $A\mathbf{x} = \mathbf{b}$), but it takes facts from advanced calculus to prove this.

For the last step, we pick any starting vector $\mathbf{x}_0$. To know that the sequence $\mathbf{x}_0$, $\mathbf{x}_1 = F(\mathbf{x}_0)$, $\mathbf{x}_2 = F(\mathbf{x}_1), \ldots$ converges to the fixed point $\mathbf{r}$, it is sufficient to show the norms $\|\mathbf{x}_n - \mathbf{r}\| \to 0$ as $n \to \infty$. Now for any $k \ge 1$, $\mathbf{x}_k = F(\mathbf{x}_{k-1})$, and we know that $\mathbf{r} = F(\mathbf{r})$. Thus

$$\begin{aligned} \|\mathbf{x}_k - \mathbf{r}\|_* &= \|F(\mathbf{x}_{k-1}) - F(\mathbf{r})\|_* = \|B\mathbf{x}_{k-1} + \mathbf{c} - (B\mathbf{r} + \mathbf{c})\|_* \\ &= \|B(\mathbf{x}_{k-1} - \mathbf{r})\|_* \le \|B\|_* \|\mathbf{x}_{k-1} - \mathbf{r}\|_* \end{aligned}$$

Applying this over and over, we see that

$$\begin{aligned} \|\mathbf{x}_n - \mathbf{r}\|_* &\le \|B\|_* \|\mathbf{x}_{n-1} - \mathbf{r}\|_* \le \|B\|_*^2 \|\mathbf{x}_{n-2} - \mathbf{r}\|_* \\ &\le \cdots \le \|B\|_*^n \|\mathbf{x}_0 - \mathbf{r}\|_* \end{aligned}$$

Since $\|B\|_* < 1$, it follows that $\|B\|_*^n \to 0$, as $n \to \infty$; hence $\|\mathbf{x}_n - \mathbf{r}\|_* \to 0$, as $n \to \infty$. This is exactly what we needed to show, so we are done.

We now turn to our next application.

Condition Numbers

Recall from Section 1.7 the difficulty that led to the brief discussion of condition numbers. We start with a system $A\mathbf{x} = \mathbf{b}$, where A is nonsingular. Thus the system has a unique solution (call it $\mathbf{x}$). Suppose we change $\mathbf{b}$ to $\mathbf{b}'$ (as, due to round-off error, we probably do just by entering $\mathbf{b}$ into a computer). Then the solution changes also (say, to $\mathbf{x}'$). The question is, how big is the change from $\mathbf{x}$ to $\mathbf{x}'$ compared to the change from $\mathbf{b}$ to $\mathbf{b}'$? In other words, what is $\|\mathbf{x} - \mathbf{x}'\|_*$ compared to $\|\mathbf{b} - \mathbf{b}'\|_*$? The answer is, sometimes the two changes are comparable, but sometimes the change in $\mathbf{x}$ is *much* greater than the change in $\mathbf{b}$ (and this can produce disastrous results). We need to be able to tell which case is which.

There is one technical detail: we want to compare the *relative* change, not the *absolute* change. Suppose that $\|\mathbf{x} - \mathbf{x}'\|_*$, which is the **absolute** change, is approximately 0.001. If $\|\mathbf{x}\|_* \approx 1000$, then the change in **x** may easily be acceptable; however, if $\|\mathbf{x}\|_* \approx 0.01$, then the change in **x** probably is not acceptable. Thus our question is, if **b** and, consequently, **x** are nonzero, then

(6.82) $$\text{“How does } \frac{\|\mathbf{x} - \mathbf{x}'\|_*}{\|\mathbf{x}\|_*} \text{ compare to } \frac{\|\mathbf{b} - \mathbf{b}'\|_*}{\|\mathbf{b}\|_*}\text{?”}$$

NOTE The fraction $\frac{\|\mathbf{x} - \mathbf{x}'\|_*}{\|\mathbf{x}\|_*}$ is called the **relative change** in **x**.

The question in (6.82) is fairly easy to answer with matrix norms. Since $A\mathbf{x} = \mathbf{b}$ and $A\mathbf{x}' = \mathbf{b}'$ and since A is invertible, we know that

(6.83) $$\|\mathbf{x} - \mathbf{x}'\|_* = \|A^{-1}\mathbf{b} - A^{-1}\mathbf{b}'\|_* = \|A^{-1}(\mathbf{b} - \mathbf{b}')\|_* \le \|A^{-1}\|_*\|\mathbf{b} - \mathbf{b}'\|_*$$

(6.84) $$\|\mathbf{b}\|_* = \|A\mathbf{x}\|_* \le \|A\|_*\|\mathbf{x}\|_* \quad \text{so} \quad \frac{1}{\|\mathbf{x}\|_*} \le \frac{\|A\|_*}{\|\mathbf{b}\|_*}$$

Putting Inequalities (6.83) and (6.84) together gives us

(6.85) $$\frac{\|\mathbf{x} - \mathbf{x}'\|_*}{\|\mathbf{x}\|_*} \le \|A\|_*\|A^{-1}\|_* \frac{\|\mathbf{b} - \mathbf{b}'\|_*}{\|\mathbf{b}\|_*}$$

Now let us define the *condition number* of an invertible matrix:

(6.86)

DEFINITION If A is an invertible matrix and $\|\cdot\|_*$ is a matrix norm, then the **condition number** of A, denoted by $c_*(A)$, is defined by

$$c_*(A) = \|A\|_*\|A^{-1}\|_*$$

THEOREM If A is an invertible matrix and $\mathbf{b} \neq \mathbf{0}$, then

$$\frac{\|\mathbf{x} - \mathbf{x}'\|_*}{\|\mathbf{x}\|_*} \le c_*(A) \frac{\|\mathbf{b} - \mathbf{b}'\|_*}{\|\mathbf{b}\|_*}$$

Of course, the proof follows immediately from the definition and from (6.85). Several comments should be made at this point. First, different norms yield different values for the condition number, but for a given matrix, *all condition numbers are roughly about the same.* Recall from Section 1.7 that what we need to know is the k for which $c \approx 10^k$; what we are saying is that for a given matrix A, the k's are all about the same for different norms. From Section 1.7, the bigger k is, the more likely that it will be troublesome.

Our main problem is how to compute $c_*(A)$. After all, if we use the definition $\|A\|_*\|A^{-1}\|_*$, then we must compute A^{-1} first. However, we want to avoid that much work (and we may not be able to find A^{-1} accurately anyway). Thus we do not compute $c_*(A)$ exactly; we only *estimate* it. A fairly sophisticated but computationally efficient way of estimating $c_\infty(A)$ is used in MATLAB and other computer packages.* For classroom purposes, however, there is a simpler method. Recall from Example 2 that if A is symmetric, then $\|A\| = |\lambda_1|$, where $|\lambda_1| \geq |\lambda_2| \geq \cdots \geq |\lambda_n|$ (and the λ_i's are the eigenvalues of A). We know that the eigenvalues of A^{-1} are $\lambda_1^{-1}, \ldots, \lambda_n^{-1}$. Thus we have $|\lambda_n^{-1}| \geq |\lambda_{n-1}^{-1}| \geq \cdots \geq |\lambda_1^{-1}|$, and hence $\|A^{-1}\| = |\lambda_n^{-1}|$. Taken together,

(6.87) **THEOREM** If A is a symmetric, nonsingular matrix, then $c(A) = |\lambda_1/\lambda_n|$.

In the same way, we can see from Theorem (6.72) that $c(A) = \sigma_1/\sigma_n$ for a nonsymmetric matrix A. If the matrices are small, you may be able to compute (or estimate) the eigenvalues easily; singular values tend to be harder to estimate. If A is large, then the methods described in Section 5.8 yield reasonable rough estimates for λ_1 and λ_n, but again, singular values are much harder to approximate.

Examples are given in the exercises.

Exercise 6.4

In Exercises 1–6, find $\|A\|$ and $\|A^{-1}\|$ for the given matrix A.

1. $\begin{bmatrix} -3 & \\ & -2 \end{bmatrix}$

2. $\begin{bmatrix} 12 & \\ & \frac{2}{3} \end{bmatrix}$

3. $\begin{bmatrix} 5 & -2 \\ -2 & 2 \end{bmatrix}$

4. $\begin{bmatrix} -1 & -6 \\ -6 & 4 \end{bmatrix}$

5. $\begin{bmatrix} 2 & 1 & 1 \\ 1 & 2 & 1 \\ 1 & 1 & 2 \end{bmatrix}$

6. $\begin{bmatrix} 0 & -2 & -2 \\ -2 & 0 & 2 \\ -2 & 2 & 0 \end{bmatrix}$

In Exercises 7–10, find the eigenvalues of the given matrix A. Then find an $\mathbf{x}$ such that $\|A\mathbf{x}\|/\|\mathbf{x}\| > |\lambda_{\max}|$.

7. $\begin{bmatrix} 3 & 1 \\ & -2 \end{bmatrix}$

8. $\begin{bmatrix} 1 & 10 \\ & -2 \end{bmatrix}$

9. $\begin{bmatrix} 1 & -2 \\ -3 & 0 \end{bmatrix}$

10. $\begin{bmatrix} 3 & 1 \\ -3 & 7 \end{bmatrix}$

11. In $\mathbb{R}^n$, show that $\|\cdot\|_\infty$ satisfies properties (a) and (b) of Theorem (6.74).

* *See Cline, Moler, Stewart, Wilkinson,* An estimate for the condition number of a matrix, *SIAM Num. Anal.* ***16****(1979), pp. 368–75.*

12. In $\mathbb{R}^n$, show that $\|\cdot\|_\infty$ satisfies property (c) of Theorem (6.74). (Careful; this is tricky.)

In Exercises 13–18, find $\|\mathbf{x}\|$ and $\|\mathbf{x}\|_\infty$, for the given $\mathbf{x}$.

13. $[3 \quad -4]^T$
14. $[-5 \quad 12]^T$
15. $[2 \quad 1 \quad -2]^T$
16. $[5 \quad -3 \quad 4]^T$
17. $[2 \quad 6 \quad -3]^T$
18. $[6 \quad 2 \quad -9]^T$
19. In $\mathbb{R}^2$, sketch $\{\mathbf{x} \mid \|\mathbf{x}\| = 1\}$ and $\{\mathbf{x} \mid \|\mathbf{x}\|_\infty = 1\}$ on the same axes.
20. In general, is either of the inequalities $\|\mathbf{x}\|_\infty \le \|\mathbf{x}\|$ or $\|\mathbf{x}\| \le \|\mathbf{x}\|_\infty$ true for all $\mathbf{x}$ in $\mathbb{R}^n$? Give proofs or counterexamples to support your answers.

It is true that

$$\|A\|_\infty = \max_i \{|a_{i1}| + |a_{i2}| + \cdots + |a_{in}|\}, \text{ where } 1 \le i \le n \tag{6.88}$$

The one-page proof can be found in almost any numerical analysis textbook. For example,

$$\left\| \begin{bmatrix} 1 & -2 \\ 3 & 1 \end{bmatrix} \right\|_\infty = \max\{3, 4\} = 4$$

In Exercises 21–26, use Equation (6.88) to compute $\|A\|_\infty$ for the given matrix A.

21. $\begin{bmatrix} -3 & 15 \\ 12 & -9 \end{bmatrix}$
22. $\begin{bmatrix} 10 & 10 \\ -7 & -14 \end{bmatrix}$
23. $\begin{bmatrix} -7 & 12 \\ -2 & 15 \end{bmatrix}$
24. $\begin{bmatrix} 6 & -8 \\ 10 & -4 \end{bmatrix}$
25. $\begin{bmatrix} -3 & 16 & -4 \\ 11 & -9 & 2 \\ 8 & -5 & 12 \end{bmatrix}$
26. $\begin{bmatrix} 10 & -9 & 13 \\ -3 & 0 & 22 \\ -7 & 11 & 10 \end{bmatrix}$
27. Show that the properties in (6.79) hold: [that is, suppose A is diagonally dominant, suppose $A = L + D + U$, as described in (6.59), Section 6.3, and suppose $B = -D^{-1}(L + U)$. Then $b_{ii} = 0$ and $b_{ij} = -a_{ij}/a_{ii}$ for $i \ne j$.
28. *(continuation of Exercise 27)* For each i, show that $|b_{i1}| + \cdots + |b_{in}| < 1$. (Hence, from Equation (6.88), we have an alternative proof of (6.80): if A is diagonally dominant, then $\|B\|_\infty < 1$.)

In Exercises 29–32, compute $c(A)$ and $c_\infty(A)$ for the given matrix A. Feel free to use Equation (6.88).

29. $\begin{bmatrix} 1 & 3 \\ 3 & 1 \end{bmatrix}$
30. $\begin{bmatrix} 3 & -2 \\ -2 & 0 \end{bmatrix}$
31. $\begin{bmatrix} 6 & -2 \\ -2 & 3 \end{bmatrix}$
32. $\begin{bmatrix} 0 & -6 \\ -6 & 5 \end{bmatrix}$

In Exercises 33–36, solve the given system. Then perform the indicated change to the system and solve the new system. Finally, compute c_∞ of the coefficient matrix and evaluate the inequality given in Theorem (6.86). (You will have to compute A^{-1} to find $\|A^{-1}\|_\infty$.)

33.
$$\begin{aligned} 2x + y &= 3 \\ 2.001x + y &= 3.002 \end{aligned}$$

Change the second equation to: $= 3.001$

34.
$$\begin{aligned} 3x + \quad 2y &= -1 \\ 3x + 2.001y &= -1.002 \end{aligned}$$

Change the second equation to: $= -1.005$

35.
$$\begin{aligned} 4x - \quad 3y &= 18 \\ 4x - 3.001y &= 18.002 \end{aligned}$$

Change the second equation to: $= 17.998$

36.
$$\begin{aligned} 2x - \quad 5y &= -3 \\ 2x - 5.002y &= -3.002 \end{aligned}$$

Change the second equation to: $= -3.006$

37. (a) What is the name of the inequality that gives us

$$\|(A + B)\mathbf{x}\|_* \leq \|A\mathbf{x}\|_* + \|B\mathbf{x}\|_*$$

(b) Use (a) to show that $\|A + B\|_* \leq \|A\|_* + \|B\|_*$.

38. (a) Explain why $\|AB\mathbf{x}\|_* \leq \|A\|_*\|B\|_*\|\mathbf{x}\|_*$.
(b) Use (a) to show that $\|AB\|_* \leq \|A\|_*\|B\|_*$.
(c) Use (b) to show that $c_*(AB) \leq c_*(A)c_*(B)$.

39. Show that if λ is any eigenvalue of A, then $|\lambda| \leq \|A\|_*$.

6.5 General Vector Spaces and Linear Transformations Over an Arbitrary Field

This section is an introduction to some of the subjects found in the next level of abstraction in linear algebra. You will meet such subjects if you study a variety of topics ranging from abstract algebra to applied coding theory. We begin by discussing fields, which are really abstract number systems.

Fields

Most of the time in this text we have been using the real number system. In a few places in Chapter 5, we had to make reference to the complex number system (see Sections 5.2 and 5.4). For coding theory (Section 3.9), we used the finite number system $Z_2 = \{0, 1\}$. All of these are examples of fields.

(6.89)

DEFINITION A **field** F is a set on which addition and multiplication are defined so that for all x, y, and z in F,

(a) $x + (y + z) = (x + y) + z$
Associative law of addition

(b) $x + y = y + x$
Commutative law of addition

(c) There is a **zero**, 0, in F such that $x + 0 = x$
Additive identity

(d) For all x in F, there is **negative**, $-x$, in F such that $x + (-x) = 0$.
Additive inverse

(e) $x(y + z) = xy + xz$ Distributive law

(f) $x(yz) = (xy)z$
Associative law of multiplication

(g) $xy = yx$
Commutative law of multiplication

(h) There is a **one**, 1, in F, $1 \neq 0$, such that $x1 = x$. Multiplicative identity

(i) For all *nonzero* x in F, there is a **(multiplicative) inverse**, x^{-1}, in F such that $xx^{-1} = 1$. Multiplicative inverse

Example 1 You are already acquainted with at least three fields

(a) The set of all real numbers, $\mathbb{R}$.
(b) The set of all complex numbers $\mathbb{C} = \{a + bi\}$, briefly discussed in Section 5.4
(c) The set of all rational numbers $Q = \left\{\frac{a}{b} \,\middle|\, a, b \text{ are integers}, b \neq 0\right\}$.

We leave it to the exercises to verify that these three examples satisfy the axioms of a field, as given in Definition (6.89). ■

Example 2 The set of all integers $Z = \{\ldots, -2, -1, 0, 1, 2, \ldots\}$ is *not* a field.* It satisfies (a)–(h) of (6.89) but not (*i*); it is called a **commutative ring with one**. ■

Example 3 In Section 3.9, we introduced $Z_2 = \{0, 1\}$, which is a finite field. ■

* *The letter "Z" comes from the German word for "number," Zahlen.*

Example 3 is a special case of a whole family of finite fields, which we are now going to discuss briefly.

Pick an integer $n \geq 2$ and fix it throughout this discussion. We are going to describe Z_n, and some of its properties will depend upon this n.

In preparation we consider the rational numbers. Recall that many different ratios represent the same rational numbers. For example,

$$\frac{2}{3}, \frac{4}{6}, \frac{-10}{-15}, \ldots$$

are all different ratios, but they all represent the same rational number. To represent that number, we usually pick the ratio that is reduced to lowest terms, in this case $\frac{2}{3}$.

In Z_n elements are expressed as $\bar{k}$, where k is an integer. We say $\bar{k}$ and $\bar{\ell}$ are *equal*, $\bar{k} = \bar{\ell}$, if $k - \ell$ is a multiple of n. When this happens, $\bar{k}$ and $\bar{\ell}$ represent the same "number" in Z_n, and to represent that number we usually pick the k between 0 and $n - 1$.

NOTE In this notation, $Z_2 = \{\bar{0}, \bar{1}\}$, whereas previously we wrote $Z_2 = \{0, 1\}$.

Example 4 In Z_6

$$\begin{aligned} \bar{0} &= \bar{6} = \overline{-6} = \overline{12} = \overline{-12} = \cdots \\ \bar{1} &= \bar{7} = -5 = \overline{13} = \overline{-11} = \cdots \\ &\vdots \\ \bar{5} &= 11 = \overline{-1} = \overline{17} = \overline{-7} = \cdots \end{aligned}$$

and $Z_6 = \{\bar{0}, \bar{1}, \ldots, \bar{5}\}$. ■

Example 5 In Z_7

$$\begin{aligned} \bar{0} &= \bar{7} = -7 = 14 = \overline{-14} = \cdots \\ \bar{1} &= \bar{8} = \overline{-6} = \overline{15} = \overline{-13} = \cdots \\ &\vdots \\ \bar{6} &= \overline{13} = \overline{-1} = \overline{20} = \overline{-8} = \cdots \end{aligned}$$

and $Z_7 = \{\bar{0}, \bar{1}, \ldots, \bar{6}\}$. ■

You can see we must be careful what value n has; for example, $\bar{1}$ in Z_6 and $\bar{1}$ in Z_7 are quite different. We define **addition** and **multiplication** in Z_n by

$$\bar{k} + \bar{\ell} = \overline{k + \ell} \quad \text{and} \quad \bar{k} \cdot \bar{\ell} = \overline{k\ell} \tag{6.90}$$

We leave it to the exercises to show:

(6.91)

> THEOREM $Z_n = \{\bar{0}, \bar{1}, \ldots, \overline{n-1}\}$, together with the definitions of addition and multiplication given in (6.90), satisfy axioms (a)–(h) of the definition of a field (6.89).

Thus Z_n is at least a commutative ring with one, just like the integers described in Example 2. We shall also leave it to the exercises to show:

(6.92)

> THEOREM Z_n is a field if and only if n is a prime.

This discussion has been quite brief, and there are *many* other kinds of fields around. Many of them are useful in both theory and applications, and this is why the next topic is important.

Vector Spaces Over Fields

In Section 3.3, we defined a vector space to be a set whose elements you can add and multiply by real numbers so as to satisfy certain axioms. We now replace the real numbers, $\mathbb{R}$, with any field, F.

(6.93)

> DEFINITION Let F be a field. A set V is a **vector space over** F if addition is defined on V and scalar multiplication by elements of F is defined on V, so that all the axioms of a vector space [Definition (3.19), Section 3.3] still hold (if you replace the word "number" with "element of the field F" in the definition).

We have essentially been studying vector spaces over the reals, $\mathbb{R}$. In Chapter 5 we saw a little of vector spaces over the complex numbers, $\mathbb{C}$, and in Section (3.9) we worked in vector spaces over Z_2. The next example is very important no matter what the field is.

Example 6 Let F be any field. Let F^n be the set of all n-tuples of elements of F,

$$F^n = \{(x_1, x_2, \ldots, x_n) \mid x_i \in F\}$$

We define addition and scalar multiplication in the usual way,

$$(x_1, \ldots, x_n) + (y_1, \ldots, y_n) = (x_1 + y_1, \ldots, x_n + y_n)$$
$$r(x_1, \ldots, x_n) = (rx_1, \ldots, rx_n), \qquad \text{for } r \text{ in } F$$

Then $V = F^n$ is a vector space over F. ■

(6.94) REMARKABLE FACT 1 Most of the concepts in Chapter 3, such as linear independence, dependence, span, basis, and dimension, go through when working in vector spaces over an arbitrary field. The same proofs work, and it is straightforward, though time consuming, to verify this.

(6.95) REMARKABLE FACT 2 The concepts of norm and inner product do *not* hold in the same way for some fields. For example, if we try $\|\mathbf{v}\| = x_1^2 + \cdots + x_1^2$ in Z_2^n, then for $v = (\bar{1}, \bar{1}) \in Z_2^2$,

$$\|\mathbf{v}\| = \bar{1}^2 + \bar{1}^2 = \bar{1} + \bar{1} = \bar{0}$$

Thus $\|\mathbf{v}\| = 0$ but $\mathbf{v} \neq 0$, a most undesirable result.

As stated in (6.94), there is a tremendous amount of material that holds in vector spaces over arbitrary fields. This revelation leads us to the next topic.

Matrices and Linear Transformations

We now briefly develop the relationship between matrices, coordinates, and linear transformations. By a **matrix over a field** F we simply mean a matrix whose entries are from F. We add, scalar multiply, and multiply matrices over F exactly the way we have been doing (the scalars also come from F). We also define null space, column space, row space, rank, and so on, in exactly the same way. What is more, all the theorems, such as

$$rk(A) + \dim NS(A) = \text{no. of cols. of } A$$

hold for matrices A over F.

Suppose now V is an n-dimensional vector space over F and $\alpha = \{\boldsymbol{\alpha}_1, \ldots, \boldsymbol{\alpha}_n\}$ is a basis for V.

Example 7 Let $V = Z_2^3$, which is three dimensional. If

$$\boldsymbol{\alpha}_1 = (\bar{1}, \bar{0}, \bar{0}), \qquad \boldsymbol{\alpha}_2 = (\bar{1}, \bar{1}, \bar{0}), \qquad \text{and} \qquad \boldsymbol{\alpha}_3 = (\bar{1}, \bar{1}, \bar{1})$$

then $\alpha = \{\boldsymbol{\alpha}_1, \boldsymbol{\alpha}_2, \boldsymbol{\alpha}_3\}$ is a basis for V, as you can check. ■

Let V and $\alpha = \{\boldsymbol{\alpha}_1, \cdots, \boldsymbol{\alpha}_n\}$ be as above. If $\mathbf{v}$ is a vector in V, then $\mathbf{v}$ is a linear combination of the basis α, $\mathbf{v} = a_1\boldsymbol{\alpha}_1 + a_2\boldsymbol{\alpha}_2 + \cdots + a_n\boldsymbol{\alpha}_n$. By the **coordinate matrix** of $\mathbf{v}$ in the basis α, which we denote by $[\mathbf{v}]_\alpha$, we mean the

$n \times 1$ matrix

$$[\mathbf{v}]_\alpha = \begin{bmatrix} a_1 \\ a_2 \\ \vdots \\ a_n \end{bmatrix}$$

Example 8 If $\mathbf{v} = (\bar{0}, \bar{1}, \bar{1})$ in Z_2^3 and α is the basis given in Example 7, then

$$\mathbf{v} = \boldsymbol{\alpha}_1 + \boldsymbol{\alpha}_3 = \bar{1}\boldsymbol{\alpha}_1 + \bar{0}\boldsymbol{\alpha}_2 + \bar{1}\boldsymbol{\alpha}_3 \qquad \text{so} \qquad [\mathbf{v}]_\alpha = \begin{bmatrix} \bar{1} \\ \bar{0} \\ \bar{1} \end{bmatrix}$$ ■

Suppose now that V and W are vector spaces over the *same* field F. Then we can easily extend the definition of a linear transformation $T: V \to W$ from Definition (4.3), Section 4.1, to this situation.

(6.96)

DEFINITION Let V and W be vector spaces over the same field F. A function $T: V \to W$ is called a **linear transformation** if for all $\mathbf{u}$ and $\mathbf{v}$ in V and for all scalars r in F
(a) $T(\mathbf{u} + \mathbf{v}) = T(\mathbf{u}) + T(\mathbf{v})$
(b) $T(r\mathbf{u}) = rT(\mathbf{u})$

Suppose V and W are finite-dimensional vector spaces over the same field F, $\alpha = \{\boldsymbol{\alpha}_1, \ldots, \boldsymbol{\alpha}_n\}$ is a basis for V, and $\beta = \{\boldsymbol{\beta}_1, \ldots, \boldsymbol{\beta}_m\}$ is a basis for W. Let $T: V \to W$ be a linear transformation. Then we can associate with T, α, and β an $m \times n$ matrix called (logically) the **matrix associated with** T (in the bases α and β) and denoted by $[T]_{\beta\alpha}$ as follows:

1. For each $\boldsymbol{\alpha}_i$, $T(\boldsymbol{\alpha}_i)$ is a vector in W. Find $[T(\boldsymbol{\alpha}_i)]_\beta$, the coordinate matrix of $T(\boldsymbol{\alpha}_i)$ in the basis β, which is an $m \times 1$ matrix.
2. Let $[T]_{\beta\alpha}$ be the $m \times n$ matrix with $[T(\alpha_i)]_\beta$ as its ith column.

Example 9 Let $V = Z_2^3$ and let α be the basis described in Example 7. Let $W = Z_2^2$. Let $\boldsymbol{\beta}_1 = (\bar{0}, \bar{1})$ and $\boldsymbol{\beta}_2 = (\bar{1}, \bar{1})$, so that $\beta = \{\boldsymbol{\beta}_1, \boldsymbol{\beta}_2\}$ is a basis for W. Let $T: V \to W$ be the linear transformation given by

$$T(x_1, x_2, x_3) = (x_1 + x_3, x_2 + x_3)$$

Find $[T]_{\beta\alpha}$.

Solution By Example 7, $\alpha = \{\boldsymbol{\alpha}_1, \boldsymbol{\alpha}_2, \boldsymbol{\alpha}_3\}$, where $\boldsymbol{\alpha}_1 = (\bar{1}, \bar{0}, \bar{0})$, $\boldsymbol{\alpha}_2 = (\bar{1}, \bar{1}, \bar{0})$, and $\boldsymbol{\alpha}_3 = (\bar{1}, \bar{1}, \bar{1})$. We first compute the $T(\boldsymbol{\alpha}_i)$'s:

$$\begin{aligned} T(\boldsymbol{\alpha}_1) &= T(\bar{1}, \bar{0}, \bar{0}) = (\bar{1}, \bar{0}) = \boldsymbol{\beta}_1 + \boldsymbol{\beta}_2 = \bar{1}\boldsymbol{\beta}_1 + \bar{1}\boldsymbol{\beta}_2 \\ T(\boldsymbol{\alpha}_2) &= T(\bar{1}, \bar{1}, \bar{0}) = (\bar{1}, \bar{1}) = \boldsymbol{\beta}_2 = \bar{0}\boldsymbol{\beta}_1 + \bar{1}\boldsymbol{\beta}_2 \\ T(\boldsymbol{\alpha}_3) &= T(\bar{1}, \bar{1}, \bar{1}) = (\bar{0}, \bar{0}) = 0 = \bar{0}\boldsymbol{\beta}_1 + \bar{0}\boldsymbol{\beta}_2 \end{aligned}$$

Then

$$[T(\boldsymbol{\alpha}_1)]_\beta = \begin{bmatrix} \bar{1} \\ \bar{1} \end{bmatrix}, \qquad [T(\boldsymbol{\alpha}_2)]_\beta = \begin{bmatrix} \bar{0} \\ \bar{1} \end{bmatrix}, \qquad [T(\boldsymbol{\alpha}_3)]_\beta = \begin{bmatrix} \bar{0} \\ \bar{0} \end{bmatrix}$$

Thus

$$[T]_{\beta\alpha} = \left[[T(\boldsymbol{\alpha}_1)]_\beta \,\vdots\, [T(\boldsymbol{\alpha}_2)]_\beta \,\vdots\, [T(\boldsymbol{\alpha}_3)]_\beta\right] = \begin{bmatrix} \bar{1} & \bar{0} & \bar{0} \\ \bar{1} & \bar{1} & \bar{0} \end{bmatrix}$$ ■

We can now state the very nice relationship between linear transformations, bases, and matrices.

(6.97) **THEOREM** Let V and W be finite-dimensional vector spaces over F and let $T: V \to W$ be a linear transformation. Let α be a basis for V and β be a basis for W. Then for any $\mathbf{v}$ in V,

$$[T]_{\beta\alpha}[\mathbf{v}]_\alpha = [T(\mathbf{v})]_\beta$$

We first give an example to illustrate this relationship, and then we prove the theorem.

Example 10 Let V, W, T, α, and β be as in Example 9. Let $\mathbf{v} = (\bar{0}, \bar{1}, \bar{1})$ be as in Example 8. Then by Examples 8 and 9,

(6.98)
$$[\mathbf{v}]_\alpha = \begin{bmatrix} \bar{1} \\ \bar{0} \\ \bar{1} \end{bmatrix}, \qquad [T]_{\beta\alpha} = \begin{bmatrix} \bar{1} & \bar{0} & \bar{0} \\ \bar{1} & \bar{1} & \bar{0} \end{bmatrix}$$

By Example 9, $T(x_1, x_2, x_3) = (x_1 + x_3, x_2 + x_3)$, so

$$T(\mathbf{v}) = T(\bar{0}, \bar{1}, \bar{1}) = (\bar{1}, \bar{0}) = \boldsymbol{\beta}_1 + \boldsymbol{\beta}_2 \qquad \text{and hence} \qquad [T(\mathbf{v})]_\beta = \begin{bmatrix} \bar{1} \\ \bar{1} \end{bmatrix}$$

We now compute

$$[T]_{\beta\alpha}[\mathbf{v}]_\alpha = \begin{bmatrix} \bar{1} & \bar{0} & \bar{0} \\ \bar{1} & \bar{1} & \bar{0} \end{bmatrix} \begin{bmatrix} \bar{1} \\ \bar{0} \\ \bar{1} \end{bmatrix} = \begin{bmatrix} \bar{1} \\ \bar{1} \end{bmatrix} = [T(\mathbf{v})]_\beta$$

which illustrates Theorem (6.97). ■

Proof of Theorem (6.97) Any $\mathbf{v}$ in V is a linear combination of the $\boldsymbol{\alpha}_i$'s, and that will give us $[\mathbf{v}]_\alpha$,

(6.99)
$$\mathbf{v} = c_1\boldsymbol{\alpha}_1 + c_2\boldsymbol{\alpha}_2 + \cdots + c_n\boldsymbol{\alpha}_n \qquad \text{so} \qquad [\mathbf{v}]_\alpha = \begin{bmatrix} c_1 \\ \vdots \\ c_n \end{bmatrix}$$

Next,

$$T(\mathbf{v}) = T(c_1\boldsymbol{\alpha}_1 + \cdots + c_n\boldsymbol{\alpha}_n) = c_1T(\boldsymbol{\alpha}_1) + \cdots + c_nT(\boldsymbol{\alpha}_n)$$

since T is linear. Therefore, by elementary properties of matrices,

$$\begin{aligned}[T(\mathbf{v})]_\beta &= [c_1T(\boldsymbol{\alpha}_1) + \cdots + c_nT(\boldsymbol{\alpha}_n)]_\beta \\ &= c_1[T(\boldsymbol{\alpha}_1)]_\beta + \cdots + c_n[T(\boldsymbol{\alpha}_n)]_\beta \\ &= \left[[T(\boldsymbol{\alpha}_1)]_\beta \vdots \cdots \vdots [T(\boldsymbol{\alpha}_n)]_\beta\right]\begin{bmatrix} c_1 \\ \vdots \\ c_n \end{bmatrix} \\ &= [T]_{\beta\alpha}[\mathbf{v}]_\alpha\end{aligned}$$

by the definitions of $[T]_{\beta\alpha}$ and $[\mathbf{v}]_\alpha$. ■

The correspondence discussed in Theorem (6.97) goes both ways. Suppose we are given V with $\alpha = \{\boldsymbol{\alpha}_1, \ldots, \boldsymbol{\alpha}_n\}$ as a basis, W with $\beta = \{\boldsymbol{\beta}_1, \ldots, \boldsymbol{\beta}_m\}$ as a basis, and an $m \times n$ matrix A. Then for any $\mathbf{v}$ in V,

1. Find $A[\mathbf{v}]_\alpha = \begin{bmatrix} b_1 \\ \vdots \\ b_m \end{bmatrix}$.
2. Let $T(\mathbf{v}) = b_1\boldsymbol{\beta}_1 + \cdots + b_m\boldsymbol{\beta}_m$.

You can verify that T is linear and $[T]_{\beta\alpha} = A$. Therefore,

(6.100) **THEOREM** There is a one-to-one correspondence between linear transformations from V to W and $m \times n$ matrices over F.

This concludes our brief introduction to general vector spaces. As you have seen, all the basic relationships hold in general vector spaces. The material relating to eigenvalues and inner product spaces does, also. The end result is a beautiful and powerful theory that is extremely useful to both theoretical and applied mathematicians.

Exercise 6.5

1. Verify that the real numbers, $\mathbb{R}$, satisfy the axioms of a field [Definition (6.89)].
2. Verify that the complex numbers, $\mathbb{C}$, satisfy the axioms of a field. [Make sure you see that it is important to be able to express $(a + bi)^{-1}$ in the form $c + di$ *and* that you know how to do it.]

3. Verify that the rational numbers, Q, satisfy the axioms of a field.
4. Verify that Z_n satisfies axioms (a), (c), (e), and (g) of a field.
5. Verify that Z_n satisfies axioms (b), (d), (f), and (h) of a field.
6. Show that if n is not a prime, there are nonzero elements $\bar{a}, \bar{b}$ in Z_n such that $\bar{a}\bar{b} = \bar{0}$. Why does this imply that Z_n cannot satisfy axiom (i)? (HINT If $\bar{a}^{-1}$ exists, multiply $\bar{a}\bar{b} = \bar{0}$ by $\bar{a}^{-1}$ and get a contradiction.)

In Exercises 7–10, assume n is a prime.

7. Suppose a is not a multiple of n. Show that if ab is a multiple of n, then b must be a multiple of n.
8. Suppose $\bar{a} \neq \bar{0}$ in Z_n. Use Exercise 7 to show that if $\bar{a}\bar{b} = \bar{a}\bar{c}$, then $\bar{b} = \bar{c}$.
9. Suppose $\bar{a} \neq \bar{0}$ in Z_n. Use Exercise 8 to show $\bar{a} \cdot \bar{0}, \bar{a} \cdot \bar{1}, \ldots, \bar{a} \cdot \overline{n-1}$ are all distinct elements of Z_n.
10. Suppose $\bar{a} \neq \bar{0}$ in Z_n. Use Exercise 9 to show there is a $\bar{b}$ in Z_n such that $\bar{a}\bar{b} = \bar{1}$ in Z_n. Thus Z_n satisfies axiom (i) of a field.

In Exercises 11–16, we work in the field $Z_3 = \{\bar{0}, \bar{1}, \bar{2}\}$.

11. Find the multiplicative inverses for each of the nonzero elements of Z_3.
12. In Z_3^2 let $\boldsymbol{\alpha}_1 = (\bar{1}, \bar{1})$ and $\boldsymbol{\alpha}_2 = (\bar{1}, \bar{2})$. Show that $\alpha = \{\boldsymbol{\alpha}_1, \boldsymbol{\alpha}_2\}$ is a basis for Z_3^2.
13. In Z_3^3 let $\boldsymbol{\beta}_1 = (\bar{1}, \bar{0}, \bar{0})$, $\boldsymbol{\beta}_2 = (\bar{2}, \bar{1}, \bar{0})$, and $\boldsymbol{\beta}_3 = (\bar{0}, \bar{2}, \bar{1})$. Show that $\beta = \{\boldsymbol{\beta}_1, \boldsymbol{\beta}_2, \boldsymbol{\beta}_3\}$ is a basis for Z_3^3.
14. Let $T(x, y) = (\bar{2}x, x + \bar{2}y, y)$. Show that $T: Z_3^2 \to Z_3^3$ is linear.
15. Find $[T]_{\beta\alpha}$.
16. Let $\mathbf{x} = (\bar{2}, \bar{1})$. Find $[\mathbf{x}]_\alpha$, $[T(\mathbf{x})]_\beta$ and verify that $[T(\mathbf{x})]_\beta = [T]_{\beta\alpha}[\mathbf{x}]_\alpha$.

In Exercises 17–22 we work in the field $Z_5 = \{\bar{0}, \bar{1}, \bar{2}, \bar{3}, \bar{4}\}$.

17. Find the multiplicative inverses for each nonzero element of Z_5.
18. In Z_5^3 let $\boldsymbol{\alpha}_1 = (\bar{4}, \bar{0}, \bar{0})$, $\boldsymbol{\alpha}_2 = (\bar{3}, \bar{2}, \bar{0})$, and $\boldsymbol{\alpha}_3 = (\bar{0}, \bar{1}, \bar{1})$. Show that $\alpha = \{\boldsymbol{\alpha}_1, \boldsymbol{\alpha}_2, \boldsymbol{\alpha}_3\}$ is a basis for $\mathbf{Z}_5^3$.
19. In Z_5^2 let $\boldsymbol{\beta}_1 = (\bar{2}, \bar{1})$ and $\bar{\boldsymbol{\beta}}_2 = (\bar{1}, \bar{4})$. Show that $\beta = \{\boldsymbol{\beta}_1, \boldsymbol{\beta}_2\}$ is a basis for Z_5^2.
20. Let $T(x, y, z) = (\bar{3}x + \bar{2}z, y + \bar{4}z)$. Show that $T: Z_5^3 \to Z_5^2$ is linear.
21. Find $[T]_{\beta\alpha}$.
22. Let $\mathbf{x} = (\bar{4}, \bar{3}, \bar{2})$. Find $[\mathbf{x}]_\alpha$, $[T(\mathbf{x})]_\beta$ and verify $[T(\mathbf{x})]_\beta = [T]_\beta = [T]_{\beta\alpha}[\mathbf{x}]_\alpha$.

Review Exercises

In Exercises 1 and 2, use $f \cdot g = \int_{-\pi}^{\pi} f(t)g(t)\,dt$.

1. Show that $\sin nx \cdot \sin mx = 0$.

2. Find the Fourier series for $F(x) = \sin 2x$.

3. If we used $f \cdot g = \int_0^{\pi} f(t)g(t)\,dt$, for which m and n do we obtain $\sin nx \cdot \sin mx = 0$?

4. Let

$$F(x) = \begin{cases} x, & 0 \le x \le \pi \\ 2\pi - x, & \pi < x \le 2\pi \end{cases}$$

and use $f \cdot g = \int_0^{2\pi} f(t)g(t)\,dt$.

(a) Using the formulas from Theorem (6.9), Section 6.1, compute a_n and b_n for $n \le 2$. (HINT First compute $\int_0^{\pi}$ and then use symmetry for $\int_{\pi}^{2\pi}$.)
(b) Graph $y = F(x)$ and $y = a_0 + a_1 \cos x + b_1 \sin x + a_2 \cos 2x + b_2 \sin 2x$.

5. Apply the Gram–Schmidt process to $f_0(x) = 1$, $f_1(x) = x$, and $f_2(x) = x^2$ on $[0, 2]$.

6. Use the results of Exercise 5 to find the equation of the parabola that best fits $y = \sqrt{x}$ on $[0, 2]$.

7. Find the pseudoinverse of

$$A = \begin{bmatrix} 1 & 1 & 1 \\ 1 & 1 & 1 \end{bmatrix}$$

8. Use the pseudoinverse to solve the least-squares problem for $x + y = 2$. (NOTE Here $A = [1 \quad 1]$.)

9. Determine which of Z_7 or Z_8 is a field.

10. How many vectors are there in Z_5^3? How many "lines," that is, one-dimensional subspaces, are there in Z_5^3?

11. In Z_3^3 show that $\boldsymbol{\alpha}_1 = (\bar{1}, \bar{2}, \bar{1})$ and $\boldsymbol{\alpha}_2 = (\bar{2}, \bar{1}, \bar{2})$ are linearly dependent. If $\boldsymbol{\alpha}_1$ and $\boldsymbol{\alpha}_2$ were vectors in Z_5^3, would they be linearly dependent or independent?

12. For $\boldsymbol{\alpha}_1$ and $\boldsymbol{\alpha}_2$ of Exercise 11, find an $\boldsymbol{\alpha}_3$ so that $\boldsymbol{\alpha}_1$, $\boldsymbol{\alpha}_2$, and $\boldsymbol{\alpha}_3$ is a basis for Z_5^3. How many such $\boldsymbol{\alpha}_3$'s are there?

13. In Z_3^3 let $\boldsymbol{\beta}_1 = (\bar{1}, \bar{2}, \bar{1})$, $\boldsymbol{\beta}_2 = (\bar{1}, \bar{1}, \bar{2})$, $\boldsymbol{\beta}_3 = (\bar{1}, \bar{1}, \bar{0})$. Show that $\boldsymbol{\beta}_1$, $\boldsymbol{\beta}_2$, and $\boldsymbol{\beta}_3$ form a basis for Z_3^3.

14. Let $\boldsymbol{\varepsilon}_1 = (\bar{1}, \bar{0})$ and $\boldsymbol{\varepsilon}_2 = (\bar{0}, \bar{1})$ be the standard basis for Z_3^2 and let $T: Z_3^3 \to Z_3^2$ by $T(x, y, z) = (x - y + \bar{2}z, \bar{2}x + y - z)$. Find $[T]_{\varepsilon\beta}$.

15. The matrix $A = \begin{bmatrix} 2 & -1 & 0 \\ -1 & 2 & -1 \\ 0 & -1 & 2 \end{bmatrix}$ has eigenvalues of 2, $2 \pm \sqrt{2}$. Find the Jacobi matrix $-D^{-1}(L + U)$ and its eigenvalues and the Gauss–Seidel matrix $-(L + D)^{-1}U$ and its eigenvalues.

16. What is the relationship between the condition numbers of A and of A^{-1}?

17. Let $A = \begin{bmatrix} 4 & 3 \\ 2 & -1 \end{bmatrix}$, $\mathbf{x} = \begin{bmatrix} 3 \\ -4 \end{bmatrix}$, and $\mathbf{y} = A\mathbf{x}$. Find $\|A\|_*$, $\|\mathbf{x}\|_*$, and $\|\mathbf{y}\|_*$, and verify $\|A\mathbf{x}\|_* \le \|A\|_*\|\mathbf{x}\|_*$ for $\|\cdot\|_* = \|\cdot\|$ and $\|\cdot\|_\infty$.

APPENDIXES

Appendix A More on LU Decompositions

It is the purpose of this appendix to show how to prove some of the statements and theorems in Section 1.5, particularly those involving interchanges while computing LU decompositions. We begin by taking a slightly different approach to the construction of an LU decomposition.

We begin with the matrix

$$A = \begin{bmatrix} 2 & 8 & 1 & 1 \\ 4 & 13 & 3 & -1 \\ -2 & -5 & -3 & 3 \\ -6 & -18 & -1 & 1 \end{bmatrix}$$

given in Example 1 of Section 1.5. We showed there that if

$$U = \begin{bmatrix} 2 & 8 & 1 & 1 \\ 0 & -3 & 1 & -3 \\ 0 & 0 & -1 & 1 \\ 0 & 0 & 0 & 2 \end{bmatrix}$$

then

$$U = E_{34}E_{24}E_{23}E_{14}E_{13}E_{12}A \tag{A.1}$$

where

$$E_{12} = \begin{bmatrix} 1 & & & \\ -2 & 1 & & \\ & & 1 & \\ & & & 1 \end{bmatrix}, \quad E_{13} = \begin{bmatrix} 1 & & & \\ & 1 & & \\ 1 & & 1 & \\ & & & 1 \end{bmatrix}, \quad E_{14} = \begin{bmatrix} 1 & & & \\ & 1 & & \\ & & 1 & \\ 3 & & & 1 \end{bmatrix}$$

$$E_{23} = \begin{bmatrix} 1 & & & \\ & 1 & & \\ & 1 & 1 & \\ & & & 1 \end{bmatrix}, \quad E_{24} = \begin{bmatrix} 1 & & & \\ & 1 & & \\ & & 1 & \\ & 2 & & 1 \end{bmatrix}, \quad E_{34} = \begin{bmatrix} 1 & & & \\ & 1 & & \\ & & 1 & \\ & & 4 & 1 \end{bmatrix}$$

Equation (A.1) can be condensed by letting $L_1 = E_{14}E_{13}E_{12}$, $L_2 = E_{24}E_{23}$, and (to keep up the pattern) $L_3 = E_{34}$. Then Equation (A.1) becomes

(A.2)
$$U = L_3L_2L_1A$$

where

$$L_1 = \begin{bmatrix} 1 & & & \\ -2 & 1 & & \\ 1 & & 1 & \\ 3 & & & 1 \end{bmatrix}, \quad L_2 = \begin{bmatrix} 1 & & & \\ & 1 & & \\ & 1 & 1 & \\ & 2 & & 1 \end{bmatrix}, \quad L_3 = \begin{bmatrix} 1 & & & \\ & 1 & & \\ & & 1 & \\ & & 4 & 1 \end{bmatrix}$$

Note that L_i has 1's on the diagonal and the only other nonzero entries are in the ith column. For our purposes another of the important properties the L's have is that if

$$A \to A_1 \to A_2 \to A_3 = U$$

is the sequence obtained for A in Example 1 of Section 1.5, then

$$L_1A = A_1, \qquad L_2A_1 = A_2, \qquad \text{and} \qquad L_3A_2 = A_3 = U$$

We can now describe the general situation.

(A.3)

THEOREM If Gaussian elimination can be performed on A, yielding nonzero pivots without any row interchanges, then

$$U = L_{n-1} \cdots L_2L_1A$$

where U is either upper triangular if A is square, $n \times n$, or U is the echelon form obtained from A if A is rectangular, $n \times p$. Each L_i is of the form

$$L_i = \begin{bmatrix} 1 & & & & \\ & \ddots & & & \\ & & 1 & & \\ & & m_{i+1} & \ddots & \\ & & \vdots & & \\ & & m_n & & 1 \end{bmatrix}$$

where the m's are in the ith column and are the multiples used in Gaussian elimination when working with the pivot in the ith row of A.

In addition, if $L = L_1^{-1}L_2^{-1} \cdots L_{n-1}^{-1}$, then $A = LU$ is the LU decomposition for A.

Theorem (A.3) is the general case if there are no row interchanges. Example 1 illustrates how to modify the above procedure when there are row interchanges.

Example 1 Let

$$A = \begin{bmatrix} 0 & 0 & 0 & 3 \\ 2 & 8 & -1 & 1 \\ -2 & -4 & -3 & -1 \\ 6 & 0 & 5 & 1 \end{bmatrix}$$

Then without augmenting A, Gaussian elimination could proceed as follows:

$$\begin{bmatrix} 0 & 0 & 0 & 3 \\ 2 & 8 & -1 & 1 \\ -2 & -4 & -3 & -1 \\ 6 & 0 & 5 & 1 \end{bmatrix} \to \begin{bmatrix} 2 & 8 & -1 & 1 \\ 0 & 0 & 0 & 3 \\ -2 & -4 & -3 & -1 \\ 6 & 0 & 5 & 1 \end{bmatrix} \quad \text{Interchange rows 1 and 2.}$$

$$\to \begin{bmatrix} 2 & 8 & -1 & 1 \\ 0 & 0 & 0 & 3 \\ 0 & 4 & -4 & 0 \\ 0 & -24 & 8 & -2 \end{bmatrix} \quad \text{Add row 1 to row 3 and add } -3 \text{ times row 1 to row 4.}$$

$$\to \begin{bmatrix} 2 & 8 & -1 & 1 \\ 0 & 4 & -4 & 0 \\ 0 & 0 & 0 & 3 \\ 0 & -24 & 8 & -2 \end{bmatrix} \quad \text{Interchange rows 2 and 3.}$$

$$\to \begin{bmatrix} 2 & 8 & -1 & 1 \\ 0 & 4 & -4 & 0 \\ 0 & 0 & 0 & 3 \\ 0 & 0 & -16 & -2 \end{bmatrix} \quad \text{Add 6 times row 2 to row 4.}$$

$$\to \begin{bmatrix} 2 & 8 & -1 & 1 \\ 0 & 4 & -4 & 0 \\ 0 & 0 & -16 & -2 \\ 0 & 0 & 0 & 3 \end{bmatrix} \quad \text{Interchange rows 3 and 4.}$$

Denote this sequence by

$$A \to A_1 \to A_2 \to A_3 \to A_4 \to A_5$$

The $A_1 = P_1A$, $A_2 = L_1A_1$, $A_3 = P_2A_2$, $A_4 = L_2A_3$, $A_5 = P_3A_4$, and $A_5 = U$, where

$$P_1 = \begin{bmatrix} 0 & 1 & & \\ 1 & 0 & & \\ & & 1 & \\ & & & 1 \end{bmatrix}, \quad P_2 = \begin{bmatrix} 1 & & & \\ & 0 & 1 & \\ & 1 & 0 & \\ & & & 1 \end{bmatrix}, \quad P_3 = \begin{bmatrix} 1 & & & \\ & 1 & & \\ & & 0 & 1 \\ & & 1 & 0 \end{bmatrix}$$

$$L_1 = \begin{bmatrix} 1 & & & \\ 0 & 1 & & \\ 1 & & 1 & \\ -3 & & & 1 \end{bmatrix}, \quad L_2 = \begin{bmatrix} 1 & & & \\ & 1 & & \\ & 0 & 1 & \\ & 6 & & 1 \end{bmatrix}$$

Thus

(A.4)
$$U = A_5 = P_3L_2P_2L_1P_1A$$ ■

The general situation for this case is summarized in Theorem **(A.5)**. Note that in Example 1, $L_3 = I$.

(A.5)

(A.6)

> THEOREM If A is an $n \times m$ matrix, then
> $$U = L_{n-1}P_{n-1} \cdots L_2P_2L_1P_1A$$
> where the P's are elementary permutation matrices (or the identity I_n) and the L's are as in Theorem (A.3).

Example 2 illustrates how to obtain a $PA = LU$ decomposition from equations such as (A.4) and (A.6).

Example 2 In Equation (A.4), consider the product P_2L_1, using the fact that for elementary permutation matrices $P_i^{-1} = P_i$:

(A.7)
$$P_2L_1 = P_2L_1P_2^{-1}P_2 = P_2L_1P_2P_2$$

and

$$P_2L_1P_2 = \begin{bmatrix} 1 & & & \\ & 0 & 1 & \\ & 1 & 0 & \\ & & & 1 \end{bmatrix} \begin{bmatrix} 1 & & & \\ 0 & 1 & & \\ 1 & & 1 & \\ -3 & & & 1 \end{bmatrix} \begin{bmatrix} 1 & & & \\ & 0 & 1 & \\ & 1 & 0 & \\ & & & 1 \end{bmatrix}$$

$$= \begin{bmatrix} 1 & & & \\ 1 & 1 & & \\ 0 & & 1 & \\ -3 & & & 1 \end{bmatrix} = L_1'$$

We see that P_2 interchanges rows 2 and 3 and that $L_2' = P_2L_1P_2$ is identical to L_1, except that the entries in rows 2 and 3 of column 1 are interchanged. If we now substitute $L_1' = P_2 L_1 P_2$ into Equation (A.7), we see that $P_2 L_1 = L_1' P_2$. This is now substituted, in turn, into Equation (A.4) to obtain

$$U = P_3 L_2 P_2 L_1 P_1 A = P_3 L_2 L_1' P_2 P_1 A \tag{A.8}$$

Thus we have "slid" P_2 past L_1, changing L_1 into L_1'. We proceed in a similar way with P_3. As above, $P_3 L_2 = P_3 L_2 P_3^{-1} P_3 = P_3 L_2 P_3 P_3 = L_2' P_3$, where

$$P_3 L_2 P_3 = \begin{bmatrix} 1 & & & \\ & 1 & & \\ & & 0 & 1 \\ & & 1 & 0 \end{bmatrix} \begin{bmatrix} 1 & & & \\ & 1 & & \\ & 0 & 1 & \\ & 6 & & 1 \end{bmatrix} \begin{bmatrix} 1 & & & \\ & 1 & & \\ & & 0 & 1 \\ & & 1 & 0 \end{bmatrix}$$

$$= \begin{bmatrix} 1 & & & \\ & 1 & & \\ & 6 & 1 & \\ & 0 & & 1 \end{bmatrix} = L_2'$$

Also, $P_3 L_1' = P_3 L_1' P_3^{-1} P_3 = P_3 L_1' P_3 P_3 = L_1'' P_3$, where

$$P_3 L_1' P_3 = \begin{bmatrix} 1 & & & \\ & 1 & & \\ & & 0 & 1 \\ & & 1 & 0 \end{bmatrix} \begin{bmatrix} 1 & & & \\ 1 & 1 & & \\ 0 & & 1 & \\ -3 & & & 1 \end{bmatrix} \begin{bmatrix} 1 & & & \\ & 1 & & \\ & & 0 & 1 \\ & & 1 & 0 \end{bmatrix}$$

$$= \begin{bmatrix} 1 & & & \\ 1 & 1 & & \\ -3 & & 1 & \\ 0 & & & 1 \end{bmatrix} = L_1''$$

Therefore we see $P_3 L_2 = L_2' P_3$ and $P_3 L_1' = L_1'' P_3$. Substituting these into Equation (A.8), we obtain

$$U = P_3 L_2 L_1' P_2 P_1 A = L_2' P_3 L_1' P_2 P_1 A = L_2' L_1'' P_3 P_2 P_1 A \tag{A.9}$$

or equivalently,

$$P_3 P_2 P_1 A = L_1''^{-1} L_2'^{-1} U \tag{A.10}$$

We can now check that

$$L_1''^{-1} = \begin{bmatrix} 1 & & & \\ -1 & 1 & & \\ 3 & & 1 & \\ 0 & & & 1 \end{bmatrix}, \qquad L_2'^{-1} = \begin{bmatrix} 1 & & & \\ & 1 & & \\ & -6 & 1 & \\ & 0 & & 1 \end{bmatrix}$$

In other words the inverse of a matrix L_i in this form is another L_i of exactly the same form: the diagonal remains all 1's, all other entries are zero except for one column below the diagonal, and these entries have been replaced by their negatives. So, next we compute

$$P = P_3P_2P_1 = \begin{bmatrix} & 1 & & \\ & & 1 & \\ & & & 1 \\ 1 & & & \end{bmatrix},$$

$$L = L_1''^{-1}L_2'^{-1} = \begin{bmatrix} 1 & & & \\ -1 & 1 & & \\ 3 & -6 & 1 & \\ 0 & 0 & 0 & 1 \end{bmatrix}$$

Finally we have constructed the $PA = LU$ decomposition for A:

$$\begin{bmatrix} & 1 & & \\ & & 1 & \\ & & & 1 \\ 1 & & & \end{bmatrix}\begin{bmatrix} 0 & 0 & 0 & 3 \\ 2 & 8 & -1 & 1 \\ -2 & -4 & -3 & -1 \\ 6 & 0 & 5 & 1 \end{bmatrix}$$

$$= \begin{bmatrix} 1 & & & \\ -1 & 1 & & \\ 3 & -6 & 1 & \\ 0 & 0 & 0 & 1 \end{bmatrix}\begin{bmatrix} 2 & 8 & -1 & 1 \\ & 4 & -4 & 0 \\ & & -16 & -2 \\ & & & 3 \end{bmatrix}$$ ■

This example, then, is a general illustration for an arbitrary $n \times n$ matrix A, where Gaussian elimination naturally produces a factorization

$$L_{n-1}P_{n-1} \cdots L_1P_1A = U$$

If the P's are moved to the right past the L's, the L's are changed only by interchanging elements in the single nontrivial column, yielding

$$L'_{n-1} \cdots L'_1P_{n-1} \cdots P_1A = U$$

When the L's are brought to the other side and replaced by their inverses, the only changes made concern the elements in the single nontrivial column below the diagonal, which are replaced by their negatives. To finish the process, we then set $P = P_{n-1} \cdots P_1$ and $L = L_1'^{-1} \cdots L_{n-1}'^{-1}$, obtaining the $PA = LU$ decomposition.

Exercise Appendix A

In Exercises 1–8, for the given $n \times n$ matrix A, find matrices $L_1, \ldots, L_{n-1}$, and U so that $L_{n-1} \cdots L_1A = U$ as described in Theorem (A.3).

1. $\begin{bmatrix} 2 & 4 \\ 6 & 3 \end{bmatrix}$

2. $\begin{bmatrix} 4 & -6 \\ -3 & 5 \end{bmatrix}$

3. $\begin{bmatrix} 2 & 1 & 3 \\ -2 & 5 & 1 \\ 4 & 2 & 4 \end{bmatrix}$

4. $\begin{bmatrix} 5 & 1 & 2 \\ 10 & 3 & 8 \\ -10 & 4 & -7 \end{bmatrix}$ 5. $\begin{bmatrix} 3 & -2 & -1 \\ -6 & -4 & 2 \\ -3 & -2 & 4 \end{bmatrix}$ 6. $\begin{bmatrix} 4 & -2 & 4 \\ 2 & 3 & -1 \\ -8 & 2 & 5 \end{bmatrix}$

7. $\begin{bmatrix} 1 & 3 & 2 & -1 \\ 2 & 5 & 3 & 2 \\ -3 & 2 & -1 & 2 \\ 1 & 1 & 3 & 1 \end{bmatrix}$ 8. $\begin{bmatrix} 2 & -2 & 6 & -4 \\ 2 & -5 & 2 & 2 \\ -4 & 1 & 3 & 2 \\ 1 & 5 & 1 & -2 \end{bmatrix}$

In Exercises 9–12, find L_i^{-1} and show directly, by multiplying $L_iL_i^{-1}$, that $L_iL_i^{-1} = I$.

9. $L_1 = \begin{bmatrix} 1 & \\ -3 & 1 \end{bmatrix}$ 10. $L_1 = \begin{bmatrix} 1 & & & \\ 2 & 1 & & \\ -3 & & 1 & \\ -2 & & & 1 \end{bmatrix}$

11. $L_2 = \begin{bmatrix} 1 & & & \\ & 1 & & \\ & 3 & 1 & \\ & -3 & & 1 \end{bmatrix}$ 12. $L_2 = \begin{bmatrix} 1 & & & & \\ & 1 & & & \\ & 4 & 1 & & \\ & -2 & & 1 & \\ & 5 & & & 1 \end{bmatrix}$

13. Let

$$L_1 = \begin{bmatrix} 1 & & \\ 2 & 1 & \\ 3 & & 1 \end{bmatrix}, \quad L_2 = \begin{bmatrix} 1 & & \\ & 1 & \\ & 4 & 1 \end{bmatrix}$$

Compute L_1L_2 and L_2L_1. Describe L_1 and L_2 in terms of elementary row operations, and use this to explain why one product equals

$$\begin{bmatrix} 1 & & \\ 2 & 1 & \\ 3 & 4 & 1 \end{bmatrix}$$

and the other product does not.

14. Let

$$L_1 = \begin{bmatrix} 1 & & & \\ 2 & 1 & & \\ 3 & & 1 & \\ 4 & & & 1 \end{bmatrix}, \quad L_2 = \begin{bmatrix} 1 & & & \\ & 1 & & \\ & 5 & 1 & \\ & 6 & & 1 \end{bmatrix}, \quad L_3 = \begin{bmatrix} 1 & & & \\ & 1 & & \\ & & 1 & \\ & & 7 & 1 \end{bmatrix}$$

Compute L_1L_2, L_2L_1, L_1L_3, L_3L_1, L_2L_3, L_3L_1. For which possible products is the following true?

$$L_iL_jL_k = \begin{bmatrix} 1 & & & \\ 2 & 1 & & \\ 3 & 5 & 1 & \\ 4 & 6 & 7 & 1 \end{bmatrix}$$

Explain your answer in terms of elementary row operations.

In Exercises 15–18, find the LU decomposition for A from the given information.

15. $L_1 = \begin{bmatrix} 1 & \\ 3 & 1 \end{bmatrix}$, $U = \begin{bmatrix} 2 & 3 \\ & 4 \end{bmatrix}$

16. $L_1 = \begin{bmatrix} 1 & \\ -\frac{1}{2} & 1 \end{bmatrix}$, $U = \begin{bmatrix} 2 & 1 \\ 0 & 3 \end{bmatrix}$

17. $L_1 = \begin{bmatrix} 1 & & \\ 2 & 1 & \\ -3 & & 1 \end{bmatrix}$, $L_2 = \begin{bmatrix} 1 & & \\ & 1 & \\ & 1 & 1 \end{bmatrix}$, $U = \begin{bmatrix} 2 & -1 & 0 \\ & 3 & 1 \\ & & 4 \end{bmatrix}$

18. $L_1 = \begin{bmatrix} 1 & & & \\ 3 & 1 & & \\ -1 & & 1 & \\ 4 & & & 1 \end{bmatrix}$, $L_2 = \begin{bmatrix} 1 & & & \\ & 1 & & \\ & 0 & 1 & \\ & 2 & & 1 \end{bmatrix}$, $L_3 = \begin{bmatrix} 1 & & & \\ & 1 & & \\ & & 1 & \\ & & -5 & 1 \end{bmatrix}$,

$$U = \begin{bmatrix} 2 & -1 & 3 & 1 \\ & 4 & -1 & 2 \\ & & 5 & 1 \\ & & & -1 \end{bmatrix}$$

In Exercises 19–24, find L_i's and P_i's as in Example 2 to obtain the factorization described in Theorem (A.5).

19. $A = \begin{bmatrix} 0 & 2 \\ 3 & 4 \end{bmatrix}$

20. $A = \begin{bmatrix} 0 & 3 & 2 \\ 2 & 4 & 1 \\ 6 & 3 & 1 \end{bmatrix}$

21. $A = \begin{bmatrix} 1 & -4 & 3 \\ -2 & 8 & 5 \\ 3 & 1 & 3 \end{bmatrix}$

22. $A = \begin{bmatrix} 0 & 0 & 2 \\ 4 & 2 & 6 \\ 2 & 3 & 1 \end{bmatrix}$

23. $A = \begin{bmatrix} 0 & 0 & 0 & 3 \\ 1 & 2 & -3 & 1 \\ 3 & 1 & 2 & 1 \\ -5 & 5 & 4 & 1 \end{bmatrix}$

24. $\begin{bmatrix} 0 & 1 & -2 & 3 \\ 2 & 0 & 1 & 4 \\ 2 & -1 & 3 & 5 \\ 4 & 2 & 1 & 2 \end{bmatrix}$

Appendix B Counting Operations and Gauss–Jordan Elimination

In this appendix we shall count and compare the number of arithmetic operations required to solve linear systems. Comparisons between different solving methods are usually made on the basis of how many operations are performed with each method. For the sake of simplicity, we shall assume that our computer takes essentially the same amount of time to compute multiplications and divisions and that all other operations (additions, subtractions, interchanges) are performed much more quickly. (This is a reasonable assumption.) Thus we will count only multiplications and divisions.

We begin by counting the number of operations it takes to compute $L_{n-1}P_{n-1} \cdots L_1P_1A = U$, as described in Appendix A.

To compute P_1 or P_1A takes no multiplications/divisions.

To compute L_1 takes $n - 1$ divisions, one for each $m_{i1} = -\dfrac{a_{i1}}{a_{11}}$, $i \geq 2$.

Now the product L_1P_1A has $a_{i1} = 0$ for $i \geq 2$. Since we know these are going to be zero when we are done, there is *no need to have the computer compute these entries.* The only entries we have to compute in going from A to L_1P_1A are a_{ij} for $i, j \geq 2$. Since each one of these requires one multiplication (the formula is $a_{ij} + m_{i1}a_{1j}$) and there are $(n-1)^2$ entries,

to go from A to L_1P_1A takes $(n-1)^2 + (n-1)$ multiplications/divisions

That is, $n - 1$ to compute L_1 and $(n-1)^2$ to compute the product L_1P_1A.

To compute P_2 or $P_2(L_1P_1A)$ takes no operations.

To compute L_2 takes $n - 2$ divisions, one for each m_{i2}, $i \geq 2$.

Likewise to compute $L_2(P_2L_1P_1A)$, we have no need to compute the entries a_{ij} for $i = 1$ or 2. We need only compute a_{ij} for $i, j \geq 3$. Each one of these takes one multiplication $(a_{ij} + m_{i2}a_{2j})$ and there are $(n-2)^2$ such entries. Thus

to go from L_1P_1A to $L_2P_2L_1P_1A$ takes $(n-2)^2 + (n-2)$ multiplications/divisions

By continuing like this, we can see that it takes

$$(n-1)^2 + (n-1) + (n-2)^2 + (n-2) + \cdots + 1^2 + 1 \tag{B.1}$$

multiplications to compute the factorization $L_{n-1} \cdots P_1A = U$. We now need to use two well-known summation formulas

$$1 + 2 + \cdots + k = \frac{k(k+1)}{2}, \qquad 1^2 + 2^2 + \cdots + k^2 = \frac{k(k+1)(2k+1)}{6} \tag{B.2}$$

Applying them to (B.1), we have

$$\begin{aligned}
&1^2 + \cdots + (n-1)^2 + 1 + \cdots + (n-1) \\
&= \frac{(n-1)[(n-1)+1][2(n-1)+1]}{6} + \frac{(n-1)[(n-1)+1]}{2} \\
&= \frac{(n-1)n(2n-1)}{6} + \frac{(n-1)n}{2} \cdot \frac{3}{3} \\
&= \frac{(n-1)n}{6}[2n-1+3] = \frac{(n-1)n(n+1)}{3}
\end{aligned}$$

As described in Appendix A, it takes no operations to go from the factorization $L_{n-1} \cdots P_1 A = U$ to the $PA = LU$ decomposition. Thus when n is large, *it takes approximately* $\frac{n^3}{3}$ *operations* to compute the $PA = LU$ decomposition.

To solve $AX = B$ given $PA = LU$, we find $PAX = PB$, substitute $B' = PB$ and $LU = PA$, obtaining $LUX = B'$; substitute $Y = UX$; solve $LY = B'$ using forward substitution; and finish by solving $UX = Y$ using backsubstitution. Only the last two steps involve any operations. It takes $\frac{(n-1)n}{2}$ operations to solve $LY = B'$ using forward substitution (see Exercise 6) and $\frac{(n+1)n}{2}$ operations to solve $UX = Y$ using backsubstitution (see Exercise 5). Therefore,

(B.3)

> Once you have the $PA = LU$ decomposition, it takes only
>
> $$\frac{(n-1)n}{2} + \frac{(n+1)n}{2} = n^2$$
>
> operations to solve the equation $AX = B$.

Therefore, if we wish to solve $AX = B$ for several different B's, we can now easily see that there is considerable savings by computing the $PA = LU$ decomposition once and storing it, and then using only the factorization when the need arises.

For the complete picture, it takes approximately $\frac{n^3}{3}$ operations to find the $PA = LU$ decomposition and n^2 operations to solve it from there. Since n^2 is small compared with $\frac{n^3}{3}$ (when n is at all large), we have

(B.4)

> To solve the system $AX = B$ of n equations in n unknowns by Gaussian elimination, it takes approximately $\frac{n^3}{3}$ multiplications/divisions.

Gauss–Jordan Elimination

We now turn to an alternative method for solving linear systems, **Gauss–Jordan elimination**. This method leads to the row-reduced echelon form mentioned in Section 1.2. This form requires that the entries above and below the diagonal equal zero and that the pivots equal one. The next example illustrates how to do this with augmented matrices; it can also be done using a factorizaton of P_i's and L_i's.

Example 1 Solve the system

$$\begin{bmatrix} 1 & 1 & 1 & 1 \\ 2 & 4 & 8 & 16 \\ 3 & 9 & 27 & 81 \\ 4 & 16 & 64 & 256 \end{bmatrix} \begin{bmatrix} x_1 \\ x_2 \\ x_3 \\ x_4 \end{bmatrix} = \begin{bmatrix} 0 \\ 26 \\ 144 \\ 468 \end{bmatrix}$$

using the Gauss–Jordan method on the augmented matrix and no unnecessary interchanges.

Solution Start with the augmented matrix

$$\begin{bmatrix} 1 & 1 & 1 & 1 & 0 \\ 2 & 4 & 8 & 16 & 26 \\ 3 & 9 & 27 & 81 & 144 \\ 4 & 16 & 64 & 256 & 468 \end{bmatrix}$$

Add -2 times row 1 to row 2.
Add -3 times row 1 to row 3.
Add -4 times row 1 to row 4.

$$\begin{bmatrix} 1 & 1 & 1 & 1 & 0 \\ 0 & 2 & 6 & 14 & 26 \\ 0 & 6 & 24 & 78 & 144 \\ 0 & 12 & 60 & 252 & 468 \end{bmatrix}$$

Add $-\frac{1}{2}$ times row 2 to row 1.
Add -3 times row 2 to row 3.
Add -6 times row 2 to row 4.

$$\begin{bmatrix} 1 & 0 & -2 & -6 & -13 \\ 0 & 2 & 6 & 14 & 26 \\ 0 & 0 & 6 & 36 & 66 \\ 0 & 0 & 24 & 168 & 312 \end{bmatrix}$$

Add $\frac{1}{3}$ times row 3 to row 1.
Add -1 times row 3 to row 2.
Add -4 times row 3 to row 4.

$$\begin{bmatrix} 1 & 0 & 0 & 6 & 9 \\ 0 & 2 & 0 & -22 & -40 \\ 0 & 0 & 6 & 36 & 66 \\ 0 & 0 & 0 & 24 & 48 \end{bmatrix}$$

Add $-\frac{1}{4}$ times row 4 to row 1.
Add $\frac{11}{12}$ times row 4 to row 2.
Add $-\frac{3}{2}$ times row 4 to row 3.

$$\begin{bmatrix} 1 & 0 & 0 & 0 & -3 \\ 0 & 2 & 0 & 0 & 4 \\ 0 & 0 & 6 & 0 & -6 \\ 0 & 0 & 0 & 24 & 48 \end{bmatrix}$$

Divide row 1 by 1.
Divide row 2 by 2.
Divide row 3 by 6.
Divide row 4 by 24.

$$\begin{bmatrix} 1 & 0 & 0 & 0 & -3 \\ 0 & 1 & 0 & 0 & 2 \\ 0 & 0 & 1 & 0 & -1 \\ 0 & 0 & 0 & 1 & 2 \end{bmatrix}$$

The associated system

$$\begin{aligned} x_1 \qquad\qquad\qquad &= -3 \\ x_2 \qquad\quad &= 2 \\ x_3 \quad &= -2 \\ x_4 &= 2 \end{aligned}$$

clearly gives the solution. ■

The final form of the reductions seems to avoid backsubstitution. However, there is no such thing as a free lunch; in fact more work is done in making the terms above the diagonal zero. The bottom line is that the Gauss–Jordan method takes approximately $\frac{n^3}{2}$ operations. (See Exercise 14.)

Exercise Appendix B

In Exercises 1–6, count the number of operations (multiplications or divisions) it takes to solve the systems by forward or backward substitution, whichever is appropriate. (Just count the operations; *do not solve.*)

In Exercises 5 and 6 $a_{ii} \neq 0$.

1. $$\begin{aligned} 2x + 3y + 4z &= 5 \\ 4y + 5z &= 6 \\ 6z &= 7 \end{aligned} \qquad \begin{aligned} x + 3y + 4z &= 5 \\ y + 5z &= 6 \\ z &= 7 \end{aligned}$$

2. $$\begin{aligned} 2x \qquad\qquad &= 3 \\ 3x + 4y \qquad &= 5 \\ 4x + 5y + 6z &= 7 \end{aligned} \qquad \begin{aligned} x \qquad\qquad &= 3 \\ 3x + y \qquad &= 5 \\ 4x + 5y + z &= 7 \end{aligned}$$

3. $$\begin{aligned} 2x_1 + 3x_2 + 4x_3 + \cdots + 21x_{20} &= 22 \\ 4x_2 + 5x_3 + \cdots + 22x_{20} &= 23 \\ 6x_3 + \cdots + 23x_{20} &= 24 \\ &\vdots \\ 40x_{20} &= 41 \end{aligned} \qquad \begin{aligned} x_1 + 3x_2 + 4x_3 + \cdots + 21x_{20} &= 22 \\ x_3 + 5x_4 + \cdots + 22x_{20} &= 23 \\ x_4 + \cdots + 23x_{20} &= 24 \\ &\vdots \\ x_{20} &= 41 \end{aligned}$$

4. $$\begin{aligned} 2x_1 \qquad\qquad\qquad\qquad\qquad &= 3 \\ 3x_1 + 4x_2 \qquad\qquad\qquad\qquad &= 5 \\ 4x_1 + 5x_2 + 6x_3 \qquad\qquad\quad &= 7 \\ &\vdots \\ 21x_1 + 22x_2 + 23x_3 + \cdots + 40x_{20} &= 41 \end{aligned}$$

$$\begin{aligned} x_1 \qquad\qquad\qquad\qquad\qquad &= 3 \\ 3x_1 + x_2 \qquad\qquad\qquad\qquad &= 5 \\ 4x_1 + 5x_2 + x_3 \qquad\qquad\quad &= 7 \\ &\vdots \\ 21x_1 + 22x_2 + 23x_3 + \cdots + x_{20} &= 41 \end{aligned}$$

5. $$\begin{aligned} a_{11}x_1 + a_{12}x_2 + \cdots + a_{1n}x_n &= b_1 \\ a_{22}x_2 + \cdots + a_{2n}x_n &= b_2 \\ &\vdots \\ a_{nn}x_n &= b_n \end{aligned} \qquad \begin{aligned} x_1 + a_{12}x_2 + \cdots + a_{1n}x_n &= b_1 \\ x_2 + \cdots + a_{2n}x_n &= b_2 \\ &\vdots \\ x_n &= b_n \end{aligned}$$

6. $$\begin{aligned} a_{11}x_1 \qquad\qquad\qquad &= b_1 \\ a_{21}x_1 + a_{22}x_2 \qquad &= b_2 \\ &\vdots \\ a_{n1}x_1 + a_{n2}x_2 + \cdots + a_{nn}x_n &= b_n \end{aligned} \qquad \begin{aligned} x_1 \qquad\qquad\qquad &= b_1 \\ a_{21}x_1 + x_2 \qquad &= b_2 \\ &\vdots \\ a_{n1}x_1 + a_{n2}x_2 + \cdots + x_n &= b_n \end{aligned}$$

In Exercises 7–12, solve by Gauss–Jordan elimination.

7. $$\begin{aligned} 2x + 3y &= 5 \\ 3x + 4y &= 6 \end{aligned}$$

8. $$\begin{aligned} 2x - 3y &= 0 \\ 2x + 3y &= 2 \end{aligned}$$

9. $$\begin{aligned} x + 2y + z &= 8 \\ -x + 3y - 2z &= 1 \\ 3x + 4y - 7z &= 10 \end{aligned}$$

10. $$\begin{aligned} x_1 + 7x_2 - 7x_3 &= 9 \\ 2x_1 + 3x_2 + x_3 &= -1 \\ x_1 - 4x_2 + 3x_3 &= 0 \end{aligned}$$

11. $$\begin{aligned} 2x_1 + 3x_2 - x_3 + x_4 &= 6 \\ 4x_1 + 5x_2 + 2x_3 - x_4 &= 7 \\ -2x_1 - x_2 - x_3 - x_4 &= -1 \\ 6x_1 + 7x_2 + x_3 - 4x_4 &= 2 \end{aligned}$$

12. $$\begin{aligned} 2x - y &= 11 \\ x + z &= 5 \\ -2x + w &= -10 \\ 4x + 3y - z + w &= 4 \end{aligned}$$

13. In Example 1:

(a) Explain why it takes 3×5 multiplications to go from step 1 to step 2. Be sure to count computing the multiples -2, -3, and -4, but do not count making the a_{i1}'s zero.

(b) Explain why it takes 3×4 multiplications to go from step 2 to step 3. [Same instructions as for (a).]

(c) Explain why it takes 3×3 multiplications to go from step 3 to step 4.

(d) Explain why it takes 3×2 multiplications to go from step 4 to step 5.

(e) Explain why it takes 4 multiplications (or divisions) to go from step 5 to step 6.

14. Use Exercise 13 as a guide to count the number of multiplications required by Gauss–Jordan elimination.

Appendix C Another Application

The purpose of this appendix is to introduce an application that is vital to fields as diverse as engineering, chemistry, physics, economics, and biology, as well as to mathematics—the solving of differential equations. Linear algebra plays a crucial role throughout the spectrum of pure and applied mathematics, but unfortunately it would lead us far astray to give much of a description of a biological or economic or engineering system to which linear algebra would be applied. So we shall begin with a brief description of how one type of differential equation problem arises and then solve a similar but slightly simpler problem.

The example is included in this text not only because it is a beautiful and typical application of linear algebra, but also because it shows how linear algebra is part of the bridge between continuous and discrete mathematics.

Example 1 Suppose heat is applied to a metal bar at a known rate. Find the temperature at each point of the bar if the ends of the bar are kept at a fixed temperature and the system is in equilibrium (which means the temperature at each point does not change). To keep the numbers simple, we assume the temperature at each end of the bar is 0°C (say they are attached to reservoirs of ice water), and the bar has length one. See Figure C.1. Let $f(x)$ represent the amount of heat applied to the bar at the point x, $0 \leq x \leq 1$. Then a basic physics course would tell us the mathematical problem is to find a function $y = g(x)$ satisfying

$$y'' - cy = f(x), \qquad 0 \leq x \leq 1, \qquad g(0) = 0, \qquad g(1) = 0 \tag{C.1}$$

where c is a constant.

Figure C.1

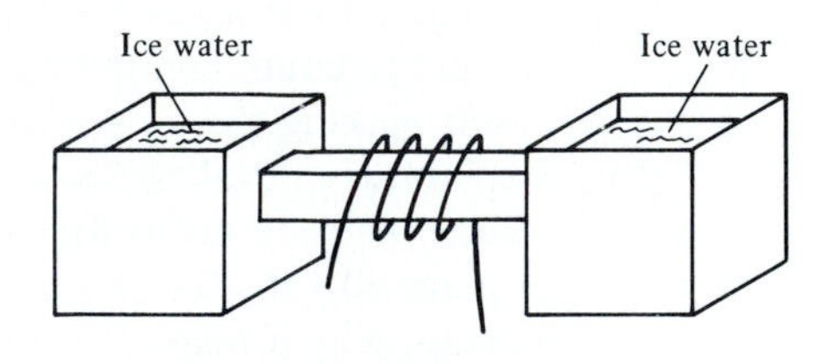

Wire gives heat source ■

Thus the problem is to find a function that satisfies a certain differential equation over an interval *and* that also takes on fixed values at two points of the interval. Since those two points are the boundary of the interval, the problem is called a *two-point boundary value problem* and it has a unique solution if f is continuous. Example 2 shows how to solve a simpler version of such a problem.

Example 2 Solve the problem

$$y''(x) = f(x), \qquad 0 \leq x \leq 1, \qquad y(0) = 0, \qquad y(1) = 0 \tag{C.2}$$

To **solve** this problem means to find a function $y = y(x)$ that is defined and twice differentiable on the interval $0 \leq x \leq 1$; that has a second derivative,

y'', that satisfies the equation $y''(x) = f(x)$, $0 \le x \le 1$, where f is given; and finally that satisfies the equation $y(0) = 0 = y(1)$. For example, if $f(x) = x$, then $y = \frac{1}{6}x^3 - \frac{1}{6}x$ is the solution, as you can easily check. Unfortunately it is only for very simple functions f that you can write down the solution so nicely. Usually the best we can hope to do is to find the values $y(x)$ for a finite number of points (and perhaps interpolate in between them).

We first approximate the derivatives in terms of y. Recall that the definition of $y'(x)$ is

$$y'(x) = \lim_{h \to 0} \frac{y(x+h) - y(x)}{h}$$

(You may have seen it with "Δx" in place of h, but it makes no difference, and the h takes less writing.) Thus

$$y'(x) \approx \frac{y(x+h) - y(x)}{h} \qquad \text{or} \qquad y'(x) \approx \frac{y(x) - y(x-h)}{h} \tag{C.3}$$

are good approximations if h is small. Since $y'' = (y')'$, using the first approximation in (C.3), we see

$$y''(x) \approx \frac{y'(x+h) - y'(x)}{h}$$

Using the second approximation in (C.3), we get

$$y''(x) \approx \frac{\dfrac{y(x+h) - y(x)}{h} - \dfrac{y(x) - y(x-h)}{h}}{h}$$

Simplifying we get

$$y''(x) \approx \frac{y(x+h) - 2y(x) + y(x-h)}{h^2} \tag{C.4}$$

which is a nicely symmetric formula expressing y'' in terms of y.

We now break the interval $[0, 1]$ up into n equal subintervals using $h = \dfrac{1}{n}$ and

$$0 = x_0, x_1, \ldots, x_{n-1}, x_n = 1, \qquad \text{where} \qquad x_i - x_{i-1} = h, \qquad 1 \le i \le n$$

Thus $x_i = ih$. We assume that $y(x_0) = 0$ and $y(x_n) = 0$. The problem is to find the values $y(x_i)$, $1 \le i \le n - 1$. By (C.2), $y''(x) = f(x)$, $0 \le x \le 1$. Thus

$$y''(x_i) = f(x_i), \qquad 1 \le i \le n - 1$$

Since $x_{i+1} = x_i + h$ and $x_{i-1} = x_i - h$, we have by (C.4),

$$\frac{y(x_{i-1}) - 2y(x_i) + y(x_{i+1})}{h^2} = f(x_i), \qquad 1 \le i \le n - 1$$

Multiplying by h^2 and using the abbreviation $y_i = y(x_i)$, we get

$$y_{i-1} - 2y_i + y_{i+1} = h^2 f(x_i), \qquad 1 \le i \le n-1^* \tag{C.5}$$

Since $y_0 = y(x_0) = 0$ and $y_n = y(x_n) = 0$, we thus have the following system of $n-1$ equations in $n-1$ unknowns.

$$\begin{aligned}
-2y_1 + y_2 \qquad\qquad\qquad &= h^2 f(x_1)\\
y_1 - 2y_2 + y_3 \qquad\qquad &= h^2 f(x_2)\\
y_2 - 2y_3 + y_4 \qquad &= h^2 f(x_3)\\
&\ \ \vdots\\
y_{n-2} - 2y_{n-1} &= h^2 f(x_{n-1})
\end{aligned}$$

or

$$\begin{bmatrix} -2 & 1 & & & \\ 1 & -2 & 1 & & \\ & 1 & -2 & 1 & \\ & & \ddots & \ddots & \ddots \\ & & & 1 & -2 \end{bmatrix} \begin{bmatrix} y_1 \\ y_2 \\ y_3 \\ \vdots \\ y_{n-1} \end{bmatrix} = h^2 \begin{bmatrix} f(x_1) \\ f(x_2) \\ f(x_3) \\ \vdots \\ f(x_{n-1}) \end{bmatrix} \tag{C.6}$$

We shall denote this by $AY = B$. We now set out to find an approximation to our problem given in (C.2). Let us pick $n = 5$ and $f(x) = x$ (since we know the exact solution, $y = \frac{1}{6}x^3 - \frac{1}{6}x$). We find the LU decomposition for A as we did in Section 1.5,

$$\begin{bmatrix} -2 & 1 & & \\ 1 & -2 & 1 & \\ & 1 & -2 & 1 \\ & & 1 & -2 \end{bmatrix} = \begin{bmatrix} 1 & & & \\ -\frac{1}{2} & 1 & & \\ & -\frac{2}{3} & 1 & \\ & & -\frac{3}{4} & 1 \end{bmatrix} \begin{bmatrix} -2 & 1 & & \\ & -\frac{3}{2} & 1 & \\ & & -\frac{4}{3} & 1 \\ & & & -\frac{5}{4} \end{bmatrix} \tag{C.7}$$

With $n = 5$, $h = \frac{1}{5} = 0.2$ and the partition of $[0, 1]$ is

$$0 = x_0, \quad 0.2 = x_1, \quad 0.4 = x_2, \quad 0.6 = x_3, \quad 0.8 = x_4, \quad 1 = x_5$$

We wish to solve $AY = B$ with

$$B = h^2 \begin{bmatrix} f(x_1) \\ f(x_2) \\ f(x_3) \\ f(x_4) \end{bmatrix} = 0.04 \begin{bmatrix} 0.2 \\ 0.4 \\ 0.6 \\ 0.8 \end{bmatrix} = 10^{-3} \begin{bmatrix} 8 \\ 16 \\ 24 \\ 32 \end{bmatrix}$$

since $f(x) = x$. Arguing as in Section 1.5, to solve $AY = B$, we substitute $A = LU$ to obtain $LUY = B$. We set $Z = UY$ and first solve $LZ = B$ by

* *This equation is a type of* ***difference equation****. The resulting matrix equation, Equation (C.6), is called a* ***finite difference matrix equation****. Difference equations arise in many different contexts; see Section 4.5, for example.*

forward substitution, obtaining $(z_1, z_2, z_3, z_4) = 10^{-3}(8, 20, 37\frac{1}{3}, 60)$. We then solve $UY = Z$ by backsubstitution, obtaining $(y_1, y_2, y_3, y_4) = (-0.032, -0.056, -0.064, -0.048)$. This happens to be the correct answer, which can be easily checked, using $y = \frac{1}{6}x^3 - \frac{1}{6}x$, since the function $f(x) = x$ is so simple. However, see Exercise 2. ■

Exercise Appendix C

1. Suppose the problem in Example 2 is $y''(x) = 4\pi^2 \sin 2\pi x$, $0 \le x \le 1$, $y(0) = 0$, $y(1) = 0$. Use $n = 4$, so that $h = \frac{1}{4}$.

   (a) Explain why the difference equations yield the system

$$\begin{bmatrix} 2 & -1 & \\ -1 & 2 & -1 \\ & -1 & 2 \end{bmatrix}\begin{bmatrix} y_1 \\ y_2 \\ y_3 \end{bmatrix} = \frac{\pi^2}{4}\begin{bmatrix} 1 \\ 0 \\ -1 \end{bmatrix}$$

   (b) Solve the system in (a) and compare with the true solution $u = \sin 2\pi x$ at $x = \frac{1}{4}, \frac{1}{2}$, and $\frac{3}{4}$.

2. The same question in Exercise 1, except this time take $f(x) = x^2$ and $n = 5$. Here the exact solution is $y = \frac{1}{12}x^4 - \frac{1}{12}x$.

3. Write down the 3×3 finite difference matrix equation if $n = 4$ (so $h = \frac{1}{4}$) and the problem is $y''(x) - y(x) = x$, $0 \le x \le 1$, $y(0) = 0$, $y(1) = 0$.

Appendix D Software and Codes for Linear Algebra

This appendix gives a brief discussion of some of the software and codes available to solve linear algebra problems. Software and codes are available in abundance for use in many types of computing situations; some are well written and some are not. What is presented here has been professionally written, well tested, and designed to be used effectively in both classroom and professional environments. The discussion is broken up into two parts, software for mainframe and minicomputers, and software for personal computers.

Mainframe and Minicomputer Software

There are two collections of subroutines aimed specifically at linear algebraic problems, EISPACK and LINPACK. Any technical computer installation should have them both.

Historically, EISPACK came first. It is a collection of subroutines for solving various aspects of the eigenvalue problem. The codes are based on many man-years of theoretical development and practical experience. The theory is developed in Wilkinson's *The Algebraic Eigenvalue Problem* (OUP, 1965), programming considerations and codes are discussed in *Handbook for Automatic Computation II, Linear Algebra* (Springer, 1971), and EISPACK's documentation (i.e., the user's manual) is Volumn 6 of Springer-Verlag's *Lecture Notes in Computer Science.* A magnetic tape containing the FORTRAN source code is incredibly inexpensive, a mere \$75, really provided as a public service by IMSL, International Mathematical and Statistical Libraries, 2500 ParkWest Tower One, 2500 CityWest Boulevard, Houston, TX 77042.

Stimulated by the success of EISPACK, LINPACK was developed in the 1970's to solve many variations of the problem $AX = B$ and a few other related problems. Prepared by Dongarra, Bunch, Moler, and Stewart, LINPACK is designed to be machine independent, efficient, and simple. The user's guide is excellent; the main part of it can be read and understood by anyone who has gone through the first chapter of this text. The guide is available (for about \$24) from SIAM, Society for Industrial and Applied Mathematics, 3600 University City Science Center, Philadelphia, PA 19104. As with EISPACK, a magnetic tape containing the FORTRAN source code is available for \$75, also from IMSL.

Large collections of subroutines to attack a whole spectrum of mathematical problems are also available, and they contain subroutines to solve linear algebraic problems. IMSL provides one of the most extensive of such collections, but another excellent one is NAG, whose home office is Oxford, England, but with U.S. address, Numerical Algorithm Group, Inc., 1101 31st Street, Suite 100, Downers Grove, IL 60515.

Personal Computers

There is a wide range of technical software available for a variety of machines. Some is very good, but some is not. Unfortunately, it is impossible to know and list all that is good, so the following short list is in no way comprehensive, and in any case, new software is being developed all the time. In the long run, the professional user will just have to try different packages and keep his or her own benchmarks for the utilities which she or he finds valuable.

For educational purposes (as well as professional purposes) the package PC-Mathlab (designed by The MathWorks, 21 Eliot Street, South Natick, MA 01760) seems very accurate and efficient, and it has nice graphics integrated with the numerical programs. With the proper hardware, its speed and accuracy rivals a VAX. However, other packages which should also be considered are IMSL's version for PC's and The Scientific Disk (produced by C. Abaci, 208 St. Mary's Street, Raleigh, NC 27605), among several others.

Bibliography and Further Readings

Senior Level Abstract Linear Algebra Textbooks

S. H. Friedberg, A. J. Insel, and L. E. Spence, *Linear Algebra*, Prentice-Hall, Inc., Englewood Cliffs, N.J., 1979.

P. R. Halmos, *Finite-Dimensional Vector Spaces*, Van Nostrand-Reinhold, Princeton, N.J., 1958.

K. Hoffman and R. Kunze, *Linear Algebra*, Prentice-Hall, Inc., Englewood Cliffs, N.J., 1971.

Junior-Senior Level Applied Linear Algebra Textbook

G. Strang, *Linear Algebra and Its Applications*, 3rd ed., Harcourt Brace Jovanovich, San Diego, 1988.

Senior-Graduate Level Applied or Numerical Linear Algebra Textbooks

G. H. Golub and C. F. Van Loan, *Matrix Computations*, 2nd ed., John Hopkins University Press, Baltimore, 1989.

A. R. Magid, *Applied Matrix Models*, John Wiley & Sons, New York, 1985.

B. Noble, *Applied Linear Algebra*, Prentice-Hall, Inc., Englewood Cliffs, N.J., 1969.

G. W. Stewart, *Introduction to Matrix Computations*, Academic Press, Inc., New York, 1973.

Further Readings

A. Ben-Israel and T. N. E. Greville, *Generalized Inverses: Theory and Applications*, John Wiley & Sons, New York, 1974.

I. S. Duff, A. M. Erisman, and J. K. Reid, *Direct Methods for Sparse Matrices*, Oxford University Press, Oxford, 1989.

L. A. Hageman and D. M. Young, *Applied Iterative Methods*, Academic Press, New York, 1981.

C. L. Lawson and R. J. Hanson, *Solving Least Squares Problems*, Prentice-Hall, Inc., Englewood Cliffs, N.J., 1975.

J. M. Ortega, *Numerical Analysis, A Second Course*, Academic Press, Inc., New York, 1972.

B. N. Parlett, *The Symmetric Eigenvalue Problem*, Prentice-Hall, Inc., Englewood Cliffs, N.J., 1980.

R. S. Varga, *Matrix Iterative Analysis*, Prentice-Hall, Inc., Englewood Cliffs, N.J., 1962.

J. M. Wilkinson, *Rounding Errors in Algebraic Processes*, Prentice-Hall, Inc., Englewood Cliffs, N.J., 1963.

J. M. Wilkinson, *The Algebraic Eigenvalue Problem*, Oxford University Press, Oxford, 1965.

D. M. Young, *Iterative Solutions of Large Linear Systems*, Academic Press, Inc., New York, 1971.

Answers to Odd-Numbered Exercises

Section 1.1

1. Not 3. Linear 5. Not 7. $(x, y) = (\frac{5}{2} - 2t, t)$ or $(s, \frac{5}{4} - \frac{1}{2}s)$
9. $(x, y, z) = (-1 - 2s + \frac{3}{7}t, s, t)$ or $= (s, -\frac{1}{2} - \frac{1}{2}s + \frac{3}{14}t, t)$
11. $(x_1, x_2, x_3, x_4) = (2 + \frac{2}{3}s_1 - \frac{5}{3}s_2 + \frac{1}{3}s_3, s_1, s_2, s_3)$
 or $= (t_1, t_2, t_3, 3t_1 - 2t_2 + 5t_3 - 6)$
13. Infinitely many 15. None

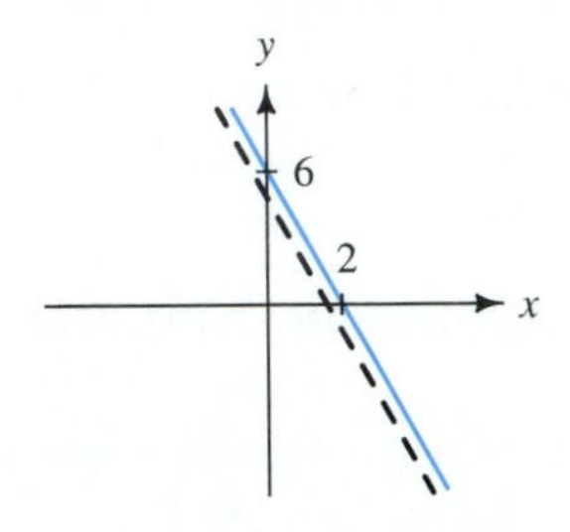

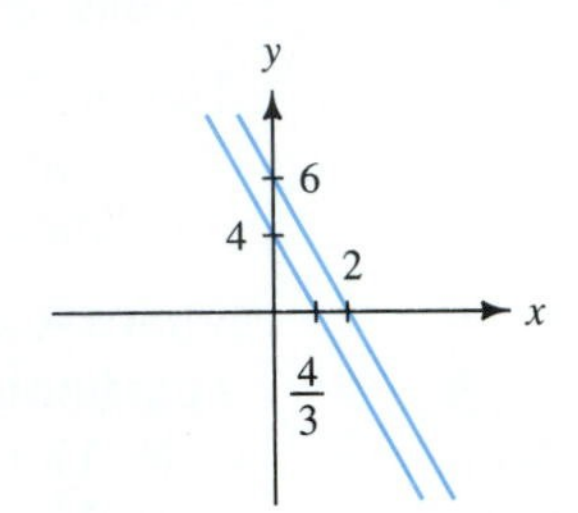

17. One, $(x, y) = (1, -1)$

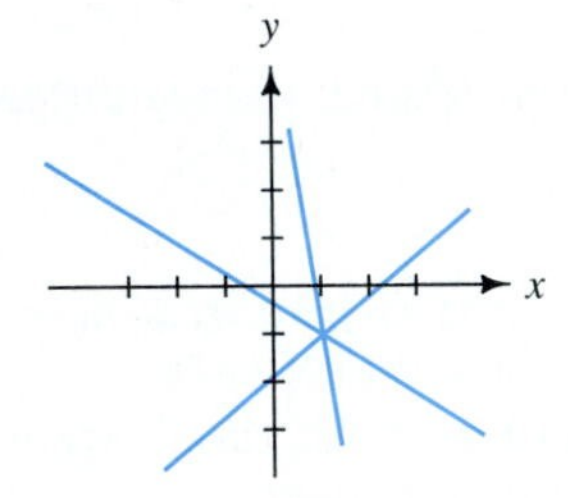

19. Change $2x - y = 2$ to $2x - y = -4$ or to $-4x - y = 2$.

21. $(-1, 2, -2)$ 23. $(\frac{7}{8}, -\frac{3}{2}, 0)$ 25. $(-\frac{1}{6}, \frac{3}{2}, 1, 2)$ 27. None
29. (a) None (b) $k \neq 4$ (c) 4 31. (a) $k \neq -6$ (b) None (c) -6

Section 1.2

1. Y 3. Y 5. N 7. (0, 3) 9. $(t, \frac{5}{3}, \frac{1}{2})$
11. $(-2, 1, 3)$ 13. $(\frac{1}{2}t - 8, t, -2, 3)$ 15. $(3, 2, 1)$

17. $(-2, 1, 3, -1)$ 19. $(\frac{3}{2}s - \frac{1}{2}t + \frac{3}{2}, s, 2 - 3t, 5 - 3t, t)$
21. Inconsistent 23. $(3, \frac{1}{2}, 2)$ 25. $(-\frac{2}{3}, 1)$ 27. $(\frac{55}{16}, \frac{43}{32})$
29. (a) T (b) F (c) T

Section 1.3

1. $A = \begin{bmatrix} 2 & 3 & 4 \\ 3 & 4 & 5 \end{bmatrix}, B = \begin{bmatrix} 1 & -1 & 1 \\ -1 & 1 & -1 \end{bmatrix}$

3. $\begin{bmatrix} 7 & 6 & 0 \\ -3 & 5 & -3 \end{bmatrix}, \begin{bmatrix} 6 & 8 & 2 \\ -2 & 0 & -12 \end{bmatrix}, \begin{bmatrix} -4 & -2 & 1 \\ 2 & -5 & -3 \end{bmatrix}$

5. $\begin{bmatrix} 10 & 0 & -1 & -3 \\ 3 & 4 & -5 & -6 \end{bmatrix}, \begin{bmatrix} 14 & 4 & 0 & -6 \\ -4 & 0 & -10 & 4 \end{bmatrix}, \begin{bmatrix} -3 & 2 & 1 & 0 \\ -5 & -4 & 0 & 8 \end{bmatrix}$

7. $\begin{bmatrix} 0 & 1 \\ -4 & -1 \end{bmatrix}, \begin{bmatrix} -13 & -20 \\ 8 & 12 \end{bmatrix}$ 9. $\begin{bmatrix} 6 & -10 \\ 0 & 1 \end{bmatrix}, \begin{bmatrix} 9 & 2 & -9 \\ 0 & 1 & 0 \\ 3 & 2 & -3 \end{bmatrix}$

11. $AB = \begin{bmatrix} 7 & -3 & 1 \\ 6 & 0 & -3 \end{bmatrix}$, BA undefined 13. Both undefined

17. $\begin{bmatrix} 8 & 0 \\ 11 & -9 \end{bmatrix}\begin{bmatrix} x \\ y \end{bmatrix} = \begin{bmatrix} -9 \\ 3 \end{bmatrix}$ 19. $\begin{bmatrix} 5 & 1 & -9 \\ 10 & 5 & -1 \end{bmatrix}\begin{bmatrix} x \\ y \\ z \end{bmatrix} = \begin{bmatrix} 10 \\ -8 \end{bmatrix}$

21. $\begin{bmatrix} -7 & -11 \\ 36 & 8 \end{bmatrix}\begin{bmatrix} s \\ t \end{bmatrix} = \begin{bmatrix} 2 \\ 4 \end{bmatrix}$ 23. $\begin{bmatrix} 4 & 22 & -12 \\ 33 & 3 & 3 \end{bmatrix}\begin{bmatrix} p \\ q \\ r \end{bmatrix} = \begin{bmatrix} 10 \\ -2 \end{bmatrix}$

25.

	Dec.	Apr.	Aug.
Total sales	11240	6620	7840
Total cost	7930	4680	5540

33. $\begin{bmatrix} -6 \\ -6 \\ 0 \end{bmatrix} = -2\begin{bmatrix} 9 \\ 0 \\ 3 \end{bmatrix} + 3\begin{bmatrix} 4 \\ -2 \\ 2 \end{bmatrix}$

35. $\begin{bmatrix} 8 \\ 21 \end{bmatrix} = 3\begin{bmatrix} 1 \\ 8 \end{bmatrix} + (-1)\begin{bmatrix} 3 \\ -1 \end{bmatrix} + 2\begin{bmatrix} -4 \\ 2 \end{bmatrix}$ 37. $\begin{bmatrix} -2 & 5 \\ 11 & -5 \end{bmatrix}$

39. $\begin{bmatrix} 7 & 6 & 5 \\ 19 & 2 & 5 \end{bmatrix}$ 41. $\begin{bmatrix} 5 & -4 \\ 9 & 4 \end{bmatrix}$

Section 1.4

1. $e^{-1} = e, \begin{bmatrix} 0 & 0 & 1 & \\ 0 & 1 & 0 & \\ 1 & 0 & 0 & \\ & & & 1 \end{bmatrix}$

3. Multiply the third row by $\frac{3}{2}$, $\begin{bmatrix} 1 & & \\ & 1 & \\ & & \frac{2}{3} \end{bmatrix}, \begin{bmatrix} 1 & & \\ & 1 & \\ & & \frac{3}{2} \end{bmatrix}$

5. $E = \begin{bmatrix} 1 & & \\ & 0 & 1 \\ & 1 & 0 \end{bmatrix}, eA = EA = \begin{bmatrix} 1 & -2 & 3 & 0 \\ 3 & -1 & 1 & -1 \\ 4 & 1 & 1 & 7 \end{bmatrix}$

7. $E = \begin{bmatrix} 1 & 0 & -2 \\ & 1 & 0 \\ & & 1 \end{bmatrix}, eA = EA = \begin{bmatrix} -4 & -6 & 1 \\ 4 & -1 & 1 \\ 3 & 3 & 1 \end{bmatrix}$

9. Yes, add 2 times row 3 to row 1. 11. N 13. N

15. $E_1 = E_2 = \begin{bmatrix} 0 & 1 & \\ 1 & 0 & \\ & & 1 \end{bmatrix}$

17. Impossible because E_3E_1 is not an elementary matrix; that is, E_3E_1 requires *two* row exchanges to achieve this exchange of rows.

19. (a) E_1 *or* $E_2 = I$; (b) $E_1 = E_2 = P_{ij}$; (c) E_1 and E_2 multiply the same row by possibly different numbers; (d) E_1 and E_2 add (possibly different) multiples of the same first row to the same second row.

21. $DA = \begin{bmatrix} 15 & -10 & 30 \\ 26 & 8 & 20 \end{bmatrix}, BD = \begin{bmatrix} 30 & 2 \\ -60 & 30 \\ 35 & -4 \end{bmatrix}$

23. $DA = \begin{bmatrix} 0 & 0 & 0 \\ 4 & -4 & 6 \\ 5 & 3 & -12 \end{bmatrix}, BD = \begin{bmatrix} 0 & -16 & 2 \\ 0 & -4 & -3 \\ 0 & -6 & -12 \end{bmatrix}$

25. $\begin{bmatrix} 5 & -2 \\ 3 & -1 \end{bmatrix}$ 27. Does not exist 29. $\begin{bmatrix} \frac{1}{2} & \frac{1}{2} & -\frac{1}{2} \\ 0 & 0 & 1 \\ \frac{1}{2} & -\frac{1}{2} & \frac{1}{2} \end{bmatrix}$

31. $\begin{bmatrix} 1 & -2 & 1 & 0 \\ 1 & -2 & 2 & -3 \\ 0 & 1 & -1 & 1 \\ -2 & 3 & -2 & 3 \end{bmatrix}$ 33. $\begin{bmatrix} 1 & 0 & 1 \\ 0 & 1 & 0 \\ 1 & 0 & -1 \end{bmatrix}$

35. $\begin{bmatrix} \frac{1}{3} & & \\ & 4 & \\ & & \frac{1}{5} \end{bmatrix}$ 37. $\begin{bmatrix} A^{-1} & \\ & B^{-1} \end{bmatrix}$

39. $\begin{bmatrix} & & & d^{-1} \\ & & c^{-1} & \\ & b^{-1} & & \\ a^{-1} & & & \end{bmatrix}$ 41. $\begin{bmatrix} 1 & & & \\ -a & 1 & & \\ a^2 & -a & 1 & \\ -a^3 & a^2 & -a & 1 \end{bmatrix}$

Section 1.5

1. $\begin{bmatrix} 1 & \\ 3 & 1 \end{bmatrix}\begin{bmatrix} 2 & 4 \\ & -9 \end{bmatrix}$ 3. $\begin{bmatrix} 1 & & \\ -1 & 1 & \\ 2 & 0 & 1 \end{bmatrix}\begin{bmatrix} 2 & 1 & 3 \\ & 6 & 4 \\ & & -2 \end{bmatrix}$

5. $\begin{bmatrix} 1 & & \\ -2 & 1 & \\ -1 & \frac{1}{2} & 1 \end{bmatrix}\begin{bmatrix} 3 & -2 & -1 \\ & -8 & 0 \\ & & 3 \end{bmatrix}$

7. $\begin{bmatrix} 1 & & & \\ 2 & 1 & & \\ -3 & -11 & 1 & \\ 1 & 2 & -\frac{1}{2} & 1 \end{bmatrix}\begin{bmatrix} 1 & 3 & 2 & -1 \\ & -1 & -1 & 4 \\ & & -6 & 43 \\ & & & \frac{31}{2} \end{bmatrix}$

9. $\begin{bmatrix} -5 \\ -11 \end{bmatrix}$

11. $\begin{bmatrix} -2 \\ -1 \\ 3 \end{bmatrix}$ 13. $\begin{bmatrix} -2 \\ 3 \\ 2 \end{bmatrix}$ 15. $\begin{bmatrix} 2 \\ -1 \\ 3 \\ -2 \end{bmatrix}$ 17. $\begin{bmatrix} 1 \\ -1 \end{bmatrix}$

19. $\begin{bmatrix} 2 \\ -3 \\ -1 \end{bmatrix}$ 21. $\begin{bmatrix} 1 \\ 1 \end{bmatrix}$ 23. $\begin{bmatrix} -\frac{1}{2} \\ -1 \\ 1 \end{bmatrix}$

25. $\begin{bmatrix} 0 & 1 \\ 1 & 0 \end{bmatrix}\begin{bmatrix} 0 & 2 \\ -3 & 4 \end{bmatrix} = \begin{bmatrix} 1 & \\ 0 & 1 \end{bmatrix}\begin{bmatrix} -3 & 4 \\ & 2 \end{bmatrix}$

27. $\begin{bmatrix} 0 & 1 & \\ 1 & 0 & \\ & & 1 \end{bmatrix}\begin{bmatrix} 0 & 2 & 3 \\ -2 & 1 & 4 \\ 6 & -7 & 2 \end{bmatrix} = \begin{bmatrix} 1 & & \\ 0 & 1 & \\ -3 & -2 & 1 \end{bmatrix}\begin{bmatrix} -2 & 1 & 4 \\ & 2 & 3 \\ & & 20 \end{bmatrix}$

29. $\begin{bmatrix} 1 & & \\ & 0 & 1 \\ & 1 & 0 \end{bmatrix}\begin{bmatrix} 2 & 3 & -1 \\ -4 & -6 & 5 \\ 4 & 1 & 2 \end{bmatrix} = \begin{bmatrix} 1 & & \\ 2 & 1 & \\ -2 & 0 & 1 \end{bmatrix}\begin{bmatrix} 2 & 3 & -1 \\ & -5 & 4 \\ & & 3 \end{bmatrix}$

31. $\begin{bmatrix} 1 & & & \\ & 0 & 1 & 0 \\ & 0 & 0 & 1 \\ & 1 & 0 & 0 \end{bmatrix}\begin{bmatrix} -2 & 3 & -1 & 2 \\ 4 & -6 & 2 & 3 \\ 6 & -1 & -2 & 3 \\ -8 & -4 & -3 & -1 \end{bmatrix}$

$= \begin{bmatrix} 1 & & & \\ -3 & 1 & & \\ 4 & -2 & 1 & \\ -2 & 0 & 0 & 1 \end{bmatrix}\begin{bmatrix} -2 & 3 & -1 & 2 \\ & 8 & -5 & 9 \\ & & -9 & 9 \\ & & & 7 \end{bmatrix}$

33. $\begin{bmatrix} 3 \\ -2 \end{bmatrix}$ 35. $\begin{bmatrix} 3 \\ -2 \\ -1 \end{bmatrix}$ 39. HINT If $i > j$, $u_{ij} = 0$.

Section 1.6

1. $\begin{bmatrix} 2 & 4 \\ -3 & 1 \end{bmatrix}$ 3. $\begin{bmatrix} 5 & 8 & -4 \\ 7 & -1 & 2 \\ 2 & -3 & 4 \end{bmatrix}$ 5. $\begin{bmatrix} 2 & -1 & 3 & 0 \\ 4 & 7 & -3 & -2 \\ 9 & 1 & 3 & -1 \end{bmatrix}$

7. $\begin{bmatrix} 2 & -5 \\ -1 & 3 \end{bmatrix}, \begin{bmatrix} 3 & 5 \\ 1 & 2 \end{bmatrix}, \begin{bmatrix} 3 & 1 \\ 5 & 2 \end{bmatrix}, \begin{bmatrix} 3 & 5 \\ 1 & 2 \end{bmatrix}$

9. $\begin{bmatrix} 1 & 2 & -3 & 4 \\ & 1 & & \\ & & 1 & \\ & & & 1 \end{bmatrix}, \begin{bmatrix} 1 & -2 & 3 & -4 \\ & 1 & & \\ & & 1 & \\ & & & 1 \end{bmatrix}, \begin{bmatrix} 1 & & & \\ -2 & 1 & & \\ 3 & & 1 & \\ -4 & & & 1 \end{bmatrix},$
$\begin{bmatrix} 1 & -2 & 3 & -4 \\ & 1 & & \\ & & 1 & \\ & & & 1 \end{bmatrix}$

11. $\begin{bmatrix} 1 & & \\ 3 & 1 & \\ 2 & & 1 \end{bmatrix}, \begin{bmatrix} 1 & & \\ -3 & 1 & \\ -2 & & 1 \end{bmatrix}, \begin{bmatrix} 1 & -3 & -2 \\ & 1 & \\ & & 1 \end{bmatrix}, \begin{bmatrix} 1 & & \\ -3 & 1 & \\ -2 & & 1 \end{bmatrix}$

13. $D(C^T)^2B^T$

15. (a) $\begin{bmatrix} 1 & & \\ 1 & 1 & \\ & 1 & 1 \end{bmatrix}\begin{bmatrix} 1 & 1 & \\ & 1 & 1 \\ & & 2 \end{bmatrix}$ (b) $\begin{bmatrix} \frac{5}{2} & -\frac{3}{2} & \frac{1}{2} \\ -\frac{3}{2} & \frac{3}{2} & -\frac{1}{2} \\ \frac{1}{2} & -\frac{1}{2} & \frac{1}{2} \end{bmatrix}$ (c) $\begin{bmatrix} 2 \\ -1 \\ -3 \end{bmatrix}$

17. (a) $\begin{bmatrix} 1 & & & \\ 2 & 1 & & \\ & 4 & 1 & \\ & & -2 & 1 \end{bmatrix}\begin{bmatrix} 2 & \frac{1}{2} & & \\ & 1 & 1 & \\ & & -2 & 1 \\ & & & 4 \end{bmatrix}$

(b) $\begin{bmatrix} \frac{1}{2} & 0 & -\frac{1}{16} & \frac{1}{32} \\ 0 & 0 & \frac{1}{4} & -\frac{1}{8} \\ -2 & 1 & -\frac{1}{4} & \frac{1}{8} \\ 4 & -2 & \frac{1}{2} & \frac{1}{4} \end{bmatrix}$ (c) $\begin{bmatrix} 1 \\ -2 \\ 3 \\ -4 \end{bmatrix}$

23. $\begin{bmatrix} 1 & & \\ & 1 & \\ & & 1 \end{bmatrix}$, (itself); $\begin{bmatrix} 0 & 1 & 0 \\ 1 & 0 & 0 \\ 0 & 0 & 1 \end{bmatrix}$, (itself); $\begin{bmatrix} & & 1 \\ & 1 & \\ 1 & & \end{bmatrix}$, (itself);
$\begin{bmatrix} 1 & 0 & 0 \\ 0 & 0 & 1 \\ 0 & 1 & 0 \end{bmatrix}$, (itself); $\begin{bmatrix} 0 & 1 & 0 \\ 0 & 0 & 1 \\ 1 & 0 & 0 \end{bmatrix} \leftrightarrow \begin{bmatrix} 0 & 0 & 1 \\ 1 & 0 & 0 \\ 0 & 1 & 0 \end{bmatrix}$

Section 1.7

1. (a) (0, 2.000) (b) (1.000, 2.000) 3. (2, −3) 5. (3, −2, 0)
7. ($\frac{1}{2}$, −2, 3) 9. (−2, 3, −1, 4)
11. $\begin{bmatrix} 0 & 1 \\ 1 & 0 \end{bmatrix}\begin{bmatrix} 2 & 1 \\ 4 & -3 \end{bmatrix} = \begin{bmatrix} 1 & \\ \frac{1}{2} & 1 \end{bmatrix}\begin{bmatrix} 4 & -3 \\ & \frac{5}{2} \end{bmatrix}$

13. $\begin{bmatrix} 0 & 1 & 0 \\ 0 & 0 & 1 \\ 1 & 0 & 0 \end{bmatrix}\begin{bmatrix} -2 & 3 & 1 \\ 4 & -8 & 3 \\ 1 & 1 & 1 \end{bmatrix} = \begin{bmatrix} 1 & & \\ \frac{1}{4} & 1 & \\ -\frac{1}{2} & -\frac{1}{3} & 1 \end{bmatrix}\begin{bmatrix} 4 & -8 & 3 \\ & 3 & \frac{1}{4} \\ & & \frac{31}{12} \end{bmatrix}$

15. $\begin{bmatrix} 0 & 0 & 1 \\ 1 & 0 & 0 \\ 0 & 1 & 0 \end{bmatrix}\begin{bmatrix} 4 & -2 & 1 \\ -2 & 3 & -1 \\ 6 & 12 & 6 \end{bmatrix} = \begin{bmatrix} 1 & & \\ \frac{2}{3} & 1 & \\ -\frac{1}{3} & -\frac{7}{10} & 1 \end{bmatrix}\begin{bmatrix} 6 & 12 & 6 \\ & -10 & -3 \\ & & -\frac{11}{10} \end{bmatrix}$

17. $\begin{bmatrix} 0 & 0 & 1 & 0 \\ 1 & 0 & 0 & 0 \\ 0 & 1 & 0 & 0 \\ 0 & 0 & 0 & 1 \end{bmatrix}\begin{bmatrix} 3 & -2 & 3 & 1 \\ 2 & 1 & 1 & -4 \\ 4 & 8 & 4 & -1 \\ 1 & 1 & 1 & 1 \end{bmatrix}$

$= \begin{bmatrix} 1 & & & \\ \frac{3}{4} & 1 & & \\ \frac{1}{2} & \frac{3}{8} & 1 & \\ \frac{1}{4} & \frac{1}{8} & 0 & 1 \end{bmatrix}\begin{bmatrix} 4 & 8 & 4 & -1 \\ & -8 & 0 & \frac{7}{4} \\ & & -1 & -\frac{133}{32} \\ & & & \frac{33}{32} \end{bmatrix}$

19.

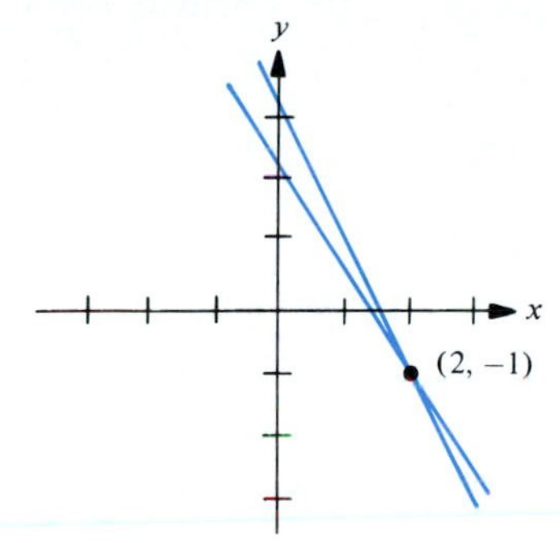

Ill conditioned,
new solution: (1, 1)

21.

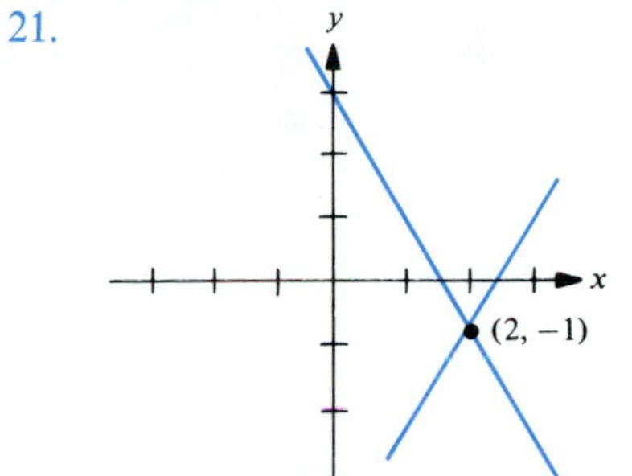

Not ill conditioned
new solution: (2.001, −1.002)

23.

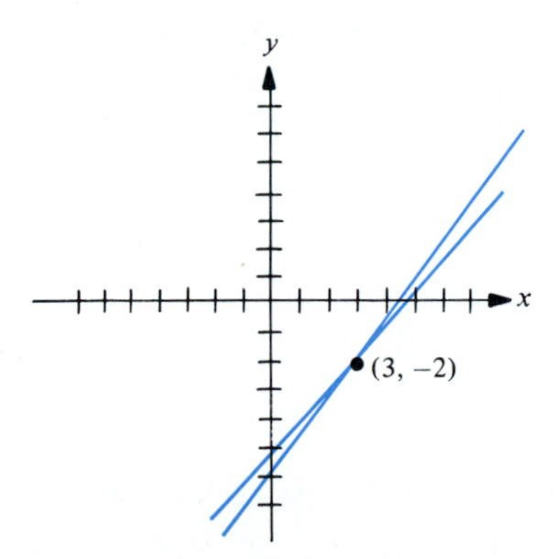

Ill conditioned,
new solution: (6, 2)

Chapter 1 Review

1. (a) $\begin{bmatrix} 2 & 3 & -1 \\ 4 & -1 & -2 \\ -2 & -1 & 1 \end{bmatrix}\begin{bmatrix} x \\ y \\ z \end{bmatrix} = \begin{bmatrix} -5 \\ 4 \\ 1 \end{bmatrix}$ (b) $\begin{bmatrix} \frac{1}{2} + \frac{1}{2}t \\ -2 \\ t \end{bmatrix}$

3. (a) $\begin{bmatrix} 1 & 0 & 0 & 1 \\ 1 & -1 & 0 & 0 \\ 1 & 0 & 1 & 0 \\ 0 & 1 & -1 & 0 \end{bmatrix}\begin{bmatrix} x \\ y \\ z \\ w \end{bmatrix} = \begin{bmatrix} 4 \\ 3 \\ 4 \\ -3 \end{bmatrix}$ (b) $\begin{bmatrix} 2 \\ -1 \\ 2 \\ 2 \end{bmatrix}$

5. $\begin{bmatrix} -5 & 2 \\ -3 & 1 \end{bmatrix}, \begin{bmatrix} 2 & 2 \\ -3 & 2 \end{bmatrix}$ 7. $\begin{bmatrix} 1 & -2 \\ & 1 \end{bmatrix}, \begin{bmatrix} 1 & 0 \\ 3 & 1 \end{bmatrix}, \begin{bmatrix} 1 & -2 \\ 3 & -5 \end{bmatrix}$

9. $\begin{bmatrix} 1 & 0 & 0 \\ 0 & 0 & 1 \\ 0 & 1 & 0 \end{bmatrix}$ 11. (a) A with 20 times row 1 added to row 2
(b) $\begin{bmatrix} 1 & \\ 2 & 1 \end{bmatrix}, \begin{bmatrix} 1 & \\ 20 & 1 \end{bmatrix}, \begin{bmatrix} 10 & \\ 20 & 10 \end{bmatrix}$

13. 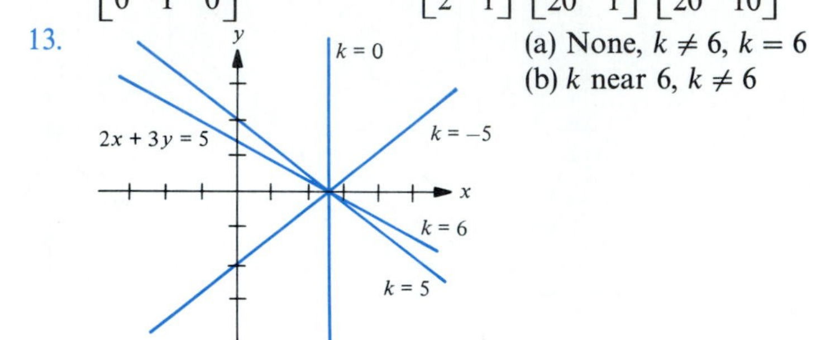

(a) None, $k \neq 6$, $k = 6$
(b) k near 6, $k \neq 6$

15. (a) $\begin{bmatrix} 0 & -1 \\ 1 & 0 \end{bmatrix}$ (b) $\begin{bmatrix} 0 & 1 \\ 0 & 0 \end{bmatrix}$ (c) $C = \begin{bmatrix} 0 & 1 \\ 1 & 0 \end{bmatrix}, D = \begin{bmatrix} 0 & -1 \\ 1 & 0 \end{bmatrix}$

17. (a), (b) a, b, c all nonzero 19. $\begin{bmatrix} -\frac{7}{2} & 2 \\ 2 & -1 \end{bmatrix}, \begin{bmatrix} 1 & \\ 2 & 1 \end{bmatrix}\begin{bmatrix} 2 & 4 \\ & -1 \end{bmatrix}$

21. Column 1 of A = Column 1 of B. *That's all!*

23. $\begin{bmatrix} 0 & 1 & 0 \\ 0 & 0 & 1 \\ 1 & 0 & 0 \end{bmatrix}\begin{bmatrix} 0 & 2 & 3 \\ 2 & 1 & 2 \\ 0 & 3 & 4 \end{bmatrix} = \begin{bmatrix} 1 & & \\ 0 & 1 & \\ 0 & \frac{2}{3} & 1 \end{bmatrix}\begin{bmatrix} 2 & 1 & 2 \\ & 3 & 4 \\ & & \frac{1}{3} \end{bmatrix}; \begin{bmatrix} 2 \\ -1 \\ -2 \end{bmatrix}$

25. (a) If the single 1 in column 1 of P is in the ith row, let P_1 interchange rows 1 and i. Next, look at column 2 of P_1P, and so on. (b) $P = P_1 \cdots P_{n-1}$

29. $[A^{-1} + B^{-1}]^{-1} = B(A + B)^{-1}A$ 31. (a) T (b) F (c) F (d) T

Section 2.1

1. 3 3. 5 5. 3 7. Y, + 9. N 11. Y, −
13. 0 15. −5 17. 21 19. 27 21. −6 23. $x + 2$
25. $x^3 + 2x^2 - 3x$ 27. −12 29. 1 31. 0
33. Each elementary product has 0 as a factor. 35. $n(n - 1)/2$

Section 2.2

1. 0 3. 0 5. -8 7. 0 9. -1 11. 0
13. 6 15. 100 17. -1284 19. 420 25. 3 27. 30

Section 2.3

1. $-1, -13, 13$ 3. $2, -20, -40$ 5. $1, -5, -5$
7. $30, -72, -2160$ 9. N 11. N 13. Y
17. $-24, -24$ 19. $3, -1, -3$ 21. $6, -2, -12$ 23. $13, 1, 13$

Section 2.4

1. $M_{11} = 4, M_{12} = 3, M_{21} = 1, M_{22} = -2$
   $C_{11} = 4, C_{12} = -3, C_{21} = -1, C_{22} = -2$
3. $M_{11} = -2, M_{12} = 0, M_{13} = 1, M_{21} = 0, M_{22} = 0, M_{23} = 2$
   $M_{31} = 2, M_{32} = 4, M_{33} = -3$
   $C_{11} = -2, C_{12} = 0, C_{13} = 1, C_{21} = 0, C_{22} = 0, C_{23} = -2$
   $C_{31} = 2, C_{32} = -4, C_{33} = -3$
5. -11 7. -4 9. -20 11. 28 13. 27 15. 6
17. 1 19. -48 21. $(b - a)(c - a)(c - b)$
23. $\begin{bmatrix} 1 & -4 \\ 1 & 2 \end{bmatrix}, \begin{bmatrix} 1 & 1 \\ -4 & 2 \end{bmatrix}, \frac{1}{6}\begin{bmatrix} 1 & 1 \\ -4 & 2 \end{bmatrix}$
25. $\begin{bmatrix} 0 & 2 & -1 \\ -1 & 11 & -7 \\ -1 & 8 & -5 \end{bmatrix}, \begin{bmatrix} 0 & -1 & -1 \\ 2 & 11 & 8 \\ -1 & -7 & -5 \end{bmatrix}, \frac{1}{1}\begin{bmatrix} 0 & -1 & -1 \\ 2 & 11 & 8 \\ -1 & -7 & -5 \end{bmatrix}$
27. $\begin{bmatrix} -8 & 4 & 0 \\ -1 & -2 & 0 \\ -26 & 8 & 20 \end{bmatrix}, \begin{bmatrix} -8 & -1 & -26 \\ 4 & -2 & 8 \\ 0 & 0 & 20 \end{bmatrix}, -\frac{1}{20}\begin{bmatrix} -8 & -1 & -26 \\ 4 & -2 & 8 \\ 0 & 0 & 20 \end{bmatrix}$
29. $(3, -2)$ 31. $(-2, \frac{1}{2}, 3)$ 33. $(3, -2, 0)$

Chapter 2 Review

1.,3. -1
5. Yes. Use the definition, and a sum of products of integers is an integer.
9. $[\det(A + B)]^2 = \det(A + B)^2 = \det[A^2 + AB + BA + B^2] = \det[A^2 + B^2]$
11. -12 13. No, $\begin{bmatrix} 1 & 1 & 0 \\ 1 & 0 & 1 \\ 0 & 1 & 1 \end{bmatrix}$

Section 3.1

1. (a) 7 mph, S (b) 1 mph, S (c) 5 mph, 143.130° (d) 2.83363 mph, 131.529° (e) 6.47847 mph, 160.886°
3. (a) 7.21110 mph, 56.3099° to AB (b) 1.5 mi (c) 15 min (d) 15 min
5. 322.360°, 405.449 mph 7. 128.176°, 398.765 km/h 9. $(-1, 1)$
11. $(-5, 11, -2)$ 13. $(-1, 4, 6)$ 15. $(-1, 0, -3)$

17. $(-8, 20, 40)$ 19. $2\sqrt{2}$ 21. $\sqrt{73}$ 23. 5 25. $6\sqrt{14}$
27. $1, \|r\mathbf{w}\| = |r|\,\|\mathbf{w}\|$ 29. $9, 0°$ 31. $0, 90°$ 33. $7, 69.4859°$
35. $\mathbf{0}$ 37. $\frac{13}{10}\mathbf{v} = (-\frac{13}{5}, \frac{26}{5})$ 39. $\frac{11}{13}\mathbf{v} = (\frac{33}{13}, \frac{44}{13}, -\frac{11}{13})$ 41. -136
43. $3\sqrt{13}$ 45. scalar · vector is undefined 47. ||scalar|| is undefined

Section 3.2

1. $(-2, 6, -8, 6, -5)$ 3. $(-9, -8, -7, -5, 3, 9)$
5. $(5, -3, 16, 3, 5)$ 7. $\begin{bmatrix} -1 \\ 0 \\ -3 \\ -4 \end{bmatrix}$ 9. $\begin{bmatrix} -8 \\ 20 \\ 40 \\ 28 \end{bmatrix}$ 11. $\sqrt{21}$
13. $\sqrt{22}$ 15. 3 17. $\sqrt{493}$ 19. $1, \|r\mathbf{w}\| = |r|\,\|\mathbf{w}\|$
21. $31, 0°$ 23. $0, 90°$ 25. $2, 83.9576°$ 27. $\mathbf{0}$
29. $\frac{5}{11}\mathbf{v} = [\frac{20}{11} \ \ \frac{10}{11} \ \ \frac{5}{11} \ \ 0 \ \ \frac{5}{11}]$ 31. $\frac{12}{19}\mathbf{v} = (\frac{36}{19}, \frac{24}{19}, -\frac{12}{19}, \frac{12}{19}, \frac{24}{19})$

Section 3.3

1. (b) is $((r + s)x_1, 2x_2)$ 17. N. Axiom (h) 19. N, Axiom (f)
21. Y 23. N, closure of scalar multiplication 25. Y
27. N, Axiom (c) 29. Y 37. $\int_0^{2\pi} \sin x \cos x \, dx = 0, \mathbf{0}$ 39. π

Section 3.4

1. Y, Y, Y 3. Y, N, N 5. Y, Y, Y 7. Y, Y, Y
9. Y, Y, Y 11. Y, Y, Y 13. Y, Y, Y 15. Y, Y, Y
17. Y, Y, Y 19. N, Y, N 21. Y, Y, Y 23. Y, Y, Y
25. $\left\{ x\begin{bmatrix} 3 \\ 1 \end{bmatrix} \middle| \, x \text{ in } \mathbb{R} \right\}$, 2,

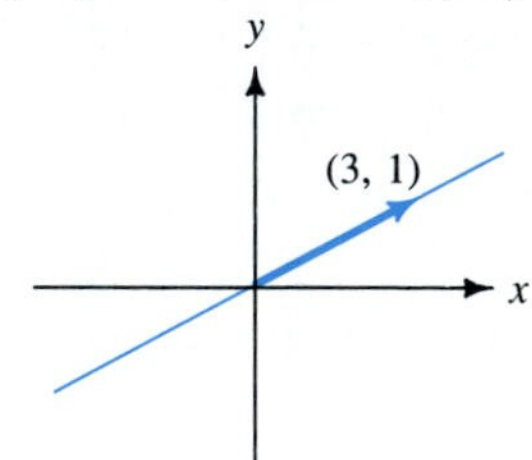

27. $\mathbb{R}^3$, 3,

29. 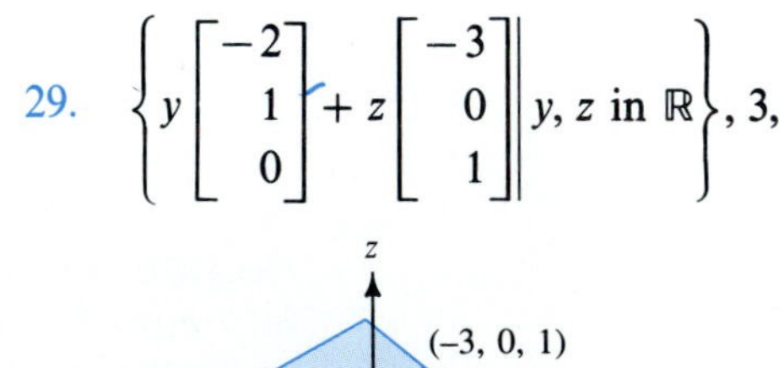
$\left\{ y\begin{bmatrix} -2 \\ 1 \\ 0 \end{bmatrix} + z\begin{bmatrix} -3 \\ 0 \\ 1 \end{bmatrix} \middle| \, y, z \text{ in } \mathbb{R} \right\}$, 3,

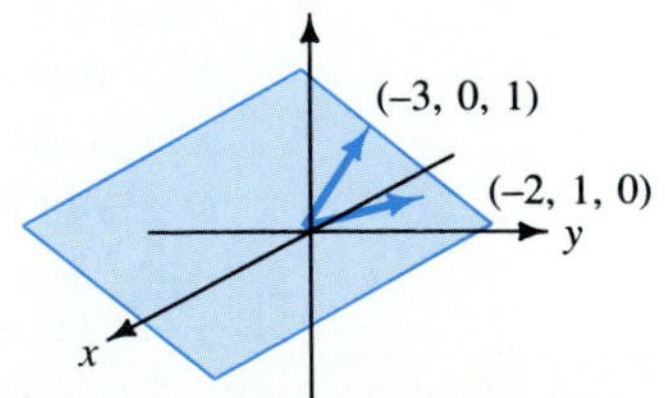

31. $\left\{ x\begin{bmatrix} 1 \\ 2 \end{bmatrix} \,\middle|\, x \text{ in } \mathbb{R} \right\}$, 2,

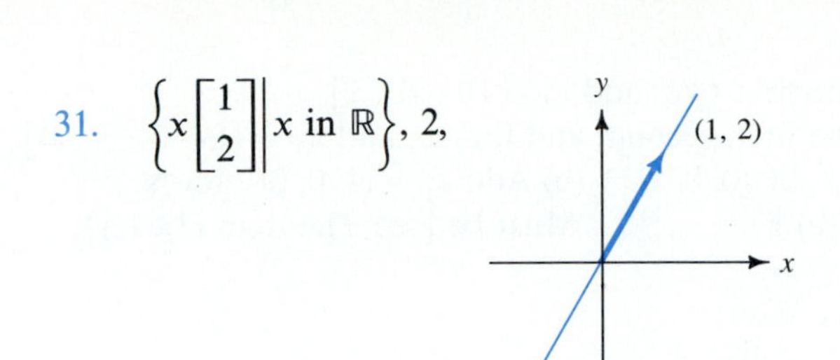

33. (a) Y, -3, 0 (b) N (c) Y, 0, 0 (d) Y, 5, 0
35. (a) N (b) Y, 0, 0 (c) Y, 3, -2 (d) Y, 2, -5
37. (a) Y, -1, 1 (b) N (c) Y, 0, 0 (d) Y, 1, 1 (e) N
39. The main diagonal in $\mathbb{R}^2$ 41. The xy-plane in $\mathbb{R}^3$
43. All diagonal 2×2 matrices 45. All polynomials of degree ≤ 1

Section 3.5

1. $\mathbf{u}_2 = 3\mathbf{u}_1$ 3. $\mathbf{w}_3 = 0\mathbf{w}_1 + 0\mathbf{w}_2$ 5. $A_3 = 2A_1 - A_2$
7. Line, dep. 9. Line, dep. 11. Larger, indep.
13. Dep., Theorem (3.46) 15. Indep., by inspection
17. Dep., by inspection 19. Dep., $p_3 = -2p_1 + 3p_2$ 21. Indep.
23. Indep. 25. Dep.
31. Suppose $c_1\mathbf{x}_1 + \cdots + c_k\mathbf{x}_k = \mathbf{0}$. Then $c_1A\mathbf{x}_1 + \cdots + c_kA\mathbf{x}_k = A\mathbf{0}$ or $c_1\mathbf{b}_1 + \cdots + c_k\mathbf{b}_k = \mathbf{0}$. Thus all $c_i = 0$ since the $\mathbf{b}$'s are linearly independent.
33. Definitely not; five vectors in $\mathbb{R}^3$.

Section 3.6

1. $3 > \dim \mathbb{R}^2$ 3. $3 < \dim \mathbb{R}^4$ 5. $2 < \dim P_2$ 7. Y
9. N 11. Y 13. N 15. Y 17. Y 19. N

21. $\begin{bmatrix} 1 & 0 \\ 0 & 0 \\ 0 & 0 \end{bmatrix}$, etc.; 6 23. $\begin{bmatrix} 1 & 0 & \cdots & 0 \\ 0 & 0 & \cdots & 0 \\ \vdots & \vdots & & \vdots \\ 0 & 0 & \cdots & 0 \end{bmatrix}$, etc.; mn 25. N

27. $\left\{ \begin{bmatrix} 2 \\ -\frac{3}{2} \\ 1 \end{bmatrix} \right\}$, 1 29. $\left\{ \begin{bmatrix} -\frac{4}{5} \\ \frac{6}{5} \\ 1 \end{bmatrix} \right\}$, 1 31. No basis, 0

33. $\left\{ \begin{bmatrix} 1 \\ -1 \end{bmatrix} \right\}$, 1 35. $\left\{ \begin{bmatrix} 1 & & \\ & 0 & \\ & & 0 \end{bmatrix}, \begin{bmatrix} 0 & 1 & \\ 1 & 0 & \\ & & 0 \end{bmatrix}, \text{etc.} \right\}$, 6

37. Infinite dimensional
39. (b) $\{(1, -1, 0, 0), (1, 0, -1, 0), (1, 0, 0, -1)\}$, 3

41. (a) $\begin{bmatrix} 1 & 1 & 0 & -1 \\ 1 & -1 & -1 & 0 \end{bmatrix}$ (b) $\begin{bmatrix} 1 \\ 0 \\ 1 \\ 1 \end{bmatrix}, \begin{bmatrix} 0 \\ 1 \\ -1 \\ 1 \end{bmatrix}$, 2

43. Take the first two; add $\varepsilon_3 = [0 \quad 0 \quad 1]$
45. Take the first, second, and fourth; add $\varepsilon_3 = [0 \quad 0 \quad 1 \quad 0]$
47. (a) $\{(1, 1, 0), (0, 0, 1)\}$ (b) Add $\varepsilon_1 = (1, 0, 0)$ (c) N
49. (a) T (b) F 51. Must be [see Theorem (3.64c)]

Section 3.7

1. $[2 \quad 4 \quad 3], [0 \quad -2 \quad 4]$, etc., $\begin{bmatrix} 2 \\ 0 \\ 3 \\ -1 \end{bmatrix}$, etc.
3. $[1 \quad -2 \quad 5 \quad 2 \quad 1 \quad 3], [0 \quad 0 \quad 2 \quad 3 \quad 0 \quad -1], [0 \quad 0 \quad 0 \quad 0 \quad 4]$; $\begin{bmatrix} 1 \\ 0 \\ 0 \end{bmatrix}, \begin{bmatrix} 5 \\ 2 \\ 0 \end{bmatrix}, \begin{bmatrix} 3 \\ -1 \\ 4 \end{bmatrix}$; $rk = 3$
5. Every row; every column; $rk = 4$
7. $[1 \quad 2 \quad -3], [0 \quad 0 \quad 11]$; $\begin{bmatrix} 1 \\ 2 \\ 3 \end{bmatrix}, \begin{bmatrix} -3 \\ 5 \\ 0 \end{bmatrix}$; $rk = 2$
9. $[0 \quad 2 \quad -3 \quad 12], [0 \quad 0 \quad 0 \quad 4 \quad 3]$; $\begin{bmatrix} 2 \\ -2 \\ 4 \end{bmatrix}, \begin{bmatrix} 1 \\ 3 \\ 6 \end{bmatrix}$; $rk = 2$
11. $(2, -1, 6), (0, 5, -14)$ 13. $\begin{bmatrix} 2 \\ 3 \\ 2 \\ -4 \end{bmatrix}, \begin{bmatrix} 0 \\ -5 \\ -8 \\ 10 \end{bmatrix}$
15. $2 - 3x + 4x^2 - 5x^3, 5x + 7x^2 + 8x^3$
17. $\{[2 \quad -1 \quad 3], [0 \quad 4 \quad -5]\}$; $\left\{\begin{bmatrix} -\frac{7}{8} \\ \frac{5}{4} \\ 1 \end{bmatrix}\right\}$; 2; 1
19. All 3 rows; no basis; 3; 0 21. $\mathbf{b} = 5\mathbf{c}_1 + 1\mathbf{c}_2$ 23. N
25. $\mathbb{R}^3, \mathbb{R}^3, \{\mathbf{0}\}$ 27. yz-plane, xy-plane, x-axis
35. $[1 \quad 3 \quad -2 \quad 4], [0 \quad 0 \quad 5 \quad 1]$; $\begin{bmatrix} -3 \\ 1 \\ 0 \\ 0 \end{bmatrix}, \begin{bmatrix} 22 \\ 0 \\ 1 \\ -5 \end{bmatrix}$; $\begin{bmatrix} 1 \\ 0 \\ 0 \end{bmatrix}, \begin{bmatrix} -2 \\ 5 \\ 0 \end{bmatrix}$; $[0 \quad 0 \quad 1]$
39. (a), (b) F, $\begin{bmatrix} 0 & 1 \\ 0 & 0 \end{bmatrix}$ (c) T, $\dim[NS(A)] = 0$

(d) T, four vectors in $\mathbb{R}^3$ (e) F, $\begin{bmatrix}1&1&1\\0&0&0\\0&0&0\\0&0&0\end{bmatrix}$

Section 3.8

1. $\begin{bmatrix}3\\-2\end{bmatrix}, \begin{bmatrix}-2\\3\end{bmatrix}$ 3. $\begin{bmatrix}-\frac{9}{4}\\-\frac{13}{4}\end{bmatrix}, \begin{bmatrix}-\frac{13}{4}\\-\frac{9}{4}\end{bmatrix}$ 5. $\begin{bmatrix}2\\-2\\3\end{bmatrix}, \begin{bmatrix}-2\\3\\2\end{bmatrix}$

7. $\begin{bmatrix}4\\3\\-2\end{bmatrix}, \begin{bmatrix}-2\\4\\3\end{bmatrix}$ 9. $\begin{bmatrix}2\\\frac{4}{3}\\-\frac{1}{6}\end{bmatrix}, \begin{bmatrix}-\frac{1}{6}\\\frac{4}{3}\\2\end{bmatrix}$ 11. $\begin{bmatrix}2\\-1\\3\\-6\end{bmatrix}, \begin{bmatrix}3\\-1\\2\\-6\end{bmatrix}$

13. $\mathbf{x} = \frac{2}{5}\mathbf{v}_1 - \frac{1}{5}\mathbf{v}_2 + 0\mathbf{v}_3 = -\frac{4}{5}\mathbf{v}_1 + \frac{7}{5}\mathbf{v}_2 + 2\mathbf{v}_3 = -\frac{1}{5}\mathbf{v}_1 + \frac{3}{5}\mathbf{v}_2 + \mathbf{v}_3$
    $\mathbf{y} = \frac{4}{5}\mathbf{v}_1 - \frac{7}{5}\mathbf{v}_2 - \mathbf{v}_3 = \frac{1}{5}\mathbf{v}_1 - \frac{3}{5}\mathbf{v}_2 + 0\mathbf{v}_3 = -\frac{2}{5}\mathbf{v}_1 + \frac{1}{5}\mathbf{v}_2 + \mathbf{v}_3$

15. $[a_1 \cdots a_n]^T$, $[b_1 \cdots b_n]^T$, $[a_1 + b_1 \cdots a_n + b_n]^T$

17. $\begin{bmatrix}\frac{1}{4}&-\frac{1}{4}\\\frac{1}{8}&\frac{3}{8}\end{bmatrix}, \begin{bmatrix}-\frac{3}{4}\\\frac{5}{8}\end{bmatrix}$ 19. $\begin{bmatrix}2&-1\\3&-2\end{bmatrix}, \begin{bmatrix}1\\3\end{bmatrix}$

21. $\begin{bmatrix}-7&-6&12\\6&5&-10\\-4&-4&7\end{bmatrix}, \begin{bmatrix}7\\-6\\3\end{bmatrix}$ 23. $\begin{bmatrix}1&0&0\\-1&1&0\\0&-1&1\end{bmatrix}, \begin{bmatrix}2\\-3\\2\end{bmatrix}$

25. $\begin{bmatrix}1&1&0\\1&1&1\\1&0&1\end{bmatrix}, \begin{bmatrix}5\\3\\0\end{bmatrix}$ 27. $\begin{bmatrix}\frac{1}{4}&-\frac{1}{4}\\\frac{1}{8}&\frac{3}{8}\end{bmatrix}, \begin{bmatrix}3&2\\-1&2\end{bmatrix}$

29. $\begin{bmatrix}-7&-6&12\\6&5&-10\\-4&-4&7\end{bmatrix}, \begin{bmatrix}5&6&0\\2&1&-2\\4&4&1\end{bmatrix}$ 31. QP

Section 3.9

1. (0, 1, 1, 0, 0, 1, 1) 3. (0, 1, 1, 1, 1, 0, 0) 5. Y; (0, 1, 1, 0)
7. N; (0, 1, 1, 0) 9. N; (0, 0, 0, 1) 11. N; (1, 1, 0, 0)
13. (0, 0, 0), (0, 0, 1), (0, 1, 0), (0, 1, 1), (1, 0, 0), (1, 0, 1), (1, 1, 0), (1, 1, 1); 8
15. 2; 2 17. 4; 4 19. 7(3) = 21; $(2^n - 1)(2^n - 2)/2$

Chapter 3 Review

1. $\begin{bmatrix}5\\0\\0\\2\end{bmatrix}, \begin{bmatrix}-5\\5\\-15\\14\end{bmatrix}$ 3. $\frac{1}{\sqrt{10}}\begin{bmatrix}0\\3\\1\\0\end{bmatrix}$, many others

5. $\{(1, 0, 0, 0), (0, 1, 1, 1)\}$
7. $\begin{bmatrix} 2 & 1 & -1 & 0 & 3 \\ 0 & 0 & 1 & 1 & 2 \\ 0 & 0 & 0 & 0 & 0 \end{bmatrix}$; 2; 2; 3

11. Y; $A_3 = 2A_1 - A_2$
13. $\begin{bmatrix} 2 \\ 0 \\ 1 \end{bmatrix} + t\begin{bmatrix} -2 \\ 1 \\ 0 \end{bmatrix}$; none

15. HINT A^{-1} exists.
17. (a) F (b) T (c) F (d) T (e) F

19. (b), (d), (e)
21. (c), (d)

23. (a) $\{[1 \quad 0 \quad 0]^T, [2 \quad 3 \quad 0]^T\} = \{\mathbf{c}_2, \mathbf{c}_3\}$
(b) $\mathbf{c}_1 = 0\mathbf{c}_2 + 0\mathbf{c}_3$, $\mathbf{c}_2 = 1\mathbf{c}_2 + 0\mathbf{c}_3$, $\mathbf{c}_3 = 0\mathbf{c}_2 + 1\mathbf{c}_3$, $\mathbf{c}_4 = 2\mathbf{c}_2 + 1\mathbf{c}_3$
(c) $\begin{bmatrix} 0 & 1 & 2 & 4 \\ 0 & 0 & 3 & 3 \\ 0 & 1 & 2 & 4 \end{bmatrix}$

25. $\left\{\begin{bmatrix} 1 & 0 \\ 0 & 1 \end{bmatrix}, \begin{bmatrix} 0 & 1 \\ 1 & 0 \end{bmatrix}\right\}$

27. (a), (b) $m = n = r$ (c) $r = n < m$ (d) $m = r < n$

Section 4.1

1. $f_A(x_1, x_2) = (x_1 + 2x_2, -2x_1, 3x_1 - x_2)$; $f_A: \mathbb{R}^2 \to \mathbb{R}^3$
3. $f_A(x_1, x_2) = (4x_1 - x_2, x_1 + 2x_2, 3x_1, -x_1 + x_2)$; $f_A: \mathbb{R}^2 \to \mathbb{R}^4$
5. $f_A(x_1, x_2, x_3, x_4) = x_1 + 2x_2 - x_3 + 3x_4$; $f_A: \mathbb{R}^4 \to \mathbb{R}^1$
7. Expands the x-axis (multiples it by 2), contracts the y-axis (multiplies it by $\frac{1}{2}$)
9. Interchanges the x- and y-axes
11. Rotates both axes by 45° (see Example 5)

13. $\begin{bmatrix} \frac{1}{2}\sqrt{3} & -\frac{1}{2} & \\ \frac{1}{2} & \frac{1}{2}\sqrt{3} & \\ & & 1 \end{bmatrix}$
15. $\begin{bmatrix} -1 & & \\ & \frac{1}{2}\sqrt{2} & -\frac{1}{2}\sqrt{2} \\ & \frac{1}{2}\sqrt{2} & \frac{1}{2}\sqrt{2} \end{bmatrix}$

17. Y; $T: \mathbb{R}^2 \to \mathbb{R}^2$; $\begin{bmatrix} 2 & 0 \\ 0 & 1 \end{bmatrix}$
19. N; $T: \mathbb{R}^2 \to \mathbb{R}^2$

21. Y; $T: \mathbb{R}^3 \to \mathbb{R}^2$; $\begin{bmatrix} 2 & 1 & 0 \\ 1 & -1 & 1 \end{bmatrix}$
23. Y; $T: \mathbb{R}^3 \to \mathbb{R}^3$; $\begin{bmatrix} 0 & 0 & 0 \\ 0 & 0 & 0 \\ 0 & 0 & 0 \end{bmatrix}$

25. Y; $T: \mathbb{R}^2 \to \mathbb{R}^3$; $\begin{bmatrix} 1 & -1 \\ 1 & 1 \\ 2 & -3 \end{bmatrix}$
27. Y
29. N

31. Y
33. Y
35. Y
37. Y
39. N

41. (a) Y (b) $J: C[a, b] \to \mathbb{R}$

Section 4.2

1. 1, 1; 7, 7
3. 5, 5; -10, -10
5. $\sqrt{13} \approx 3.6$, $\sqrt{6} + \sqrt{3} \approx 4.2$; $\sqrt{33} \approx 5.7$, $\sqrt{10} + \sqrt{13} \approx 6.8$

7. $-1, [-1], \begin{bmatrix} 2 & 1 & -1 \\ -2 & -1 & 1 \\ 4 & 2 & -2 \end{bmatrix}$, (a) = (b);

$2, \begin{bmatrix} 2 & -2 & -2 & 2 \\ -1 & 1 & 1 & -1 \\ 2 & -2 & -2 & 2 \\ 1 & -1 & -1 & 1 \end{bmatrix}, [2]$, (a) = (c).

9. Y 11. N, (a) 13. Y 15. N, (d) 17. $-14, [-14]$

19. $3, [3]$ 21. (a) 19, 19 (b) $-49, -49$

23. (a) $-82, -82$ (b) 20, 20

25. (a) $\begin{bmatrix} 38 & -76 \\ -76 & 152 \end{bmatrix}$ (b), (c) $\left\{ x\begin{bmatrix} 2 \\ 1 \end{bmatrix} \middle| x \text{ in } \mathbb{R} \right\}$ (d), (e) 1

27. (a) $\begin{bmatrix} 15 & -30 & 0 \\ -30 & 60 & 0 \\ 0 & 0 & 74 \end{bmatrix}$ (b), (c) $\left\{ x\begin{bmatrix} 2 \\ 1 \\ 0 \end{bmatrix} \middle| x \text{ in } \mathbb{R} \right\}$ (d), (e) 2

41. Only for $(a, a, \ldots, a)$ 45. $\begin{bmatrix} 1 \\ 0 \end{bmatrix}$

Section 4.3

1. (a) N (b) Y, $y = -3x$ 3. $10, \frac{5}{4}, \frac{2}{3}, \frac{13}{16}; a = \frac{1}{3}$ 5. $1, 4, 4, \frac{2}{3}$

7. $y = \frac{1}{3}x$ 9. $y = \frac{2}{3}x$ 11. $y = 1$ 13. $y = 2x + \frac{8}{3}$

15. $y = 1.3x + 1.8$ 17. $p = \frac{1}{3}\begin{bmatrix} 2 \\ 2 \\ 2 \end{bmatrix}, P = \frac{1}{3}\begin{bmatrix} 1 & 1 & 1 \\ 1 & 1 & 1 \\ 1 & 1 & 1 \end{bmatrix}$

19. $p = \begin{bmatrix} -1 \\ 0 \\ -1 \end{bmatrix}, P = \frac{1}{3}\begin{bmatrix} 2 & -1 & 1 \\ -1 & 2 & 1 \\ 1 & 1 & 2 \end{bmatrix}$

21. $p = (1, 2, 3, 4, 0, 0), P = \begin{bmatrix} 1 & & & & & \\ & 1 & & & & \\ & & 1 & & & \\ & & & 1 & & \\ & & & & 0 & \\ & & & & & 0 \end{bmatrix}$

23. $\begin{bmatrix} 4 & 2 & 6 \\ 2 & 6 & 8 \\ 6 & 8 & 18 \end{bmatrix}\begin{bmatrix} a \\ b \\ c \end{bmatrix} = \begin{bmatrix} 6 \\ 5 \\ 11 \end{bmatrix}$ 25. $\begin{bmatrix} 4 & 2 & 2 \\ 2 & 2 & 1 \\ 2 & 1 & 2 \end{bmatrix}\begin{bmatrix} a \\ b \\ c \end{bmatrix} = \begin{bmatrix} 9 \\ 3 \\ 3 \end{bmatrix}$

27. $\begin{bmatrix} 6 & 6 & 3 & 3 \\ 6 & 10 & 3 & 5 \\ 3 & 3 & 3 & 3 \\ 3 & 5 & 3 & 5 \end{bmatrix}\begin{bmatrix} a \\ b \\ c \\ d \end{bmatrix} = \begin{bmatrix} 20 \\ 27 \\ 11 \\ 14 \end{bmatrix}$ 29. $\begin{bmatrix} 4 & 6 \\ 6 & 14 \end{bmatrix}\begin{bmatrix} \ln a \\ \ln b \end{bmatrix} = \begin{bmatrix} 33 \\ 74 \end{bmatrix}$

Section 4.4

1. Y 3. N 7. (a) 6, −4, 1 (b) $\frac{5}{3}, 5\sqrt{2}, -\frac{1}{3}\sqrt{2}$

9. (a) 1, 2, 3, 4, 5, 6 (b) 6, 5, 4, 3, 2, 1 11. $y = \frac{3}{2}x + \frac{2}{3}$; $\begin{bmatrix} 1 & -1 \\ 1 & 0 \\ 1 & 1 \end{bmatrix}$

13. $y = -1.5x - 3.8$; $\begin{bmatrix} 1 & -2 \\ 1 & -1 \\ 1 & 0 \\ 1 & 1 \\ 1 & 2 \end{bmatrix}$ 15. $y = 0.4x + 0.5$; $\begin{bmatrix} 1 & -\frac{3}{2} \\ 1 & -\frac{1}{2} \\ 1 & \frac{1}{2} \\ 1 & \frac{3}{2} \end{bmatrix}$

17. $(1/\sqrt{2}, 1/\sqrt{2}), (-1/\sqrt{2}, 1/\sqrt{2})$ 19. $\begin{bmatrix} 1/\sqrt{3} \\ 1/\sqrt{3} \\ 1/\sqrt{3} \end{bmatrix}, \begin{bmatrix} -2/\sqrt{6} \\ 1/\sqrt{6} \\ 1/\sqrt{6} \end{bmatrix}, \begin{bmatrix} 0 \\ -1/\sqrt{2} \\ 1/\sqrt{2} \end{bmatrix}$

21. $\begin{bmatrix} \frac{2}{3} \\ 0 \\ \frac{2}{3} \\ -\frac{1}{3} \end{bmatrix}, \begin{bmatrix} \frac{1}{3} \\ 0 \\ -\frac{2}{3} \\ -\frac{2}{3} \end{bmatrix}, \begin{bmatrix} 2/3\sqrt{10} \\ 3/\sqrt{10} \\ -1/3\sqrt{10} \\ 2/3\sqrt{10} \end{bmatrix}$ 23. $\frac{1}{\sqrt{2}}, \sqrt{\frac{3}{2}}\,x, \frac{1}{4}\sqrt{10}(3x^2 - 1)$

25. $\begin{bmatrix} \frac{1}{2} & \frac{1}{2} \\ \frac{1}{2} & \frac{1}{2} \end{bmatrix}, \begin{bmatrix} -2/\sqrt{6} & 1/\sqrt{6} \\ 1/\sqrt{6} & 0 \end{bmatrix}, \begin{bmatrix} 1/\sqrt{30} & 4/\sqrt{30} \\ -2/\sqrt{30} & -3/\sqrt{30} \end{bmatrix}$

Section 4.5

1. $\begin{bmatrix} 2/\sqrt{5} \\ -1/\sqrt{5} \end{bmatrix}$ 3. $\begin{bmatrix} \frac{4}{5} \\ -\frac{3}{5} \\ 0 \end{bmatrix}$ 5. $[\frac{2}{7} \;\; -\frac{6}{7} \;\; \frac{3}{7}]$ 7. Y, $\begin{bmatrix} 0 & 1 \\ -1 & 0 \end{bmatrix}$

9. N 11. Y, $\begin{bmatrix} 1 & 0 & 0 & 0 \\ 0 & 1/\sqrt{2} & 0 & -1/\sqrt{2} \\ 0 & 1/\sqrt{3} & 1/\sqrt{3} & 1/\sqrt{3} \\ 0 & 1/\sqrt{6} & -2/\sqrt{6} & 1/\sqrt{6} \end{bmatrix}$

13. $\begin{bmatrix} \frac{4}{5} & \frac{3}{5} \\ \frac{3}{5} & -\frac{4}{5} \end{bmatrix}\begin{bmatrix} 5 & \frac{1}{5} \\ & \frac{7}{5} \end{bmatrix}$ 15. $\begin{bmatrix} \frac{1}{3} & 8/3\sqrt{26} \\ \frac{2}{3} & 7/3\sqrt{26} \\ -\frac{2}{3} & 11/3\sqrt{26} \end{bmatrix}\begin{bmatrix} 3 & \frac{1}{3} \\ & \sqrt{26}/3 \end{bmatrix}$

17. $y = \frac{3}{2}x - 2$ 19. $y = -\frac{61}{114}x + \frac{35}{114}$

Section 4.6

1. (a) The plane spanned by the x-axis and the line $z = y$ in the yz-plane (b) (1, 3, 2), (4, 1, 1), (0, 2, −2)

3. (a) $Q\begin{bmatrix} x_1 \\ x_2 \end{bmatrix} = \begin{bmatrix} -x_2 \\ -x_1 \end{bmatrix}$ (b) $-\boldsymbol{\varepsilon}_2, -\boldsymbol{\varepsilon}_1 - \boldsymbol{\varepsilon}_2, \boldsymbol{\varepsilon}_1 - \boldsymbol{\varepsilon}_2$

5. (a) $Q\begin{bmatrix} x_1 \\ x_2 \\ x_3 \end{bmatrix} = \begin{bmatrix} x_1 \\ \frac{3}{5}x_2 + \frac{4}{5}x_3 \\ \frac{4}{5}x_2 - \frac{3}{5}x_3 \end{bmatrix}$ (b) $\boldsymbol{\varepsilon}_1$, $\begin{bmatrix} 1 \\ \frac{3}{5} \\ \frac{4}{5} \end{bmatrix}$, $\begin{bmatrix} 1 \\ -\frac{3}{5} \\ -\frac{4}{5} \end{bmatrix}$

7. (a) $Q\begin{bmatrix} x_1 \\ x_2 \\ x_3 \end{bmatrix} = \begin{bmatrix} \frac{2}{3}x_1 + \frac{1}{3}x_2 + \frac{2}{3}x_3 \\ \frac{1}{3}x_1 + \frac{2}{3}x_2 - \frac{2}{3}x_3 \\ \frac{2}{3}x_1 - \frac{2}{3}x_2 - \frac{1}{3}x_3 \end{bmatrix}$, $\begin{bmatrix} \frac{2}{3} \\ \frac{1}{3} \\ \frac{2}{3} \end{bmatrix}$ (b) $\boldsymbol{\varepsilon}_1 + \boldsymbol{\varepsilon}_2$, $\begin{bmatrix} \frac{1}{3} \\ -\frac{1}{3} \\ \frac{4}{3} \end{bmatrix}$

9. (a) $Q\begin{bmatrix} x_1 \\ x_2 \\ x_3 \\ x_4 \end{bmatrix} = \begin{bmatrix} \frac{1}{2}x_1 - \frac{1}{2}x_2 - \frac{1}{2}x_3 + \frac{1}{2}x_4 \\ -\frac{1}{2}x_1 + \frac{1}{2}x_2 - \frac{1}{2}x_3 + \frac{1}{2}x_4 \\ -\frac{1}{2}x_1 - \frac{1}{2}x_2 + \frac{1}{2}x_3 + \frac{1}{2}x_4 \\ \frac{1}{2}x_1 + \frac{1}{2}x_2 + \frac{1}{2}x_3 + \frac{1}{2}x_4 \end{bmatrix}$, $\begin{bmatrix} \frac{1}{2} \\ -\frac{1}{2} \\ -\frac{1}{2} \\ \frac{1}{2} \end{bmatrix}$ (b) $-\boldsymbol{\varepsilon}_3 + \boldsymbol{\varepsilon}_4$, $\boldsymbol{\varepsilon}_1 - \boldsymbol{\varepsilon}_2$

11. $\begin{bmatrix} 0 & -1 \\ -1 & 0 \end{bmatrix}$ 13. $\begin{bmatrix} \frac{1}{2} & \frac{1}{2} & \frac{1}{2}\sqrt{2} \\ \frac{1}{2} & \frac{1}{2} & -\frac{1}{2}\sqrt{2} \\ -\frac{1}{2}\sqrt{2} & \frac{1}{2}\sqrt{2} & 0 \end{bmatrix}$

15. $\begin{bmatrix} \frac{1}{2} & -\frac{1}{2} & -\frac{1}{2} & -\frac{1}{2} \\ -\frac{1}{2} & \frac{1}{2} & -\frac{1}{2} & -\frac{1}{2} \\ -\frac{1}{2} & -\frac{1}{2} & \frac{1}{2} & -\frac{1}{2} \\ -\frac{1}{2} & -\frac{1}{2} & -\frac{1}{2} & \frac{1}{2} \end{bmatrix}$ 17. $\begin{bmatrix} 2/\sqrt{5} \\ -1/\sqrt{5} \end{bmatrix}$, $\begin{bmatrix} \frac{1}{5} \\ \frac{7}{5} \end{bmatrix}$

19. $\begin{bmatrix} 2/\sqrt{6} \\ 1/\sqrt{6} \\ -1/\sqrt{6} \end{bmatrix}$, $\begin{bmatrix} -1 \\ 0 \\ 1 \end{bmatrix}$ 21. $\begin{bmatrix} -3/2\sqrt{3} \\ -1/2\sqrt{3} \\ 1/2\sqrt{3} \\ -1/2\sqrt{3} \end{bmatrix}$, $\begin{bmatrix} -1 \\ \frac{1}{3} \\ \frac{2}{2} \\ -\frac{2}{3} \end{bmatrix}$

23. (a) $\begin{bmatrix} -2 & -1 \\ 1/\sqrt{2} & 3 \\ 1/\sqrt{2} & -1 \end{bmatrix}$ (b) $\begin{bmatrix} 0 & -1 \\ -1 & 0 \end{bmatrix}\begin{bmatrix} -2 & -1 \\ & -3 \end{bmatrix}$

25. (a) $\begin{bmatrix} 3 & -\frac{5}{3} \\ -2/\sqrt{6} & 3.350 \\ 1/\sqrt{6} & 1.026 \\ -1/\sqrt{6} & 0.050 \end{bmatrix}$ 27. (a) $\begin{bmatrix} \sqrt{2} & -\sqrt{2}/2 & -\sqrt{2}/2 \\ \sqrt{2}/2 & 1.225 & 0.408 \\ \frac{1}{2} & -0.996 & -1.155 \\ \frac{1}{2} & 0.0848 & -1 \end{bmatrix}$

29. (a) $\begin{bmatrix} -2 & 2 \\ 3/2\sqrt{3} & -\sqrt{2} \\ 1/2\sqrt{3} & (1 + \sqrt{2})/\sqrt{4 + 2\sqrt{2}} \\ 1/2\sqrt{3} & -1/\sqrt{4 + 2\sqrt{2}} \\ 1/2\sqrt{3} & 0 \end{bmatrix}$

31. (a) $\begin{bmatrix} -2 & -2 & -1 \\ 3/\sqrt{12} & 1 & -1 \\ \sqrt{2}/\sqrt{12} & 1/\sqrt{2} & \sqrt{3} \\ 1/\sqrt{12} & 0 & (-1 - \sqrt{3})/\sqrt{6 + 2\sqrt{3}} \\ 0 & 1/\sqrt{2} & \sqrt{2}/\sqrt{6 + 2\sqrt{3}} \end{bmatrix}$

Section 4.7

1. (a) $\begin{bmatrix} 1 & 1 \\ 1 & -1 \\ 1 & 0 \end{bmatrix}$ (b) $\begin{bmatrix} -\frac{3}{2} & 1 \\ \frac{1}{2} & -1 \\ -\frac{1}{2} & 0 \end{bmatrix}$ (c) $\begin{bmatrix} -2 \\ 2 \end{bmatrix}$ (d) $\begin{bmatrix} -2 \\ 6 \\ 2 \end{bmatrix}$

3. (a) $\begin{bmatrix} -1 & 0 & 1 \\ -1 & 2 & 0 \end{bmatrix}$ (b) $\begin{bmatrix} \frac{1}{3} & -1 & -\frac{2}{3} \\ \frac{1}{3} & 1 & \frac{1}{3} \end{bmatrix}$ (c) $\begin{bmatrix} -2 \\ -3 \\ 2 \end{bmatrix}$ (d) $\begin{bmatrix} -5 \\ -4 \end{bmatrix}$

5. (a) $\begin{bmatrix} 1 & -1 \\ 3 & 0 \\ 0 & -2 \\ 1 & 1 \end{bmatrix}$ (b) $\begin{bmatrix} -1 & 2 \\ 4 & 4 \\ -7 & -6 \\ 7 & 2 \end{bmatrix}$ (c) $\begin{bmatrix} -3 \\ 4 \end{bmatrix}$ (d) $\begin{bmatrix} 7 \\ 3 \\ 12 \\ -5 \end{bmatrix}$

7. (a) $\begin{bmatrix} 1 & -1 \\ 1 & 1 \\ 2 & 0 \\ 0 & 3 \end{bmatrix}$ (b) $\begin{bmatrix} 3 & -3 \\ -1 & 5 \\ 0 & -2 \\ -2 & 2 \end{bmatrix}$ (c) $\begin{bmatrix} 2 \\ 3 \end{bmatrix}$ (d) $\begin{bmatrix} 6 & 4 \\ 10 & -3 \end{bmatrix}$

9. (a) $\begin{bmatrix} 0 & 0 \\ 0 & 0 \\ 1 & 0 \\ 0 & 1 \end{bmatrix}$ (b) $\begin{bmatrix} 1 & 2 \\ 0 & -2 \\ -1 & 0 \\ 0 & 0 \end{bmatrix}$ (c) $\begin{bmatrix} 3 \\ -\frac{5}{2} \end{bmatrix}$ (d) $3x^2 - 2x^3$

11. (a) $\begin{bmatrix} 2 \\ 3 \\ -1 \\ -2 \end{bmatrix}$ (b) $\begin{bmatrix} 0 & 1 & 0 & 0 \\ 0 & 0 & 0 & 0 \\ 0 & 0 & 1 & 1 \\ 0 & 0 & 0 & 1 \end{bmatrix}$ (c) $\begin{bmatrix} 3 \\ 0 \\ -3 \\ -2 \end{bmatrix}$

13. See 1(b), $f_A(x, y) = (2x + y, -x, -3y)$
15. See 3(b), $f_A(x, y, z) = (-x + z, x + y)$
17. See 5(b), $f_A(x, y) = (3x, -x + y, 2y, -y)$

23. $\begin{bmatrix} 1 & -1 \\ 1 & 1 \end{bmatrix}, \begin{bmatrix} 0 & -1 \\ 2 & 2 \end{bmatrix}$ 25. $\begin{bmatrix} -1 & 2 \\ 1 & -1 \end{bmatrix}, \begin{bmatrix} -1 & 1 \\ 2 & -1 \end{bmatrix}$

27. $\begin{bmatrix} 1 & 0 & -1 \\ 1 & -1 & 1 \\ 0 & 2 & 0 \end{bmatrix}, \begin{bmatrix} 0 & 1 & -1 \\ 1 & -2 & -1 \\ 0 & 2 & 2 \end{bmatrix}$

29. $\begin{bmatrix} 1 & -1 & 0 \\ 0 & 1 & -1 \\ 1 & 0 & 1 \end{bmatrix}, \begin{bmatrix} 2 & 2 & 2 \\ 0 & 1 & -1 \\ -1 & -1 & 0 \end{bmatrix}$

31. Use $P = I$ 35. Same P 41. Use $P = B$

Chapter 4 Review

1. Y, $\begin{bmatrix} 2 & -1 \\ 3 & 0 \\ 1 & 4 \end{bmatrix}$ 7. (a) $\frac{2}{3}\begin{bmatrix} 1 \\ 1 \\ 1 \end{bmatrix}$ (b) $\frac{1}{14}\begin{bmatrix} -8 \\ 5 \\ 31 \end{bmatrix}$

9. $\frac{3}{2}$,

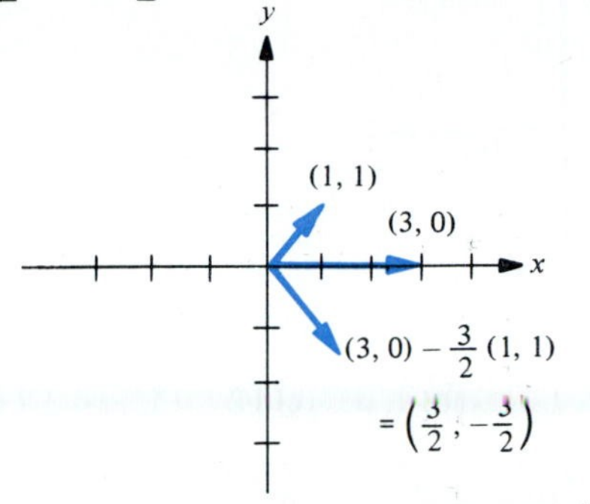

11. $(0, 0, 1/\sqrt{2}, 1/\sqrt{2})$ 13. $\begin{bmatrix} \frac{3}{5} & \frac{4}{5} \\ \frac{4}{5} & -\frac{3}{5} \end{bmatrix}\begin{bmatrix} 5 & \frac{6}{5} \\ & \frac{8}{5} \end{bmatrix}$

15. (a) $\begin{bmatrix} 0 & 0 & 0 & 1 \\ 1 & 0 & 0 & 0 \\ 0 & 1 & 0 & 0 \\ 0 & 0 & 1 & 0 \end{bmatrix}$ (b) Interchange $x_1 \leftrightarrow x_3$, $x_2 \leftrightarrow x_4$

17. 3; (0, 1, 1), (4, 1, −1) 19. Y; A, B orthogonal $\Rightarrow$ AB orthogonal

21. 16×16 23. $[1 \quad 2 \quad -3]$; does not exist

25. $\mathbf{0} = 0\mathbf{u}_1 + 0\mathbf{u}_2$ 27. $\begin{bmatrix} 0 & -1 & 0 \\ -1 & 0 & 0 \\ 0 & 0 & 1 \end{bmatrix}$

29. HINT First show that $CS(AA^T) \subset CS(A)$. Then use dimensions.

31. $\|A\mathbf{x} - \mathbf{b}\|$ or $\|A\mathbf{x} - \mathbf{b}\|^2$

33. (a) HINT You can solve $A\mathbf{x} = \mathbf{b}$; can you also solve $AA^T\mathbf{y} = \mathbf{b}$? See Exercise 29.
(b) HINT If $A\mathbf{x} = \mathbf{b}$ and $A\mathbf{x}' = \mathbf{b}$, where is $\mathbf{x} - \mathbf{x}'$? Now use Exercise 30.

Section 5.1

1. (a) 3.1 (b) Y 3. (a) 0 (b) N 5. (a) 7 (b) Y
7. (a) 0 (b) N 9. (a) 4 (b) Y 11. (a) 25 (b) Y
13. (a) 0 (b) N 15. (a) −8 (b) Y 17. (a) −12 (b) Y
19. (a) 0 (b) N 21. (a) x (b) 0 23. (a) $x^2 - 2x - 3$ (b) 3, −1
25. (a) $x + 3$ (b) −3 27. (a) $x^2 + x - 12$ (b) 3, −4
29. (a) $24x - 13$ (b) $\frac{13}{24}$
31. (a) $(x - 2)(x + 3)(x + 1)(x - 5)$ (b) −3, −1, 2, 5 33. ab
35. $125a$ 37. c

Section 5.2

1. 6, 4 3. 2, 0 5. 1, . . . , 1 7. $A\mathbf{v} = \mathbf{0} = 0\mathbf{v}$; 0

11. (a) $(\lambda - 2)(\lambda + 3)$ (b) 2, -3 (c) $\boldsymbol{\varepsilon}_1$, $\boldsymbol{\varepsilon}_2$
(d)

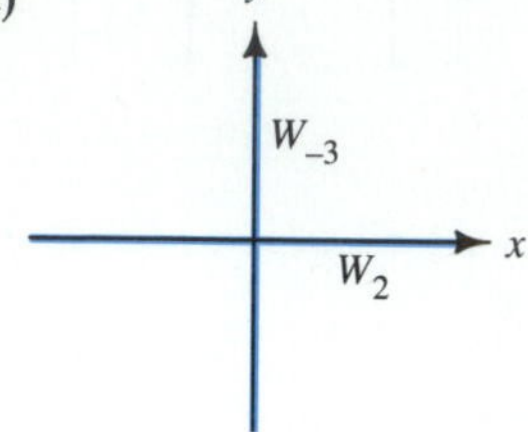

(f) $2 + (-3) = 2 + (-3)$

13. (a) $(\lambda - 1)(\lambda - 3)$ (b) 1, 3 (c) $[2 \quad 1]^T$, $[4 \quad 1]^T$
(d)

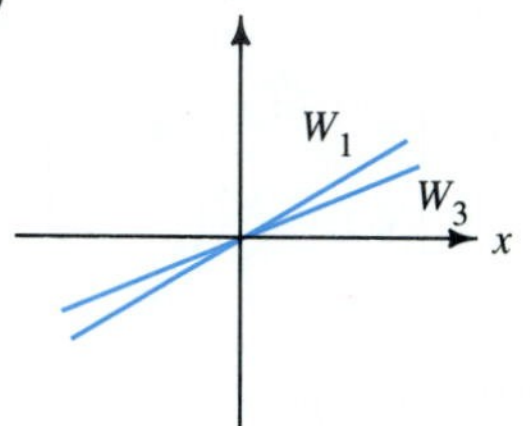

(f) $5 + (-1) = 1 + 3$

15. (a) $(\lambda - 3)(\lambda + 5)$ (b) 3, -5 (c) $[3 \quad 2]^T$, $[-1 \quad 2]^T$
(d)

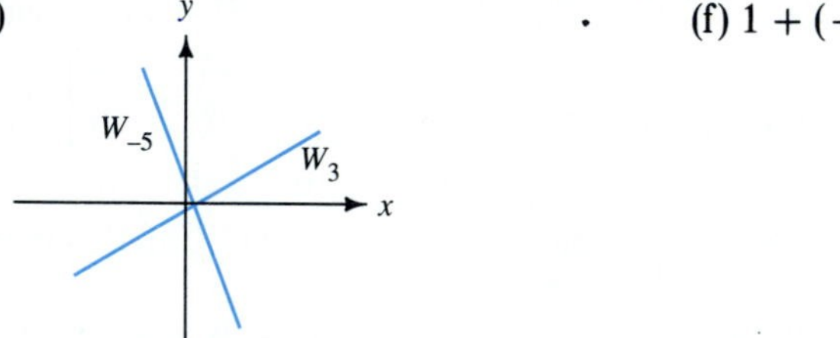

(f) $1 + (-3) = 3 + (-5)$

17. (a) $\lambda(\lambda - 1)(\lambda - 3)$ (b) 0, 1, 3 (c) $[1 \quad 1 \quad 1]^T$, $[1 \quad 0 \quad -1]^T$, $[1 \quad -2 \quad 1]^T$
(d)

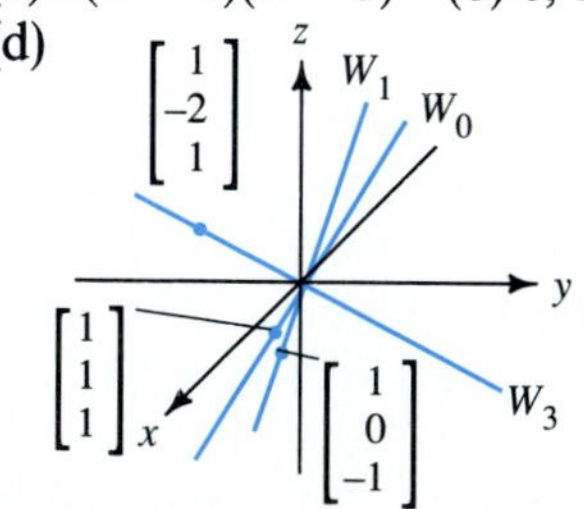

(f) $1 + 2 + 1 = 0 + 1 + 3$

19. (a) $(\lambda - 2)^2(\lambda - 4)$ (b) 2, 2, 4 (c) $[1 \quad 0 \quad 0]^T$, $[0 \quad 1 \quad 1]^T$, $[5 \quad -1 \quad 1]^T$

(d)

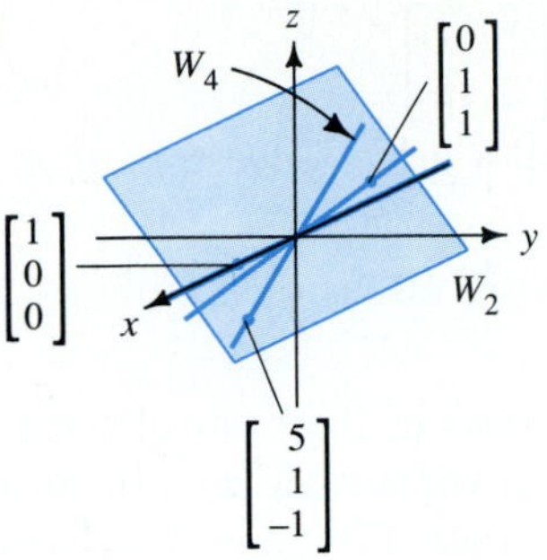

(f) $2 + 3 + 3 = 2 + 2 + 4$

21. (a) $(\lambda + 2)(\lambda - 3)^2(\lambda + 6)$ (b) $-2, 3, 3, -6$
(c) $[2 \;\; 1 \;\; 0 \;\; 0]^T, [1 \;\; 3 \;\; 0 \;\; 0]^T, [0 \;\; 0 \;\; 1 \;\; -2]^T, [0 \;\; 0 \;\; 4 \;\; 1]^T$
(f) $-3 + 4 + (-5) + 2 = -2 + 3 + 3 + (-6)$

23. λ^2 25. λ^{-1} 27. $\lambda - 7$

31. (a) $\{\mathbf{u}_1\}, \{\mathbf{u}_1, \mathbf{u}_2\}$ (b) $\mathbf{x} = \frac{1}{2}\mathbf{u}_2 + \frac{1}{4}\mathbf{u}_3$
(c) $\mathbf{u}_1, \mathbf{u}_2, \mathbf{u}_3$ are linearly independent, and $\mathbf{u}_1$ is not in $CS(A)$, by (a).

35. (a), (b) $A = \begin{bmatrix} 2 & 1 \\ 0 & 1 \end{bmatrix}, B = \begin{bmatrix} 2 & 0 \\ 1 & 1 \end{bmatrix}$
(d) $\det(AB) = \det(A)\det(B)$, $tr(A + B) = tr(A) + tr(B)$

Section 5.3

1. (a) $\begin{bmatrix} 1 & 1 \\ 1 & 0 \end{bmatrix}, \begin{bmatrix} 2 & \\ & -3 \end{bmatrix}$ (b) $\begin{bmatrix} -3 & 5 \\ 0 & 2 \end{bmatrix}$

3. (a) $\begin{bmatrix} 1 & 0 & 1 \\ 0 & 1 & 0 \\ 1 & 0 & -1 \end{bmatrix}, \begin{bmatrix} -2 & & \\ & -1 & \\ & & 0 \end{bmatrix}$ (b) $\begin{bmatrix} -1 & 0 & -1 \\ 0 & -1 & 0 \\ -1 & 0 & -1 \end{bmatrix}$

5. (a) $\begin{bmatrix} 1 & & & & \\ & 1 & & & \\ & & 1 & & \\ & & & 1 & \\ & & & & 1 \end{bmatrix}, \begin{bmatrix} 2 & & & & \\ & 3 & & & \\ & & 4 & & \\ & & & 5 & \\ & & & & 6 \end{bmatrix}$ (b) $A = \Lambda$

13. $\begin{bmatrix} 1 & 1 \\ -1 & 2 \end{bmatrix}, \begin{bmatrix} 3 & \\ & 2 \end{bmatrix}$ 15. $\begin{bmatrix} 1 & 1 & 1 \\ 1 & 0 & 1 \\ 0 & 0 & 1 \end{bmatrix}, \begin{bmatrix} 3 & & \\ & 0 & \\ & & -2 \end{bmatrix}$

17. N 19. $\begin{bmatrix} 1 & 4 \\ -1 & 1 \end{bmatrix}, \begin{bmatrix} -3 & \\ & 2 \end{bmatrix}$ 21. $\begin{bmatrix} 0 & 1 & 0 \\ 1 & 0 & 0 \\ 0 & -4 & 1 \end{bmatrix}, \begin{bmatrix} 0 & & \\ & 0 & \\ & & 1 \end{bmatrix}$

23. N 25. $\begin{bmatrix} 1 & 0 & 0 & 0 \\ 0 & 1 & 1 & 1 \\ 0 & 0 & -1 & 0 \\ 0 & 0 & 0 & 1 \end{bmatrix}, \begin{bmatrix} 3 & & & \\ & 3 & & \\ & & -2 & \\ & & & -2 \end{bmatrix}$

29. $\frac{1}{3}\begin{bmatrix} 62 & -19 \\ -38 & 43 \end{bmatrix}, \frac{1}{18}\begin{bmatrix} 7 & 1 \\ 2 & 8 \end{bmatrix}$

31. Show the columns of S are eigenvectors of A.

33. Yes; the eigenvalues are all distinct; $\begin{bmatrix} 1 & & \\ & 3 & \\ & & -6 \end{bmatrix}$

35. From the standard basis to the basis of eigenvectors.

37. $f_A(\mathbf{v}_i) = \lambda_i \mathbf{v}_i$, so $\mathbf{v}_i$ is moved further from the origin if $\lambda_i > 1$; left fixed if $\lambda_i = 1$; moved closer to the origin if $0 < \lambda_i < 1$; collapsed to $\mathbf{0}$ if $\lambda_i = 0$; and reflected through the origin if $\lambda_i < 0$.

Section 5.4

1. (a) $\begin{bmatrix} 1 \\ 2 \end{bmatrix}, \begin{bmatrix} 2 \\ -1 \end{bmatrix}$ (b) $\begin{bmatrix} 1/\sqrt{5} & 2/\sqrt{5} \\ 2/\sqrt{5} & -1/\sqrt{5} \end{bmatrix}, \begin{bmatrix} 5 & \\ & 0 \end{bmatrix}$

3. (a) $\begin{bmatrix} 2 \\ -1 \end{bmatrix}, \begin{bmatrix} 1 \\ 2 \end{bmatrix}$ (b) $\begin{bmatrix} 2/\sqrt{5} & 1/\sqrt{5} \\ -1/\sqrt{5} & 2/\sqrt{5} \end{bmatrix}, \begin{bmatrix} 6 & \\ & 1 \end{bmatrix}$

5. (a) $\begin{bmatrix} 1 \\ 1 \end{bmatrix}, \begin{bmatrix} 1 \\ -1 \end{bmatrix}$ (b) $\begin{bmatrix} 1/\sqrt{2} & 1/\sqrt{2} \\ 1/\sqrt{2} & -1/\sqrt{2} \end{bmatrix}, \begin{bmatrix} 2 & \\ & 0 \end{bmatrix}$

7. (a) $\begin{bmatrix} 1 \\ 1 \\ 1 \end{bmatrix}, \begin{bmatrix} 1 \\ -1 \\ 0 \end{bmatrix}, \begin{bmatrix} 1 \\ 1 \\ -2 \end{bmatrix}$ (b) $\begin{bmatrix} 1/\sqrt{3} & 1/\sqrt{2} & 1/\sqrt{6} \\ 1/\sqrt{3} & -1/\sqrt{2} & 1/\sqrt{6} \\ 1/\sqrt{3} & 0 & -2/\sqrt{6} \end{bmatrix}, \begin{bmatrix} 3 & & \\ & 0 & \\ & & 0 \end{bmatrix}$

9. (a) $\begin{bmatrix} 1 \\ -2 \\ 2 \end{bmatrix}, \begin{bmatrix} 2 \\ 1 \\ 0 \end{bmatrix}, \begin{bmatrix} 2 \\ -4 \\ -5 \end{bmatrix}$ (b) $\begin{bmatrix} \frac{1}{3} & 2/\sqrt{5} & 2/3\sqrt{5} \\ -\frac{2}{3} & 1/\sqrt{5} & -4/3\sqrt{5} \\ \frac{2}{3} & 0 & -5/3\sqrt{5} \end{bmatrix}, \begin{bmatrix} 6 & & \\ & -3 & \\ & & -3 \end{bmatrix}$

11. (a) $\begin{bmatrix} 2 \\ -1 \\ 0 \end{bmatrix}, \begin{bmatrix} 1 \\ 2 \\ 0 \end{bmatrix}, \begin{bmatrix} 0 \\ 0 \\ 1 \end{bmatrix}$ (b) $\begin{bmatrix} 2/\sqrt{5} & 1/\sqrt{5} & 0 \\ -1/\sqrt{5} & 2/\sqrt{5} & 0 \\ 0 & 0 & 1 \end{bmatrix}, \begin{bmatrix} 4 & & \\ & -1 & \\ & & 1 \end{bmatrix}$

13. (a) $\begin{bmatrix} 1 \\ 1 \\ 1 \\ 1 \end{bmatrix}, \begin{bmatrix} 1 \\ -1 \\ 0 \\ 0 \end{bmatrix}, \begin{bmatrix} 1 \\ 1 \\ -2 \\ 0 \end{bmatrix}, \begin{bmatrix} 1 \\ 1 \\ 1 \\ -3 \end{bmatrix}$

(b) $\begin{bmatrix} \frac{1}{2} & 1/\sqrt{2} & 1/\sqrt{6} & 1/2\sqrt{3} \\ \frac{1}{2} & -1/\sqrt{2} & 1/\sqrt{6} & 1/2\sqrt{3} \\ \frac{1}{2} & 0 & -2/\sqrt{6} & 1/2\sqrt{3} \\ \frac{1}{2} & 0 & 0 & -3/2\sqrt{3} \end{bmatrix}, \begin{bmatrix} 4 & & & \\ & 0 & & \\ & & 0 & \\ & & & 0 \end{bmatrix}$

15. $\begin{bmatrix} 2 & \\ & -3 \end{bmatrix}, \begin{bmatrix} 2/\sqrt{5} & 1/\sqrt{5} \\ 1/\sqrt{5} & -2/\sqrt{5} \end{bmatrix}$

17. $\begin{bmatrix} 3 & & \\ & -4 & \\ & & -1 \end{bmatrix}, \begin{bmatrix} \frac{1}{3} & \frac{2}{3} & -\frac{2}{3} \\ \frac{2}{3} & \frac{1}{3} & \frac{2}{3} \\ -\frac{2}{3} & \frac{2}{3} & \frac{1}{3} \end{bmatrix}$

19. $\begin{bmatrix} 2 & & \\ & 3 & \\ & & 3 \end{bmatrix}, \begin{bmatrix} 1/\sqrt{3} & 1/\sqrt{2} & 1/\sqrt{6} \\ 1/\sqrt{3} & -1/\sqrt{2} & 1/\sqrt{6} \\ 1/\sqrt{3} & 0 & -2/\sqrt{6} \end{bmatrix}$

Section 5.5

1. 1, 1, 2, 3, 5, 8   3. 1, 3, 4, 7, 11, 18
9. $S_{k+1} = \frac{1}{2}F_k$, $T_{k+1} = \frac{1}{3}S_k$, $F_{k+1} = 6T_k$
21. Half in, half out; only the fixed population
23. $\begin{bmatrix} \frac{1}{2} & 0 & \frac{1}{2} \\ 0 & \frac{1}{2} & \frac{1}{2} \\ \frac{1}{2} & \frac{1}{2} & 0 \end{bmatrix}$
25. (a) $\begin{bmatrix} 64 \\ 64 \\ 64 \end{bmatrix}$ (b) $\begin{bmatrix} 32 \\ 32 \\ 128 \end{bmatrix} \to \begin{bmatrix} 80 \\ 80 \\ 32 \end{bmatrix} \to \begin{bmatrix} 56 \\ 56 \\ 80 \end{bmatrix} \to \begin{bmatrix} 68 \\ 68 \\ 56 \end{bmatrix} \to \begin{bmatrix} 62 \\ 62 \\ 68 \end{bmatrix} \to \begin{bmatrix} 65 \\ 65 \\ 62 \end{bmatrix}$

Section 5.6

1. (a) $\begin{bmatrix} u \\ v \end{bmatrix} = ae^{-t}\begin{bmatrix} 1 \\ 1 \end{bmatrix} + be^{-2t}\begin{bmatrix} 3 \\ 4 \end{bmatrix}$ (b) $a = -11$, $b = 6$
3. (a) $\begin{bmatrix} u \\ v \end{bmatrix} = ae^{t}\begin{bmatrix} 2 \\ 1 \end{bmatrix} + be^{3t}\begin{bmatrix} 4 \\ 1 \end{bmatrix}$ (b) $a = \frac{5}{2}$, $b = -\frac{1}{2}$
5. (a) $\begin{bmatrix} u \\ v \end{bmatrix} = ae^{-t}\begin{bmatrix} 1 \\ -1 \end{bmatrix} + be^{7t}\begin{bmatrix} 3 \\ 1 \end{bmatrix}$ (b) $a = -2$, $b = 3$
7. (a) $\begin{bmatrix} u \\ v \\ w \end{bmatrix} = a\begin{bmatrix} 1 \\ 1 \\ 1 \end{bmatrix} + be^{t}\begin{bmatrix} 1 \\ 0 \\ -1 \end{bmatrix} + ce^{3r}\begin{bmatrix} 1 \\ -2 \\ 1 \end{bmatrix}$ (b) $a = 4$, $b = -1$, $c = 0$
9. (a) $\begin{bmatrix} u \\ v \\ w \end{bmatrix} = ae^{-6t}\begin{bmatrix} 1 \\ 0 \\ 0 \end{bmatrix} + be^{2t}\begin{bmatrix} 0 \\ 1 \\ 1 \end{bmatrix} + ce^{4t}\begin{bmatrix} 1 \\ -1 \\ 1 \end{bmatrix}$ (b) $a = 2$, $b = 3$, $c = -1$
11. (a) $y = ae^{2t} + be^{-t}$ (b) $a = 3$, $b = 2$
13. (a) $y = ae^{t} + be^{2t} + ce^{3t}$ (b) $a = \frac{13}{2}$, $b = -9$, $c = \frac{7}{2}$

Section 5.7

1. $2x^2 + 4xy + y^2$ 3. $3x^2 + xy - 2y^2 - 8yz + 2z^2 - 5xz$
5. Ellipse   7. Parabola   9. Hyperbola

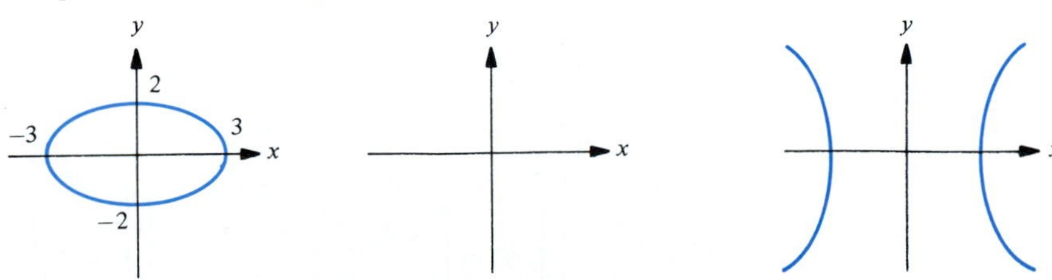

11. $[x \quad y]\begin{bmatrix} 3 & 2 \\ 2 & -5 \end{bmatrix}\begin{bmatrix} x \\ y \end{bmatrix}$

13. $[x \quad y]\begin{bmatrix} 5 & -1 \\ -1 & 2 \end{bmatrix}\begin{bmatrix} x \\ y \end{bmatrix}$

15. $[x \quad y \quad z]\begin{bmatrix} 2 & -4 & \frac{1}{2} \\ -4 & -5 & -3 \\ \frac{1}{2} & -3 & -7 \end{bmatrix}\begin{bmatrix} x \\ y \\ z \end{bmatrix}$

17. $[x \quad y \quad z]\begin{bmatrix} 2 & 0 & -\frac{1}{2} \\ 0 & 9 & \frac{3}{2} \\ -\frac{1}{2} & \frac{3}{2} & 0 \end{bmatrix}\begin{bmatrix} x \\ y \\ z \end{bmatrix}$

19. $6x'^2 + 2y'^2$

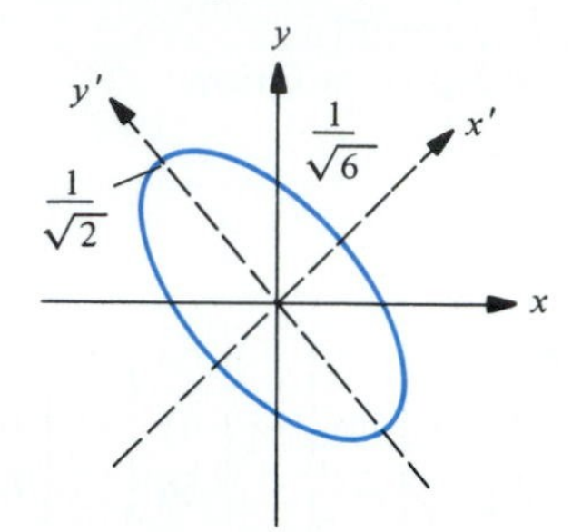

$\mathbf{x}' = \begin{bmatrix} 1/\sqrt{2} \\ 1/\sqrt{2} \end{bmatrix}, \mathbf{y}' = \begin{bmatrix} -1/\sqrt{2} \\ 1/\sqrt{2} \end{bmatrix}$

21. $x'^2 + 4y'^2$

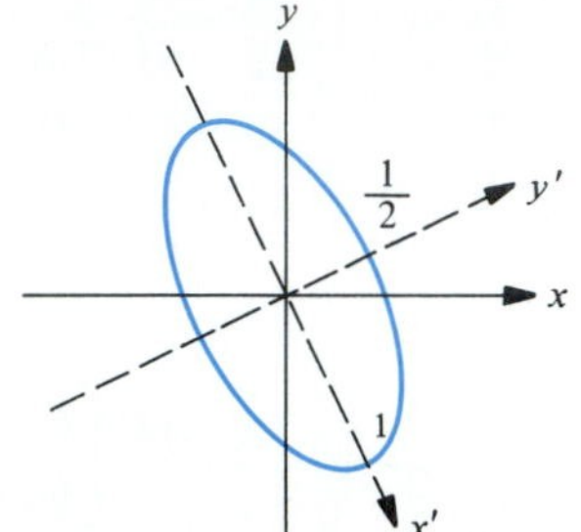

$\mathbf{x}' = \begin{bmatrix} 1/\sqrt{3} \\ -\sqrt{2}/\sqrt{3} \end{bmatrix}, \mathbf{y}' = \begin{bmatrix} \sqrt{2}/\sqrt{3} \\ 1/\sqrt{3} \end{bmatrix}$

23. $\frac{5}{2}x'^2 - \frac{5}{2}y'^2$

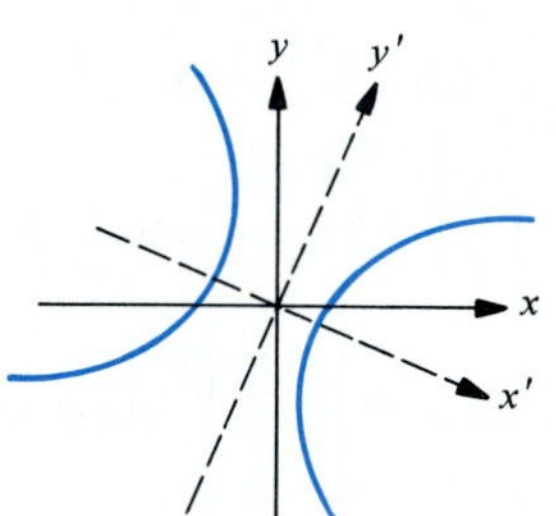

$\mathbf{x}' = \begin{bmatrix} 3/\sqrt{10} \\ -1/\sqrt{10} \end{bmatrix}, \mathbf{y}' = \begin{bmatrix} 1/\sqrt{10} \\ 3/\sqrt{10} \end{bmatrix}$

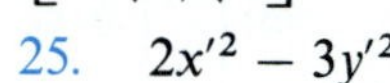

25. $2x'^2 - 3y'^2$

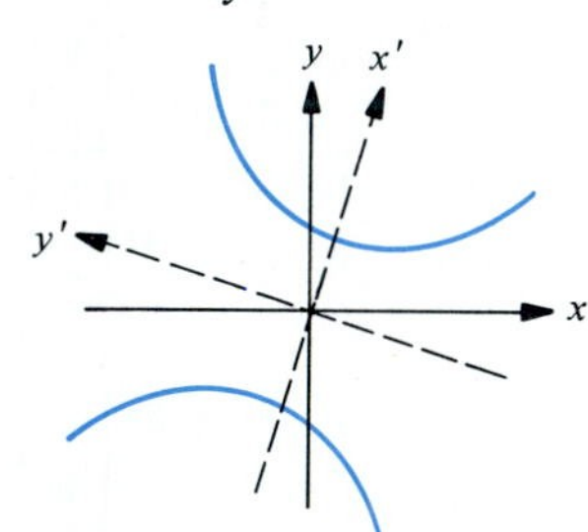

$\mathbf{x}' = \begin{bmatrix} 1/\sqrt{5} \\ 2/\sqrt{5} \end{bmatrix}, \mathbf{y}' = \begin{bmatrix} -2/\sqrt{5} \\ 1/\sqrt{5} \end{bmatrix}$

27. $\frac{1}{4}x'^2 - y'^2$

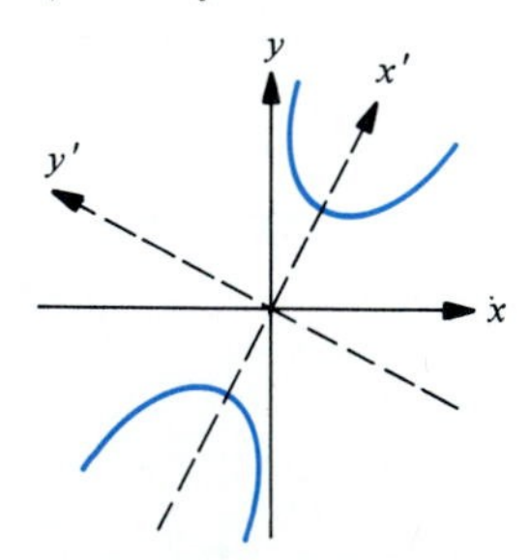

$\mathbf{x}' = \begin{bmatrix} 1/\sqrt{5} \\ 2/\sqrt{5} \end{bmatrix}, \mathbf{y}' = \begin{bmatrix} -2/\sqrt{5} \\ 1/\sqrt{5} \end{bmatrix}$

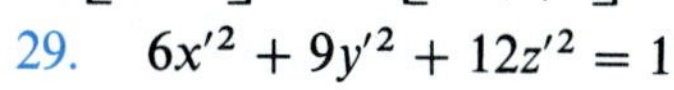

29. $6x'^2 + 9y'^2 + 12z'^2 = 1$

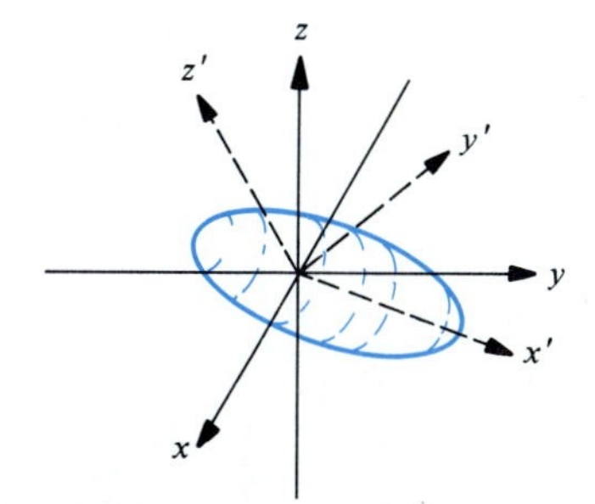

$\mathbf{x}' = \begin{bmatrix} \frac{2}{3} \\ \frac{2}{3} \\ \frac{1}{3} \end{bmatrix}, \mathbf{y}' = \begin{bmatrix} -\frac{2}{3} \\ \frac{1}{3} \\ \frac{2}{3} \end{bmatrix}, \mathbf{z}' = \begin{bmatrix} \frac{1}{3} \\ -\frac{2}{3} \\ \frac{2}{3} \end{bmatrix}$

Section 5.8

1. $B\mathbf{w} = (S^{-1}AS)(S^{-1}\mathbf{v}) = S^{-1}A\mathbf{v} = S^{-1}\lambda\mathbf{v} = \lambda\mathbf{w}$

3. $3, \begin{bmatrix} 4/\sqrt{17} \\ 1/\sqrt{17} \end{bmatrix}; 1 \begin{bmatrix} 2/\sqrt{5} \\ 1/\sqrt{5} \end{bmatrix}$ 5. $-5, \begin{bmatrix} 1/\sqrt{5} \\ -2/\sqrt{5} \end{bmatrix}; 3 \begin{bmatrix} 3/\sqrt{13} \\ 2/\sqrt{13} \end{bmatrix}$

7. $6, \begin{bmatrix} 4/\sqrt{17} \\ 1/\sqrt{17} \end{bmatrix}; 4 \begin{bmatrix} 2/\sqrt{5} \\ 1/\sqrt{5} \end{bmatrix}$ 9. 4.5, 1.4 11. 1.7, 3.3 13. 0.5, 0.5

15. 10, 0.6; 1, 2.1 17. 5, 0.6; -2, 1

Chapter 5 Review

1. 1 3. $x^2 - 4x + 11$ 5. $\begin{bmatrix} 0 & 1 \\ 3 & -2 \end{bmatrix}$

7. $S = \begin{bmatrix} 2 & 2 \\ 3 & -1 \end{bmatrix}, \Lambda = \begin{bmatrix} 7 & \\ & -1 \end{bmatrix}$

11. (a) The eigenvectors are orthogonal. (b) 0 (c) 1, $\frac{1}{10}\begin{bmatrix} 9 & -3 \\ -3 & 1 \end{bmatrix}$

13. F; identity matrix 15. (a) ± 1 (b) 0, -1 (c) $\begin{bmatrix} 3 & -1 \\ 8 & -3 \end{bmatrix}$

17. (a) T (b), (c) May not; $\begin{bmatrix} 2 & 1 & \\ & 2 & \\ & & -3 \end{bmatrix}$ 19. $F_{-k} = (-1)^{k+1}F_k$

21. $\begin{bmatrix} u \\ v \end{bmatrix} = 2e^{2t}\begin{bmatrix} 1 \\ 1 \end{bmatrix} - e^{-3t}\begin{bmatrix} 1 \\ -4 \end{bmatrix}$ 23. $S = \begin{bmatrix} 1/\sqrt{5} & 1/\sqrt{5} \\ -2/\sqrt{5} & 2/\sqrt{5} \end{bmatrix}, \Lambda = \begin{bmatrix} 1 & \\ & 5 \end{bmatrix}$

Section 6.1

5. f is g_0 of Equation (6.4) 7. $b_n = \frac{2}{n}, n \geq 1$

9. $f(x) = \pi + 2 \sin x + \sin 2x + \frac{1}{2} \sin 4x + \cdots$
$$= \pi + 2 \sum_{n=1}^{\infty} \frac{1}{n} \sin nx, \; 0 \leq x \leq 2\pi$$

11. $f(x) = \frac{4}{3}\pi^2 + 4 \sum_{n=1}^{\infty} \left(\frac{1}{n^2} \cos nx - \frac{\pi}{n} \sin nx \right), \; 0 \leq x \leq 2\pi$

13. $b_n = \frac{4}{n\pi}$, n odd; $b_n = 0$, n even; $f(x) = \frac{4}{\pi}\{\sin x + \frac{1}{3} \sin 3x + \cdots\}$,

15. $q_3(x) = x^3 - \frac{3}{5}x$, $q_4(x) = x^4 - \frac{6}{7}x^2 + \frac{3}{35}$ 17.,19. $y = x - \frac{1}{6}$

21.,23. $y = \frac{9}{10}x - \frac{1}{5}$

Section 6.2

1. 2, 1 3. 2, 0 5. $\sqrt{5}$, 0 7. $\sqrt{(9 \pm \sqrt{17})/2}$

9. $\begin{bmatrix} -1 & 0 \\ 0 & 1 \end{bmatrix}\begin{bmatrix} 3 & \\ & 1 \end{bmatrix}\begin{bmatrix} 1 & 0 \\ 0 & 1 \end{bmatrix}$ 11. $[1][\sqrt{5} \;\; 0]\begin{bmatrix} 1/\sqrt{5} & 2/\sqrt{5} \\ -2/\sqrt{5} & 1/\sqrt{5} \end{bmatrix}^T$

13. $\begin{bmatrix} 1/\sqrt{5} & 2/\sqrt{5} \\ -2/\sqrt{5} & 1/\sqrt{5} \end{bmatrix} \begin{bmatrix} 5 & \\ & 0 \end{bmatrix} \begin{bmatrix} 1/\sqrt{5} & 2/\sqrt{5} \\ -2/\sqrt{5} & 1/\sqrt{5} \end{bmatrix}^T$

15. $\begin{bmatrix} \sqrt{2}/\sqrt{3} & 1/\sqrt{3} \\ 1/\sqrt{3} & -\sqrt{2}/\sqrt{3} \end{bmatrix} \begin{bmatrix} 2 & 0 & 0 \\ 0 & 1 & 0 \end{bmatrix} \begin{bmatrix} 1/\sqrt{3} & \sqrt{2}/\sqrt{3} & 0 \\ \sqrt{2}/\sqrt{3} & -1/\sqrt{3} & 0 \\ 0 & 0 & 1 \end{bmatrix}^T$

17. $\begin{bmatrix} 1/\sqrt{5} & 2/\sqrt{5} & 0 \\ 2/\sqrt{5} & -1/\sqrt{5} & 0 \\ 0 & 0 & -1 \end{bmatrix} \begin{bmatrix} 3 & & \\ & 2 & \\ & & 2 \end{bmatrix} \begin{bmatrix} 2/\sqrt{5} & 1/\sqrt{5} & 0 \\ 1/\sqrt{5} & -2/\sqrt{5} & 0 \\ 0 & 0 & 1 \end{bmatrix}^T$

19. $\begin{bmatrix} & & & 1 \\ & & 1 & \\ & 1 & & \\ \cdot^{\cdot^{\cdot}} & & & \\ 1 & & & \end{bmatrix} \begin{bmatrix} n & & & & \\ & n-1 & & & \\ & & n-2 & & \\ & & & \ddots & \\ & & & & 1 \end{bmatrix} \begin{bmatrix} & & & 1 \\ & & 1 & \\ & 1 & & \\ \cdot^{\cdot^{\cdot}} & & & \\ 1 & & & \end{bmatrix}^T$

21. $\begin{bmatrix} -\frac{1}{3} & \\ & 1 \end{bmatrix}$ 23. $\begin{bmatrix} \frac{1}{5} \\ -\frac{2}{5} \end{bmatrix}$ 25. $\frac{1}{25}\begin{bmatrix} 1 & -2 \\ -2 & 4 \end{bmatrix}$

27. $\begin{bmatrix} \sqrt{2}/2 & -\frac{1}{2} \\ 0 & \sqrt{2}/2 \\ 0 & 0 \end{bmatrix}$ 35. $\begin{bmatrix} x \\ y \\ z \end{bmatrix} = \begin{bmatrix} \frac{1}{6} & \frac{1}{6} \\ \frac{1}{6} & \frac{1}{6} \\ \frac{1}{6} & \frac{1}{6} \end{bmatrix} \begin{bmatrix} 1 \\ 0 \end{bmatrix} = \begin{bmatrix} \frac{1}{6} \\ \frac{1}{6} \\ \frac{1}{6} \end{bmatrix}$

37. $\begin{bmatrix} x \\ y \end{bmatrix} = \frac{1}{25}\begin{bmatrix} 1 & 2 \\ 2 & 4 \end{bmatrix}\begin{bmatrix} 2 \\ 3 \end{bmatrix} = \frac{1}{25}\begin{bmatrix} 8 \\ 16 \end{bmatrix}$

39. (a) $\begin{bmatrix} -99 \\ 100 \end{bmatrix}, \begin{bmatrix} -1000 \\ 1000 \end{bmatrix}$ (b) $\begin{bmatrix} 0.70709 & * \\ 0.70728 & * \end{bmatrix} \begin{bmatrix} 2.0005 & \\ & 0.0007 \end{bmatrix} \begin{bmatrix} 0.70744 & * \\ 0.70709 & * \end{bmatrix}^T$

(c) **b** is close to $CS(A)$ but **c** is not.

Section 6.3

1. Solve $x^2 = x$ 5. Try $x_0 \geq \frac{4\pi}{9}$ 15. Y 17. N 19. N

Section 6.4

1. $3, \frac{1}{2}$ 3. 6, 1 5. 4, 1 7. $3, -2; [1 \quad 1]^T$
9. $3, -2; [2 \quad -1]^T$ 13. 5, 4 15. 3, 2 17. 7, 6
19.

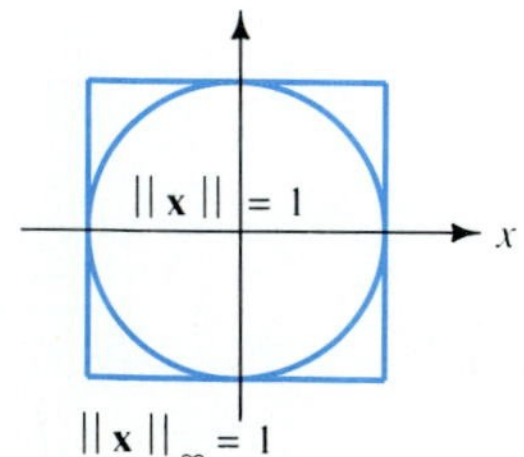

21. 21 23. 19 25. 25 29. 8, 8

31. 14, 22 33. (2, −1), (1, 1), 750 35. (3, −2), (6, 2), 2800
37. (a) Triangle inequality 39. $\|\lambda \mathbf{x}\|_* = \|A\mathbf{x}\|_* \le \|A\|_* \|\mathbf{x}\|_*$

Section 6.5

11. $\bar{1}^{-1} = \bar{1}, \bar{2}^{-1} = \bar{2}$ 15. $\begin{bmatrix} \bar{0} & \bar{0} \\ \bar{1} & \bar{1} \\ \bar{1} & \bar{2} \end{bmatrix}$

17. $\bar{1}^{-1} = \bar{1}, \bar{2}^{-1} = \bar{3}, \bar{3}^{-1} = \bar{2}, \bar{4}^{-1} = \bar{4}$ 21. $\begin{bmatrix} \bar{4} & \bar{2} & \bar{4} \\ \bar{4} & \bar{0} & \bar{4} \end{bmatrix}$

Chapter 6 Review

3. n even and m odd, or vice versa 5. $1, x - 1, x^2 - 2x + \frac{2}{3}$

7. $\begin{bmatrix} \frac{1}{6} & \frac{1}{6} \\ \frac{1}{6} & \frac{1}{6} \\ \frac{1}{6} & \frac{1}{6} \end{bmatrix}, \begin{bmatrix} 1/\sqrt{2} & 1/\sqrt{2} \\ 1/\sqrt{2} & -1/\sqrt{2} \end{bmatrix} \begin{bmatrix} \sqrt{6} & 0 & 0 \\ 0 & 0 & 0 \end{bmatrix} \begin{bmatrix} 1/\sqrt{3} & 1/\sqrt{2} & 1/\sqrt{6} \\ 1/\sqrt{3} & -1/\sqrt{2} & 1/\sqrt{6} \\ 1/\sqrt{3} & 0 & -2/\sqrt{6} \end{bmatrix}$

9. Z_7 11. Independent

15. $\begin{bmatrix} 0 & \frac{1}{2} & 0 \\ \frac{1}{2} & 0 & \frac{1}{2} \\ 0 & \frac{1}{2} & 0 \end{bmatrix}, 0, \pm\sqrt{2}/2; \begin{bmatrix} 0 & \frac{1}{2} & 0 \\ 0 & \frac{1}{4} & \frac{1}{2} \\ 0 & \frac{1}{8} & \frac{1}{4} \end{bmatrix}, 0, 0, \frac{1}{2}$

17. $\|\cdot\|_2 : \sqrt{10 \dfrac{3 + \sqrt{5}}{2}} \approx 5.11667, 5, 10; \|\cdot\|_\infty : 7, 4, 10$

Appendix A

1. $\begin{bmatrix} 1 & \\ -3 & 1 \end{bmatrix}, \begin{bmatrix} 2 & 4 \\ & -9 \end{bmatrix}$ 3. $\begin{bmatrix} 1 & & \\ 1 & 1 & \\ -2 & 0 & 1 \end{bmatrix}, I, \begin{bmatrix} 2 & 1 & 3 \\ & 6 & 4 \\ & & -2 \end{bmatrix}$

5. $\begin{bmatrix} 1 & & \\ 2 & 1 & \\ 1 & & 1 \end{bmatrix}, \begin{bmatrix} 1 & & \\ & 1 & \\ & -\frac{1}{2} & 1 \end{bmatrix}, \begin{bmatrix} 3 & -2 & -1 \\ & -8 & 0 \\ & & 3 \end{bmatrix}$

7. $\begin{bmatrix} 1 & & & \\ -2 & 1 & & \\ 3 & & 1 & \\ -1 & & & 1 \end{bmatrix}, \begin{bmatrix} 1 & & & \\ & 1 & & \\ & 11 & 1 & \\ & -2 & & 1 \end{bmatrix}, \begin{bmatrix} 1 & & & \\ & 1 & & \\ & & 1 & \\ & & \frac{1}{2} & 1 \end{bmatrix},$
$\begin{bmatrix} 1 & 3 & 2 & -1 \\ & -1 & -1 & 4 \\ & & -6 & 43 \\ & & & \frac{31}{2} \end{bmatrix}$

9. $\begin{bmatrix} 1 & \\ 3 & 1 \end{bmatrix}$ 11. $\begin{bmatrix} 1 & & & \\ & 1 & & \\ & -3 & 1 & \\ & 3 & & 1 \end{bmatrix}$

13. L_1 adds 2 times row 1 to row 3 and 3 times row 1 to row 4. L_2 adds 4 times row 2 to row 3. In computing L_2L_1, the (2, 1) entry of L_1 is nonzero, so 4 times this entry is added to the (3, 1) entry.

15. $\begin{bmatrix} 1 & \\ -3 & 1 \end{bmatrix}\begin{bmatrix} 2 & 3 \\ & 4 \end{bmatrix}$ 17. $\begin{bmatrix} 1 & & \\ -2 & 1 & \\ 3 & -1 & 1 \end{bmatrix}\begin{bmatrix} 2 & -1 & 0 \\ & 3 & 1 \\ & & 4 \end{bmatrix}$

19. $\begin{bmatrix} 3 & 4 \\ & 2 \end{bmatrix} = \begin{bmatrix} 0 & 1 \\ 1 & 0 \end{bmatrix} A$

21. $\begin{bmatrix} 1 & -4 & 3 \\ & 13 & -6 \\ & & 11 \end{bmatrix} = \begin{bmatrix} 1 & & \\ & 0 & 1 \\ & 1 & 0 \end{bmatrix}\begin{bmatrix} 1 & & \\ 2 & 1 & \\ -3 & & 1 \end{bmatrix} A$

23. $$\begin{bmatrix} 1 & 2 & -1 & 3 \\ & -5 & 11 & -2 \\ & & 22 & 0 \\ & & & 3 \end{bmatrix} = \begin{bmatrix} 1 & & & \\ & 1 & & \\ & & 0 & 1 \\ & & 1 & 0 \end{bmatrix}\begin{bmatrix} 1 & & & \\ & 1 & & \\ & 0 & 1 & \\ & 3 & & 1 \end{bmatrix}\begin{bmatrix} 1 & & & \\ & 0 & 1 & \\ & 1 & 0 & \\ & & & 1 \end{bmatrix}$$
$$\times \begin{bmatrix} 1 & & & \\ 0 & 1 & & \\ -3 & & 1 & \\ 5 & & & 1 \end{bmatrix}\begin{bmatrix} 0 & 1 & & \\ 1 & 0 & & \\ & & 1 & \\ & & & 1 \end{bmatrix} A$$

Appendix B

1. 6, 3 3. 210, 190 5. $\dfrac{n(n+1)}{2}, \dfrac{(n-1)n}{2}$ 7. −2, 3

9. 3, 2, 1 11. −2, 3, 1, 2

Appendix C

1. $\dfrac{\pi^2}{4}\begin{bmatrix} \frac{1}{2} \\ 0 \\ -\frac{1}{2} \end{bmatrix}$

3. $$\begin{bmatrix} -(2+h^2) & 1 & \\ 1 & -(2+h^2) & 1 \\ & 1 & -(2+h^2) \end{bmatrix}\begin{bmatrix} y_1 \\ y_2 \\ y_3 \end{bmatrix} = h^3\begin{bmatrix} 1 \\ 2 \\ 3 \end{bmatrix}, h = \tfrac{1}{4}$$

INDEX

B C D E F G H I J 3 4 5 6 7 8 9